Statistics and Data Visualization Using R

For Michelle, Tim, and Jessica, who taught me to always leave a note.

Sara Miller McCune founded SAGE Publishing in 1965 to support the dissemination of usable knowledge and educate a global community. SAGE publishes more than 1000 journals and over 800 new books each year, spanning a wide range of subject areas. Our growing selection of library products includes archives, data, case studies and video. SAGE remains majority owned by our founder and after her lifetime will become owned by a charitable trust that secures the company's continued independence.

Los Angeles | London | New Delhi | Singapore | Washington DC | Melbourne

Statistics and Data Visualization Using R

The Art and Practice of Data Analysis

David S. Brown
University of Colorado Boulder

Los Angeles | London | New Delhi
Singapore | Washington DC | Melbourne

FOR INFORMATION:

SAGE Publications, Inc.
2455 Teller Road
Thousand Oaks, California 91320
E-mail: order@sagepub.com

SAGE Publications Ltd.
1 Oliver's Yard
55 City Road
London, EC1Y 1SP
United Kingdom

SAGE Publications India Pvt. Ltd.
B 1/I 1 Mohan Cooperative Industrial Area
Mathura Road, New Delhi 110 044
India

SAGE Publications Asia-Pacific Pte. Ltd.
18 Cross Street #10-10/11/12
China Square Central
Singapore 048423

Acquisitions Editor: Leah Fargotstein
Product Associate: Ivey Mellem
Production Editor: Rebecca Lee
Copy Editor: Christina West
Typesetter: Hurix Digital
Indexer: Integra
Cover Designer: Gail Buschman
Marketing Manager: Victoria Velasquez

Copyright © 2022 by SAGE Publications, Inc.

All rights reserved. Except as permitted by U.S. copyright law, no part of this work may be reproduced or distributed in any form or by any means, or stored in a database or retrieval system, without permission in writing from the publisher.

All third party trademarks referenced or depicted herein are included solely for the purpose of illustration and are the property of their respective owners. Reference to these trademarks in no way indicates any relationship with, or endorsement by, the trademark owner.

Printed in the United States of America

Library of Congress Control Number: 2021907690

ISBN 978-1-5443-3386-1

This book is printed on acid-free paper.

21 22 23 24 25 10 9 8 7 6 5 4 3 2 1

Brief Contents

Preface — xxi
Acknowledgments — xxv
About the Author — xxvii

Chapter 1 • Getting Started — 1
Chapter 2 • An Introduction to Data Analysis — 29
Chapter 3 • Describing Data — 67
Chapter 4 • Central Tendency and Dispersion — 99
Chapter 5 • Univariate and Bivariate Descriptions of Data — 131
Chapter 6 • Transforming Data — 175
Chapter 7 • Some Principles of Displaying Data — 205
Chapter 8 • The Essentials of Probability Theory — 237
Chapter 9 • Confidence Intervals and Testing Hypotheses — 271
Chapter 10 • Making Comparisons — 307
Chapter 11 • Controlled Comparisons — 347
Chapter 12 • Linear Regression — 391
Chapter 13 • Multiple Regression — 427
Chapter 14 • Dummies and Interactions — 459
Chapter 15 • Diagnostics I: Is Ordinary Least Squares Appropriate? — 485
Chapter 16 • Diagnostics II: Residuals, Leverages, and Measures of Influence — 513

Chapter 17 • Logistic Regression	537
Appendix • Developing Empirical Implications	569

Glossary	573
References	581
Index	583

Detailed Contents

Preface	xxi
Acknowledgments	xxv
About the Author	xxvii

Chapter 1 • Getting Started — 1

Learning Objectives	1
Overview	1
R, RStudio, and R Markdown	2
ART AND PRACTICE OF DATA VISUALIZATION: Copy Relevant Bits of Code	3
KNOWLEDGE CHECK: Explain the difference between R, RStudio, and R Markdown.	3
Objects and Functions	4
KNOWLEDGE CHECK: Distinguish between functions and objects.	6
Getting Started in RStudio	6
ART AND PRACTICE OF DATA VISUALIZATION: Keep It Together	6
KNOWLEDGE CHECK: Download and install both R and RStudio.	12
Navigating RStudio With R Markdown	12
ART AND PRACTICE OF DATA VISUALIZATION: A Multitude of Options	17
Using R Markdown Files Versus R-Scripts	18
KNOWLEDGE CHECK: Identify the main features of R Markdown.	18
A Little Practice	19
KNOWLEDGE CHECK: Create a Word or HTML file from R Markdown.	23
Summary	24
ART AND PRACTICE OF DATA VISUALIZATION: Curiosity and Humor	24
Common Problems	24
Review Questions	25
Practice on Analysis and Visualization	25
Annotated R Functions	26
Answers	27

Chapter 2 • An Introduction to Data Analysis — 29

Learning Objectives	29
Overview	29
Motivating Data Analysis	30
Big Data Are Getting Bigger	30

Data Analysis Is a Marketable Skill	*31*
Data Analysis Is a Public Good	*31*
KNOWLEDGE CHECK: Explain the importance of acquiring skills in data analysis.	32
The Main Components of Data Analysis	33
KNOWLEDGE CHECK: List the components of data analysis and how they fit together.	34
Developing Hypotheses by Describing Data	35
ART AND PRACTICE OF DATA VISUALIZATION: Describing Data and Formulating Hypotheses	35
Hypothesis I: Ethnolinguistic Fractionalization	*36*
Hypothesis II: Women's Suffrage	*42*
Hypothesis III: Human Capital	*48*
Hypothesis IV: Political Stability	*50*
KNOWLEDGE CHECK: Form hypotheses from descriptions of data.	53
Model Building and Estimation	54
KNOWLEDGE CHECK: Explain the connection between hypotheses, models, and estimates.	55
Diagnostics	56
Stability of the Results	*56*
Residual Plots	*58*
KNOWLEDGE CHECK: Define diagnostics and explain their role in data analysis.	60
Next Questions	61
KNOWLEDGE CHECK: Formulate new questions.	61
Summary	62
Common Problems	62
Review Questions	63
Practice on Analysis and Visualization	63
Annotated R Functions	65
Answers	65

Chapter 3 • Describing Data 67

Learning Objectives	67
Overview	67
Data Sets and Variables	69
KNOWLEDGE CHECK: Explain the difference between data sets and variables.	70
Different Kinds of Variables	71
Continuous Variables	*71*
Categorical Variables	*72*
Ordered Categorical Variables	*72*
ART AND PRACTICE OF DATA VISUALIZATION: Resist Taking Data as Given	76
KNOWLEDGE CHECK: Identify continuous, categorical, and ordered categorical variables.	77
Describing Data Saves Time and Effort	77
Shape of the Data	*78*
Range of the Data	*82*

ART AND PRACTICE OF DATA VISUALIZATION: Small Changes for the Same View of Different Data	82
KNOWLEDGE CHECK: Discuss how describing data saves time and effort.	83
Identify Puzzles, Questions, Hypotheses, and Clues	84
Puzzles and Questions: An Important Difference	*84*
ART AND PRACTICE OF DATA VISUALIZATION: A Quick Way to Distinguish Questions From Puzzles	85
Describing Data to Refine the Question	*85*
Describing Data Reveals Additional Clues	*88*
ART AND PRACTICE OF DATA VISUALIZATION: Maximize Space	90
KNOWLEDGE CHECK: Connect data description to answering questions and identifying puzzles.	90
Measurement	91
Validity	*91*
Reliability	*92*
KNOWLEDGE CHECK: Create valid and reliable measures.	93
Summary	93
Common Problems	94
Review Questions	94
Practice on Analysis and Visualization	94
Annotated R Functions	95
Answers	96

Chapter 4 • Central Tendency and Dispersion   99

Learning Objectives	99
Overview	99
Measures of Central Tendency: The Mode, Mean, and Median	100
Mode	*100*
Mean	*102*
ART AND PRACTICE OF DATA VISUALIZATION: Lines Raise the Bar	106
Median	*106*
KNOWLEDGE CHECK: Calculate the mode, mean, and median.	108
Mean Versus Median	109
ART AND PRACTICE OF DATA VISUALIZATION: Option *Alpha*	109
KNOWLEDGE CHECK: Explain the difference between the mean and median.	112
Measures of Dispersion: The Range, Interquartile Range, and Standard Deviation	112
Range	*113*
Interquartile Range	*114*
Standard Deviation	*115*
KNOWLEDGE CHECK: Calculate the range, interquartile range, and standard deviation.	121
Interquartile Range Versus Standard Deviation	122
KNOWLEDGE CHECK: Explain the difference between the interquartile range and the standard deviation.	123
A Note on Variance	*123*

Summary	124
Common Problems	124
Review Questions	124
Practice on Analysis and Visualization	124
Annotated R Functions	127
Answers	127

Chapter 5 • Univariate and Bivariate Descriptions of Data 131

Learning Objectives	131
Overview	131
The Good, the Bad, and the Outlier	131
ART AND PRACTICE OF DATA VISUALIZATION: Using Different Packages in R	132
KNOWLEDGE CHECK: Ask the questions univariate views answer.	133
Five Views of Univariate Data	133
Frequency Table	*134*
Bar Plot	*135*
Boxplot (or Box-and-Whisker Plot)	*136*
Histogram	*139*
Stem-and-Leaf Plot	*143*
KNOWLEDGE CHECK: Create and analyze univariate views of the data.	146
Are They in a Relationship?	146
KNOWLEDGE CHECK: Ask the questions bivariate views answer.	147
Scatter Plot	*148*
ART AND PRACTICE OF DATA VISUALIZATION: Dependent and Independent Variables	149
ART AND PRACTICE OF DATA VISUALIZATION: Caution With Lines and Curves	152
Boxplot (Bivariate)	*154*
Mosaic Plot	*159*
Cross-Tab	*162*
Bubble Plot	*163*
KNOWLEDGE CHECK: Create and analyze bivariate views of the data.	169
Summary	170
Common Problems	170
Review Questions	171
Practice on Analysis and Visualization	171
Annotated R Functions	172
Answers	173

Chapter 6 • Transforming Data 175

Learning Objectives	175
Overview	175
Theoretical Reasons for Transforming Data	176
Transform the Data to Fit the Theory	*176*
Transform Both Data and Question to Match Each Other	*177*
KNOWLEDGE CHECK: Connect hypotheses to the data.	178

Transforming Data for Practical Reasons	178
KNOWLEDGE CHECK: Create views that uncover patterns by transforming data.	181
Transforming Data—Continuous to Categorical Variables	182
ART AND PRACTICE OF DATA VISUALIZATION: Defining New Variables	182
KNOWLEDGE CHECK: Explain why and how to create categories from continuous data.	187
Transforming Data—Changing Categories	187
ART AND PRACTICE OF DATA VISUALIZATION: Collect More Data Than You Need	191
KNOWLEDGE CHECK: Create categorical variables to connect hypotheses to the data.	193
Box-Cox Transformations	194
ART AND PRACTICE OF DATA VISUALIZATION: Transforming Variables Changes Their Units	194
ART AND PRACTICE OF DATA VISUALIZATION: A Tip for Taking the Log of Variables With Zeros	195
KNOWLEDGE CHECK: Describe the Box-Cox ladder of transformation.	201
Summary	201
Common Problems	202
Review Questions	202
Practice on Analysis and Visualization	202
Annotated R Functions	203
Answers	204

Chapter 7 • Some Principles of Displaying Data — 205

Learning Objectives	205
Overview	205
Some Elements of Style	206
Eliminate the Clutter	207
Focus	209
Integrate Words and Pictures	211
ART AND PRACTICE OF DATA VISUALIZATION: Figure and Table Proximity	214
Pictures Should Be Worth a Thousand Words	215
ART AND PRACTICE OF DATA VISUALIZATION: Blinded by the Light	216
Know Your Audience	220
Know Your Purpose: Explanatory, Exploratory, or Informational	222
KNOWLEDGE CHECK: List some elements of style.	227
The Basic Elements of a Story	228
KNOWLEDGE CHECK: Tell your story.	229
Documentation (Establishing Credibility as a Storyteller)	229
KNOWLEDGE CHECK: Describe the elements and importance of documentation.	230
Build an Intuition (Setting the Context)	231
KNOWLEDGE CHECK: Explain how to set the context.	231
Show Causation (The Journey)	231
KNOWLEDGE CHECK: Establish causation.	232
From Causation to Action (The Resolution)	232
KNOWLEDGE CHECK: Identify the actions implied by the analysis.	233

Summary 233
Common Problems 233
Review Questions 234
Practice on Analysis and Visualization 234
Annotated R Functions 235
Answers 236

Chapter 8 • The Essentials of Probability Theory 237

Overview 237
Learning Objectives 237
Populations and Samples 238
    ART AND PRACTICE OF DATA VISUALIZATION: Be Precise With Language 239
    KNOWLEDGE CHECK: Explain the distinction between populations and samples. 239
Sample Bias and Random Samples 240
    KNOWLEDGE CHECK: Recognize potential sources of bias and why randomization helps. 241
The Law of Large Numbers 241
    Visualizing the Law of Large Numbers 242
    KNOWLEDGE CHECK: Identify uses of the law of large numbers. 246
The Central Limit Theorem 246
    Sampling Distribution of Means Approaches Normality as n Increases 248
    ART AND PRACTICE OF DATA VISUALIZATION: Statistic's Golden Rule 250
    Sampling Distribution of Sums Is Normal 250
    Number of Observations When Drawing From a Normal Distribution 251
    A Useful Property of the Central Limit Theorem 253
    Sampling From Different Distributions 253
    KNOWLEDGE CHECK: Explain why the central limit theorem is so important. 259
The Standard Normal Distribution 260
    The Standard Normal Distribution and Critical z-Scores 263
    KNOWLEDGE CHECK: Connect sampling distributions to the standard normal distribution. 265
Summary 266
Common Problems 266
Review Questions 267
Practice on Analysis and Visualization 267
Annotated R Functions 269
Answers 270

Chapter 9 • Confidence Intervals and Testing Hypotheses 271

Learning Objectives 271
Overview 271

Confidence Intervals With Large Samples	272
Finding the Population Proportion	*272*
ART AND PRACTICE OF DATA VISUALIZATION: Decimals and Percentages	273
Finding the Population Mean	*277*
KNOWLEDGE CHECK: Calculate confidence intervals around proportions and means.	280
Small Samples and the t-Distribution	280
Degrees of Freedom	*283*
Sample Standard Deviation With Small Samples	*284*
Constructing Confidence Intervals With Small Samples	*289*
ART AND PRACTICE OF DATA VISUALIZATION: The Magic Number	289
Example: Women's Pay Versus Men's	*290*
KNOWLEDGE CHECK: Use the t-distribution to calculate confidence intervals.	291
Comparing Two Sample Means	292
Example: Two Groups and Two Incomes	*293*
Example: Race and Attitudes Toward the Police	*295*
Example: Income and Support for Trump	*296*
KNOWLEDGE CHECK: Determine if the means of two populations are different.	297
Confidence Levels	297
KNOWLEDGE CHECK: Explain what confidence means in statistics.	299
A Brief Note on Statistical Inference and Causation	300
KNOWLEDGE CHECK: Connect confidence intervals to statistical estimates.	301
Summary	301
Common Problems	302
Review Questions	303
Practice on Analysis and Visualization	303
Annotated R Functions	304
Answers	305

Chapter 10 • Making Comparisons   307

Overview	307
Learning Objectives	307
Why Do We Make Comparisons?	308
ART AND PRACTICE OF DATA VISUALIZATION: Is It a Lot and Why?	308
KNOWLEDGE CHECK: Discuss the reasons why we compare things.	309
Questions That Beg Comparisons	309
KNOWLEDGE CHECK: Identify questions that imply comparison.	310
Comparing Two Categorical Variables	310
Example: Attitudes Toward Police	*310*
Example: Religion and Politics	*313*
KNOWLEDGE CHECK: Construct and interpret figures when both variables are categorical.	315
ART AND PRACTICE OF DATA VISUALIZATION: Axis Labels	315
Comparing Continuous and Categorical Variables	315
Example: Obama Feeling Thermometer	*315*
KNOWLEDGE CHECK: Construct and interpret figures with continuous and categorical variables.	317

Comparing Two Continuous Variables ... 318
    Example: Gender and Education ... 318
    Example: Gender and Policymaking ... 319
    ART AND PRACTICE OF DATA VISUALIZATION: Pick Representative Points ... 321
    KNOWLEDGE CHECK: Construct and interpret figures with continuous variables. ... 321
Exploratory Data Analysis: Investigating Abortion Rates in the United States ... 322
    Recap ... 324
    KNOWLEDGE CHECK: Connect exploring data with generating hypotheses to make discoveries. ... 328
Good Analysis Generates Additional Questions ... 329
    KNOWLEDGE CHECK: Compose the next question after your analysis. ... 329
Summary ... 329
Common Problems ... 330
Review Questions ... 330
Practice on Analysis and Visualization ... 330
Annotated R Functions ... 331
Answers ... 332

Chapter 11 • Controlled Comparisons ... 347

Learning Objectives ... 347
Overview ... 347
What Is a Controlled Comparison? ... 348
    KNOWLEDGE CHECK: Explain the difference between a comparison and a controlled comparison. ... 348
Comparing Two Categorical Variables, Controlling for a Third ... 349
    Example: Attitudes Toward the Police ... 349
    Example: Attitudes Toward Immigration ... 355
    ART AND PRACTICE OF DATA VISUALIZATION: Reliability and Validity ... 360
    KNOWLEDGE CHECK: Create a controlled comparison between categorical variables. ... 362
Comparing Two Continuous Variables, Controlling for a Third ... 363
    Example: Infant Mortality ... 363
    Example: Homicide Rates ... 365
    ART AND PRACTICE OF DATA VISUALIZATION: Dichotomous Variables ... 368
    KNOWLEDGE CHECK: Create a controlled comparison between continuous variables. ... 369
Arguments and Controlled Comparisons ... 370
    KNOWLEDGE CHECK: Discuss how arguments and controlled comparisons are related. ... 370
Summary ... 371
Common Problems ... 371
Review Questions ... 372
Practice on Analysis and Visualization ... 372
Annotated R Functions ... 374
Answers ... 375
Practice on Analysis and Visualization ... 380

Chapter 12 • Linear Regression — 391

Learning Objectives — 391
Overview — 391
The Advantages of Linear Regression — 392
    KNOWLEDGE CHECK: Summarize the advantages of fitting a line to data. — 392
The Slope and Intercept in Linear Regression — 393
    ART AND PRACTICE OF DATA VISUALIZATION: The Language of Regressing Y on X — 393
    Interpretation of the Slope and Intercept — 394
    Example: Voter Turnout and Education — 394
    KNOWLEDGE CHECK: Identify the main features of linear regression. — 398
Goodness of Fit (R^2 Statistic) — 399
    KNOWLEDGE CHECK: Describe model fit. — 402
Statistical Significance — 403
    Calculating the t-Ratio — 403
    ART AND PRACTICE OF DATA VISUALIZATION: With More Information, Confidence Increases — 404
    KNOWLEDGE CHECK: Describe statistical significance. — 405
Examples of Bivariate Regressions — 405
    Does the Level of Religiosity in a State Influence Abortion Rates? — 405
    ART AND PRACTICE OF DATA VISUALIZATION: Standard Errors of the Estimate — 407
    Does Religiosity Affect Gun Laws? — 408
    Are Ethnically Heterogeneous Countries Violent? — 409
    ART AND PRACTICE OF DATA VISUALIZATION: The .lm Suffix for Linear Models — 410
    Does Violence Cause Political Instability? — 412
    Is Income (GDP per Capita) Related to Turnout? — 412
    KNOWLEDGE CHECK: Interpret coefficients in a variety of contexts. — 415
    ART AND PRACTICE OF DATA VISUALIZATION: Bivariate Regressions and Multiple Regression — 416
Summary — 416
Common Problems — 416
Review Questions — 417
Practice on Analysis and Visualization — 417
Annotated R Functions — 418
Answers — 419

Chapter 13 • Multiple Regression — 427

Learning Objectives — 427
Overview — 427
What Is Multiple Regression? — 428
    Why Multiple Regression? — 429
    KNOWLEDGE CHECK: Define multiple regression. — 429

Regression Models and Arguments 430
    KNOWLEDGE CHECK: Identify the parallels between a model and an argument. 431
Regression Models, Theory, and Evidence 431
    ART AND PRACTICE OF DATA VISUALIZATION: Honesty 433
    KNOWLEDGE CHECK: Discuss the relationship between theory and evidence 434
Interpreting Estimates in Multiple Regression 435
    Substantive Significance 435
    Statistical Significance 436
    ART AND PRACTICE OF DATA VISUALIZATION: Levels of Significance 437
    Goodness of Fit: R^2 437
    KNOWLEDGE CHECK: Interpret coefficients, t-ratios, and measures of fit. 438
Example: Homicide Rate and Education 438
    Theory 439
    Description of Data 439
    Estimation 443
    Empirical Implication 445
    Discussion 448
    KNOWLEDGE CHECK: Employ multiple regression analysis to examine a problem. 448
Summary 449
Common Problems 449
Review Questions 450
Practice on Analysis and Visualization 450
Annotated R Functions 451
Answers 452
Practice on Analysis and Visualization 452

Chapter 14 • Dummies and Interactions   459

Learning Objectives 459
Overview 459
What Is a Dummy Variable? 460
    KNOWLEDGE CHECK: Describe a dummy variable. 460
Additive Models and Interactive Models 461
    KNOWLEDGE CHECK: Discuss the difference between interactive and additive models. 461
Bivariate Dummy Variable Regression 461
    KNOWLEDGE CHECK: Explain what questions a dummy variable helps answer. 463
Multiple Regression and Dummy Variables 464
    ART AND PRACTICE OF DATA VISUALIZATION: Dummy Variables in R 465
    KNOWLEDGE CHECK: Interpret dummies in multiple regression. 467
Interactions in Multiple Regression 468
    Example: Bernie Sanders, Education, and Income 470
    Example: Foreign Aid, GDP per Capita, and Democracy 474
    KNOWLEDGE CHECK: Interpret results from a multiple regression with an interactive term. 478
Summary 479

Common Problems	479
Review Questions	480
Practice on Analysis and Visualization	480
Annotated R Functions	481
Answers	482

Chapter 15 • Diagnostics I: Is Ordinary Least Squares Appropriate? 485

Learning Objectives	485
Overview	485
Diagnostics in Regression Analysis	486
KNOWLEDGE CHECK: Describe why diagnostics are so important.	487
Properties of Statistics and Estimators	487
KNOWLEDGE CHECK: Identify properties of estimators.	491
The Gauss-Markov Assumptions	492
KNOWLEDGE CHECK: Discuss the Gauss-Markov assumptions.	497
The Residual Plot	498
ART AND PRACTICE OF DATA VISUALIZATION: Color Transparency With *alpha*	503
KNOWLEDGE CHECK: Connect residual plots to Gauss-Markov assumptions.	506
Summary	507
Common Problems	507
Review Questions	508
Practice on Analysis and Visualization	508
Annotated R Functions	511
Answers	512

Chapter 16 • Diagnostics II: Residuals, Leverages, and Measures of Influence 513

Learning Objectives	513
Overview	513
Outliers	514
KNOWLEDGE CHECK: Describe what we can learn from residuals.	516
Leverages	517
KNOWLEDGE CHECK: Explain the difference between residuals and leverages.	522
Measures of Influence	523
Cook's Distance	523
ART AND PRACTICE OF DATA VISUALIZATION: One Command, Multiple Plots	523
Dfbetas	524
ART AND PRACTICE OF DATA VISUALIZATION: Three Dimensions	527
KNOWLEDGE CHECK: Employ measures of influence to make discoveries.	528

Added Variable Plots	529
KNOWLEDGE CHECK: Identify potential cases that influence regression results.	530
Summary	531
Common Problems	532
Review Questions	532
Practice on Analysis and Visualization	532
Annotated R Functions	535
Answers	535

Chapter 17 • Logistic Regression 537

Learning Objectives	537
Overview	537
Questions and Problems That Require Logistic Regression	538
KNOWLEDGE CHECK: Discuss problems best summarized by a nonlinear framework.	538
Logistic Regression Violates Gauss-Markov Assumptions	539
KNOWLEDGE CHECK: Explain how logistic relationships violate the Gauss-Markov assumptions.	542
Working With Logged Odds	542
KNOWLEDGE CHECK: Calculate a logistic regression.	544
Working With Predicted Probabilities	544
Bivariate Logistic Regression	*545*
Multivariate Logistic Regression	*547*
Example: Obama's Electoral College Wins in 2012	*548*
KNOWLEDGE CHECK: Explain how to convert logged odds to predicted probabilities.	551
Model Fit With Logistic Regression	551
Example: Obama, Income, and Education	*551*
Receiver Operating Characteristic Curves and the Area Under the Curve	*553*
ART AND PRACTICE OF DATA VISUALIZATION: Calculate the AUC	559
KNOWLEDGE CHECK: Evaluate the ability of the logistic model to predict outcomes.	560
Summary	560
Common Problems	561
Review Questions	561
Practice on Analysis and Visualization	562
Annotated R Functions	562
Answers	563

Appendix • Developing Empirical Implications 569

Overview	569
Developing Empirical Implications	569
Testing Additional Dependent Variables	569

Testing Additional Independent Variables	570
Using Information on Cases	570
Causal Mechanisms	570
The Rabbit Hole	570

Glossary	573
References	581
Index	583

Preface

Infiltrate the dealers… find the suppliers.

Ice Cube in *21 Jump Street* (Sony Pictures, 2012)

This book was written to encourage, inspire, and excite students about data analysis in the social sciences. Its fundamental premise is that students *learn* data analysis by *doing* data analysis. Toward that end, it starts with simple graphical techniques used to explore data and to ask interesting questions of the data. Emphasis is placed on methods used to identify problems that lurk far beneath the clean veneer of a regression table. In the end, readers will be conversant in basic data analytic techniques and will have developed an approach to data analysis, understanding the conceptual, analytical, and even philosophical choices made. Crucial in my view, an important goal is to excite the reader about the enterprise. The material is designed to engage as we confront real-world issues and problems with real data. Readers are encouraged to play with the data the book is based on by downloading the data file on the SAGE companion website. After the data are downloaded, be sure to execute the *installD()* and *libraries()* commands before starting. The first command installs all the necessary packages, while the second loads them. The *installD()* command only needs to be executed once; the *libraries()* command needs to be executed whenever R is restarted.

Who Is This Book For?

This book is for several audiences. Primarily, the book is for the beginner. It assumes no prior knowledge of statistics or calculus, yet having a solid background in either does not make the exercise unprofitable. The book stems from a large "Introduction to Quantitative Methods" course I taught at the University of Colorado. It is a required class for political science majors who need to read, understand, and critically examine the growing amount of quantitative evidence. Our sincere hope in the class is to arm students with a set of skills to help solve problems.

Like a biologist uses an electron microscope, the data analyst uses R, an object-oriented statistical language that has established a strong foothold in private industry, primarily among data scientists. While learning statistics with pencil and paper is admirable and advantageous pedagogically, students must be armed with a state-of-the-art tool in this era of Big Data. The book is designed for the reader to download the accompanying data and follow along. The code provides, in my experience, an excellent set of commands that many beginning, intermediate, and advanced analysts can use.

For those with more experience, the book presents an approach that emphasizes how simple analyses can generate better questions through the back-and-forth between description, theory, and evidence. The book encourages formulating hypotheses, looking at the evidence, then generating additional hypotheses from that evidence. A deep understanding of the material is demonstrated by the reader, in my opinion, when a hypothesis is constructed to ask the next question. More than learning code, more than understanding probability theory, this book seeks to foster a never-ending cycle of discovery characterized by describing what we see, formulating a hypothesis, testing it empirically, then generating the next question or hypothesis. In that sense, even those with more technical facility can benefit.

Organization

Most statistics texts start with the basics of probability theory, followed by sampling and hypothesis testing, and end with correlation and regression analysis. While conceptually coherent, the student first must begin the semester overcoming a fear of probability theory, must decipher a standard normal or t-table, then in the last two weeks (usually after Thanksgiving or spring break) master bivariate or multiple regression analysis. There are two schools of thought here and the book accommodates both. Some insist that the probabilistic underpinnings of regression theory must come before the first line is fit to the data. Others like to start fitting lines and constructing models on day 1. They argue that only after the student has confronted the challenge of building models, generating estimates, and evaluating the model's fit is there a strong motivation to understand the probabilistic machinery used to generate t-ratios, R^2 statistics, and confidence intervals. The book is designed so that after describing data and making comparisons, you can skip Chapters 8 and 9 and proceed to regression analysis. The chapters on diagnostics provide an intuitive sense of the Gauss-Markov assumptions that underpin linear regression.

There is also a chapter dedicated to the presentation of data (Chapter 7). Too often, no time is spent even with the most elemental principles of how to present the findings we generate to an audience. I mix the pioneering work by Edward Tufte with some more recent contributions that emphasize the storytelling aspects of the enterprise.

Armed with the mechanics of multiple regression analysis, the book ends with logistic regression—many questions in the social sciences involve binary choices (to vote or not), binary states of being (college graduate or not), and binary contexts (whether you live in a city). Given the importance and ubiquity of these questions in the social sciences, the student is introduced to logistic regression, how to analyze it, and how to report its result.

Underlying Philosophy

Early in my academic career I was influenced by an argument between two senior scholars who I deeply respect. I can't remember the subject, but it was during a graduate seminar taught by the two professors. As we went about evaluating and examining an assigned article, one of the professors was clearly bothered by its lack of theoretical rigor. The paper's argument seemed to change with each new piece of evidence. Confronted with a fairly obvious example of post hoc reasoning in the paper, the two professors started arguing about whether the article made any contribution at all. While the two were good friends, the argument got very heated and reached its boiling point when one declared emphatically, "I don't trust work in which the theory keeps changing in order to fit the facts!" The room became silent as we waited for the response. After a brief moment of silence, the other professor smiled wryly and stated, "I don't trust work in which the facts keep changing to fit the theory!"

There it was. A clear line was drawn that day in my mind in terms of how I wanted to approach my craft. Recognizing that "the facts" are never independent of our understanding, backgrounds, and life experiences, they are for me on somewhat more solid ground than are the theories we use to explain them.

Having said that, we do have to start somewhere and the best data sets in my mind are constructed with theories and questions that are interesting, powerful, and important. So, much like the police captain (played by Ice Cube in the movie *21 Jump Street*) admonishes his underlings (played by Channing Tatum and Jonah Hill) to first infiltrate the dealers and then find the suppliers, the book encourages students to first generate a hypothesis and then test it empirically. In the end, the book is no soldier in the wars between social science as a deductive enterprise or an exercise in data mining. (Channing Tatum does, after all, ask his captain whether they can simply just find the suppliers first.) In the intervening pages my hope is to demonstrate that the best empirical work relies on an honest discussion between the two.

 Access Instructor Resources, including PowerPoint® slides and test banks, at **http://edge.sagepub.com/brownstats1e**.

Acknowledgments

I would like to thank, first and foremost, my PSCI 2075 students at the University of Colorado Boulder who did not allow my attempts at humor and outdated cultural references get in their way of learning. Their good humor, spirit, and grit made teaching the class a true pleasure. I also want to thank my wonderful colleagues at CU—in particular, Andy Philips, who gave important feedback during the early stages of this enterprise. Finally, I want to thank the team at SAGE. Leah Fargotstein gave me the opportunity to write this book and patiently helped me transform some casual course notes into this final product. Chelsea Neve was also an important part of that process, providing help on how to transform prose into pedagogy. My copyeditor, Christina West, did heroic work as well. Her patience, skill, and kindness made the last part of the journey an absolute pleasure. Ivey Mellem and Rebecca Lee made a very complex process on the back end seem simple and easy on the front end. Despite all of the inspiration, support, and help from others, the remaining mistakes are mine.

SAGE and the author would also like to thank the following reviewers for their comments on the development of this book:

Zachary Albert, Brandeis University

Shavonne Arthurs, Seton Hall University

Hunter Bacot, University of North Carolina at Greensboro

Nathanial Bastian, Northwestern University

Salem Boumediene, University of Illinois Springfield

Scott Comparato, Southern Illinois University

Renato Corbetta, University of Alabama at Birmingham

Sarah Crocco, University of Maryland–College Park

Todd Daniel, Missouri State University

Eric Dunford, Georgetown University

Catherine Garcia, University of Nebraska–Lincoln

Jonathan Hack, Social Science Research Council

Troy Hooper, Texas Tech University Health Sciences Center

Ahmed Ibrahim, Johns Hopkins University

Whitt Kilburn, Grand Valley State University

David Lamb, University of Southern Florida

Alice Long, Penn State Shenango

Matt Miles, Brigham Young University–Idaho

Joseph Nedelec, University of Cincinnati

Carl Palmer, Illinois State University

Galen Papkov, Florida Gulf Coast University

Esther Pearson, Lasell University

Peter Peregrine, Lawrence University

Catherine Persall, St. Joseph's College

Matthew Phillips, University of North Carolina at Charlotte

Chris Prener, Saint Louis University

Alessandro Quartiroli, University of Wisconsin–La Cross

Jason Renn, Utah State University

Matthew Risler, Loras College

George Robinson, North Carolina A&T University

Andrew Rosenberg, University of Florida

Shayna Rusticus, Kwantlen Polytechnic University

Josh Ryan, Utah State University

Jennifer Samson, Arkansas Tech University

Jeffery Stone, University of California, Los Angeles

Joseph Szmania, Moravian College

Bradly Theissen, St. Ambrose University

Ches Thurber, Northern Illinois University

James Walke, Alabama A&M University

Lili Wang, Arizona State University

Kyle Woosnam, University of Georgia

Jingshun Zhang, Florida Gulf Coast University

About the Author

David S. Brown is Professor of Political Science and Divisional Dean of the Social Sciences at the University of Colorado Boulder. He received his PhD in political science from the University of California, Los Angeles and was the inaugural Kenneth Boulding Post-Doctoral Fellow at the Institute for Behavioral Science at CU Boulder. He then served as Assistant Professor at Rice University before joining the faculty in political science at CU. He studies comparative politics with a focus on institutions and their impact on economic development. His work has been published in the *American Political Science Review, American Journal of Political Science, British Journal of Political Science,* and *Annals of the Association of American Geographers.*

1 Getting Started

CHAPTER OUTLINE

Learning Objectives
Overview
R, RStudio, and R Markdown
Objects and Functions
Getting Started in RStudio
Navigating RStudio With R Markdown
Using R Markdown Files Versus R-Scripts
A Little Practice
Summary
Common Problems
Review Questions
Practice on Analysis and Visualization
Annotated R Functions
Answers

LEARNING OBJECTIVES

- Explain the difference between R, RStudio, and R Markdown.
- Distinguish between functions and objects.
- Download, install, and navigate both R and RStudio.
- Identify the main features of R Markdown.
- Create a Word or HTML file from R Markdown.

Overview

Learning how to use a high-powered microscope in biology class is a useful analogy to our task in this chapter. Learning how to manipulate and examine data in RStudio is analogous to positioning biological material on a slide, correctly placing it underneath the lens, and bringing it into focus. This analogy can be taken one step further. In biology class, you also learn how to take lab notes and present your findings. RStudio makes it easy to record your analysis and to present your results with professional-looking documents.

The goal of this chapter is to introduce you to R, RStudio, and R Markdown. We will be using RStudio to execute commands in R. RStudio makes many tasks in R easier since it provides an intuitive interface between you and R. Within RStudio, you will use R Markdown to take notes and generate professional-looking documents.

Mastering statistical software is an important part of analyzing data. Good analysis follows from scientists' skillful use of their lab equipment; scientists skilled in the lab are more likely to make interesting discoveries. The same holds true for you.

R, RStudio, and R Markdown

R is a statistical programming language based on **S**, a programming language developed at Bell Labs in the 1960s. R is an open-source language, administered by the R Core Development Team and the R Foundation. The program is free. Thousands of users contribute by writing their own R code and making it available in **packages**. Thousands of packages exist, all designed to help with specific aspects of data analysis. Packages are add-on features that you install to help make your analysis easier and more effective. Users submit packages to the R online community to test the code. Once testing is complete, the code is made available to all (https://www.r-project.org).

The interface for R, what the user actually sees when running it, is rather austere and based primarily on typing instructions to a prompt. In 2010, J. J. Allaire developed RStudio, an integrated development environment (IDE) that makes using R much easier. While the work you'll be doing in this book will technically be in R, you will be executing the commands, taking notes, and producing reports using RStudio.

Within RStudio, there is a note-taking and document-generating feature called **R Markdown**. R Markdown is a lightweight markup language for working with R based on Markdown, a more general lightweight markup language. Why bother with R Markdown? Why not simply cut and paste graphics and statistical output into a Word file? While you can simply cut and paste graphs and statistical output into Word, I recommend using R Markdown. R Markdown serves two important purposes. First, it makes combining data graphics with text extremely easy, generating documents in HTML, Word, or LaTeX.[1] Second, R Markdown encourages the careful practice of taking notes so that you can easily replicate your results. Communicating results, both to yourself and to a wider audience, is an important but often overlooked part of data analysis. Not here.

This book is designed to be read with your computer at your side. Once you download R and RStudio onto your computer (which we will describe in more detail next), you will be using RStudio to do all of the analysis. You do not need to open R, just RStudio. Almost every table and figure in this book was created with RStudio. There are exceptions, in which case I began with RStudio and edited the figure with Adobe Illustrator. To follow along with this book, you will only need RStudio.

This book is designed so that the examples can be easily replicated. The instructions you type into RStudio (the R code) for each table and figure are presented before the actual table and figure appear in the text so that you can take the code, implement it, and reproduce the same tables and figures. The code will always appear in what will be referred to as a "code chunk" just above the table or figure produced. Just to give an example, the code in Code Chunk 1-1 produces a figure called a scatter plot. I'll go into more detail on scatter plots in this and subsequent chapters. For now, it's important to establish the direct connection between the code chunk and the figure that follows. In Figure 1-1, I plot the relationship between state spending on education per student and the state's median income.

---

[1] PDF files can only be produced if the appropriate LaTeX files are loaded onto your machine. LaTeX accomplishes the same thing as R Markdown but is a much more complete and therefore complicated markup language. If you do have the proper LaTeX files on your computer, when you generate a PDF file RStudio converts the R Markdown markup language to LaTeX. While files generated this way are of exceptional quality since they use the formatting language based on TeX, it can open up a whole new world of syntax that requires mastery. To produce PDF documents, a suite of files must be downloaded. The required software can be found at https://www.latex-project.org/get/#tex-distributions and is available for Mac, Windows, and Linux platforms.

Code Chunk 1-1

```
plot(states$medinc, states$stuspend,
     main = "Figure 1-1: Spending is Higher in Wealthier States",
     col = "#bf0000", pch = 20, font.main = 1)
```

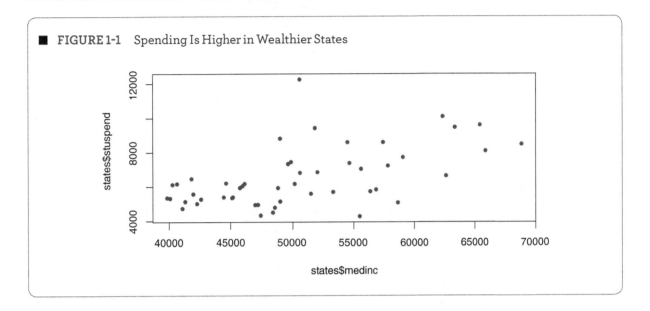

■ FIGURE 1-1 Spending Is Higher in Wealthier States

 Art and Practice of Data Visualization
COPY RELEVANT BITS OF CODE

Experienced practitioners rely on several pieces of code that they know produce specific graphs and statistical analysis. Having a file where that code can be found, copied, and edited is highly recommended. RStudio makes that easy. Rarely are entire analyses created by typing code from scratch (there's simply too much to remember and therefore there are too many possibilities for mistakes).

KNOWLEDGE CHECK: Explain the difference between R, RStudio, and R Markdown.

---

1. Which of the following statements accurately describes R?
    a. A statistical programming language based on S
    b. An integrated development environment (IDE)
    c. A lightweight markup language
    d. A statistical programming language based on RStudio

2. Which of the following statements accurately describes RStudio?
    a. A statistical programming language based on S
    b. An integrated development environment for R
    c. A lightweight markup language
    d. A statistical programming language based on R
3. Which of the following statements accurately describes R Markdown?
    a. A statistical programming language based on RStudio
    b. A lightweight markup language
    c. A feature in RStudio that helps replicate results
    d. A feature in RStudio that helps presents results

Objects and Functions

While this is not a book dedicated solely to learning R, there are a couple of concepts worth mentioning before we embark on our journey. For comprehensive treatments of the R language, I find Kurt Gaubatz's *A Survivor's Guide to R* the most helpful (Gaubatz, 2015). Instead of a full-blown chapter (or two or three) on R, I want to briefly discuss two features of R that I think will help make sense of the instructions you'll be typing or copying into RStudio.

Let's start with functions. In R, the words function and command are interchangeable. **Functions** tell R to take specific actions. They are like the functions you learned in high school math or like the verbs you use every day: *open* the door, *eat* your broccoli, or worse, *finish* your homework. In these examples, the commands or functions are *open*, *eat*, or *finish*. Also note that each function is paired with an object: *door, broccoli, homework*. In R, functions are denoted by a word directly followed by a set of parentheses that contain an object. R code for the examples above would look like *open(door)*, *eat(broccoli)*, or *finish(homework)*. Since R really doesn't physically open doors, eat broccoli, or finish homework (although one could argue it does open doors and it at least helps finish homework), let's use a more realistic example from data we will be using in the book. Here is an actual example from R:

$$\text{hist(states\$hsdiploma)}$$

Here *hist* is the function and *states$hsdiploma* is the object. In this simple example, we're telling R to create a histogram (a simple picture of the data we'll use in our first session). I will provide all of the data we use in this book, which includes information on the 50 U.S. states, information about respondents from a survey, and information collected from countries throughout the world. In this example, the data we will use is the percentage of each U.S. state's adult population that finished high school. In the data, it is called *states$hsdiploma*. Whenever you see a word before parentheses, you'll know that word is a function. There are lots of functions in R, and you can define your own.

The other basic concept is the **object**. In the everyday examples above, the objects are *door*, *broccoli*, and *homework*. While functions might be considered verbs, objects would be nouns. In the R example above, *states$hsdiploma* is the object on which the function *hist* operates. R is what we call an object-oriented language. That means R is designed to work with objects. In the example above, the object *states$hsdiploma* is a list of numbers. An object can be a list of numbers, several lists of numbers (what we'll call a data set), or a variety of other things. You can define your own objects. R code that defines objects can be recognized by the combination of the *lesser than* sign and the dash ("<-"), which essentially means "is equal to." Consider the following example:

myobject <- hist (states$hsdiploma)

In this example, I am defining an object named *myobject* as a histogram of the measure named *states$hsdiploma*, the percentage of each state's adult population with a high school diploma. The structure of *states$hsdiploma* is simply a list of numbers, one number for each state. To help solidify the concept of functions and objects, let's define a function and an object with some numbers we'll simply type into the computer. Let's start by defining an object:

newobject <- c(1,1,1,1,3,4,5,5,5,5,6,6)

The code above tells R to create an object called *newobject* and to define it as a list of 12 numbers. Here I decided to make a list of 12 numbers ranging from 1 to 6. The list of numbers has the same structure as *states$hsdiploma*, only we're using numbers we entered ourselves. The list called *states$hsdiploma* contains 50 numbers (one number for each state). The *c* in this instance tells R to combine the numbers into a list. Once we have *newobject* defined as a list of 12 numbers, we can use a function to make a histogram of that list, which is now an object called *myobject*. The command is written as follows:

hist(newobject)

This barely scratches the surface as an introduction to the object-oriented programming language R. The primary reason we're not dwelling too long on the basics of R is simple: the best way to learn data analysis using R is *to do* data analysis using R. Second, current practice in coding, regardless of the language (R, C++, JavaScript, Python, etc.), involves identifying useful chunks of code (often from a variety of different sites on the Internet), copying them, and using them in your work. Being able to recognize a function and an object will carry you a long way. Rest assured, this book contains all of the chunks of code you'll need to do sophisticated, effective, and insightful work.

There is one more note before we proceed. Most of the code you'll see in R is a function (or command) with a set of parentheses. As I noted, in those parentheses is an object of some sort. Just like our example of a door, broccoli, and homework, the object can be many different things. You'll notice as we proceed that in the parentheses there are often additional instructions we give, always separated by commas. Those extra commands allow us to be more specific. In our example of *open(door)*, we could have *open(door, slowly)* or *open(door, just a crack)*. These commands can get fairly long, but they give you that much more control over how the data are manipulated, providing even more insight.

> **KNOWLEDGE CHECK: Distinguish between functions and objects.**

4. Given the analogy between verbs and functions in R, which of the following makes sense?
   a. *homework()*
   b. *finish()*
   c. *combine()*
   d. *numbers()*
5. Given the analogy between nouns and objects in R, which of the following makes sense?
   a. *finish(homework)*
   b. *numbers(combine)*
   c. *homework(finish)*
   d. *combine(numbers)*

Getting Started in RStudio

To follow the material in this book, you need to download four separate things: R, RStudio, and two files from SAGE. Once you've loaded the correct version of R onto your computer (https://www.r-project.org), you can download and use RStudio (https://www.rstudio.com). Once R and RStudio have been installed on your computer, two files are available from the SAGE server that are required: *Getting_Started_with_R.Rmd* and *Art_and_Practice.RData*. Both files can be found at **edge.sagepub.com/brownstats1e**. The first file contains material that will help you get started with RStudio and R Markdown. The second file contains the data you'll be using and a few functions I've created.

>
> **Art and Practice of Data Visualization**
> **KEEP IT TOGETHER**
>
> It's best to create a file folder somewhere on your computer to keep your files together in one place. In terms of best practices, if you keep this folder somewhere in the Cloud (using Dropbox, iCloud, or any equivalent service), your work will always be available from any computer and won't be lost if something happens to your computer.

- Step 1. Open the RStudio application. It should look like Figure 1-2.

There are a number of ways to work with R. In this book, we will use R Markdown so we can create professional-looking documents in Word, HTML, or LaTeX.

- Step 2. Navigate to the upper-left pane and select the yellow folder button . This step is illustrated in Figure 1-2a.

Chapter 1 Getting Started   7

■ **FIGURE 1-2**   View When RStudio Is Opened

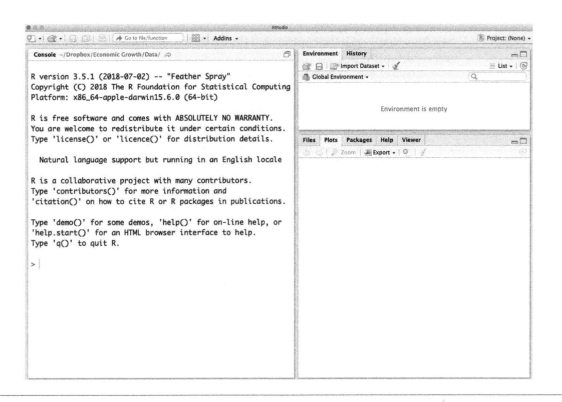

■ **FIGURE 1-2A**   Open the R Markdown File

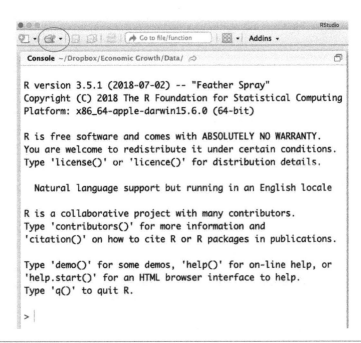

It will open a browsing window where you can navigate to the R Markdown file *Getting_Started_with_R.Rmd*. When you've opened the file, you should see a window pane open up in the upper-left corner of RStudio. Figure 1-3 illustrates what you should see.

Notice there's a gray area within the R Markdown file. The gray area is called a **code chunk**, which is where we place our instructions (the actual R code). Since this is the first code chunk that appears in the window, I've entitled it *setup* and it should include instructions to load the data and the address of where they reside on your computer.

- Step 3. Navigate to the upper-right pane and click on the yellow folder , which will allow you to browse for the data file (Figure 1-4a).[2] Find *Art_and_Practice.RData* on your computer and then select it. If you complete this step correctly, you should see three data sets in the upper-right pane: *nes*, *states*, and *world* (Figure 1-4a).

Note that when you select the file, the load command with the actual address of that file appears in the lower-left pane (the console pane). It will look similar to what is depicted in Figure 1-4b but will vary depending on where you placed the file on your computer: load ("~/Art_and_Practice.RData"). You can copy that command starting with the "l" in *load*.

Once you've copied that command, paste it on any line in the code chunk (Figure 1-4c). In Figure 1-4c, I drew a red ellipse that indicates where you can paste the code. That tells R Markdown where your data are so that you can **knit**[3] this file into a Word, HTML, or LaTeX document.

■ **FIGURE 1-3** R Markdown File Appears

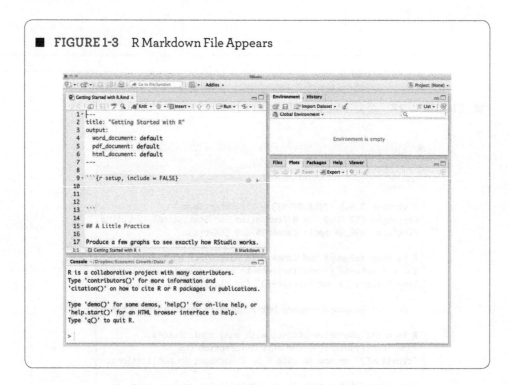

---

[2] The data file *Art_and_Practice.RData* includes data but also some functions. By data, I mean the actual numbers you'll need to make graphs, perform statistical tests, and start exploring important subjects (e.g., gun control laws in the United States, democracy across the world). Defining data and data sets more fully is saved for Chapter 2.

[3] "Knitting" is the term we use when converting the R Markdown file to a document.

■ **FIGURE 1-4A** Opening the Data File

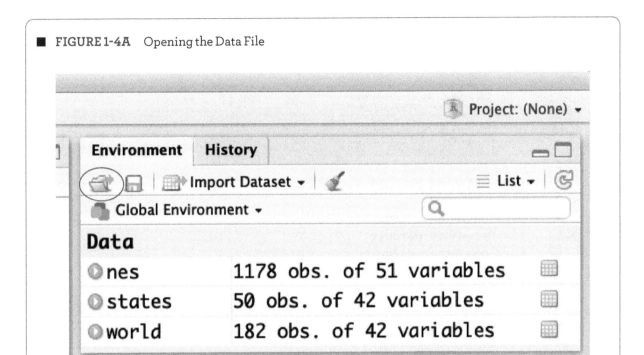

■ **FIGURE 1-4B** Copying the *load* Command

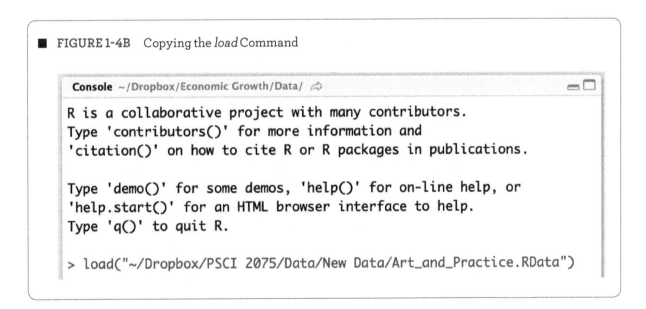

- Step 4. Install the packages you'll need for this book. Execute the function I defined as *installD()* by typing it at the prompt in the console window (see Figure 1-5a). Included in the data file are a few functions written for this book. The function *installD()* is one of them and it installs all of the packages you need.

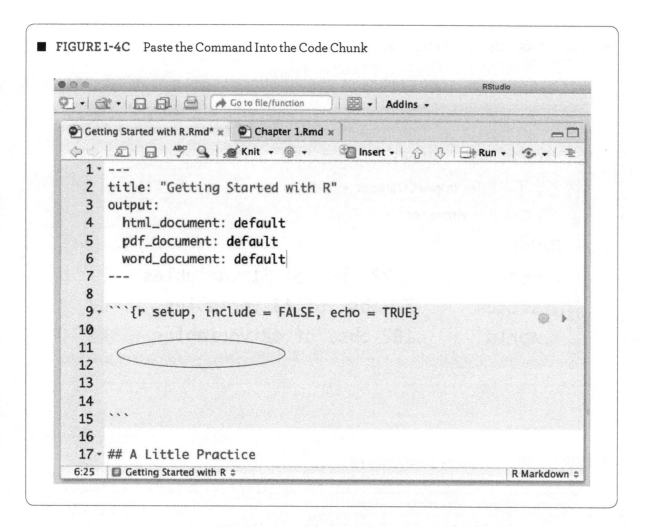

FIGURE 1-4C Paste the Command Into the Code Chunk

Once you've typed *installD()* and pressed enter to the prompt in the console pane (lower-left), the computer will start installing the packages. This might take 3–5 minutes or a bit longer depending on the speed of your computer. Once all of the packages are installed, the console pane will return you to the prompt.

- Step 5. Type *libraries()* at the prompt in the console pane (lower-left) and press enter to load all of the packages you just installed into RStudio so that you can use them (see Figure 1-5b). This is a specific command written for this book.

The *libraries()* command is similar to the *library()* command, which you can use to load one package at a time. You'll see some activity in the console as each package is loaded. This procedure usually lasts only a manner of seconds. Once you're back to the prompt in the console, you're ready to go!

To review, the first time you open RStudio, you need to type *installD()* in the console in order to install the packages for the book. The *installD()* command grabs all of the packages you need from the Internet. As you progress beyond this book, installing additional packages will be necessary. Fortunately, the RStudio interface makes installing packages easy. The process involves nothing more than clicking the *Packages* tab in the lower-right pane, clicking the

FIGURE 1-5A Install the Packages for the Book

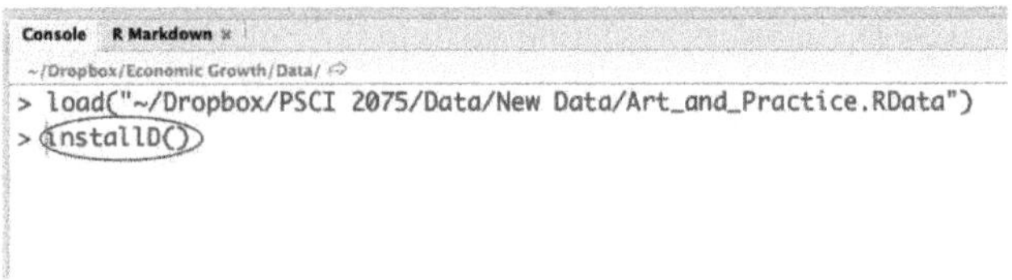

```
> load("~/Dropbox/PSCI 2075/Data/New Data/Art_and_Practice.RData")
> installD()
```

FIGURE 1-5B Loading Packages

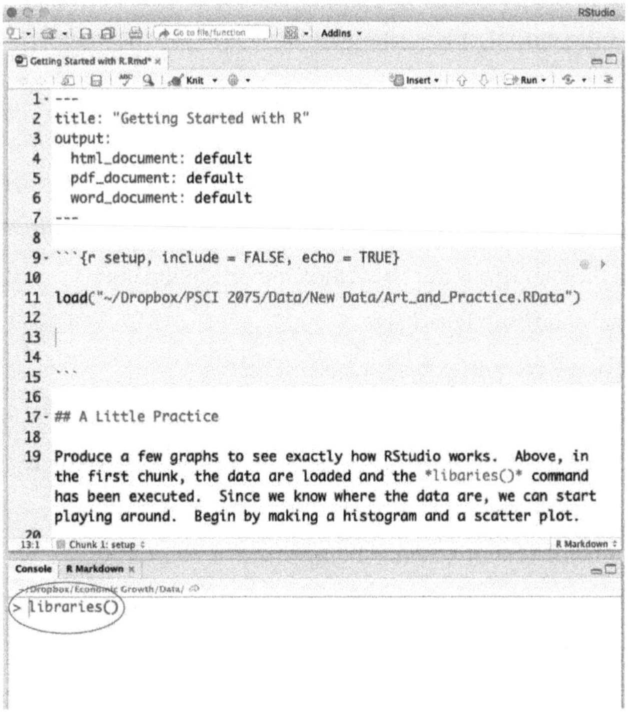

```
1  ---
2  title: "Getting Started with R"
3  output:
4    html_document: default
5    pdf_document: default
6    word_document: default
7  ---
8
9  ```{r setup, include = FALSE, echo = TRUE}
10
11 load("~/Dropbox/PSCI 2075/Data/New Data/Art_and_Practice.RData")
12
13
14
15 ```
16
17 ## A Little Practice
18
19 Produce a few graphs to see exactly how RStudio works.  Above, in
   the first chunk, the data are loaded and the *libaries()* command
   has been executed.  Since we know where the data are, we can start
   playing around.  Begin by making a histogram and a scatter plot.
```

```
> libraries()
```

install button **Install**, and typing in which package you want to install. RStudio does the rest.

Remember that once RStudio stops installing packages (the cursor should come back to the '>' prompt in the console), you need to type *libraries()* and press enter. The libraries command (a function I've defined for the book) makes the packages you've installed available. While you should only have to run *installD()* once, you'll need to type *libraries()* every time you restart RStudio. To run all of the commands in the setup chunk, click the green arrow ▶ that appears in the upper-right part of the code chunk. If everything has gone according to plan, this will load the data and execute the *libraries()* command.

Before we continue, a note about packages is in order. One of R's strengths is that there are thousands of packages available. At the same time, that is one of its weaknesses. With so many packages written by so many different contributors, some packages have functions with the same name. For example, the *select()* function exists in the 'dplyr' package and in the package called *MASS*. Sometimes one package will "mask" another. When you call a function, R will simply run the function with the package last loaded. In order to specify the function you want, sometimes it's necessary to type the package name followed by two colons ("::"). For example, in Chapter 3 you'll see that the *select()* command is preceded by *dplyr::* in order for the function to run properly. In this book, there are just a few code chunks where this happens. In each case, I will draw your attention to those instances.

> **KNOWLEDGE CHECK: Download and install both R and RStudio.**

6. Which of the following accurately describes the *installD()* command?
   a. It installs and loads R packages.
   b. It is a base R command that loads all packages.
   c. It is an author-defined command that loads all packages for this book.
   d. It is an author-defined command that installs all packages for this book.
7. Which of the following is true about the *libraries()* command?
   a. It installs and loads R packages.
   b. It is a base R command that loads all packages.
   c. It is an author-defined command that loads all packages required for this book.
   d. It is an author-defined command that installs all packages required for this book.
8. Which of the following statements are true?
   a. The *installD()* command must be typed in at the beginning of every session.
   b. The *libraries()* command must be typed in at the beginning of every session.
   c. The file *Art_and_Practice.RData* contains the data you'll need for this class.
   d. You should only need to type the *installD()* command once in this class.

Navigating RStudio With R Markdown

There are many different ways to work with RStudio. While this book shows how to work with R Markdown, you can follow along by simply copying or typing in commands to the console pane. I use R Markdown for several reasons. First, it places your code next to the images produced, providing a direct connection between your code and analysis. Consequently, it encourages good note-taking and careful analysis. Second, it helps us and others replicate our work, an important aspect of the scientific process. With the notes and code placed together, others can follow how you arrived at your conclusions. Perhaps even more importantly, it makes it easy for *you* to remember how you arrived at your conclusions. Finally, R Markdown generates professional-looking documents in a variety of formats. How you present your findings can be just as important as how you generated them in the first place. This section provides a brief introduction on how to use R Markdown in RStudio.

Before we continue, let's get oriented with the basics: the YAML, the code chunk, and how to execute commands in R Markdown. Let's pick up where we left off in the last section. Your screen should resemble Figure 1-6a. The R Markdown file *Getting_Started_with_R.Rmd* is open and we already added the *load()* command that brings in our data file and the *libraries()* command that loads the packages we'll need.

For future reference, to create a new R Markdown file, simply click on the arrow next to the file icon with the green plus sign and select R Markdown. It is located at the very top left of the RStudio window.

At the beginning of the R Markdown file, demarcated by three horizontal dashes, there are instructions that specify the main title and a list of the documents we can create: PDF, Word, and HTML (see Figure 1-6b). This section is referred to as the **YAML**, which stands for *YAML Ain't Markup Language*. It contains commands that set features for the entire document. When you open a new R Markdown file, it will have basic YAML instructions already included. We're going to keep the YAML simple for now, but if you want to take advantage of additional commands, consult the *R Markdown Cheat Sheet* available at the RStudio website (https://resources.rstudio.com/rstudio-developed/rmarkdown-2-0).

In addition to downloading the cheat sheet from the RStudio website, you can access the entire set of cheat sheets by simply going to the RStudio help menu at the very top of the screen. In Figure 1-6c, you can see how to access the various cheat sheets without leaving RStudio.

The more important part of the R Markdown document is the gray area called the *code chunk* (see Figure 1-6d). Code chunks include the actual R code that executes commands. They are created by typing three apostrophes and then a bracketed {r}, which marks the

■ **FIGURE 1-6A**   RStudio With R Markdown File

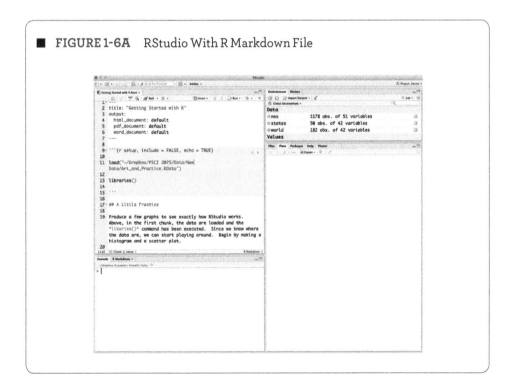

■ **FIGURE 1-6B** YAML

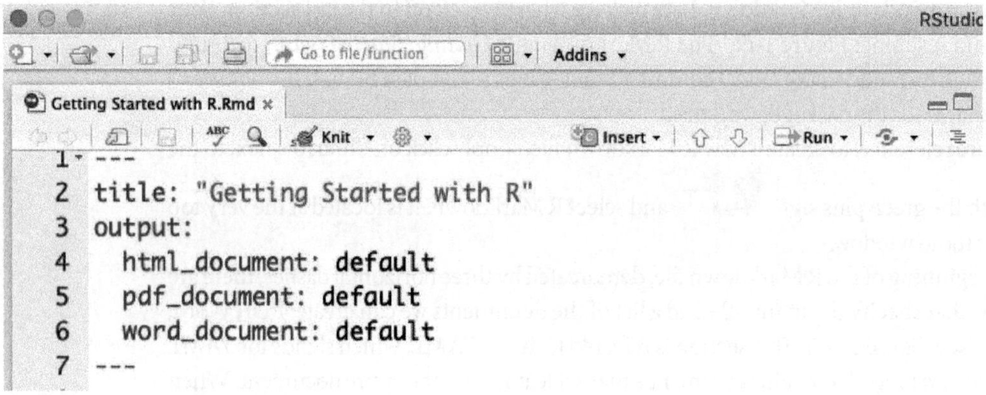

■ **FIGURE 1-6C** Accessing the Cheat Sheet for R Markdown

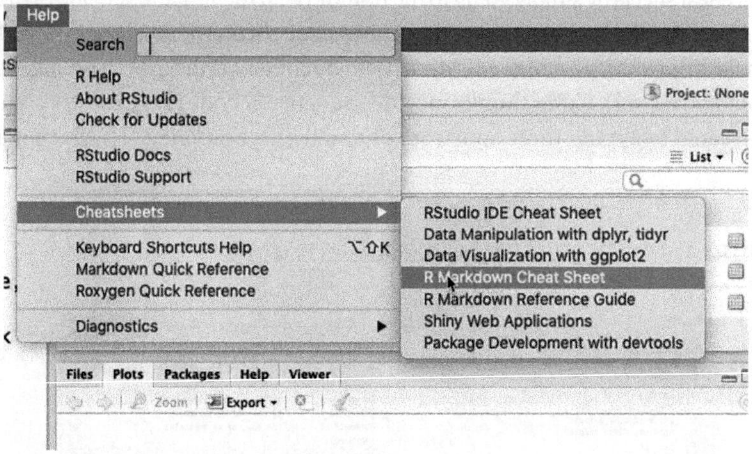

■ **FIGURE 1-6D** The Code Chunk

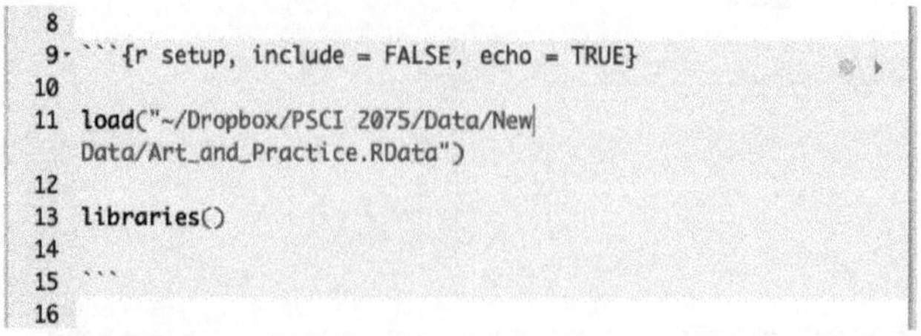

beginning. Three more apostrophes indicate where the code chunk ends. You can also create a code chunk by typing *control+command+i* on a Mac or *control+alt+i* on a Windows machine.

The code located at the top of the code chunk (in between the brackets) tells R Markdown to do several things. First, the *include = FALSE* statement tells R Markdown to not include any extraneous output from that specific code chunk in the document. For example, if the *include* statement is set on *TRUE*, the knitted document will show all of the output produced when the *libraries()* command was executed. We don't want all of that to appear in our document, so we type "FALSE".

Second, the *echo = TRUE* instruction tells R Markdown to include the code you entered in the code chunk, which is a good way to show your work (see Figure 1-6e). In each case, if you change the capitalized TRUE and FALSE, the opposite will happen.

The instructions (the actual R code) you either type or paste are placed inside the gray area (see Figure 1-6f). The command *load()* indicates, for example, what data file you are going to use. For this book, the name of the file is *Art_and_Practice.RData*. Once the command is executed, the data are loaded into RStudio.

Now that we've specified the code chunk options we want and we have the actual commands entered into the code chunk, we can execute them by clicking on the *Run* button at the top of the upper-left pane or by clicking on the green arrow in the code chunk (see Figure 1-6g). Again, I've placed red ellipses marking the two locations where you can execute the code. You can also execute each line of code separately by simply placing the cursor anywhere in the line and typing *command+enter* on a Mac or *control+enter* on a Windows machine.

■ **FIGURE 1-6E** Chunk Options

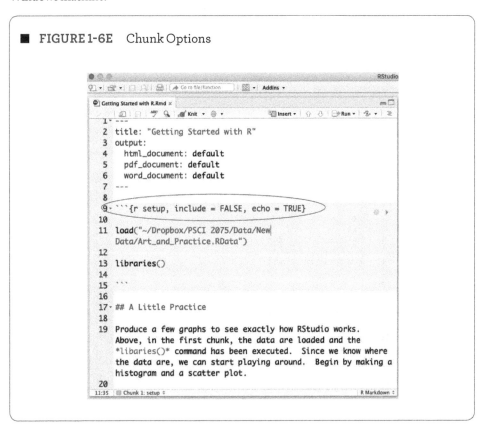

FIGURE 1-6F The R Code That Is Executed

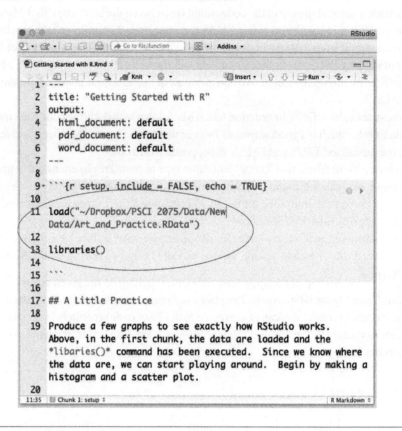

FIGURE 1-6G Executing Commands

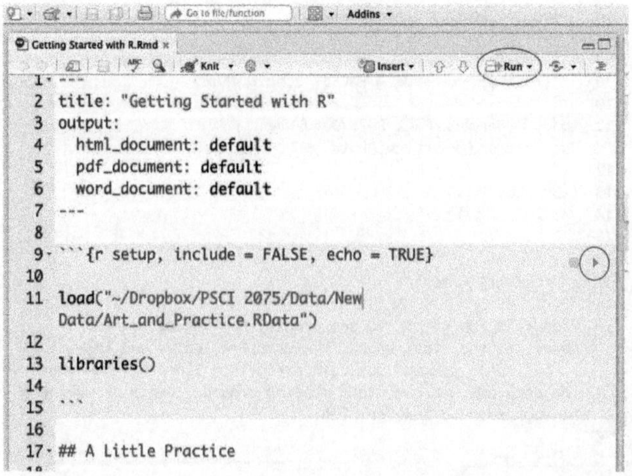

Finally, anything appearing in the white areas is simply the text you'd like to include in your document (your notes). Type anything you'd like into this space (see Figure 1-6h). R Markdown is a stripped-down version of more elaborate markup languages. This is intentional so that you don't have to spend time learning more code. All of the commands you'll need to produce a professional-looking document are available from the drop-down menu that selects the R Markdown cheat sheet (refer back to Figure 1-6c).

There are a number of ways to work with RStudio. Described below are a set of best practices to get started. Once you are more comfortable with executing commands, analyzing output, and knitting documents, develop your own workflow when analyzing data. Before we proceed, let's review the different panes in the RStudio window.

■ **FIGURE 1-6H** Where to Type Text

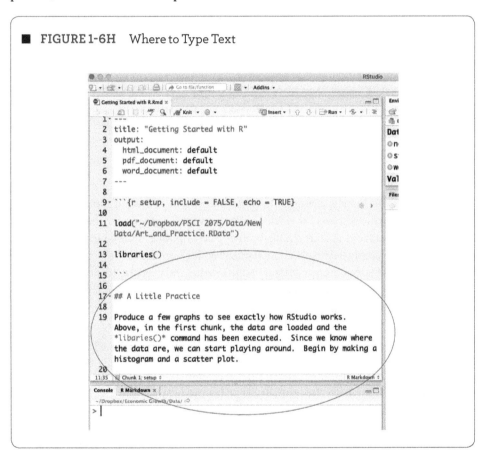

Art and Practice of Data Visualization
A MULTITUDE OF OPTIONS

There is no single correct way to operate in RStudio. There are multiple options available to achieve the same estimates, pictures, and ways to present the results. More than anything, a healthy curiosity and sense of humor are required to develop your own effective workflow that originates with a data set and ends with important insights.

- *The upper-left pane.* Most of your work will occur in the upper-left pane of RStudio. This is where you can create R Markdown files that record your work, notes, and observations. It's a good idea to develop the habit of executing commands in this manner, since they are all saved and easy to access.

- *The lower-left pane (the console).* In the console pane, you can observe the commands being executed. You can type commands directly into the console if you don't necessarily need to record them. For example, before running more sophisticated analyses, you may want to see a simple picture of the data before you proceed.[4]

- *The lower-right pane.* The pane in the lower-right corner is used to manage packages and to view the figures and tables you create. If you're using an R Markdown file and you want tables and figures to appear in the lower-right pane, find the gear wheel in the upper-left pane and select Chunk Output in Console. Once they appear in the lower-right pane, they can be enlarged, saved as files, and exported as individual items.

- *The upper-right pane (the global environment).* This pane shows what data sets are available and what functions you've defined. It helps you keep track of the data and objects you are working with.

Using R Markdown Files Versus R-Scripts

As noted earlier, this book was written in R Markdown. All of the code you'll need to execute is contained in the sections I've labeled as "code chunks" in the book. They proceed each table and figure. Before each code chunk, I provide descriptions of the code to explain what is going on. You'll simply need to become familiar with where that code is, type it into your own R Markdown file, make any necessary adjustments, and execute. To aid in locating the code you'll need, at the end of each chapter is a guide that lists all of the R functions I use in the chapter and in what labeled code chunk they first appear.

R-scripts are files that contain code you want to save. You can't generate nice-looking documents from them, but they serve as more informal ways to save your code for a specific task so that you can retrace your steps. If you don't plan on writing in-depth commentary or producing a document, operating with R-scripts is highly recommended. Note that they don't include code chunks like R Markdown. The entire R-script document can be thought of as one big code chunk in R Markdown. As long as there's no hashtag in front of the command, it can be executed as if it were in the console. As mentioned earlier, this book is structured to follow along using R Markdown rather than R-scripts. Nevertheless, operating RStudio with R-scripts and following this book is easily done if that's what you prefer.

> **KNOWLEDGE CHECK: Identify the main features of R Markdown.**

9. Indicate which statements describe the YAML section of an R Markdown document.

    a. It provides a place to execute commands in R.

    b. It contains commands that set features for the entire document.

---

[4] Note that you can generate the previous command in the console pane by typing the up arrow on your keyboard.

c. It is where you record the most often-used commands.

d. It always appears at the end of the R Markdown document.

10. Which of the following features of a code chunk is true?

    a. It provides a place to execute commands in R.

    b. Code chunks appear as gray boxes in an R Markdown file.

    c. The only way to execute a code chunk is by clicking the green arrow.

    d. You should indicate which data file to use in the first code chunk.

11. Which of the following describes navigation in RStudio?

    a. You can only execute commands in the upper-left pane.

    b. You can execute commands in the console pane.

    c. Typing *command+enter* (Mac) or *control+enter* (Windows) executes commands.

    d. Typing *control+command+i* (Mac) or *control+alt+i* (Windows) generates a code chunk.

12. What is the difference between an R Markdown file and an R-script?

    a. R Markdown files should be used for producing documents for presentation.

    b. R-scripts should be used for producing documents for presentation.

    c. R Markdown files should only be used for saving often-repeated or complex combinations of code.

    d. R-scripts files should only be used for saving often-repeated or complex combinations of code.

A Little Practice

Let's produce a picture of our data to see how everything works. To review, we loaded the data, executed the *libraries()* command, and opened up the R Markdown file *Getting_Started_with_R.Rmd*. Now we can start playing around. Let's start by making a histogram that reveals the data's distribution. Suppose we're interested in educational attainment. We have a **variable**, a list of numbers, that records the percentage of a state's adult population with a high school degree (*hsdiploma*). A collection of variables is called a **data set**. Stated properly, *hsdiploma* is a variable in the *states* data set. I give a more complete description of variables and data sets in Chapter 2. For now, let's see what the variable called *hsdiploma* actually looks like.

Code Chunk 1-2 draws a histogram, a useful picture that indicates the lowest, highest, and most common values of a variable (Figure 1-7). Code Chunk 1-2 tells R to draw a histogram. Remember, the code can either be typed into an R Markdown code chunk or executed in the console pane. Since we're using *Getting_Started_with_R.Rmd*, the code is included in the code chunk. Execute the code by clicking on the green arrow ▶ found at the upper-right corner of the code chunk to produce Figure 1-7.

Code Chunk 1-2

```
hist(states$hsdiploma, main = "Figure 1-7: This Is My First Plot!",
     font.main = 1, col = "#0000bf")
```

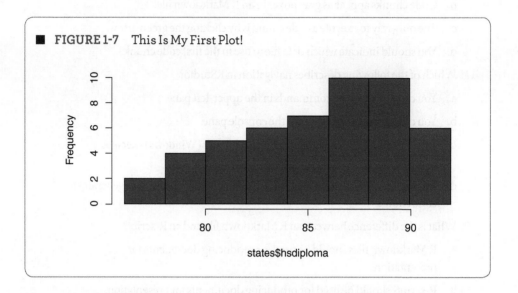

FIGURE 1-7 This Is My First Plot!

As I explained earlier when introducing functions and objects, all functions in R begin with the name of the function and a set of parentheses where you insert information that the function needs (the object). Note that we use the dollar sign ($) to separate the data set name from the variable name. By specifying the data set and the variable name, R knows which data set you're using and specifically which variable you'd like to see. In Code Chunk 1-2, we use the function *hist()* to draw the histogram. Specify which variable to use (*states$hsdiploma*) within the parentheses. You can also add a title (*main =*), specify the font for the title (*font.main =*), and indicate a color (*col =*).

By clicking on the right-pointing green arrow ▶ in the upper-right corner of the code chunk, R studio will execute the command and draw the histogram.

Let's draw another picture, a scatter plot of the two variables *murderrate* and *hsdiploma* from the *states* data set. Scatter plots show how two different variables relate to one another. In the following example, we'll look at educational attainment and murder rates (the number of homicides per 100,000 population) in the U.S. states. Note the appearance of *pch =*, which specifies the shape of the dots.

Code Chunk 1-3

```
plot(states$hsdiploma, states$murderrate,
     main = "Figure 1-8: Education Reduces Violence \n",
     font.main = 1, col = "#bf0000", pch = 19)
```

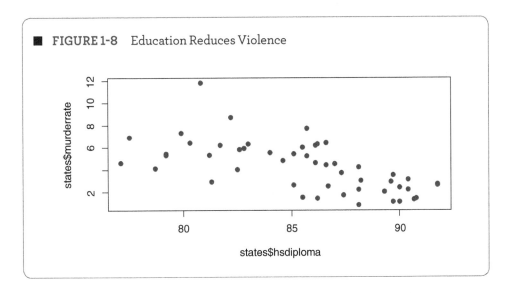

■ FIGURE 1-8   Education Reduces Violence

You can change the settings so that your output will appear in the bottom-right pane for closer inspection. To do that, go to the gear icon at the top of this window ⚙ ▼ and choose *Chunk Output in Console*. Now that we've loaded the data, loaded the packages, and produced a few plots, let's *knit* the document.

Once you've checked to make sure all of the code chunks in the R Markdown file run properly by clicking the green button in each one, choose the small arrow by the *Knit* icon 🪡 Knit ▼ at the top of this window and choose either *Word* or *HTML*. Once you've done that, you'll see RStudio going through the process of generating a document (you'll see its progress in the console window). If done correctly, you should have a nice-looking HTML or Word document of this brief analysis.

As an additional example, let's make a picture of some data we've created by hand to demonstrate the power of R as an object-oriented programming language. Earlier, we discussed the difference between executing a function and defining an object. We've executed two functions already to create a histogram and a scatter plot.

For clarity, we'll use the example from our discussion on functions and objects. In *Getting_Started_with_R.Rmd*, you'll see a code chunk that defines an object called *myobject* as a list of 12 numbers (the example we used earlier). Once that object is defined, we can put it inside the parentheses for the function to act on it. R knows whether it's the right kind of object or not. If you try to execute a function on an object that is inappropriate, R will object!

To explain why R might object, let's refer back to our example *eat(broccoli)*. R expects some kind of food when you're using the command *eat()*. If you typed in *eat(glass)*, R would spit it out (give you an error message). It knows *glass* is not the right kind of object given the function *eat*.

Since the commands are already included in the code chunk, simply click the green arrow ▶ on the upper-right side of the chunk to execute. You should see a histogram that looks like the one in Figure 1-9. This is the beauty of object-oriented languages: once we've defined an object, there are a multitude of functions we can apply to it.

In Code Chunk 1-4, there are two commands. First, an object (*myobject*) is defined. The function *c()* tells R to combine the elements separated by commas into a list (a variable in this case). Having defined an object called *myobject* that contains 12 numbers, we draw a histogram of that variable. We use the same *hist()* function as in Code Chunk 1-2, only this time we color it green.

Code Chunk 1-4

```
myobject <- c(1,1,1,1,3,4,5,5,5,5,6,6)

hist(myobject, main = "Figure 1-9: Object that You Defined",
     font.main = 1, col = "#00ff00")
```

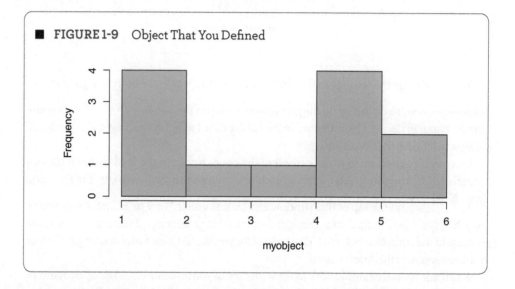

■ FIGURE 1-9   Object That You Defined

As we discussed earlier, the purpose of this chapter is to get you up and running with RStudio. Just copying the code and running it yourself is an important step. But don't stop there. See how small changes to the code change the output. Gaining the facility necessary to be a good data analyst comes with practice. Let's take the previous example and make some modifications.

Take the code in the last code chunk and alter it. Just to get you started, I copied the previous code chunk, pasted it into Code Chunk 1-5, and played around with the numbers in the variable *myobject* to change the shape of the histogram. I converted numbers to threes and fours to emphasize the middle of the range. I also took the color option and changed it from green (#00ff00) to red (#bf0000). The result is displayed in Figure 1-10.

Code Chunk 1-5

```
myobject <- c(1,2,3,3,3,4,4,4,5,5,6)

hist(myobject, main = "Figure 1-10: Object that You Defined",
     font.main = 1, col = "#bf0000")
```

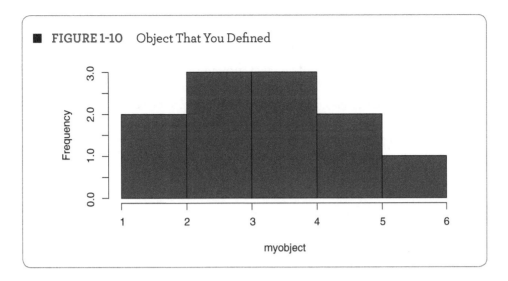

FIGURE 1-10 Object That You Defined

As you'll learn, colors and shapes can easily be overdone, so much so that they distract from understanding what the data actually say. We'll pay more attention to that as we progress through the chapters. For now, to quote famous school teacher, Miss Frizzle:

Take chances. Make mistakes. Get messy!

> **KNOWLEDGE CHECK:** Create a Word or HTML file from R Markdown.

13. Identify which commands will likely produce an error message.
    a. play(football)
    b. play(homework)
    c. eat(pizza)
    d. eat(spoon)

14. What does *knitting* do?
    a. Executes the R commands
    b. Generates an HTML document
    c. Generates a Word document
    d. Generates a PDF

15. Which of the following statements are true?
    a. Used within the *hist()* command, *main* = allows you to specify the color of a histogram.
    b. Used within the *hist()* command, *states$hsdiploma* specifies the variable to be plotted.
    c. Used within the *hist()* command, *col* = specifies the column number to use.
    d. Used within the *plot()* command, *pch* = specifies the shape of each point to be plotted.

SUMMARY

This chapter introduced R, RStudio, and R Markdown. You also learned how to load data, load the packages necessary for executing commands, and generate some simple pictures of the data. While this chapter is no substitute for helpful books devoted to learning the R programming language, you now have enough skill to start exploring.

If you play around with RStudio in these beginning chapters, you'll quickly develop the skills necessary to do some amazing work and to make some interesting discoveries. A few short commands and basic knowledge of RStudio change everything.

Art and Practice of Data Visualization
CURIOSITY AND HUMOR

Very much like writing, the way we go about coding and organizing our work reveals a lot about ourselves. Our level of patience will be on full display. One of the strengths and drawbacks of R is that there are a multitude of ways to do the same thing. While there is a lot of online help and advice, it will differ. Occasionally, since much of what we use in R is written by many different people, there are inconsistencies that can make it difficult to troubleshoot. Add to that, learning a new system of syntax can lead to despair. Learning the basics of R is not for the faint-hearted. The more you develop a sense of calm and a sense of humor throughout the enterprise, the faster you'll learn. Getting stuck and developing an effective way to get unstuck can be very rewarding and can be translated from this realm into many others. It's called critical thinking.

COMMON PROBLEMS

- Most of the problems I've seen at this stage include one of two things: (1) the versions of R, RStudio, and the computer's system software are not compatible; or (2) when the *installD()* command is executed, a clean and complete installation of the packages does not occur. So, pay attention to the version numbers of R and RStudio and make sure they are compatible with your system's software.

- Pay attention to what is happening once the *installD()* command is executed. Be on the lookout for packages that don't load or if the process ends abruptly. If the following *libraries()* command does not execute, go back and try the *installD()* command again. Sometimes, it's helpful to reinstall both R and RStudio and start again. Taking the extra 5–10 minutes to reinstall the applications and to install all of the packages properly will prevent problems down the road.

- Once you're up and running, you may find that something isn't working correctly. There are all sorts of sites to find help, but here are some of the main problems I've run across at this stage of the journey:

    ○ There was a typo in the command.

    ○ A copy of R from another class already existed on the computer.

- When installing packages, some files got corrupted.
- Not all of the packages from the *installD()* command got installed.
- Compatible versions of R and RStudio were not downloaded.
- The *libraries()* command was not executed when RStudio was opened.
- The *installD()* command was never executed.
- The versions of R and RStudio were not compatible with the computer's system software.
- The *Art_and_Practice.RData* file was not opened in the correct pane.

REVIEW QUESTIONS

1. What is RStudio?
2. What is R Markdown?
3. What does the console pane in RStudio do?
4. What is the difference between an object and a function in R?
5. How are commands executed in R Markdown?
6. When should you use an R-script?
7. What does the keystroke combination *command+enter* achieve? (Mac)
8. What are the advantages in using R Markdown when navigating RStudio?
9. What is *knitting*?
10. What color is #00ff00?

PRACTICE ON ANALYSIS AND VISUALIZATION

1. What is the difference between installing and loading packages?
    a. Loading packages has to be performed each time RStudio is reopened.
    b. Installing packages has to be performed each time RStudio is reopened.
    c. Installing packages occurs after they've been loaded.
    d. Loading packages occurs after they've been installed.
2. Which statements generally characterize navigating RStudio?
    a. There is a correct and widely accepted way to navigate RStudio.
    b. R-scripts are the best way to save your notes and commands as you are performing analysis.
    c. R Markdown files are the best way to save your notes and commands as you perform your analysis.
    d. Always execute commands in the console pane.
3. Which of the following would be likely functions?
    a. eat
    b. pray
    c. calculate
    d. Sally
4. Which of the following would be likely objects?
    a. eat
    b. pray
    c. prayer
    d. broccoli

5. Which commands make sense in the R context?
   a. eat(sandwich)
   b. broccoli(eat)
   c. eat(broccoli, steamed, everyday)
   d. eat(steamed, broccoli, everyday)

6. What is a code chunk in R Markdown?
   a. Defines features for the entire R Markdown document
   b. Contains executable code in R
   c. R code that is knitted in an R Markdown file
   d. A great place to write comments on your code

7. Which of the following statements are true?
   a. RStudio will run without having R open and running.
   b. R will run without RStudio open and running.
   c. RStudio will run if R has been installed on your computer.
   d. RStudio will run only if R is open first.

8. Which of the following statements are true?
   a. *installD()* installs all of the packages you'll need for the book.
   b. *libraries()* is an author-defined command that installs the packages you'll need for this book.
   c. *libraries()* is an author-defined command that loads the packages you'll need for this book.
   d. you need to execute *installD()* every time you open RStudio.

9. What best describes the activity in the console pane?
   a. Execution of single commands
   b. Execution of commands you don't need to save
   c. Execution of long combinations of code
   d. A good place to save commands for future use

10. What will you need to navigate RStudio successfully?
    a. A background in differential calculus
    b. At least one class in programming
    c. A previous class in statistics
    d. A sense of humor

ANNOTATED R FUNCTIONS

The following functions appear in this chapter. They are listed in order of their first appearance (with the code chunk number in parentheses) and annotated here to give a very brief description of their use. Some are not stand-alone functions and only work in combination with other commands. As a reminder, the code in every chapter will work properly if executed in the order it appears. Proper execution also depends on typing the author-defined *libraries()* command, which loads the required R packages.

plot(): basic R command to generate a plot. In this chapter, it is used to produce a scatter plot. (1-1)

hist(): basic R command to generate a histogram. (1-2)

c(): function used to combine elements separated by commas into a list. (1-4)

ANSWERS

KNOWLEDGE CHECK

1. a
2. b
3. b, c, d
4. b, c
5. a, d
6. b
7. a, b, d
8. b, c, d
9. a, d
10. b, d
11. b, c, d
12. b, d
13. b, d
14. b, c, d
15. b, d

PRACTICE ON ANALYSIS AND VISUALIZATION

1. a, d
2. c
3. a, b, c
4. c, d
5. a, c
6. b, c
7. a, b, c
8. a, c
9. a, b
10. d

 Access digital resources, including datasets, at http://edge.sagepub.com/brownstats1e.

2 An Introduction to Data Analysis

CHAPTER OUTLINE

Learning Objectives
Overview
Motivating Data Analysis
The Main Components of Data Analysis
Developing Hypotheses by Describing Data
Model Building and Estimation
Diagnostics
Next Questions
Summary
Common Problems
Review Questions
Practice on Analysis and Visualization
Annotated R Functions
Answers

LEARNING OBJECTIVES

- Discuss the goals and scope of the book.
- Explain the importance of acquiring skills in data analysis.
- List the components of data analysis and how they fit together.
- Form hypotheses from descriptions of data.
- Explain the connection between hypotheses, models, and estimates.
- Define diagnostics and explain their role in data analysis.
- Formulate new questions.

Overview

Scholars, practitioners, and policymakers interested in explaining human behavior are drowning in data. This book is designed for those who want to swim safely to shore. The goal is to introduce the method, logic, art, and practice of data analysis. Specifically, the book provides the essential skills and tools necessary to examine data in the service of solving problems. Toward that end, you will learn to marry the art and practice of data analysis with how experienced practitioners approach problems, formulate hypotheses, estimate models, and present their results. The underlying philosophy is learning by doing. Learning statistics along with the art and practice of data analysis is best achieved by *doing* data analysis.

This introductory chapter is organized as follows. First, I discuss some of the motivations behind analyzing quantitative data. Second, I provide a short exploration of an important question in the social sciences to illustrate the main components of data analysis. The example does not fully explain each component. The goal here is to provide a road map for the rest of the book.

The book is organized according to the process of data analysis laid out in this chapter: describing data and formulating hypotheses, building and estimating models, diagnostics, and generating the next question. Presenting your results also forms an important part of this book. The process of data analysis is iterative: it is an ongoing conversation that goes back and forth between all of the different components. For example, the estimates we generate from our statistical models rarely settle the matter. New and better questions are often the result.

Motivating Data Analysis

The motivations are many, but here are three of the most rehearsed. First, more things are being measured and quantified. Not only are more data available on demographic and financial trends, our daily activity (driving, dating, shopping, and listening to music) is available for analysis. Second, and partly related to the first, data analysis is in demand. Consequently, the skills associated with analyzing quantitative data are marketable. Third, data analysis helps us separate facts from fiction, a defining characteristic of successful democratic societies.

Big Data Are Getting Bigger

A significant part of our lives is spent on the Internet. The bread crumbs left behind are used by government, private industry, and others. Politicians are driven (in part) by survey results. CEOs make decisions based on data collected from their employees and their customers. Here are just a few examples of data analysis that reveal both its ubiquity and effectiveness.

1. Noticing an increasing amount of junk mail from Target advertising diapers, baby formula, and onesies, the inundated recipient calls Target to ask why they're sending him so much baby-oriented advertising (it had been years since there was a baby in the house). Target explained that recent purchasing data indicated there was a pregnant woman in his household. One week after calling Target, he discovers his daughter is pregnant.

2. By constructing a meta-analysis of polls, Nate Silver (in his FiveThirtyEight column) correctly predicted the outcome of the presidential race in 2012 in all 50 states.

3. When diseases spread (e.g., the Ebola outbreaks in Africa), the U.S. Centers for Disease Control and Prevention and other authorities previously relied on 2-week-old data collected from hospitals and clinics. In a quickly developing scenario like the H1N1 (bird flu), Ebola, Zika, or COVID-19 virus outbreaks, 2 weeks is too late. A team from Google let the Internet figure out where outbreaks occur. To develop their model, they traced the spread of H1N1 and correlated it with search terms on the Internet: high fever, cough, and aches. They eventually found 45 phrases that allowed them to track the outbreak, informing officials when and where, exactly, the newest flu outbreak was occurring.[1]

4. In an article entitled "China Invents the Digital Totalitarian State" (2016), *The Economist* details the plans of a new Chinese program called the social-credit project designed to increase trust in society by making social-credit scores available. This example, more than the others, illustrates how big data can be abused. In what is called the "judgment defaulter's list" (a tally of citizens who have defied a court order), Chinese authorities keep track of who they consider to be trustworthy in society to help spur economic cooperation and trust among its citizens.

---

[1] The examples above can be found in *Big Data: A Revolution That Will Transform How We Live, Work, and Think* by Kenneth Cukier and Viktor Mayer-Schönberger (2013). Google's attempts to track flu outbreaks have not come without criticism. For a nice discussion of the shortcomings associated with big data, see the discussion in Lazer et al. (2014).

5. To help reduce crime in Chicago, a city plagued by high homicide rates, new data and data analysis software help to reduce crime ("Violent Crime Is Down in Chicago," 2018). Equipped with sensors that locate gunshots throughout the city, along with maps of liquor stores and freeway on-ramps, analysts identify areas where crime will most likely occur. That information, combined with data on televised sporting events, increases the accuracy of locating potential problems. The new data allow police in Chicago's inner-city areas to patrol more effectively.

Whether in commerce, politics, government, health, or crime, data are the key to solving important problems. These examples show that data are used increasingly to understand and influence human behavior or in the Chinese case exert social control. Those who know how to collect, analyze, and explain data can influence important decisions.

Data Analysis Is a Marketable Skill

Quantitative data analysis is a marketable skill. Confronted by an ocean of data, companies, government offices, and nongovernmental organizations recognize the need for accurate and timely analysis. One need only to Google the phrase "jobs in R" to see the possibilities.

In the private sector, an increasing number of companies rely on surveys to understand their market, employees, and customers. The expanded use of Survey Monkey and Qualtrics adds to that reliance. Conducting surveys is an art. Done poorly, surveys either mislead or say nothing about the problem at hand. Understanding the basics of quantitative data analysis develops an appreciation and facility for survey research.

Government agencies need well-trained social scientists. Ten or 15 years ago, government agencies would only pay lip service to their need for them. Over the last several years, they've actually started to hire social scientists. Data skills will bolster a career in local, state, and federal government agencies; organizations know that a work force staffed with effective analysts produces better policy.

Postgraduate programs want students with a quantitative background. A firm grasp of quantitative skills can help gain entrance into many top 20 departments, no matter the discipline. Law school candidates with a specialty in engineering, computer science, or quantitative analysis have a unique skill that puts them head and shoulders above others, an important consideration given the competition.

In addition to law school, quantitative skills can open doors to policy schools and graduate programs in the social sciences. The number of subjects we study in the social sciences that involve examining quantitative evidence is growing. Over the last few decades, there's just too much quantitative data and evidence to be ignored. Applicants to graduate programs in these areas increase the likelihood of gaining entrance to top-ranked schools if they can demonstrate a facility with data analysis.

Data Analysis Is a Public Good

Perhaps more important than these purely instrumental reasons, society depends on good analysis. As wonderfully illustrated in the book *A Mathematician Reads the Newspaper*, we deserve the kind of public policy we get (Paulos, 1995). In one example, Paulos recounts a contested state senate race in Pennsylvania where fraud was detected by fitting a regression line to previous election results and determining that the race in question was simply too anomalous to be credible. Given the current divisiveness that characterizes U.S. politics, understanding sampling and the inferences we can draw from it can either bolster our faith in the results or suggest a new election is necessary.

Another characteristic of our political age is the growth of conspiracy theories. With so much data floating around, it is easy to connect dots that are completely unrelated yet convince many that dark forces are at work. In a prescient example, Paulos recounts an exercise undertaken by John Leavy, a computer programmer from the University of Texas. Leavy fed data into a computer to come up with similarities between different pairs of U.S. presidents. In that exercise, he found some very interesting connections between Presidents William McKinley and James Garfield who were both assassinated. Observe the similarities:

> It turns out that both of these presidents were Republicans who were born and bred in Ohio. They were both Civil War veterans, and both served in the House of Representatives. Both were ardent supporters of protective tariffs and the gold standard, and both of their last names contained eight letters. After their assassinations they were replaced by their vice presidents, Theodore Roosevelt and Chester Alan Arthur, who were both from New York City, who both sported mustaches, and who both had names containing seventeen letters. (Paulos, 1995, p. 91)

Whether citizens become data scientists or sophisticated consumers of quantitative data, a more informed public is better able to determine fact from fiction. An important goal of this book is to demonstrate how statistics are used and abused. Distinguishing between fact and fiction depends on individuals with the skill and knowledge this book hopes to provide.

> **KNOWLEDGE CHECK: Explain the importance of acquiring skills in data analysis.**

1. Given big data's growing influence on our lives, why is data analysis an important skill?
    a. It helps in our careers.
    b. We get better policy.
    c. It defines our roles as citizens and consumers.
    d. With more skills we can realize higher salaries.
2. Data analysis as a skill (a private good) is helpful with which of the following?
    a. We can realize a higher salary.
    b. Both government and the private sector need skilled individuals.
    c. It helps with gaining entrance into postgraduate degree programs (e.g., law school, business school, graduate school in general).
    d. It helps society.
3. As a public good, which of the following describe the benefits of having data analytic skills.
    a. It leads to better government policy.
    b. It helps society determine fact from fiction.
    c. It aids democracy.
    d. It helps businesses understand their customers.

The Main Components of Data Analysis

There is no shortage of good textbooks on how to conduct social science research in which the focus is on the process of hypothesis formation, theory testing, and drawing inferences. Here, the goal is the same but the approach is different. In this book we learn those foundations of inquiry through data visualization and by working on real problems with real data.

In that spirit, this section outlines four components of data analysis in the order they should occur: (1) describing data and formulating hypotheses, (2) building and estimating models, (3) diagnostics, and (4) generating the next question. There are concepts and techniques (i.e., model building and estimation, transforming variables, diagnostics, etc.) that will be new. The purpose of this chapter is to introduce the broad outlines of good data analysis with an example. When new techniques and concepts are introduced, I will indicate where they are covered in detail later in the book rather than dwell on them here. Let me define the main components of data analysis.

1. Describing data and formulating hypotheses

We describe data to better understand the problem and to ask better questions. At its base, describing data focuses primarily on identifying the typical case (central tendency) and understanding how typical that typical case is (dispersion). Describing data, however, should go much deeper than that. Observing where specific cases or entire classes of cases lie in relation to others is an important part of the enterprise. The more we know about our data, the better questions we'll generate and the better hypotheses we'll formulate. The concepts and tools necessary for describing data are found in Chapters 3–7.

Before we continue, here is a word about **theories** and **hypotheses**. Typically, hypotheses are guesses and theories are interconnected hypotheses that form a greater whole and have been tested with some success. The distinction is often fuzzy since in the vernacular theories and hypotheses are often used interchangeably. In this book, a hypothesis will refer to a specific guess about how two things are related (e.g., religiosity and political ideology). A theory will refer to a set of related hypotheses that explain successfully some empirical phenomenon.

2. Building and estimating models

Once we have some familiarity with our data and some possible explanations have been forwarded, we move on to building and estimating models. **Models** are simplified versions of reality that help us understand our complex world. Models can be thought of as arguments or explanations. They are arguments we make to explain an empirical problem or puzzle. For example, if we want to explain why some countries have high rates of homicide, we construct a model or argument that might include income, age of the population, number of police, and efficacy of the judicial system. There are a multitude of other possible causes we could include, but it helps to keep things simple. We don't want to recreate reality; we merely want to approximate it. With a good model in hand (i.e., a model that contains the main causes but not every single possible one), we can begin to understand how important each cause is and we can estimate its impact.

Estimates can be very sensitive to the model we choose. How we look at the data influences what we see. Consequently, model building and estimation is a process that should be performed as a back-and-forth between theory and evidence. The practice of model building is treated in Chapters 10–14.

3. Diagnostics

After we've constructed models and obtained some estimates, we turn to **diagnostics**. Diagnostics are a set of tools we use to determine whether we're using the right kind of model. To ascertain whether our model is appropriate, we examine how well the predictions from our model match reality. The difference between our prediction and reality is called the residual. For example, if our model does a good job of predicting infant mortality rates in all countries except for the oil-rich states of the Middle East, the resulting diagnostics will say so. That is, the residuals for these cases will be relatively large. Perhaps our model estimates are overly influenced by those Middle Eastern countries. Diagnostics help us determine whether our estimates provide a good sense of how the world really works, are the product of some strange cases, or are the result of a poorly chosen model.

It is important to keep in mind that diagnostics can both detect problems and help uncover interesting relationships, generating additional explanations or hypotheses. Diagnostics are explained in Chapters 15 and 16.

4. Generating the next question

Finally, armed with our estimates and a sense of how well our predictions fit reality, additional questions and ideas inevitably surface. A useful way to construct or identify these additional questions and ideas is to use a simple if-then statement: *if* the estimates we obtained are correct, *then* we would expect to see x. Following each set of estimates with that statement helps to unearth possible explanations and additional hypotheses to test. Since it is impossible to prove anything with complete certainty, the exercise of generating additional hypotheses to test is extremely important. Much like a prosecutor in a court of law, we must provide as much circumstantial evidence as possible to convince the jury. The more evidence we provide, the more likely the jury will find the defendant guilty. The Appendix presents a way to approach generating additional hypotheses with a few tips and tricks.

Now that we have a sense of what the main components of data analysis are, let's dive into an example of what they look like in real life. The next section serves as an introduction to data analysis by providing a very brief exploration of a problem, a puzzle that has occupied the attention of the social sciences for decades.

> **KNOWLEDGE CHECK: List the components of data analysis and how they fit together.**

4. Which are components of data analysis?
    a. Generating and forming hypotheses
    b. Diagnostics
    c. Describing the data
    d. Collecting the data
5. What is the correct order of the components of data analysis?
    a. Estimating models, forming hypotheses, diagnostics, forming new questions
    b. Describing data, forming hypotheses, estimating models, diagnostics, forming new questions

c. Forming hypotheses, describing data, estimating models, diagnostics, forming new questions

d. Describing data, forming hypotheses, forming new questions, diagnostics

6. Which of the following are accurate statements about data analysis?

    a. There is an agreed-upon method everyone follows.

    b. Hypothesis formation always comes before describing data.

    c. Describing data always comes before hypothesis formation.

    d. There is an agreed-upon order of analysis.

7. What is the difference between a theory and hypothesis?

    a. There really is no difference.

    b. Hypotheses are guesses, theories are certain.

    c. Theories are simply hypotheses that have stood the test of time.

    d. Theories define an integrated approach or set of hypotheses that have accumulated corroborating evidence over time.

8. Which statements best describe models?

    a. A model replicates reality.

    b. A model is a simplification of reality.

    c. A model is best judged by its ability to predict.

    d. A model's use is defined both by its ability to predict and its simplicity.

Developing Hypotheses by Describing Data

The best analysis starts with a puzzle or a question. To illustrate the broad outlines of data analysis, let's begin with a tried-and-true example. Why are some countries richer than others? Work drawn from development economics, political science, and sociology emphasizes education, religion, ethnic fractionalization, gender equality, and political stability.[2] These explanations represent only some of the more common hypotheses scholars have forwarded to account for the huge disparity we observe between countries. Let's explore these hypotheses in a little more detail.

Art and Practice of Data Visualization

DESCRIBING DATA AND FORMULATING HYPOTHESES

Describing data and formulating hypotheses is an interactive process. Data analysis never starts without some guess about what we'll find. An honest analysis, however, changes what we know. At its best, data analysis leads to new questions we neglected to ask in the first place.

---

[2] For an informative treatment of growth theory and the current scholarship on the empirics of growth, see Jones (2015). For a recent review of the literature on long-run growth trends, see Lloyd and Lee (2018).

Hypothesis I: Ethnolinguistic Fractionalization

What is the relationship between a country's ethnic landscape and economic growth? Some have argued there is a direct relationship between economic development and the number of ethnic groups in a country. Endowed with a multitude of different languages, religions, and beliefs, individuals in heterogeneous societies may find it more difficult to organize production and markets. Rivalries between different ethnic groups (ethnic conflict) can also inhibit economic development.

To explore the first hypothesis, we now move to more involved R code. Remember that the best way to work with R is to save chunks of code so that they can be copied and pasted when needed. Once you've established how you want a scatter plot or any other figure to look, it's easy to switch out variables, choose colors, or change a figure's underlying features.

So far, you've been introduced to just a few simple base R functions, *hist()* and *plot()*. To get you up and running with publishable figures and tables, let's dive in. Throughout the book, I use a package in R called 'ggplot2.' The "gg" in **ggplot** stands for the grammar of graphics, a way of thinking about graphs based on layers. Let me explain.

The code in Code Chunk 2-1 starts with the *ggplot()* function. Within the parentheses, we first indicate which data set to use: *world*. The next feature of the code is the *aes()* function or "aesthetics." The aesthetics indicate which variables to use and any shapes, colors, or labels to add. Once that first layer is established, note that the rest of the code is simply a series of additional layers. First, we add a layer of points with *geom_point()*. We then add a straight line with *geom_smooth()*, a title with *ggtitle()*, x- and y-axis labels with *xlab* and *ylab*, a style with *theme_minimal()* and *theme()*, and some text with *geom_text_repel()*. The code within the *geom_text_repel()* function is a little complex because we want R to only label the two data points that represent Luxembourg (LUX) and Madagascar (MDG). Notice how each layer is added by placing an addition symbol between each added layer. I encourage you to start with the *ggplot()* function, execute it, and observe the result. Then, add each additional layer one by one, observing how the plot evolves into the final product. That exercise will demonstrate the logic of the grammar of graphics: the simple addition of layer upon layer.

The code in Code Chunk 2-1 draws a nice scatter plot that illustrates the relationship between ethnic heterogeneity and the logged value of gross domestic product (GDP) per capita.

Code Chunk 2-1

```
ggplot(world, aes(ethfrac, log(gdppc), label = iso3c)) +
  geom_point(color="#bf0000") +
  geom_smooth(method="lm", se=FALSE, color="#0000bf") +
  ggtitle("Figure 2-1: ELF Lowers Income") +
  ylab("GDP per Capita (logged)") +
  xlab("Ethnolinguistic Fractionalization (ELF)") +
  theme_minimal() +
  theme(plot.title = element_text(size = 8, face = "bold"),
        axis.title = element_text(size = 8, face = "bold")) +
  geom_text_repel(size = 2.8,
      aes(label=ifelse(iso3c=="MDG" |
                       iso3c=="LUX",
                       as.character(iso3c),''),
          hjust = 0, vjust=-1), show.legend=FALSE)
```

FIGURE 2-1   ELF Lowers Income

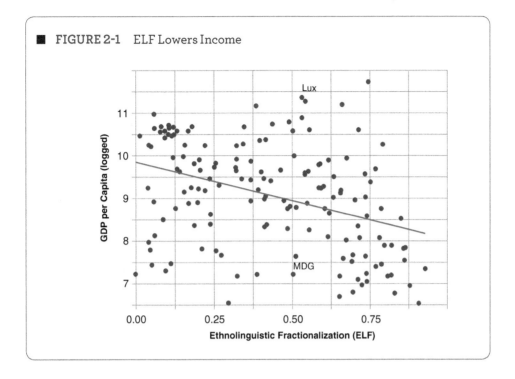

In Figure 2-1, I use GDP per capita to measure wealth (income). GDP is simply the total amount of all goods and services in a country produced in a given year. Put simply, ethnolinguistic fractionalization (ELF) indicates how likely your neighbor speaks a different language. The higher the score, the more languages spoken over a country's population. The **scatter plot** of GDP per capita and the degree of ethnic heterogeneity[3] suggests that as ethnic heterogeneity increases, GDP per capita decreases. Scatter plots show where a single observation (a country, in this case) is situated on a two-dimensional grid delineated by two variables (GDP per capita [logged] and ethnolinguistic fractionalization). Scatter plots will be discussed at greater length in Chapter 5.

Note that while there is an overall negative relationship (the line slopes downward), knowing the degree of ethnic heterogeneity in a country does little to help us predict the income level with precision. For example, Madagascar (MDG) and Luxembourg (LUX) have roughly the same ethnolinguistic heterogeneity score (ELF) but extremely different levels of per capita income. Even though knowing the ELF score can give us a rough guess, there is still considerable variation in GDP per capita left to explain.

Although there does seem to be a negative relationship between the ELF score and GDP per capita, a downward sloping line may not be the best way to summarize the relationship. A closer look at the plot indicates there is a cluster of countries that might be influencing the line (the upper-left corner). If those cases were not there, it's possible the line would have a flatter slope. That cluster of countries might be pulling the left part of the line upward, resulting in a negative slope. It turns out that those countries fall into a particular category labeled the

---

[3] Per capita income is measured by GDP per capita. The GDP per capita measure is presented in its logged form for reasons that will be explained in the chapter on transforming variables (Chapter 6). The measure of ethnicity used is the ethnolinguistic fractionalization score constructed by Alesina et al. (2003). In earlier formulations, the score was constructed to represent the probability that two randomly selected individuals in a country were from the same ethnolinguistic group. Alesina and coauthors use the same underlying construct but combine racial and linguistic characteristics. These data are from Pollock (2014).

industrial countries, a collection of countries that includes Europe, North America, Australia, New Zealand, and Japan.

In the following code, note how some changes to Code Chunk 2-1 allow me to color the industrial countries blue, turn the original line into a dashed line, and fit a separate line (using the *smooth()* function) to the nonindustrialized countries (those colored red). I can also place some words in the figure with the *annotate()* function. We're not going to get bogged down with all of the code at this point, since we want to concentrate our efforts on the components of data analysis. As the chapters unfold, you will become well acquainted with all of the code in Code Chunk 2-2.

Code Chunk 2-2

```
ggplot(world, aes(ethfrac, log(gdppc))) +
  geom_point(color=ifelse(world$aclpregion=="Industrial Countries",
                    "#0000bf", "#bf0000")) +
  geom_smooth(method="lm", col="grey", se=FALSE,
            linetype = "dashed") +
  ggtitle("Figure 2-2: Industrialized Countries Are Clustered") +
  ylab("GDP per Capita (logged)") +
  xlab("Ethnolinguistic Fractionalization") +
  theme_minimal() +
  theme(plot.title = element_text(size = 8, face = "bold"),
      axis.title = element_text(size = 8, face = "bold")) +
  geom_smooth(data=subset(world,
                        aclpregion != "Industrial Countries"),
                        color="#bf0000", se=FALSE, method = "lm") +
  annotate("text", x = .15, y = 11.5,
         label = "Industrial Countries", col="#0000bf")
```

Figure 2-2 demonstrates how sensitive the line is to the influence of the industrial countries. Note how the line fit for the rest of the world (the red line) is relatively flat compared to the gray dashed line that accounts for the industrial countries.

To further emphasize the influence of the cluster, I asked R to fit a curved line to all of the countries in the world that are not classified "industrial." Smooths will be discussed and defined in detail later (Chapter 5). For now, think of a **smooth** as simply a line or curve that R fits to the data. The curve demonstrates there may be another pattern lurking. In Figure 2-3, the black line shows that when we describe the data with a nonlinear smooth (the curve), an interesting pattern emerges. While the overall trend might be roughly negative (gray dashed line), a majority of the data (the nonindustrialized countries) tell a different story. As we move from left to right along the x axis, GDP per capita increases, reaches a maximum, then decreases. Among nonindustrialized countries, there may be an optimal degree of ethnolinguistic heterogeneity that encourages economic dynamism.

In Code Chunk 2-3, I've only replaced the straight red line in Figure 2-2 with a curved black line. The curved line is the default for *geom_smooth()*, so I simply removed the expression *method="lm"* that tells R to draw a straight line.

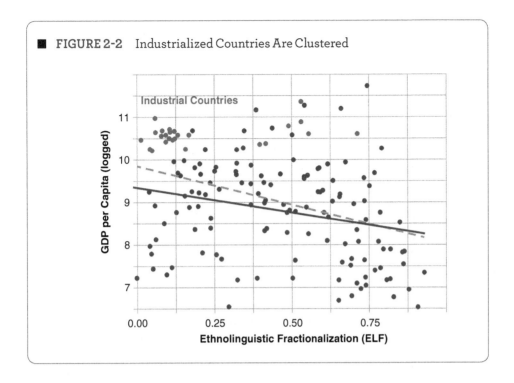

FIGURE 2-2 Industrialized Countries Are Clustered

Code Chunk 2-3
```
ggplot(world, aes(ethfrac, log(gdppc))) +
  geom_point(color=ifelse(world$aclpregion=="Industrial Countries",
                          "#0000bf", "#bf0000")) +
  geom_smooth(method="lm", col="grey", se=FALSE,
              linetype = "dashed") +
  ggtitle("Figure 2-3: The Relationship Could Be Nonlinear") +
  ylab("GDP per Capita (logged)") +
  xlab("Ethnolinguistic Fractionalization") +
  theme_minimal() +
  theme(plot.title = element_text(size = 8, face = "bold"),
        axis.title = element_text(size = 8, face = "bold")) +
  geom_smooth(data=subset(world,
                          aclpregion != "Industrial Countries"),
              color="black", se=FALSE) +
  annotate("text", x = .15, y = 11.5,
           label = "Industrial Countries", col="blue")
```

The curved pattern depicted by the black line might lead us to conclude that if there is a severe lack of ethnic heterogeneity or if there is too much, economic development suffers.

While the observation has some merit, an even closer look indicates the upside-down U-shape pattern may be somewhat overblown. Note in Figure 2-4 that the four most impoverished and ethnolinguistically homogeneous countries are all poor, very small island

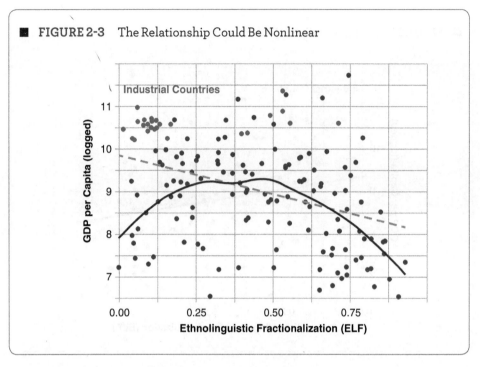

FIGURE 2-3 The Relationship Could Be Nonlinear

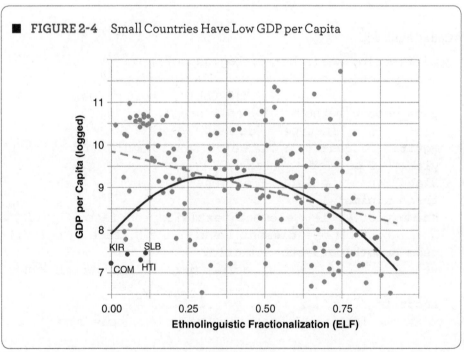

FIGURE 2-4 Small Countries Have Low GDP per Capita

nations: Comoros (COM), Haiti (HTI), Kiribata (KIR), and the Solomon Islands (SLB). Given the small size of these countries and those in close proximity, perhaps size matters.[4]

---

[4] There is a literature in economics that examines whether there is an optimal state size for economic performance. For one contribution in that debate, see Alesina (2003).

Among the developing countries, the smallest (regardless of heterogeneity) all have relatively low per capita incomes.

To highlight the four cases, I use the *ifelse()* function both to color them differently and to include their labels.

Code Chunk 2-4

```
ggplot(world, aes(ethfrac, log(gdppc))) +
  geom_point(color=ifelse(world$iso3c=="COM" |
                          world$iso3c=="HTI" |
                          world$iso3c=="KIR" |
                          world$iso3c=="SLB", "#bf0000", "grey")) +
  geom_smooth(method="lm", col="grey", se=FALSE,
              linetype = "dashed") +
  ggtitle("Figure 2-4: Small Countries Have Low GDP per Capita") +
  ylab("GDP per Capita (logged)") +
  xlab("Ethnolinguistic Fractionalization") +
  theme_minimal() +
  theme(plot.title = element_text(size = 8, face = "bold"),
        axis.title = element_text(size = 8, face = "bold")) +
  geom_smooth(data=subset(world,
                          aclpregion != "Industrial Countries"),
              color="black", se=FALSE) +
  geom_text_repel(size = 2.8, aes(label=ifelse(iso3c=="COM" |
                                               iso3c=="HTI" |
                                               iso3c=="KIR" |
                                               iso3c=="SLB",
                                               as.character(iso3c),''))) 
```

Since we may want to avoid summarizing the relationship between GDP per capita and ELF based on the small island nations of Comoros, Kiribata, Haiti, and the Solomon Islands, there does seem to be a negative relationship between GDP per capita (logged) and the ELF score as we travel from left to right across the x axis. There is, in other words, some preliminary evidence that the hypothesis has some merit. Nevertheless, a great deal of variation remains and the relationship is by no means perfectly **linear**. By linear, I mean a straight line with a single slope.

Before we move on to exploring the next hypothesis, let's take stock to highlight the process of formulating hypotheses and describing data. We started the section with a hunch about the ethnic makeup of a country and its economy. A simple scatter plot showed there might be a negative relationship between the two. We identified a cluster of countries that explains the downward sloping line. We then fit a curve to the remaining data and found there was an upside-down U-shaped pattern: countries with low and high levels of ethnolinguistic heterogeneity had lower levels of income compared to countries with medium levels. Further investigation revealed that the countries with low levels of income and low heterogeneity were small island nations. That observation suggested the geographic size of a country might matter.

Throughout this section, we saw how identifying different clusters of data led to new discoveries and suggested additional hypotheses to test.

Hypothesis II: Women's Suffrage

Another hypothesis concerns women and their role in both government and the economy. Are women being educated? Do they have equal rights? We might hypothesize that women's participation in society is essential to economic development. As a very crude measure of women's participation, we have data—a variable in the *world* data set—that record the year women obtained the right to vote. What is the relationship between the year women gained suffrage and GDP per capita? To answer the question, consider the following scatter plot that plots GDP per capita against the year women gained suffrage in the country (Figure 2-5).

Code Chunk 2-5
```
ggplot(world, aes(womyear, log(gdppc))) +
  geom_point(col="#bf0000") +
  geom_smooth(method="lm", se=FALSE, color="#0000bf") +
  ggtitle("Figure 2-5: The Later the Suffrage,
          the Poorer the Country") +
  ylab("GDP per Capita (logged)") +
  xlab("Year Women Gained Suffrage") +
  theme_minimal() +
  theme(plot.title = element_text(size = 8, face = "bold"),
        axis.title = element_text(size = 8, face = "bold"))
```

Although GDP per capita can take on very different values at each level of women's suffrage (there is considerable variation), there is a downward sloping trend. In other words, as the year when women were granted suffrage increases (suffrage was granted later), it appears that GDP per capita decreases. What is driving the result? Perhaps the downward sloping line merely describes the difference between the industrialized West and Africa. The vote was not extended to anyone in Africa, after all, until independence was achieved in the 1960s.

In Code Chunk 2-6, only a few refinements are added to the code in Code Chunk 2-5. First, we use the *ifelse()* command to color the industrial countries blue and the sub-Saharan countries red. Second, we use the *annotate()* function to place the "Industrial Countries" and "Sub-Saharan Africa" labels in the figure.

If the relationship is simply the difference between the industrial countries[5] and sub-Saharan Africa, there may be things other than women's participation that distinguish the industrial

---

[5] There is no good way to name this group of countries: the West, the industrialized West, developed countries, postindustrialized, high-income countries, and so forth. Rather than enter into a debate on nomenclature, I simply use the classification given by Przeworski et al. (2000) whose data are used here. The countries considered industrial are Australia, Austria, Belgium, Canada, Switzerland, Cyprus, Germany, Denmark, Spain, Finland, France, United Kingdom, Greece, Ireland, Italy, Japan, Lichtenstein, Luxembourg, Malta, Netherlands, Norway, New Zealand, Portugal, San Marino, Sweden, and the United States.

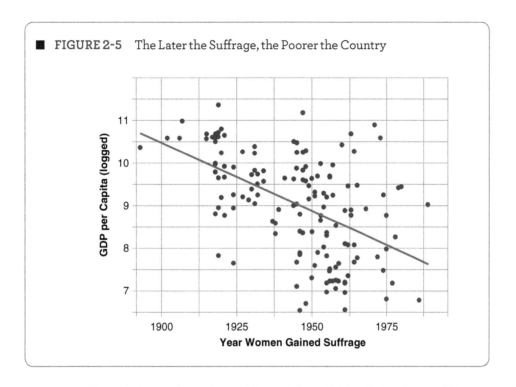

FIGURE 2-5 The Later the Suffrage, the Poorer the Country

Code Chunk 2-6

```
ggplot(world, aes(womyear, log(gdppc))) +
  geom_point(color=ifelse(world$aclpregion==
                          "Industrial Countries", "#0000bf",
                          ifelse(world$aclpregion==
                          "Sub-Saharan Africa", "#bf0000", "grey"))) +
  annotate("text", x = 1905, y = 11.75,
           label = "Industrial Countries",
           col="#0000bf") +
  annotate("text", x = 1975, y = 6.2,
           label = "Sub-Saharan Africa", col="#bf0000") +
  geom_smooth(method="lm", se=FALSE) +
  ggtitle("Figure 2-6: The Industrial Countries and
          Sub-Saharan Africa Form Clusters") +
  ylab("GDP per Capita (logged)") +
  xlab("Year Women Gained Suffrage") +
  theme_minimal() +
  theme(plot.title = element_text(size = 8, face = "bold"),
        axis.title = element_text(size = 8, face = "bold"))
```

countries and sub-Saharan Africa from each other (e.g., a colonial past). The cluster of industrial countries in the upper-left quadrant of the plot and the cluster of sub-Saharan African countries in the lower-right quadrant certainly suggests that the date of independence or colonialism is part of this story (see Figure 2-6).

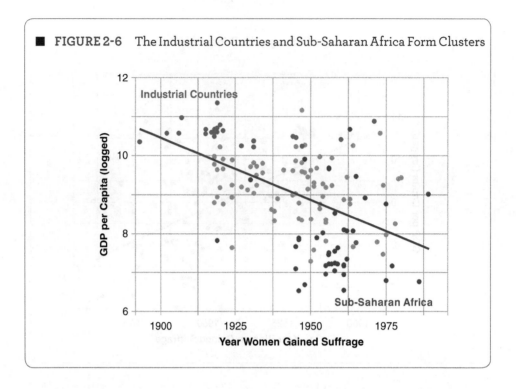

FIGURE 2-6   The Industrial Countries and Sub-Saharan Africa Form Clusters

Although we've made some important discoveries so far, let's not stop here. Part of the challenge and fun of data analysis is to develop a keen eye that notices every nook and cranny. Many times, that's how the most interesting discoveries are made. Note, for example, there are relatively few observations (dots) between 1935 and 1945 on the x axis. This implies that countries stopped granting suffrage during those years.

In Code Chunk 2-7, I introduce the *geom_vline()* function, which creates a vertical line on a specific part of the x axis. This is helpful when we want to highlight a very specific value.

Code Chunk 2-7

```
ggplot(world, aes(womyear, log(gdppc))) +
  geom_point(color=ifelse(world$womyear < 1935, "#bf0000",
                          ifelse(world$womyear > 1944,
                                 "#0000bf", "grey"))) +
  geom_smooth(method="lm", se=FALSE) +
  geom_vline(xintercept=1934, col="grey", linetype = "dashed") +
  geom_vline(xintercept=1944, col="grey", linetype = "dashed") +
  ggtitle("Figure 2-7: Gap During World War II") +
  ylab("GDP per Capita (logged)") +
  xlab("Year Women Gained Suffrage") +
  theme_minimal() +
  theme(plot.title = element_text(size = 8, face = "bold"),
        axis.title = element_text(size = 8, face = "bold"))
```

The pause between 1935 and 1945 might be a feature of the data worth exploring. The gap during the period suggests that the fight for women's rights took a back seat to fighting World War II (see Figure 2-7). To illustrate, I colored the points before 1935 red and the points after 1945 blue. After the war, a number of countries joined the community of nations

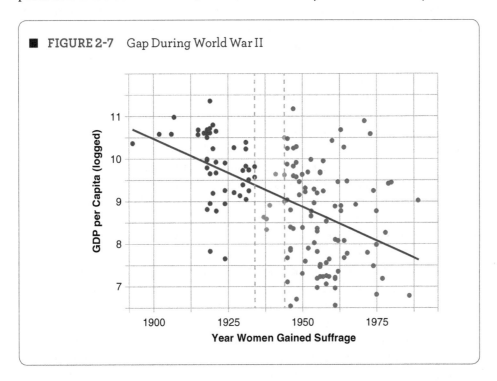

FIGURE 2-7 Gap During World War II

Code Chunk 2-8

```
ggplot(world, aes(womyear, log(gdppc))) +
  geom_point(color="#bf0000") +
  geom_smooth(method="lm", se=FALSE, color="#0000bf") +
  ggtitle("Figure 2-8: Large Variation in GDP per Capita") +
  ylab("GDP per Capita (logged)") +
  xlab("Year Women Gained Suffrage") +
  theme_minimal() +
  theme(plot.title = element_text(size = 8, face = "bold"),
        axis.title = element_text(size = 8, face = "bold")) +
  geom_text_repel(size = 2.8,
                  aes(label=ifelse(iso3c=="ETH" |
                                   iso3c=="BDI" |
                                   iso3c=="SLE" |
                                   iso3c=="HUN" |
                                   iso3c=="CYP" |
                                   iso3c=="LBY",
                                   as.character(iso3c),''),
                  hjust = 0, vjust=-1), show.legend=FALSE)
```

as independent entities. At their birth, many granted voting rights to women. Clearly, GDP per capita could be driven by the political autonomy of a country rather than women's rights. Both the two clusters of countries and the gap separating pre- and postwar cases suggest that colonialism and independence explains levels of GDP per capita.

Another feature of the relationship involves the huge variation in per capita income for countries that granted suffrage at around the same time. By huge variation, I mean that at very similar dates when suffrage was granted, levels of GDP per capita range from being fairly high to fairly low. Note that Ethiopia (ETH), Burundi (BDI), and Sierra Leone (SLE) granted suffrage at around the same time as Hungary (HUN), Cypress (CYP), and Libya (LBY), yet the two groups of countries have very different levels of per capita income (Figure 2-8). Clearly, women's suffrage alone does not explain everything. If it did, all of the countries plotted in the graph would be much closer to the line.

Refer back to Figure 2-6, which had the industrial countries in blue, the sub-Saharan African countries in red, and the rest of the world's regions in gray. Given the pattern we observed between the industrial countries and Africa in Figure 2-6, a very useful test might be to fit a line to the sub-Saharan countries separately to see if a relationship exists between women's suffrage and income per capita for sub-Saharan Africa. By doing so, we're asking whether the negative relationship between GDP per capita and the year of suffrage is indeed negative or if it is purely generated by the differences between the industrial countries and sub-Saharan Africa. If there is a strong negative relationship between GDP per capita and women's suffrage, then we would expect to find it among the sub-Saharan African countries themselves.

To distinguish different groups in a figure, it's often helpful to create a variable that indicates a 1 for a specific group and a 0 for all of the rest. The two first lines in Code Chunk 2-9 create such a variable for sub-Saharan Africa. First, I use the *ifelse()* function to state that if the variable *world$region* equals sub-Saharan Africa, then generate a 1; otherwise generate a 0. I then take that new variable *world$SSA* and convert it to a factor. I'll go into more detail on factors in Chapter 5.

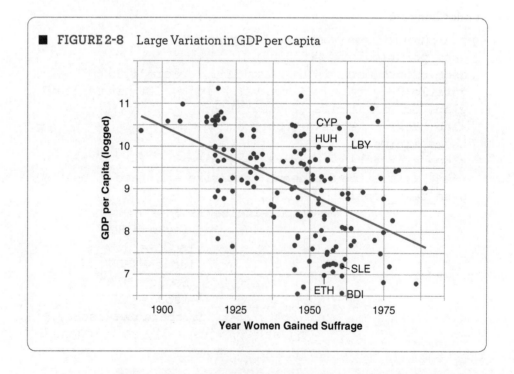

■ FIGURE 2-8   Large Variation in GDP per Capita

Code Chunk 2-9

```
world$SSA <- ifelse(world$region=="Sub-Saharan Africa", 1, 0)
world$SSA <- as.factor(world$SSA)

ggplot(world, aes(womyear, log(gdppc), col=SSA)) +
  geom_point() +
  theme_minimal() +
  theme(plot.title = element_text(size = 8, face = "bold"),
        axis.title = element_text(size = 8, face = "bold")) +
  geom_smooth(method="lm", se=FALSE) +
  ggtitle("Figure 2-9: No Relationship in Sub-Saharan Africa") +
  ylab("GDP per Capita (logged)") +
  xlab("Year Women Gained Suffrage") +
  scale_color_manual(values=c("lightblue", "#0000bf")) +
  theme(legend.position = "none") +
  annotate("text", x = 1928, y = 6.8,
           label = "Sub-Saharan Africa", col="#0000bf")
```

In Figure 2-9, the sub-Saharan countries are colored dark blue. As the figure demonstrates, there does not seem to be a relationship between women's suffrage and economic development in sub-Saharan Africa (the line is flat). In the rest of the world, however, the relationship seems strongly negative. Sub-Saharan Africa is different in this respect and accounting for it helps clarify the strong relationship that exists between GDP per capita and women's suffrage in the other countries.

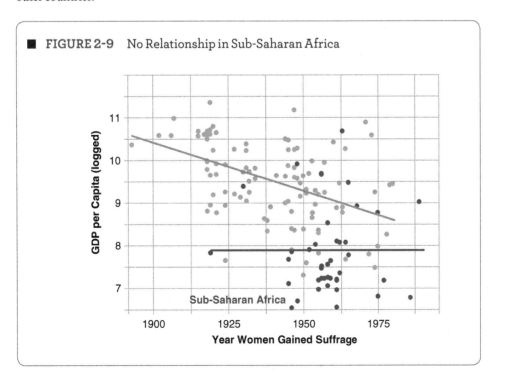

■ FIGURE 2-9   No Relationship in Sub-Saharan Africa

Think about what this implies for our investigation. Women's suffrage occurred simultaneously with independence and democratization in many countries. To the degree that other things are intertwined with women's suffrage (independence, democratization, or both), a measure that merely captures the year women were allowed to vote may only be the result of a larger process. Consequently, these other important events—independence and democratization—need to be accounted for in our explanation.

Before we examine the next hypothesis, let's pause here again to take stock of how the process of formulating hypotheses and describing data unfolds. Careful observation of simple scatter plots provides considerable information. In this example, we again took advantage of coloring the observations by region to make additional discoveries. We first explored whether the negative relationship between income and the year women were granted suffrage could be explained by the differences between industrial countries and sub-Saharan Africa. We also noticed a slight gap in the data during World War II. This suggested that other things might explain the relationship between women's suffrage and income. Finally, we examined whether the negative relationship among all the observations existed within the sub-Saharan cases. Throughout the process, we can see how describing data and formulating hypotheses interact.

Good descriptions of the data can help us evaluate hypotheses as well as formulate new ones. With one scatter plot, we went from evaluating the hypothesis that women's suffrage influences economic growth to discovering how a country's independence or move toward democracy should be considered as well. Simple views of the data may provide what might seem to be an endless number of patterns, discrepancies, or oddities that are worth investigating. As the last few figures demonstrate, important insights can be gained by thoroughly examining each plot. I hope to have demonstrated here that a simple scatter plot can tell us a lot about a relationship if we simply take the time to investigate, coloring groups of data differently to learn what cases determine the slope of a line or the shape of a curve. Spotting interesting aspects of the data through visualization in different ways is a useful skill worth developing.

Hypothesis III: Human Capital

Continuing on, an important component of economic growth is human capital: education. As a population accumulates knowledge, economists argue economic development follows. How does a measure of human capital (an index of human capital constructed by Robert Barro and Jong-Wha Lee) relate to GDP per capita? According to Figure 2-10, it appears the relationship is strong and positive: the more human capital per worker, the higher GDP per capita. Of course, a developed economy may allow more people to attend school: the direction of causality may run the other way (from wealth to education). Nevertheless, we can say there's a strong association between the two variables. Note how the observations in this figure are much closer to the line, which indicates our predictions of GDP per capita at each level of the human capital index. Also observe how the shape of this plot is better **summarized** by a line than the previous two plots. Note my use of the word "summarize." We use summaries to characterize the data. When we use a line, we summarize the data with two numbers: the intercept of a line and its slope.

The code in Code Chunk 2-10 should look familiar by now. Along with the *ggplot()* function that establishes a grid with two dimensions (GDP per capita and human capital), it specifies a layer of dots with the function *geom_point()* and a line with the function *geom_smooth()*. It also uses the *ifelse()* function to label the cases of Qatar (QAT), Kuwait (KWT), and Zimbabwe (ZWE).

Code Chunk 2-10

```
ggplot(world, aes(pwthc, log(gdppc))) +
  geom_point(col="#bf0000") +
  geom_smooth(method="lm", se=FALSE, col="#0000bf") +
  ggtitle("Figure 2-10: Strong Relationship Between
          GDP and Education") +
  ylab("GDP per Capita (logged)") +
  xlab("Barro and Lee's Human Capital Index") +
  geom_text_repel(size = 2.8,
      aes(label=ifelse(iso3c=="QAT" |
                       iso3c=="KWT" |
                       iso3c=="ZWE",
                       as.character(iso3c),''),
                       hjust = 0, vjust=-1),
    show.legend=FALSE) +
  theme_minimal() +
  theme(plot.title = element_text(size = 8, face = "bold"),
        axis.title = element_text(size = 8, face = "bold"))
```

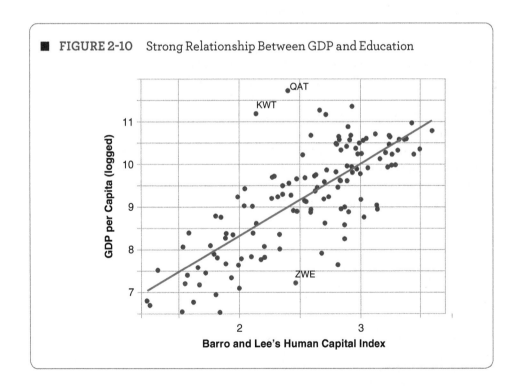

FIGURE 2-10 Strong Relationship Between GDP and Education

Besides the distinct linear pattern, a noticeable feature of the plot is the outliers, the cases far removed from the rest of the data. For example, Qatar (QAT) and Kuwait (KWT) stand out as having been severely underpredicted—they lie considerably above the line (the line represents our prediction). Given the levels of the human capital index, we would expect GDP per capita to be much lower in those countries. Of course, both have significant proven oil reserves, perhaps accounting for our inaccurate prediction. Note also one of the cases lying below the line, Zimbabwe (ZWE). At Zimbabwe's level of human capital, we would have predicted a higher level of per capita income. In Zimbabwe's case (a primarily agricultural country), growing corruption in the 1990s and 2000s along with unstable property rights—disputes over land—might help explain its underachievement. Even though the human capital index does fairly well in explaining GDP per capita, the outliers help identify other possible explanations: natural resources, property rights, and politics.

In this section, we learned that identifying specific cases suggests possible hypotheses we might want to test in the future. Outliers alert us to important causal factors that we hadn't previously considered. In this example, while investigating the importance of human capital's role in determining a country's income, we found that oil (Qatar and Kuwait) and property rights (Zimbabwe) might contribute to income levels as well. Again, we see how describing data generates additional hypotheses.

Hypothesis IV: Political Stability

What role does politics play in the development of national economies? Countries with unstable regimes are probably not the best venues for trade and investment. Is political instability associated with lower levels of wealth? To consider that possibility, GDP per capita is plotted (Figure 2-11) against the number of years since a significant change in the regime (the variable *durable*). The *durable* variable is logged for the same reasons we logged GDP per capita: transforming variables by taking logs will be explained in detail when I cover transforming data in Chapter 6. There does seem to be a positive relationship between regime stability and GDP per capita—countries with relatively stable regimes experience high levels of income (Figure 2-11). As we move from left to right along the x axis, the level of income appears to increase.

Code Chunk 2-11 lists the set of commands used previously to draw a scatter plot. The only wrinkle involves logging the *durable* variable. Since the *durable* variable contains some zeros, we add a 1 to the variable before logging since the log of 0 is undefined. There'll be more on logging and transforming variables in Chapter 6.

Code Chunk 2-11
```
ggplot(world, aes(log(durable + 1), log(gdppc))) +
  geom_point(col="#bf0000") +
  theme_minimal() +
  geom_smooth(se=FALSE, col="#0000bf", method="lm") +
  ggtitle("Figure 2-11: Political Stability Increases Wealth") +
  ylab("GDP per Capita (logged)") +
  xlab("Durable (logged)") +
  theme_minimal() +
  theme(plot.title = element_text(size = 8, face = "bold"),
        axis.title = element_text(size = 8, face = "bold"))
```

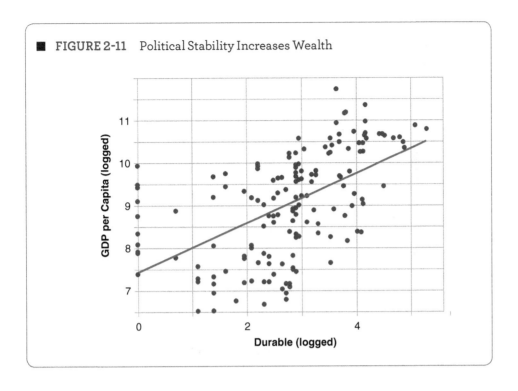

■ FIGURE 2-11   Political Stability Increases Wealth

While there does seem to be a positive relationship between stability and income, the blue line may not be the best way to summarize the relationship. In particular, notice the points at 0 on the x axis. Save for one, they are all located above the line. Also notice that as we move from left to right along the x axis between 0 and 2, most of the observations are located below the line. Then, above 3, most of the points can be found above the line.

To compare a straight line with a curved line, I added two lines to the figure with the *geom_smooth()* function. One of the layers simply draws a curved line to the data (the default), while the other specifies *method="lm"*, which tells R to fit a straight line to the data (Code Chunk 2-12).

Code Chunk 2-12

```
ggplot(world, aes(log(durable + 1), log(gdppc))) +
  geom_point(col="grey") +
  geom_smooth(se=FALSE, col="#0000bf") +
  geom_smooth(method="lm", se=FALSE,
              linetype = "dashed", col = "grey") +
  ggtitle("Figure 2-12: The Nonlinear Relationship Between
          GDP per Capita and Political Stability") +
  ylab("GDP per Capita (logged)") +
  xlab("Durable (logged)") +
  theme_minimal() +
  theme(plot.title = element_text(size = 8, face = "bold"),
        axis.title = element_text(size = 8, face = "bold"))
```

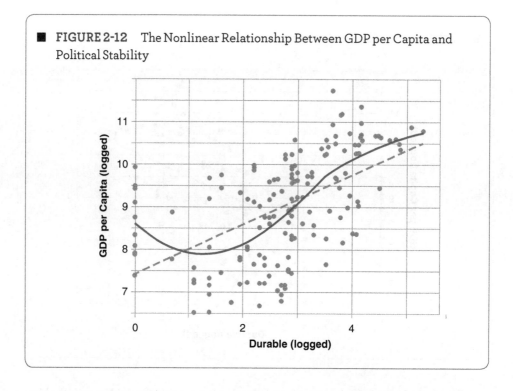

FIGURE 2-12 The Nonlinear Relationship Between GDP per Capita and Political Stability

In this example, the shape of the relationship may be best described by an s-shaped curve. At the low end of *durable*, income is high (higher than our line predicts).[6] As we move from left to right on the x axis, income per capita increases sharply and we begin to systematically underpredict the level of GDP per capita (the majority of cases lie above our prediction on the right portion of the figure).

Although the curvature of the s-shape is not severe, patterns like this suggest there are thresholds. In this example, the s-shape suggests that economic development does not take off until several years of political stability are achieved. At the high end of political stability, the s-shape indicates that each additional year of political stability change does not improve economic development appreciably. In other words, once the same set of constitutional rules have been in force for a number of years, each additional year does not have the same positive influence.

While the relationship between regime stability and GDP per capita may not be strictly linear (it may not be a straight line) and there remains plenty of variation, there is a **positive association** between the two variables. In other words, as regime stability increases, so does GDP per capita: the two variables increase together. Regime stability might be another variable we should consider when trying to explain why some countries are rich and some are poor.

To summarize, in this section we saw another example of how describing data can inform formulating hypotheses and vice versa. In earlier examples, we saw that certain techniques, such as identifying clumps of data as well as individual outliers, help generate additional hypotheses we might want to test. This example demonstrates that examining the functional form of the

---

[6] Since the log of 0 is undefined, I added 1 to the durable variable before taking logs since some of the cases registered 0, a common trick when transforming data. Notice the cases at 0 along the x axis; they represent the cases that were 0 in raw form but then registered 1 when I added the constant 1 to the variable. The log of 1 is 0.

relationship—whether the observations form a straight line or s-shaped curve—also informs our investigation. Now that we have a better sense of what the data look like and we've interrogated a few likely suspects, it's time to move on to model building and estimation.

> **KNOWLEDGE CHECK: Form hypotheses from descriptions of data.**

9. Which of the following accurately describe the process of forming hypotheses?
    a. We always enter into the exercise with some preconceived notions or priors.
    b. Never enter into the exercise with some preconceived notions or priors.
    c. Adding different colors and labeling cases can help us generate hypotheses.
    d. Don't start describing the data until you have generated a hypothesis to test.

10. When evaluating the hypothesis that ethnolinguistic heterogeneity explains income levels, what features of the data generated new hypotheses?
    a. Incomes are high in Europe.
    b. The industrialized countries of the world are clustered together.
    c. The relationship may not be linear.
    d. There are some poor island nations that have low heterogeneity.

11. What features of the data generated the hypothesis that independence has an important role to play in explaining GDP per capita?
    a. There is a gap in the data during World War II.
    b. There is a clear linear relationship between when women were allowed to vote and GDP per capita.
    c. Small island nations were early in allowing women to vote.
    d. There is a linear relationship between income and human capital.

12. What visualization techniques were used to make discoveries in this section?
    a. Representing groups of data with different shapes
    b. Representing groups of data with different colors
    c. Singling out specific cases with labels
    d. Singling out specific cases with color

13. What did you notice in the preceding section in terms of describing data and generating hypotheses?
    a. Viewing the data can prove hypotheses.
    b. Describing data usually settles the matter.
    c. Describing the data often amends the hypothesis or leads to new questions.
    d. Hypotheses are rarely proven true or completely dismissed by useful descriptions of the data.

Model Building and Estimation

Once the data are described and hypotheses have been generated, it's time to develop our argument or to construct a model. We're now moving into material that will be covered in Chapters 10–14. Regression tables will appear in this section without an in-depth discussion of their mechanics. They are presented here to help identify some important tools you will develop that figure prominently in the process of data analysis.

The scatter plots we've examined so far suggest that there are multiple causes of economic development. Ethnolinguistic heterogeneity, women's suffrage, human capital, and regime stability all seem to matter. To obtain the best estimate of their relationship with GDP per capita, we need a good model. For our purposes, let's define a model as simply a list of factors we want to include in an explanation. Well-reasoned hypotheses and a familiarity with the data should be combined toward that end. A good model identifies only the essential elements that help accurately predict outcomes—the outcome in this example is a country's GDP per capita.

Before we continue, it is important to acknowledge that both scholars and practitioners disagree on how to build models. Important philosophical debates lurk beneath the surface. Put simply, the debate concerns whether theory or the data should be our guide. One side doesn't trust the data, arguing that knowing all of the facts in the world does us no good unless we know how things fit together—the theory. The other side doesn't trust the theory, arguing that the best theory comes from knowing as much about the world as possible—the data. Devotion to one approach to the exclusion of the other, in my opinion, often misleads. Relying on a conversation between the two is more fruitful. So far, we've seen how looking at the data can inform theory. Next we find how theory can influence the conclusions drawn from the data. Recognizing how both approaches work in practice is a primary concern of this book.

Start with a simple model: a model that uses one factor—the year women were granted suffrage—to explain GDP per capita (Table 2-1). Models with two variables, a dependent variable (GDP per capita) and an independent variable (women's suffrage; *womyear*), are called **bivariate regression models**. Models with more than two variables—a dependent variable (GDP per capita) and more than one independent variable—are called **multiple regression models**. In the bivariate case, the model uses a line to summarize the relationship between GDP per capita and women's suffrage without accounting for anything else.

The code in Code Chunk 2-13 defines an object called *mod1* and then presents the results from *mod1* in a regression table using the *stargazer()* function from the 'stargazer' package. The *mod1* object is defined as a simple bivariate regression model using the *lm()* command. The *lm* in the *lm()* function stands for *linear model*. In the regression, the log of GDP per capita (*gdppc*) is regressed on the year women were granted suffrage in a country (*womyear*). An important convention in statistics is to say that the dependent variable *is regressed on* the independent variable.

Code Chunk 2-13

```
mod1 <- lm(log(gdppc)~womyear, data=world)

stargazer(mod1, header=FALSE,
          title = "Table 2-1: Estimates for a Bivariate Model",
          type = "html", out = "table3.htm")
```

TABLE 2-1 Estimates for a Bivariate Model

	DEPENDENT VARIABLE:
	log(gdppc)
womyear	−0.032***
	(0.005)
Constant	70.750***
	(8.831)
Observations	143
R^2	0.257
Adjusted R^2	0.252
Residual SE	1.046 (df = 141)
F statistic	48.874*** (df = 1; 141)

Note: ***$p < 0.01$.

The stargazer function requires the right kind of an object as an argument. In this case, the object *mod1*—defined by a linear model—is appropriate. We can also specify the title of the table, the type of output (HTML in this case), and a file where the output will appear. Note that when you knit to an HTML file or PDF, a nice table will appear. Unfortunately, as of this writing, stargazer does not work when knitting a Microsoft Word document.

With the bivariate model, we find there is a **negative association** between GDP per capita and women's suffrage—the number (−.032). In other words, as the women's suffrage variable increases (the later the date suffrage was granted), GDP per capita decreases. The number (−.032) represents the slope of the line that describes the relationship between women's suffrage and GDP per capita.

Before moving to the next section, let's recap. After describing the data and exploring a few hypotheses, we constructed a model to explain GDP per capita. We first constructed a very simple model that fit a line summarizing the relationship between GDP per capita and the year women were granted suffrage in a country. We found that the slope of that line (our estimate) was −.032, implying that women's participation in politics does play an important role in economic development.

> **KNOWLEDGE CHECK: Explain the connection between hypotheses, models, and estimates.**

14. Which of the following statements about model building are true?
    a. Theory should always come before the facts.
    b. The facts should always come before the theory.
    c. There should be a conversation, a back-and-forth between theory and facts.
    d. In the author's opinion, there should be a conversation, a back-and-forth between theory and facts.

15. Indicate which statements are true.
    a. Models with two variables are called bivariate regression models.
    b. Models with more than two variables are called bivariate models.
    c. To obtain the best estimate of a relationship between two variables, we need a good model.
    d. Models with more than two variables are called multiple regression models.

Diagnostics

Now that we've obtained estimates, we need to ascertain how much trust to place in our results. Tables with model estimates can evince an air of confidence and authority. Those estimates aligned so neatly in nice-looking columns and rows could be based on some questionable assumptions, outlying cases, or very poor measures. Much like when being confronted by a pushy salesman who leads you over to a bright, shiny, Corvette on a used-car lot, it might be good to check under the hood. Diagnostics provide the tools to determine whether our estimates are reasonable or highly misleading. Will the Corvette go from 0 to 60 in less than 4 seconds or will it die once we've driven it off the lot?

Stability of the Results

Checking the **stability** of the results and whether we've met certain assumptions are only two possible ways to perform diagnostics. Chapters 15 and 16 are devoted to the enterprise. Here, I simply want to illustrate how our choice of model (theory) can influence our estimates of the relationship (the data).

To illustrate, I estimate a multiple regression model with the *lm()* function and then construct a table displaying the results using the *stargazer()* command (Code Chunk 2-14).

Code Chunk 2-14

```
mod2 <- lm(log(gdppc)~womyear + ethfrac + durable + pwthc,
           data=world)

stargazer(mod2, header=FALSE,
          title = "Table 2-2: Estimates Are Unstable",
          type = "html", out = "table4.htm")
```

How stable are the results from the bivariate model? By stability, I mean how much the results change when we alter the model slightly. If small changes in the model (adding a few variables) produce noticeable changes, we say the results are unstable. Remember that in the bivariate regression, the slope was –.032. Once other variables are added (see Table 2-2), the slope on women's suffrage changed to –.001, decreasing in magnitude by a factor of 32! Clearly, our estimate of women's suffrage depends on the particular model we use.

To see how this translates visually, I generate a plot that shows how women's suffrage is related to GDP per capita when we account for human capital, political instability, and ethno-linguistic fractionalization. This is called an **added variable plot**, which shows the relationship

TABLE 2-2 Estimates Are Unstable

	DEPENDENT VARIABLE:
	log(gdppc)
womyear	−0.001
	(0.004)
ethfrac	−0.698**
	(0.291)
durable	0.007***
	(0.002)
pwthc	1.416***
	(0.163)
Constant	6.682
	(8.201)
Observations	105
R^2	0.725
Adjusted R^2	0.714
Residual SE	0.657 (df = 100)
F statistic	65.813*** (df = 4; 100)

Note: **$p < .05$; ***$p < 0.01$.

between two variables when other variables are accounted for. Again, you'll learn how to generate this very useful view of the data in Chapters 15 and 16. Until then, consider the relationship between GDP per capita (logged) and women's suffrage when all of the variables are included in the model (Figure 2-13).

An easy way to visualize results from regression analysis involves using an R package called 'visreg.' In Code Chunk 2-15, I use the *visreg()* function to plot the predicted values from the multiple regression model I defined in Code Chunk 2-14. In this example, I want to visualize the relationship between women's suffrage and GDP per capita, accounting for the other variables in the model. The 'visreg' package makes this easy. I simply indicate which model I want to use (*mod2*, in this case) and specify the independent variable I'm interested in (*womyear*).

Code Chunk 2-15

```
visreg(mod2, "womyear", ylab="Predicted Values GDP (logged)",
       xlab="Year Women Gained Suffrage",
       main="Figure 2-13: Women's Suffrage Has No Effect",
       band=FALSE)
```

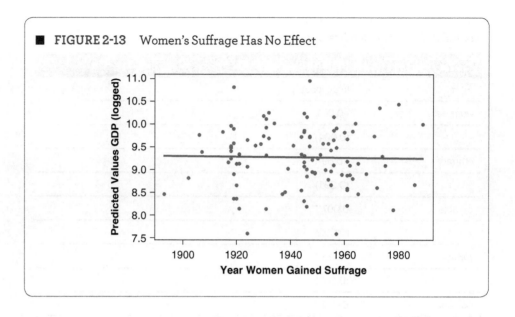

■ FIGURE 2-13   Women's Suffrage Has No Effect

The blue line in the figure represents what level of GDP per capita model 2 predicts based on women's suffrage. The dots represent the actual values as they exist for each country.

While this doesn't necessarily settle the issue, the added variable plot suggests there is not much of a relationship between model 2's prediction of income (GDP per capita) and women's suffrage when controlling for the other variables. Once human capital, political stability, and ethnolinguistic fractionalization have been accounted for, knowing when women were granted suffrage doesn't really tell us much about the level of GDP per capita. This exercise suggests that the relationship between GDP per capita and women's suffrage is *unstable with respect to* the inclusion of political stability, ethnolinguistic fractionalization, and human capital in our model. It illustrates an important lesson: our estimates depend on the model we choose.

Residual Plots

Finally, we can check whether our predictions get better or worse as we range over our predictions, from low to high levels of GDP per capita. If our predictions are systematically better or worse for rich or poor countries, we know our model could be improved. A **residual plot** (Figure 2-14) tells us very quickly if the difference between each case and our prediction (what we call the residual) follows any pattern. Figure 2-14 plots the residuals against the predictions. The horizontal dashed line represents perfect predictions. If a case rests on that line, our model has correctly predicted the actual level of income for that country. Cases further away from the dashed line, such as Singapore (SGP), are those that the model predicted poorly. This residual plot indicates that the model severely underpredicted the level of income in Singapore and Luxembourg.

To produce residuals, we first need to estimate the model and collect the residuals. In Code Chunk 2-16, you can see I use the *lm()* function again to estimate a linear model. This time it includes more than one independent variable; it includes four: *womyear*, *ethfrac*, *durable*, and *pwtch*. Once the model is estimated, I create two variables (*world$res* and *world$pred*) using the *resid()* function and the *predict()* function.

Code Chunk 2-16

```
mod2 <- lm(log(gdppc)~womyear + ethfrac +
           durable + pwthc, na.action = na.exclude,
        data=world)

world$res <- resid(mod2)
world$pred <- predict(mod2)
```

Now that I have a variable with the predicted values from the model and the residuals, I'm ready to plot the two against each other. I do this using the code in Code Chunk 2-17. Code Chunk 2-17 defines a scatter plot that colors the points red, labels the points of five countries using the *ifelse()* function, and draws a horizontal line that represents the predicted values (where the residuals equal 0).

Code Chunk 2-17

```
ggplot(world, aes(pred, res)) +
  geom_point(col="#bf0000") +
  geom_text_repel(size = 2.8,
      aes(label=ifelse(iso3c=="SGP" |
                         iso3c=="TJK" |
                         iso3c=="ZWE" |
                         iso3c=="USA" |
                         iso3c=="LUX",
                    as.character(iso3c),'')),
         hjust = 0, vjust=-1), show.legend=FALSE) +
  geom_hline(yintercept = 0, linetype = 2) +
  theme_minimal() +
  theme(plot.title = element_text(size = 8, face = "bold"),
        axis.title = element_text(size = 8, face = "bold")) +
  xlab("Predicted or 'Fitted' Values") +
  ylab("Residuals") +
  ggtitle("Figure 2-14: Some Possible Outliers")
```

Here we look for patterns. As we'll learn in subsequent chapters, an important assumption in model estimation is that our predictions are just as good or bad for low or high values. In this example we're making the assumption that the model's accuracy does not change when we're making predictions for relatively poor countries or relatively rich ones. The plot in Figure 2-14 can be characterized as a formless cloud: our predictions are just as good or bad as we make predictions for rich and poor countries. There are no obvious observable patterns. From this we would conclude that an important assumption of our analysis is met.[7]

---

[7] An explanation of the assumptions associated with ordinary least squares (OLS) regression analysis, which is used in this example, will be presented in the chapter on diagnostics (Chapter 15).

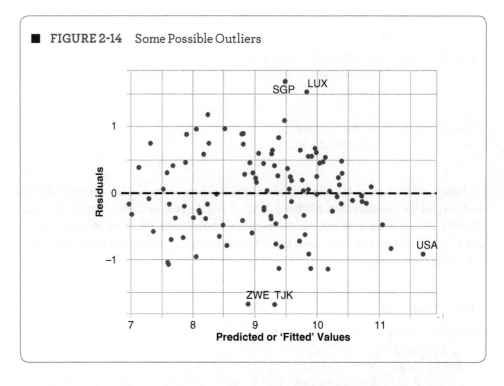

FIGURE 2-14 Some Possible Outliers

The residual plot also indicates if any outliers exist (cases far removed from the rest of the data). With this regression model, Singapore (SGP), Luxembourg (LUX), Tajikistan (TJK), the United States (USA) and Zimbabwe (ZWB) are potential outliers. These outlying cases are ones our model did not predict very well. Why? As we learned with the scatter plots in the previous section, investigating those cases may provide clues as to what other factors might be involved.

> **KNOWLEDGE CHECK: Define diagnostics and explain their role in data analysis.**

16. Which are good analogies for diagnostics?
    a. Checking underneath the hood of a car
    b. Examining the X-ray of a patient
    c. Cross-examining a witness in a trial
    d. Questioning a politician at a press conference

17. What are we looking for when we perform diagnostics?
    a. Do the results show a strong relationship?
    b. Is the hypothesis confirmed?
    c. Do the results change with small changes to the model?
    d. Are there odd cases that are unduly influencing the results?

18. What are we looking for in a residual plot?
    a. Odd cases.
    b. Odd patterns
    c. Any patterns
    d. The typical case

Next Questions

One of the most important aspects of data analysis involves formulating new questions. What do the estimates imply and how can that be tested? In the regression analysis described earlier, including human capital, ethnolinguistic fractionalization, and political stability in the regression changed the estimate of women's suffrage. Why is that? Perhaps women's political rights and the human capital index are closely related (we can test that). Women's influence on GDP might also depend on human capital. The right to vote may have little impact in countries with low levels of education. When women are educated, the right to vote could be extremely powerful. In addition to these possibilities, the model should probably contain other variables (e.g., date of independence, colonial past, and democracy). As we include those variables in our model, the estimates will change, leading to new discoveries.

Asking the next question performs two functions. First, it can help confirm our hypotheses. Generating the next question, forming a hypothesis around it, and testing it can bring new evidence to bear on our original hypothesis. Second, it can lead to new discoveries. Often, we find that the original question we asked was not the most interesting, or that it sent us barking up the wrong tree. In either case, asking the next question is a fundamental aspect of data analysis.

How do we formulate new questions? I find that a helpful way to spark new questions starts with an if-then statement: "if" the results are true, "then" what else would we expect to see in the data? This exercise tests our knowledge of the problem, our ability to think creatively, and our ability to think logically. This is surely part of what we mean by "critical thinking." As with anything, it requires practice. It is a muscle that, if used often, will grow big and strong. If seldom used, it will atrophy.

A final feature of new questions that I'll mention is that they often help us "think outside the box." While this is an overused and ill-defined directive often muttered when the boss is completely out of ideas, thinking outside the box comes directly from asking new questions. In other words, asking new questions *is* thinking outside the box. New questions force us to consider "what if"? When we think outside the box, we're simply exploring beyond our current model—the proverbial box.

> **KNOWLEDGE CHECK: Formulate new questions.**

19. Why do we formulate new questions?
    a. They challenge our old questions.
    b. They force us to reconsider our model.
    c. They provide additional evidence.
    d. They help us think "outside the box."

20. How do we formulate new questions?
    a. Get to know the subject.
    b. Use an if-then statement.
    c. Practice.
    d. Exercise.

SUMMARY

In this chapter, the goal was to motivate the enterprise and to introduce the process of data analysis. The growing amount of information and its use in our everyday lives requires we know how to use and understand quantitative data. In particular, I hope to have shown the back-and-forth between theory and evidence that undergirds the approach taken in the following pages. Although all scholars and practitioners take slightly different approaches, the process outlined in this chapter is a good way to start: (1) describing data and formulating hypotheses, (2) building and estimating models, (3) diagnostics, and (4) generating the next question.

This book is less about probability theory and statistics and more about ways to discover patterns that exist in our data and the inferences we can draw from them. There are many things we want to explain (wealth, inequality, violence, voting, etc.), and there are a lot of data to explore.

We have many tools at our disposal to understand the many characteristics of our data. For example, if our problem is to reduce homicide rates in the United States, it would be helpful to know where the lowest and highest rates are found. Do most states have similar rates or do they vary dramatically? Are similar rates clustered in the South, Northeast, and West? Answering these questions involves describing the data. The more we know about our data, the better questions we'll ask and the better models we'll construct. In Chapter 3, I discuss the importance of describing data and provide examples that illustrate why it is so important.

COMMON PROBLEMS

- *Exploration versus presentation.* Understanding the difference between exploration and presentation is usually not articulated in most books on statistics. Much of what was presented in this chapter shows what goes on behind the scenes. Following paths that ultimately lead nowhere is not usually presented. Exploring data for the purpose of discovery and presenting data for the purpose of persuasion are two very different things.

- *Understanding the residual plot.* Spend time understanding the analytics of the residual plot. The most difficult aspect is to understand that the flat horizontal line represents residuals with a value of 0, indicating that the cases were perfectly predicted. Dots above the line represent instances where the model underpredicted and dots below the horizontal line represent instances where the model overpredicted.

- *Theory versus hypothesis.* Students and practitioners often fall into using these two terms interchangeably. Theories are frameworks or systems of relationships used to explain a variety of different phenomenon. Hypotheses are more specific and tentative. Hypotheses represent our best guesses about how the world works. Those guesses are tested to confirm or challenge existing theory. As evidence accumulates indicating that the theory is incorrect, it's time to change our theory.

- *Patience.* Patience, or rather the lack thereof, is a common theme in data analysis. Students and practitioners get impatient. Their impatience manifests itself in three primary ways during the process of data analysis. First, they often jump past providing a good description of the data right to estimating models. Second, students and practitioners report their results without spending adequate time with diagnostics. Finally, relatively little time is spent thinking about how to present data effectively. As you'll see in Chapter 7, considerations concerning data presentation are crucial.

REVIEW QUESTIONS

1. Describe the struggle between theory and evidence.
2. What are the main components of data analysis?
3. Explain how colors, shapes, and labels aid in data analysis.
4. Why is generating the next question such a useful exercise?
5. What are we looking for when we examine a residual plot?
6. What is meant by stability of the results?
7. How would you describe the grammar of graphics used in ggplot?
8. What is a model?
9. What are diagnostics?
10. What purpose do diagnostics serve?

PRACTICE ON ANALYSIS AND VISUALIZATION

1. What does the *ifelse()* function below say in English?

    ifelse(world, region =="Europe", 1, 0)

    a. If the *world* variable equals "region," then assign 1.
    b. If the *world* variable equals "region," then assign 0.
    c. If *world$region* equals Europe, assign 1 otherwise 0.
    d. If *world$region* equals Europe, assign 0 otherwise 1.

2. Which of the following statements accurately describe the process of data analysis?

    a. Model estimation benefits from describing data.
    b. Diagnostics should always represent the end of our analysis.
    c. Discovery comes once model estimates have been generated.
    d. If a model is properly formulated, diagnostics are not necessary.

3. Which of the following statements accurately describe the purpose of diagnostics?

    a. They help identify important cases.
    b. They indicate whether we are using the appropriate model.
    c. They generate additional questions.
    d. They can help generate additional hypotheses.

4. When we examined the relationship between women's suffrage and income levels, what if-then statement did we generate?

    a. If the relationship between women's suffrage and GDP per capita is just the result of the important differences between Europe and sub-Saharan Africa, then we would expect to see the same relationship exist among the countries of sub-Saharan Africa.
    b. If the relationship between women's suffrage and GDP per capita is not just the result of the important differences between Europe and sub-Saharan Africa, then we would expect to see

the same relationship exist among the countries of sub-Saharan Africa.

c. If the relationship between suffrage and income is strong, the relationship should exist between Europe and Africa.

d. If the relationship between women's suffrage and GDP per capita is not just the result of the important differences between Europe and sub-Saharan Africa, then we would not expect to see the same relationship exist among the countries of sub-Saharan Africa.

5. What did the outlying cases of Kuwait, Qatar, and Zimbabwe illustrate when looking at the relationship between human capital accumulation and GDP per capita?

a. They indicated the possible importance of natural resources, property rights, and politics.

b. They should be removed from the data since they are not representative cases.

c. They all point to the importance of oil as an explanation of GDP per capita.

d. Corruption is clearly an important variable that needs to be considered.

6. What did exploring the stability of our results reveal?

a. There are important thresholds we need to account for in our model estimation.

b. The wrong estimation technique is being used.

c. Regardless of the variables we include in this model, the results always indicate the same thing.

d. Our estimates for a variable in the model can depend heavily on how we specify the model—what variables we include in the model.

7. What is an added variable plot?

a. A plot that shows the residuals against the predicted values

b. A plot that shows the predicted values against the residuals

c. A plot showing the relationship between one variable, accounting for all of the others

d. A plot showing the relationship between the dependent and independent variables

8. What observation led us to ask whether the relationship between ethnic heterogeneity and GDP per capita was a linear pattern?

a. There were many poor island nations.

b. The clump of countries we identified as being "industrialized" may be generating the linear pattern.

c. The slope of the line declined when ignoring the European cases.

d. There seems to be a nonlinear relationship between income and ethnolinguistic heterogeneity when all of the cases are considered.

9. If there is large variation in y at different levels of x in a scatter plot, what does that indicate?

a. The relationship is not linear.

b. There is no relationship between the x and y variables.

c. The x variable does a good job of explaining the variation in y.

d. There could be other variables that might help explain the variation.

10. Which of the following are true in the context of residual plots?

a. Points above the horizontal line represent cases that are underpredicted.

b. Points above the horizontal line represent cases that are overpredicted.

c. Patterns in the data suggest there is a problem with the model.

d. Patterns in the data suggest the model is appropriate.

ANNOTATED R FUNCTIONS

The following functions appear in this chapter. They are listed in order of their first appearance (with the code chunk number in parentheses) and annotated here to give a very brief description of their use. Some are not stand-alone functions and only work in combination with other commands. As a reminder, the code in every chapter will work properly if executed in the order it appears. Proper execution also depends on typing the author-defined *libraries()* command, which loads the required R packages.

ggplot(): defines the basic structure of a plot (usually the x and y variables). (2-1)

aes(): the aes (called "aesthetics") function is used in ggplot to define the basic structure of the plot, which often includes the variables you want to use and any shapes or colors. (2-1)

ylab(): labels the y axis in ggplot. (2-1)

xlab(): labels the x axis in ggplot. (2-1)

theme_minimal(): specifies a minimalist style for ggplot. (2-1)

theme(): specifies font, size, and so forth in a ggplot. (2-1)

annotate(): allows the placement of text in the figure. (2-2)

geom_vline(): places a horizontal line in the figure. (2-7)

scale_colour_manual: a function that allows the user to specify exactly what colors to use in the figure. (2-9)

stargazer(): a function from the 'stargazer' package that helps create professional-looking tables. (2-13)

lm(): a function that specifies a linear regression. The "lm" stands for linear model. (2-13)

ANSWERS

KNOWLEDGE CHECK

1. a
2. a, b, c
3. a, b, c
4. a, b, c
5. b
6. c
7. d
8. b, d
9. a, c
10. b
11. a
12. b, c, d
13. c, d
14. d
15. a, c, d
16. a, b, c, d
17. c, d
18. a, c
19. a, b, c, d
20. a, b, c, d

PRACTICE ON ANALYSIS AND VISUALIZATION

1. c
2. a
3. a, b, c, d
4. b
5. a
6. d
7. c
8. b
9. d
10. a, c

 Access digital resources, including datasets, at http://edge.sagepub.com/brownstats1e.

3 Describing Data

CHAPTER OUTLINE

Learning Objectives

Overview

Data Sets and Variables

Different Kinds of Variables

Describing Data Saves Time and Effort

Identify Puzzles, Questions, Hypotheses, and Clues

Measurement

Summary

Common Problems

Review Questions

Practice on Analysis and Visualization

Annotated R Functions

Answers

LEARNING OBJECTIVES

- Explain the difference between data sets and variables.
- Identify continuous, categorical, and ordered categorical variables.
- Discuss how describing data saves time and effort.
- Connect data description to answering questions and identifying puzzles.
- Create valid and reliable measures.

Overview

Describing data is often given short shrift, brushed aside by those eager to test their hypothesis with sophisticated econometric models. Data description, however, performs several important functions: it helps us identify questions and puzzles, it informs the construction of our hypotheses and theories, and it helps us gauge the validity and reliability of our measures. Ultimately, well-crafted descriptions can make very powerful statements. Take, for example, a figure popularized by Edward Tufte: Charles Minard's description of Napoleon's march into Russia (Figure 3-1).

This graph's beauty lies in the information conveyed along with its simple elegance. From it, we can discern the battles fought, where they were waged, and their toll on Napoleon's men. This antiwar poster reveals the horrible cost of war. The poster documents where Napoleon's troops died, and it offers an explanation. At the bottom of the figure, Minard records the temperature at specific points in the march, suggesting weather influenced the mortality rate of Napoleon's troops. Not only does the poster give an overall view of the campaign, if examined closely it reveals important battles.

On the way back from Moscow, note the crossing of the Berezina River where Napoleon's troop strength was cut in half. The Grand Army found that the usually frozen Berezina had unexpectedly melted, making for a treacherous crossing. Surrounded by Russians and reduced to crossing on two quickly assembled bridges, the French were decimated. In this battle alone, historians estimate 21,000 French soldiers lost their lives (de Fezensac, 1852, p. xviii).

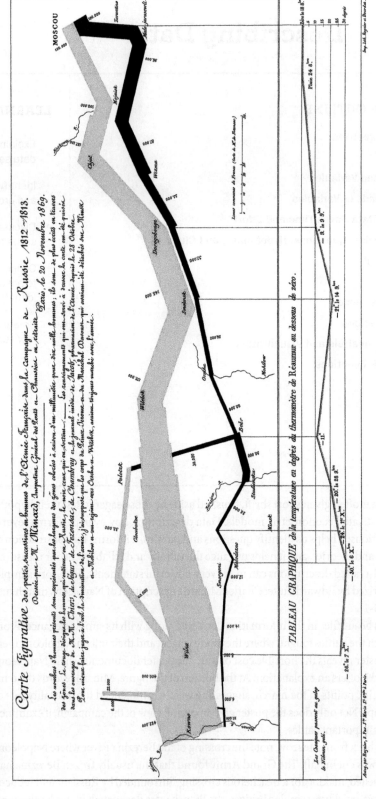

FIGURE 3-1 The Grand Army's March

Source: Charles Minard (1781–1870), via Wikimedia Commons.

While our descriptions will be rudimentary in comparison to Minard's work, the goal should be similar: to produce as much information as possible with elegance and purpose. While there are no simple rules to guide us, there are important conventions we can follow to perfect our craft.

In order to construct useful pictures, we need to know the underlying structure of the data. This chapter starts with the distinction between data sets and variables. The discussion following explains the difference between continuous and categorical variables. The chapter then illustrates how descriptions of the data can quickly help evaluate the connection between our questions, hypotheses, and the data. Finally, good analysis requires we have collected valid and reliable data. Determining the data's validity and reliability is aided by good data description.

Data Sets and Variables

To generate useful descriptions of our data, we need to understand their basic structure: the difference between **data sets** and variables. We must also become familiar with specific kinds of variables since their features determine how they are described. The following section discusses the most basic structural features of data you'll need to know in order to produce useful and informative descriptions.

To better illustrate the difference between a data set and variables, consider Table 3-1, which shows the first six rows of the *states* data set. To produce the table, I define a data set *stdat*, which contains the variables *state*, *region*, *trumpwin*, *percwom*, and *inc* from the *states* data set. The *select()* command is helpful because it indicates which variables I want to use in a data set. The *head()* command tells R I would like to list the first six rows of the variables that I have selected. (As with all of the code chunks in this book, I encourage you to copy Code Chunk 3-1a and execute it in RStudio.)

Code Chunk 3-1a
```
stdat <- head(dplyr::select(states, state, region, trumpwin, percwom, inc))
```

Code Chunk 3-1b
```
knitr::kable(stdat, format = "pandoc", caption = "Table 3-1: First Six
                    Lines of States Data")
```

TABLE 3-1 First Six Lines of *States* Data

STATE	REGION	TRUMPWIN	PERCWOM	INC
Alabama	South	1	75.49	34650
Alaska	West	1	79.02	45529
Arizona	West	1	84.00	35875
Arkansas	South	1	88.52	34014
California	West	0	89.94	44481
Colorado	West	0	79.57	44088

I then generate the table using the *kable()* function, which displays the data set *stdat* in tabular form.

A variable describes a particular characteristic of a person, place, or thing. A data set is a collection of two or more variables. In a data set, columns are variables and rows are cases. This data set contains several variables (columns) that describe different characteristics of the 50 states (rows). To illustrate, five variables are selected from the states data set: *state*, *region*, *trumpwin*, *percwom*, and *inc*. The variable *state* indicates the state's name, *region* indicates the region, the variable *trumpwin* records whether Donald Trump won the state's electoral college in the 2016 race, *percwom* indicates the average woman's salary relative to men, and *inc* records per capita income.

Before we proceed, here's a note of caution. Much of this book is organized in a way that builds knowledge cumulatively. In other words, any concepts or structures not firmly grasped in this chapter will cause problems down the road. At the same time, a solid foundation will make the going much easier in subsequent chapters. Often, one is tempted to move on to the more interesting, colorful aspects of data analysis too quickly. My advice here is to understand the concepts and ideas in this chapter before continuing.

> **KNOWLEDGE CHECK: Explain the difference between data sets and variables.**

COUNTRY	GDPPC	TURNOUT	WOMYEAR
Afghanistan	1629.167	45.83	NA
Angola	6360.849	62.77	1975
Albania	9646.582	53.31	1920
U Arab Emirates	56245.478	NA	NA
Argentina	18333.995	81.07	1947
Armenia	6376.268	62.87	1921

1. The columns in the data set above represent which of the following?
    a. Data sets
    b. Variables
    c. Cases
    d. Countries

2. The rows in the data set above represent which of the following?
    a. Variables
    b. Data sets
    c. Turnout
    d. Cases

3. Which of the following are variables?
   a. *world*
   b. *turnout*
   c. *womyear*
   d. *Afghanistan*
4. Identify which are data sets or variables.
   a. *states*
   b. *hsdiploma*
   c. *nes*
   d. *ftobama*

Different Kinds of Variables

It is better to spend time learning the different kinds of variables than countless time debugging error messages. A shaky foundation on this subject will also make matching hypotheses to data difficult. Error messages can be avoided and a tight connection between hypotheses and data can be formed if you understand the difference between continuous, categorical, and ordered categorical variables.

From a practical standpoint, many of the summaries, models, functions, and pictures we can generate in R depend on knowing what kind of variable we're using. The wrong kind of variable used with a command or function will often elicit an error message. Unfortunately, it won't always be clear what has gone wrong—R's error messages can be less than informative. Even more problematic, sometimes an error message won't appear at all when it should. An inappropriate variable used in a model or function will sometimes produce a result, and it will appear as if everything is fine even though it isn't. In such cases, it is possible to continue down the wrong path without detecting the problem. At best, a significant amount of time will be spent retracing steps already taken. At worst, the entire enterprise can be wrecked. Fortunately, good description helps prevent such mishaps. Moreover, pictures of the data readily confess to the crime, because they indicate immediately whether something strange has happened.

Knowing the difference between continuous, categorical, and ordered categorical variables not only helps avoid going down the wrong path, but it can also help us more tightly connect our hypotheses to the data. For example, if we want to measure the concept of democracy, do we use a variable that measures it on a continuum or is there a clear distinction between authoritarian and democratic regimes? If we want to understand whether retirees vote differently than the rest of the population, should we use a variable that records a person's age or one that indicates whether the respondent is retired? Continuous, categorical, and ordered categorical variables will be explored, defined, and discussed in this chapter so that we can avoid error messages and more tightly connect the data to our questions.

Continuous Variables

Continuous variables measure counts or amounts. The word "continuous" is used since it conveys the idea that the variable can take on any value between its minimum and maximum. Height, weight, and age are all concepts or things that can take on any value. They are continuous measures. They all represent some kind of count or amount.

Continuous variables, consequently, are helpful when measuring things, concepts, or ideas that increase or decrease incrementally. It's clear that populations or a person's age, weight, and height increase incrementally, but what about concepts? Take the concept of democracy referred to earlier. If democracy is something that waxes and wanes in very subtle and incremental ways, a continuous measure provides a fitting representation. If we're trying to measure happiness, we might think of it as a continuous variable, an attitudinal construct that can take on an infinite number of values varying between sadness and elation. Finally, if we're trying to measure personal wealth, it can take on thousands, millions, if not billions of different values (for each dollar or penny added or subtracted).

Note that we have some degree of choice, all of these concepts could lend themselves to being measured with categorical variables: democratic or authoritarian, happy or sad, rich or poor. In other words, we could take each variable and decide that once past a certain threshold, an authoritarian regime becomes democratic, a sad person becomes happy, or a poor person becomes rich. Ideally, how we conceptualize democracy, happiness, and wealth should determine how me measure it. Unfortunately, the kind of data we need is not always the kind of data we get. Consequently, practical matters sometimes make the decision for us.

Categorical Variables

Categorical variables indicate differences in kind rather than counts or amounts. They identify distinct characteristics of people, places, or things. In Table 3-1, the *region* variable is considered a categorical variable: the distinct characteristic of where a U.S. state is located. The variable *trumpwin* is also a categorical variable. Even though it contains numbers, it indicates two distinct possibilities: Trump won the state's electoral college or he didn't (1 or 0). Another common example would be a variable registering a respondent's party affiliation. Often surveys ask whether a respondent identifies as a Democrat, Republican, or Independent.

Note that with categorical variables, the difference between one distinct category and another is not incremental; it is a difference in kind. While the difference could be small, it is nonetheless distinct. You are either a first-year, sophomore, junior, or senior. You cannot be a sophomore and junior at the same time.

To summarize, data generally come in two forms: continuous and categorical variables. Variables with numbers of counts or amounts are continuous variables. Variables containing names or specific characteristics are categorical variables. While continuous variables use numbers (e.g., GDP per capita), categorical variables can use either numbers (*trumpwin*) or names (*region*). Whether using numbers or names, categorical variables reference distinct characteristics of people, places, or things. Understanding the difference between continuous and categorical variables is extremely important because it determines what tools we use to describe them.

Ordered Categorical Variables

Sometimes categorical variables contain classifications that can be ordered along a dimension. One example is the variable *pid7* from the National Election Studies (NES) data that records whether an individual self-identifies as a Strong Democrat, Weak Democrat, Independent, Weak Republican, or Strong Republican. Another, the *educ* variable from the NES data, records whether an individual finished elementary, high school, college, or graduate school. Note that both of these variables hold categories that can be ordered: the party identification variable can be ordered from most liberal to most conservative and the education variable from least educated to most educated. If there is a recognizable order, arranging the views of the data to reflect that order can be helpful. Variables whose categories that can be ordered according to some dimension are called **ordered categorical variables**.

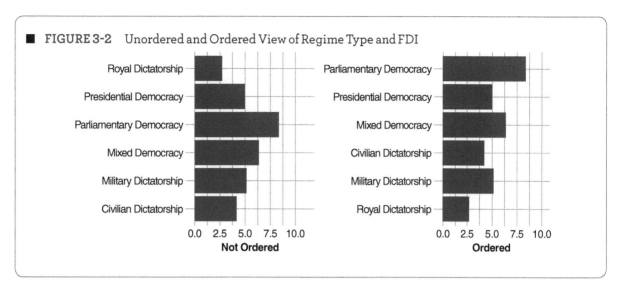

FIGURE 3-2 Unordered and Ordered View of Regime Type and FDI

Consider the example in Figure 3-2. I construct two bar plots of foreign direct investment (FDI; net inflows as a percentage of GDP) against regime type. First, a few words about the code. The first command defines a new variable *world$g* that reorders the levels (categories) of the *world$regime* variable.

Code Chunk 3-2a

```
world$g <- factor(world$regime,
                  levels = c("Royal Dictatorship",
                             "Military Dictatorship",
                             "Civilian Dictatorship",
                             "Mixed Democracy",
                             "Presidential Democracy",
                             "Parliamentary Democracy"))
```

The next two objects that I define (*fig1* and *fig2*) provide the instructions for a bar plot where the mean value of each category is calculated and then plotted. You'll notice that for each figure I use the *ggplot()* command introduced in Chapter 2 to define the basic structure of the plot. In this case, I'll be using the variables *regime* and *fdi* from the *world* data set. Since I want to plot the mean of each category, I use the *stat_summary()* command and specify using the mean. Note also the use of the command *coord_flip()*, which changes the orientation of the bar plot from being vertical to horizontal.

Code Chunk 3-2b

```
fig1 <- ggplot(world, aes(x=regime, y=(fdi))) +
  stat_summary(fun.y=mean, geom="bar", fill="#0000ff", aes(group=1)) +
  ylab("Not Ordered") +
  xlab("") +
  theme_minimal() +
```

```
    theme(plot.title = element_text(size = 8, face = "bold"),
          axis.title = element_text(size = 8, face = "bold")) +
    coord_flip(ylim=c(0,10))


fig2 <- ggplot(world, aes(x=g, y=(fdi))) +
  stat_summary(fun.y=mean, geom="bar", fill="#0000ff", aes(group=1)) +
    ylab("Ordered") +
    xlab("") +
    theme_minimal() +
    coord_flip(ylim=c(0,10)) +
    theme(plot.title = element_text(size = 8, face = "bold"),
          axis.title = element_text(size = 8, face = "bold"))
```

Finally, I use a command that indicates where to place the two figures: the *grid.arrange()* command. The *ncol=2* option tells R to place the two figures side by side.

Code Chunk 3-2c

```
grid.arrange(fig1, fig2, ncol=2,
             top = textGrob("Figure 3-2: Unordered and Ordered View of
                            Regime Type and FDI",gp=gpar(fontsize=10,
                            fontface = "bold")))
```

The bars indicate the average value of FDI across different kinds of regimes. In the first panel (left), I use the order of regime type given to us: there is no particular order. One could, however, arrange the regime categories on a spectrum according to their level of repression. Repression refers to things like the abuse of human rights, whether the right to assemble is limited, or whether members of the press are threatened by the regime. Suppose we want to know whether repression is related to FDI. To answer the question, in Code Chunk 3-2a I moved royal dictatorship to the bottom (most repressive) and switched military dictatorship and civilian dictatorship.

Note how different the two figures look. To the extent we've ordered the categories from least repressive (top) to the most repressive (bottom), the plot with ordered categories suggests FDI decreases as we move from parliamentary democracy to royal dictatorship. By ordering the categories according to repression, we can determine if interesting patterns exist in the data. In the unordered example on the left, it's hard to make any determination of what we're seeing. In the ordered example on the right, there is preliminary evidence that countries with more democratic systems (less repression) are more likely to receive FDI.

One could challenge the way I've reordered the categories. For example, it's not clear that military dictatorships are always more repressive than civilian ones. It's also not clear why one would put parliamentary democracies before presidential democracies. Regardless of my choices, it's clear that the order of the categories displayed can make a big difference in our analysis. If the variable lends itself to order, how those categories are displayed is important.

Let's consider another example that illustrates how ordering categories can change what we see (Figure 3-3). Consider the variable that records the marital status of individuals in the NES survey (*marstat*). There are six different categories for the variable listed in no apparent order:

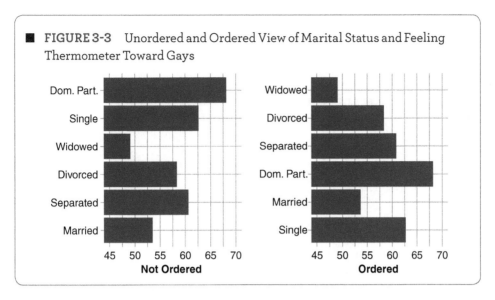

FIGURE 3-3 Unordered and Ordered View of Marital Status and Feeling Thermometer Toward Gays

married, separated, divorced, widowed, single, and domestic partnership. I first shorten the label of "Domestic Partnership" to "Dom. Part." for presentation purposes.

Code Chunk 3-3a

```
levels(nes$marstat)[levels(nes$marstat)=="Domestic Partnership"]
                        <- "Dom. Part."
```

I then create another version of the variable called *nes$g* that reorders the categories in the variable based on when they are experienced in life: single, married, domestic partnership, separated, divorced, or widowed. In the code, to illustrate another way to reorder the categories in a variable, I'm using the numbers associated with each category rather than the actual category names. I encourage you to play around with this command, changing the order of the numbers to see how it affects the order of the categories. To see how the order of the categories changes, first execute the command *levels(nes$marstat)*. Once you've created the new variable *nes$g* with the different order, issue the command *levels(nes$g)*.

Code Chunk 3-3b

```
nes$g <- factor(nes$marstat,
                levels(nes$marstat)[c(5, 1, 6, 2, 3, 4)])
```

Suppose, for purposes of this example, we're interested in the relationship between marriage status and an individual's views toward gays. As in the previous example, I define two figures that show the unordered and ordered versions of the variable *marstat*.

Code Chunk 3-3c

```
fig1 <- ggplot(nes, aes(x=marstat, y=ftgay)) +
  stat_summary(fun.y=mean, geom="bar", fill="#0000ff",
               aes(group=1)) +
```

```
    ylab("Ordered") +
    xlab("") +
    theme_minimal() +
    coord_flip(ylim=c(45,70)) +
    theme(plot.title = element_text(size = 8, face = "bold"),
          axis.title = element_text(size = 8, face = "bold"))

fig2 <- ggplot(nes, aes(x=g, y=ftgay)) +
    stat_summary(fun.y=mean, geom="bar", fill="#0000ff",
                 aes(group=1)) +
    ylab("Ordered") +
    xlab("") +
    theme_minimal() +
    coord_flip(ylim=c(45,70)) +
    theme(plot.title = element_text(size = 8, face = "bold"),
          axis.title = element_text(size = 8, face = "bold"))
```

Once the two bar plots are defined, *grid.arrange()* is used again to actually place them side by side on the page.

Code Chunk 3-3d

```
grid.arrange(fig1, fig2, ncol=2,
             top = textGrob("Figure 3-3: Unordered and Ordered View of
    Marital Status and Feeling Thermometer Toward
    Gays", gp=gpar(fontsize=10,fontface = "bold")))
```

If we simply calculated the average feeling thermometer score for each category and viewed it with a bar plot without any concern for how the categories are ordered, it's relatively difficult to understand the relationship between marital status and views toward gays and lesbians (the bar plot on the left). We see that widowers have the lowest opinion of gays while individuals in a domestic partnership have the highest, but there's really no underlying reason why one category is located next to the other.

Observe what happens if we reorder the categories according to how an individual experiences different marital states through their life span: single, married, domestic partnership,

Art and Practice of Data Visualization
RESIST TAKING DATA AS GIVEN

The previous two examples demonstrate why we must resist taking the data as given. The individuals or organizations who construct the data we use likely have different agendas and different hypotheses. As a result, categories are placed together in a way that is either easy or interesting for their analysis, but not for ours. With R, we can reorder categories and examine them, leading to discoveries that would not have otherwise been made.

separated, divorced, and widowed (setting aside for the moment getting married again after a divorce). By reordering the categories and constructing the same bar plot, we see a different pattern (the bar plot on the right). The bar plot now reveals that older respondents have lower opinions of gays and lesbians. Also revealed is a very marked change between respondents who are married and those in a domestic partnership. How we order the categories can draw our eyes to different relationships and alert us to other causes. In the case of marital status and views toward gays and lesbians, the different view suggests that age and religion (as expressed by a traditional marriage versus a domestic partnership) might matter.

> **KNOWLEDGE CHECK: Identify continuous, categorical, and ordered categorical variables.**

5. Which of the following are continuous variables?
    a. *world$gdppc*
    b. *states$inc*
    c. *nes$marstat*
    d. *states$region*
6. Which of the following are categorical variables?
    a. *world$polity2*
    b. *states$trumpwin*
    c. *nes$pid7*
    d. *nes$birthyr*
7. Which of the following are or could be ordered categorical variables?
    a. *states$region*
    b. *nes$educ*
    c. *nes$birthright_b*
    d. *states$weed*

Describing Data Saves Time and Effort

Getting familiar with your data is important for many reasons. As you'll see in the next sections, familiarity with the data leads to better questions, it helps us choose the appropriate summaries, it helps identify interesting puzzles, and it helps us spot mistakes. In this section we address two characteristics of the data—shape and range—that can help save a great deal of time and effort. First, important choices you'll make in data analysis depend on the shape of the data. If you don't know the shape of the data at the beginning, you may proceed down the wrong path, only later learning that you have to retrace your steps. Second, knowing the range of the data can help quickly determine if we're asking the right question in the first place.

As I stated at the outset of this chapter, it's hard to overemphasize the importance of describing the data as a first step in your analysis. Too often, analysts eagerly proceed to fitting models, generating estimates that give them the answer they want. Two common practical problems can easily be avoided with a cursory view of the data. First, some models and summaries of the data we use are easily influenced by the shape of the data. We need to know the shape of the data first to determine what summaries and models to use.

Second, we need to know the scale of the problem. Suppose we are interested in explaining Houston's temperature on an hourly basis. What if we find it only varies 5 degrees over a 24-hour period? Would that still be a puzzle worth exploring? We can avoid wasting a considerable amount of time if we know the degree of variation beforehand. In other words, we need to know the range—the difference between the minimum and maximum—of a variable. Knowing a variable's shape and range at the outset are just two ways to save a lot of time and effort.

Shape of the Data

Where do the majority of a variable's cases lie between its minimum and maximum values? Are values evenly distributed throughout? Do the cases congregate closer to the minimum, median, or maximum? To answer these questions, we examine the **shape** or **distribution** of the data. In this sense, the words "shape" and "distribution" are synonymous. They describe a characteristic of the data that indicates where most of the cases lie in relation to a variable's range of values. Understanding the shape of our data is important since some simple summaries of the data can be influenced by different shapes or distributions.

As will be demonstrated in Chapter 4, when a variable contains either extremely small or large values in relation to the rest of the data, some simple numerical summaries (e.g., the mean or standard deviation) of the variable will be misleading. Knowing the shape or distribution of the data helps us choose the summaries that provide the most accuracy. Let me explain what I mean by shape.

To illustrate, I created two variables that are plotted in Figure 3-4. Note that I first use the *data.frame()* function to create a data set called *mydata* that has two variables: one is "normally" distributed and approximates a bell-shaped curve (using the function *rnorm()*), and the other has an "exponential" distribution (using the function *rexp()*). The normal curve is defined by my choice of observations (1,000,000), the mean (0), and the standard deviation (1). We will tackle the mean and standard deviation in Chapter 4. The exponential curve is different from the normal curve in that it is defined by the number of observations (1,000,000) and the rate at which each of the 1,000,000 events occur (1). In order to produce a figure that overlays one curve over the other, I have to take the two variables (the two columns of the data set) and stack them on top of each other with the *melt()* function.

Code Chunk 3-4a

```
myData <- data.frame(normal=rnorm(1000000, m=0, sd=1),
                     skewed=rexp(1000000, rate=1))

data <- melt(myData)
```

I then plot the two variables using what is called a **density plot** and overlay them on top of each other for comparison. There are a couple of things to note in the *ggplot()* command I use in Code Chunk 3-4b. First, note that I use the *data* data set. If you look at that data set with the *head()* command, you'll see that there is one column labeled *variable* and another column labeled *value*; these are the two variables in the data set. This helps makes sense of what you see in the *aes()* function within the *ggplot()* and *geom_density()* functions. The other feature

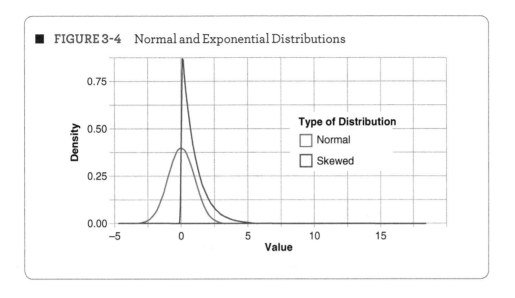

FIGURE 3-4 Normal and Exponential Distributions

Code Chunk 3-4b
```
ggplot(data, aes(x=value)) +
  geom_density(aes(group=variable, col=variable)) +
  theme_minimal() +
  scale_colour_manual(values =
      c("normal" = "#0000ff", "skewed" = "#bf0000")) +
  labs(color='Type of Distribution') +
  theme(legend.position = c(.7, .5)) +
  ggtitle("Figure 3-4: Normal and Exponential Distributions") +
  theme(plot.title = element_text(size = 8, face = "bold"),
        axis.title = element_text(size = 8, face = "bold"))
```

in the code worth noting is that I manually color the two different density plots using a very particular shade of blue (#0000ff) and red (#bf0000).

Density plots indicate where each observation is located over the range of the variable. For the blue line (the normal distribution), most of the variable's observations are at 0 on the x axis (the height of the curve), with a very few approaching 4 and −4. The red line (exponential distribution) also has most of its observations close to 0. It has, however, a few observations upward of 8.

In terms of shape, we say the blue curve approximates a bell-shaped curve: it is symmetrical since one half of the distribution (values above zero) mirrors the other half (those below zero). The variable outlined in red represents the exponential distribution. The red curve is not symmetrical, containing observations upward of 7 and 8. If it were symmetrical, it would also have observations of −7 and −8. Since the curve is not symmetrical, we say that the distribution outlined in red is **skewed**.

As we'll see in the next chapter, the shape of the variable in red warns that some simple summaries of the data will be misleading. This is good information to have at the outset since it could affect which summaries and models we decide to use.

Let's consider another example only this time with real data from the *world* data set. We're somewhat familiar with **histograms** already, so let's consider how the shape of the data is revealed by examining histograms with density plots overlaid. To understand what we mean by different shapes, we'll look at two variables: voter turnout (*world$turnout*) and infant mortality (*world$inf*).

To draw a histogram with a density plot overlay, we need to make a slight change to the code we normally use with the command *geom_histogram()*. Since histograms and density plots represent different units, they are both scaled in this example so that we can compare the curve with the bars of the histogram. The histogram and density plot are scaled by using the *..ncount..* option in the histogram and the *..scaled..* option in the density plot.

With real data, the histogram of voter turnout is pretty close to the ideal bell-shaped curve. We would consider the distribution of voting turnout to be normally distributed with the exception of one value well below 25%.

Now consider the shape of infant mortality in the world: the number of deaths in each country per 1,000 live births. Notice that in Code Chunks 3-5 and 3-6 I only specify one

Code Chunk 3-5

```
ggplot(world, aes(x = turnout)) +
    geom_histogram(aes(y = ..ncount..), fill = "#0000ff") +
    geom_density(aes(y = ..scaled..)) +
    theme_minimal() +
    ylab("Scaled Density and Observation Count") +
    xlab("Voter Turnout") +
    ggtitle("Figure 3-5: Voter Turnout Is Normally Distributed") +
    theme(plot.title = element_text(size = 8, face = "bold"),
        axis.title = element_text(size = 8, face = "bold"))
```

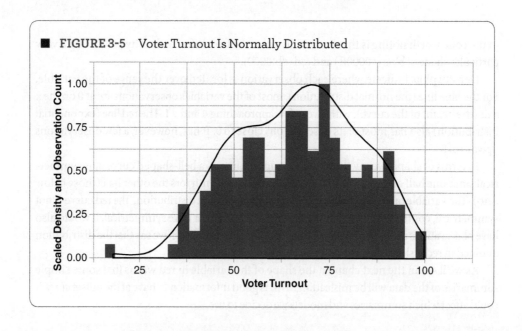

FIGURE 3-5 Voter Turnout Is Normally Distributed

variable in the aesthetics of the *ggplot()* command since I'm only graphing one variable: in Code Chunk 3-5 it's *world$turnout* and in Code Chunk 3-6 it's *world$inf*. Again, in figures where density plots are shown with histograms, we put two periods before and after "ncount" and "scaled" in the aesthetics for the *geom_histogram()* and *geom_density()* commands.

Code Chunk 3-6

```
ggplot(world, aes(x = inf)) +
  geom_histogram(aes(y = ..ncount..), fill = "#0000ff") +
  geom_density(aes(y = ..scaled..)) +
  theme_minimal() +
  ylab("Scaled Density and Observation Count") +
  xlab("Infant Mortality Rate") +
  ggtitle("Figure 3-6: Infant Mortality Is Not Normally Distributed") +
  theme(plot.title = element_text(size = 8, face = "bold"),
      axis.title = element_text(size = 8, face = "bold"))
```

Fortunately, there are many countries that have relatively low numbers of infant deaths per 1,000 live births. Unfortunately, there are some countries that witness more than 30, 60, and even 90 deaths per 1,000 live births. For our purposes, it is important to recognize that the shape of the distribution is not normal, prompting us to take certain precautions when we proceed with summarizing the data and estimating models.

To review, shape is an important characteristic of the data. As we'll see in the next chapter and throughout the book, the summaries and models we use to describe the data can be affected by their shape. If we use the wrong kind of summary or model given the data's shape, we can be misled. Consequently, describing the shape or distribution of our data should be where we start.

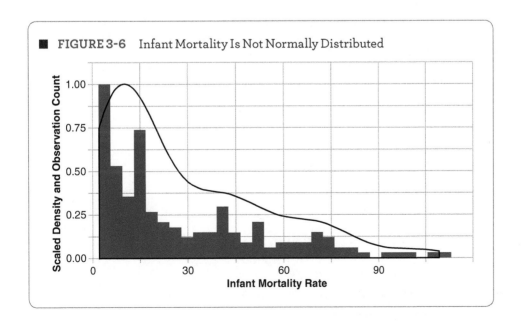

FIGURE 3-6 Infant Mortality Is Not Normally Distributed

Range of the Data

The **range** of a variable is simply the difference between the variable's minimum and maximum value. Describing data—particularly understanding a variable's range—can help us spot questions or problems that might not be worth our time and energy in the first place. For example, suppose we're interested in explaining student success. One can imagine developing a model designed to explain why some students scored better than others on an exam: hours studied, number of previous statistics courses, hours slept the night before, and so on. Imagine we go through all the trouble of building a plausible model to account for student success. Suppose after the model is estimated, we construct a simple table that lists the test scores only to find that the difference between the lowest score and the highest score is less than 3 points (say between 93% and 96%). Knowing that several points separate all students, would you have invested the time and energy to analyze the problem in the first place?

The moral of the story: a dependent variable with a small range may signal there's really not much of a puzzle to begin with. While "What explains student success?" is an intrinsically interesting question, a quick description of the data would have saved a lot of time.

Let's consider an example from our data: the percentage of a state's adult population that finished high school. Educational outcomes are clearly an important thing to track and the percentage of the adult population that completed high school is a clear and recognizable measure. Before we spend any resources investigating why some states have higher rates than others, it would be helpful to know the range of outcomes. Consider the histogram and density plot in Figure 3-7. The code is essentially the same as that found in Code Chunks 3-5 and 3-6, with the only change being the new variable name *hsdiploma*.

Art and Practice of Data Visualization
SMALL CHANGES FOR THE SAME VIEW OF DIFFERENT DATA

In the last three examples where I've generated the same view with different variables, you can see how easy it is to make small changes in the code to produce the same view of different data. Once I have settled on the code that generates the kind of figure I want to examine, it's easy to copy and paste that code and make small changes to see the same view of different data. You can save that code in an R-script or R Markdown file for future use.

Code Chunk 3-7
```
ggplot(states, aes(x = hsdiploma)) +
  geom_histogram(aes(y = ..ncount..), fill = "#0000ff") +
 geom_density(aes(y = ..scaled..)) +
  theme_minimal() +
  ylab("Scaled Density and Observation Count") +
  xlab("Percentage of Population With High School Diploma") +
  ggtitle("Figure 3-7: Not Much Separation Between States") +
  theme(plot.title = element_text(size = 8, face = "bold"),
        axis.title = element_text(size = 8, face = "bold"))
```

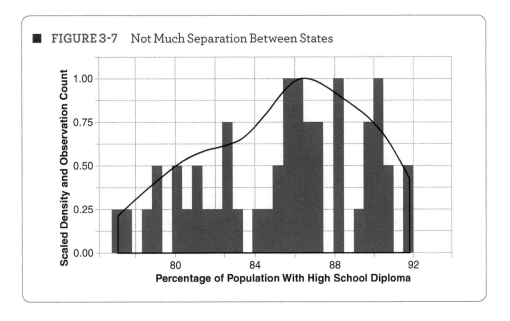

FIGURE 3-7 Not Much Separation Between States

As can be observed, most of the cases are between 80% and 90%. In effect, our study would largely involve explaining why in some states 8 out of 10 adults finished high school, while in others it is 9 out of 10. Given the relatively small range of outcomes, we might be tempted to search for different ways to measure educational outcomes. Again, spending hours examining the data without first considering the size of the problem could be a colossal waste of time. In this example, it's conceivable that in the end, our model might only account for a 2- to 3-percentage point difference. Resources might be better spent examining something else.

> **KNOWLEDGE CHECK:** Discuss how describing data saves time and effort.

8. Why is knowing the shape of the data helpful?
   a. It can help determine the scale of the problem.
   b. It can help in choosing the right summary of the data.
   c. It can help us choose which kind of model to use.
   d. It might lead to a new discovery.

9. Why is knowing the range of the data helpful?
   a. It can help determine the scale of the problem.
   b. It can help in choosing the right summary of the data.
   c. It can help us choose which kind of model to use.
   d. It might lead to a new discovery.

Identify Puzzles, Questions, Hypotheses, and Clues

A good working knowledge of the data—through rigorous data description—helps identify puzzles, formulate questions, build theories, and discover additional clues. Since each and every one of these activities is crucial to good data analysis, describing data is fundamental. Note that the phrase "a good working knowledge of the data" is synonymous with "knowing your subject." There is no substitute for an intimate knowledge of the problem at hand. The more you know your subject's history and various characteristics, the better your questions and hypotheses. This is no different than knowing your data, which makes data description so important. In this section, I'll explain and provide some examples of how taking the time and making the effort to describe the data can aid the mission.

Puzzles and Questions: An Important Difference

Describing the data in a comprehensive way helps us identify puzzles. Why is that important? True puzzles have no obvious solutions. Questions, however, can come loaded; they can reveal our bias. Starting with a true puzzle can be a helpful way to begin your analysis with a relatively high degree of objectivity. Consider the following analogy.

Questions are analogous to a police investigation where suspects have already been identified. Questions are an important part of the scientific enterprise, but they are different than puzzles. A famous question in many police investigations (at least in Victorian England) asks "Did the butler do it?" To answer the question, the investigation is conducted by determining whether there is a link between the butler and the crime. Detectives, in this case, have to be careful not to limit their investigation to evidence associated with their favorite suspect. This is sometimes difficult since the experienced detective may have seen a large number of cases in which the butler did do it. Good detectives guard against this bias all the time, since investigations focusing on the butler are more likely to find the butler did it. Similarly, analysts who begin with their favorite cause will probably find they were right all along.

While eliminating suspects one by one is an important part of any investigation, starting with a question ("Did the butler do it?") rather than the puzzle itself ("Who committed the murder?") holds important implications for the investigation. Starting with a question might unwittingly eliminate important possibilities or suspects that demand investigating. Questions don't always have to but are more likely to come with bias. Having solved previous crimes that involved a butler may cause the experienced detective to look there first. The experienced scholar is no different. The big temptation here is to conduct an investigation not to solve the crime but to show that the butler did it.

Starting with questions puts the analyst in a bind especially when the *importance of the question depends on the answer*. In other words, if the crime's importance depends on who did it, the detective—even if completely well intentioned—is more likely to arrest the wrong person. The same applies to the scholar or policy analyst. For example, if I start by asking if party identification is associated with a person's preference for dogs or cats, the scales may be inadvertently tipped by my desire to find something interesting: a strong correlation between Republicans and loving dogs. A study that finds there is no correlation between party identification and pet preference is not likely to get published or even picked up by the local news.

If the investigation or study starts with a question that is intrinsically interesting regardless of the answer, the scholar will not be tempted to tilt the scales one way or the other. To the degree we can be objective about our work, answering these kinds of questions puts the scholar in the best possible position from the outset. What's an example of a question that is intrinsically interesting regardless of the answer?

Some of the most popular articles published in the *New York Times* focus on personal well-being. It seems on a weekly basis there is a new discovery about caffeine or alcohol having a deleterious or beneficial impact on health. From the individual's point of view, it would be extremely important to know the health effects of each since alcohol and caffeine are so prevalent. Consequently, discovering whether the impact is negative, positive, or inconsequential is extremely important. Clearly, Starbucks and many distilleries in Kentucky would prefer a particular answer. Consumers would prefer the correct one.

Questions will come in all kinds of forms, the intrinsically interesting questions or ones motivated by specific interests. It is important to recognize which is which in order to know whether the investigator (a pharmaceutical firm or a graduate student) faced pressure to manufacture a conclusion.

Describing and exploring data with an open mind is more likely to produce a true puzzle that begs to be solved. While starting with a question is perfectly valid, recognize whether the answer to the question determines its importance. If it does, it might be wise to reformulate the question.

Art and Practice of Data Visualization
A QUICK WAY TO DISTINGUISH QUESTIONS FROM PUZZLES

One way to help identify questions from puzzles simply lies in the word that begins the sentence. Puzzles usually start with the words "how" and "why." Questions usually start with the words "are" and "do." This is simply a quick, although not always fool-proof, way of helping determine if the person asking a question already has an answer.

Describing Data to Refine the Question

Knowing what the data look like at an early stage may help to reformulate your question. For example, suppose we're interested in the relationship between the number of times an individual has been arrested and their views toward the judicial system. This is a typical question in jury selection: prosecutors and defense lawyers want to know if potential jurors have had a formational experience with the police. A quick description of the variable on arrests would reveal that a vast majority of respondents have never been arrested, a few individuals have one arrest, and an even smaller number have been arrested multiple times. In that case, it might make more sense to change our question slightly. Rather than asking whether the number of arrests influences attitudes, it might make more sense to ask whether being arrested at all has an effect.

Let's consider another example. Suppose we're interested in our town's police–community relations. Police conduct has become increasingly political in the last few years. What drives people's attitudes toward the police? Given the politicized nature of the subject, we might think that party identification influences people's attitudes toward those who wear the uniform. A simple table showing the average score on a feeling thermometer (0–100, with 100 being the highest rating) by party affiliation shows something very interesting (see Table 3-2).

To calculate the mean values for each category of party identification, I rely on an extremely useful package in R called **dplyr**. Dplyr is a useful way to manage and manipulate data. Many questions can be answered by a simple table. For example, which countries or states have the highest homicide rates? Are the states with the highest levels of poverty in the South? Scrolling up and down the columns of data is not a viable solution. Dplyr is the best way to search for answers. In many ways, dplyr is analogous to using the search feature of an application to find a particular word in a 100-page document. Not using dplyr is analogous to reading the entire document to find a particular word.

In the following code I've created an object called a **tibble** and named it *avgaytherm*. (For our purposes, a tibble is an object that allows for easy manipulation in dplyr.) In that tibble command, I use the "subset" feature since I only want the cases where the variable *pid7* does not have missing data (NA). I also want to eliminate the respondents who answered "not sure."

Code Chunk 3-8a

```
avgaytherm <- as_tibble(subset(nes, pid7!="NA" & pid7!="Not sure"))
```

The next set of commands takes advantage of the 'dplyr' package. I take the tibble *avgaytherm* and group it by party identification. I then summarize the feeling thermometer for each group. The result is labeled *dt*.

While I don't cover it in this book, there is ample documentation on dplyr's features elsewhere. The book, *R for Data Science*, provides an excellent treatment (Wickham & Grolemund, 2017). Also note the keystroke combination %>% in the code. This is called a pipe operator and can roughly be translated into the English word *then*.

The code used to generate Table 3-2 says "define an object called *dt* by taking the data *avgaytherm*, *then* group it by the variable *pid7*, *then* calculate the mean value on attitudes toward police for each category in *pid7*." The command *mutate_if()* simply establishes the format so that the output in the table only shows two decimal places.

Code Chunk 3-8b

```
dt <- avgaytherm %>%
  group_by(pid7) %>%
  summarise(ftpolice.mean = mean(ftpolice)) %>%
  mutate_if(is.numeric, format, digits=4)
```

Finally, with *dt* defined, I use the *kable()* function again to present the data in table form.

Code Chunk 3-8c

```
knitr::kable(dt, format = "pandoc",
       caption = "Table 3-2: Democrats Show Less Support for Police")
```

TABLE 3-2 Democrats Show Less Support for Police

PID7	FTPOLICE.MEAN
Strong Democrat	58.04
Weak Democrat	62.69
Lean Democrat	61.41
Independent	62.58
Lean Republican	73.78
Weak Republican	76.32
Strong Republican	80.73

While a relationship between conservatism and attitudes is apparent, this simple description of our data adds nuance to our hypothesis. While there is a considerable difference between how respondents rated the police at either end of the ideological spectrum, we notice that the largest jump occurs between Independents and those who lean Republican. If what drives attitudes toward the police is purely ideological, we would expect to see a bigger difference between Strong Democrat and Independent.

The data in Table 3-2 indicate that attitudes toward the police remain fairly stable among different categorizations of Democrats and Independents. Attitudes among those groups vary from a feeling thermometer of 58.04 to 62.58. There is a big jump, however, between Independents and Lean Republican. The large jump between Independents and Republicans suggests there may be something going on other than ideology. Instead, there may be influences that are strictly partisan. Or our partisan categories don't map well with political ideology (another potentially interesting finding). In either case, we may want to change our ideological theory and amend it by considering factors associated with party affiliation.

Too often, we throw a variable into the analysis, generating estimates and conclusions without realizing that such a large number of cases have missing data. One way to save time is to understand how complete the data are for a specific variable. Toward that end, a simple summary table can be very helpful.

In Table 3-3, I tallied the number of observations (nobs) and the number of missing values (NAs) for four variables in the world data set: *nourish*, *gtbeduc*, *polity2*, and *gdppc*.

To produce the table, I first create a data set that includes only the variables that I want to show in the table. I use the 'dplyr' package again. The code below says to define a data frame called *df* by taking the *world* data set and selecting *nourish*, *gtbeduc*, *polity2*, and *gdppc*.

Code Chunk 3-9a

```
df <- world %>%
   dplyr::select(nourish, gtbeduc, polity2, gdppc)
```

I then issue a command *basicStats()* that calculates the number of observations and missing values—denoted by NA—for each variable.

Code Chunk 3-9b
```
dft <- basicStats(df)[c("nobs", "NAs"),]
```

Now that we have an object I called *dft*, I can print it out using the *kable()* command, which will produce a nice table.

Code Chunk 3-9c
```
knitr::kable(dft, format = "pandoc", caption =
             "Table 3-3: Too Many Missing Cases (NAs)")
```

TABLE 3-3 Too Many Missing Cases (NAs)

	NOURISH	GTBEDUC	POLITY2	GDPPC
nobs	182	182	182	182
NAs	70	51	27	6

In Table 3-3, we see that the *nourish* variable (percentage of the population undernourished) has 70 missing values. That means roughly one-third of the 182 possible data points are not available. In this case, entire continents may be missing or severely underrepresented. Consequently, the conclusions one wishes to draw from these data may not be possible. A simple summary of the data will help avoid spending hours investigating a particular question under the mistaken assumption that a representative sample of cases is being considered. It is better to know the limitations of the data—with simple data description—before proceeding too far.

Describing Data Reveals Additional Clues

Sometimes even the simplest views or basic understanding of the data can reveal important relationships that previous theory or hypotheses had ignored. The more we describe the data, the more likely we will identify patterns that reveal possible causes for what we're trying to explain. Consider the scatter plot in Figure 3-8.

I'll formally introduce the scatter plot in Chapter 5. For now, notice that in the *ggplot()* command I am specifying two variables in the aesthetics command: *inc* and *infant*. In that same command I am telling ggplot to color the plot by *region*. Once I've specified the main structure of the plot, then I simply use *geom_point()* to indicate a layer of dots that define a scatter plot.

Code Chunk 3-10

```
ggplot(states, aes(inc, infant, col=region)) +
  geom_point() +
  theme_minimal() +
  ggtitle("Figure 3-8: Infant Mortality and Income in the United
States") +
  ylab("Deaths per 1,000 Live Births") +
  xlab("Income per Capita") +
  scale_colour_manual(values = c("South" = "#0000bf",
      "West" = "#bf0000", "Midwest" = "#00ffff",
      "Northeast" = "#04183d")) +
  labs(color="") +
  theme(legend.position = c(.75,.75)) +
  theme(plot.title = element_text(size = 8, face = "bold"),
      axis.title = element_text(size = 8, face = "bold"))
```

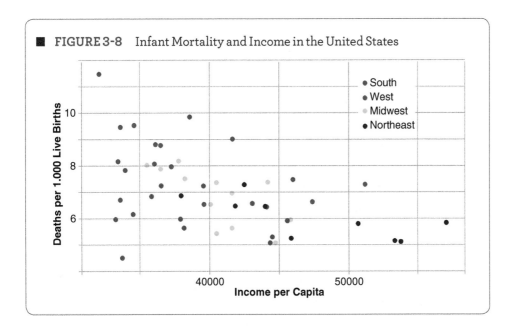

FIGURE 3-8   Infant Mortality and Income in the United States

Suppose we're interested in infant mortality, the number of deaths per 1,000 live births. Infant mortality is quite often used as a very broad gauge of a population's level of health, particularly for women. A likely cause of infant mortality is income. To investigate, we construct a scatter plot to see the relationship between infant mortality and per capita income in each of the U.S. states. We also decided to color each case according to region: Northeast, South, Midwest, and West.

While income does seem to be related to infant mortality—as income rises, infant mortality falls—notice that the states with the highest infant mortality are located in the South. While

> **Art and Practice of Data Visualization**
>
> **MAXIMIZE SPACE**
>
> When possible, place the legend of a figure within the graph itself. Placing the legend inside the grid rather than in the margins allows you to use the space available for the data rather than the key used to describe it. This is easily done by using the *legend.position* command (Code Chunk 3-10).

the southern states have low incomes in common, we know they have relatively low educational attainment and are also more likely to be conservative; their government representatives are less likely to spend on education and health. Just by describing the relationship between income and infant mortality, we've discovered a number of other factors to consider, providing additional clues that demand further investigation.

> **KNOWLEDGE CHECK: Connect data description to answering questions and identifying puzzles.**

10. Which are questions and which are puzzles?
    a. Are Democrats more likely to have dogs or cats?
    b. Why do some people have dogs rather than cats?
    c. Do minimum wage laws help raise wages?
    d. Why do minimum wage laws differ among states?
11. Indicate whether each statement is true or false?
    a. Questions are not part of the scientific process.
    b. Solving puzzles can be done without any bias.
    c. A good puzzle's importance does not depend on the answer.
    d. A question's importance can depend on the answer.
12. What other clues exist in Figure 3-8?
    a. Population density is related to infant mortality.
    b. Income doesn't matter until $40,000.
    c. Partisanship is correlated with infant mortality.
    d. Being poor guarantees a high infant mortality rate.

Measurement

Translating hypotheses into meaningful indicators is a significant challenge in data analysis. Scrutinize your measures. Discovering a strong empirical pattern is exciting. Discovering nobody believes your measure is frustrating. Take the time to develop **valid** and **reliable** measures at the outset.

Suppose we find a strong correlation between a measure of democracy and economic growth. Our discovery will only be as strong as the connection between our democracy measure and the definition of democracy. For example, a measure based on free and fair elections will not fully capture how a chief executive operates once installed in office. In addition to the difficulties associated with measuring democracy, few will trust the result unless we've accurately measured the annual change in economic activity. Accounting for the production of all goods and services within a country is no small feat, especially when a sizable portion of it operates informally in the black market. With respect to the strong correlation between democracy and economic growth, the issues are validity (mapping our measure to a definition of democracy) and reliability (accurately accounting for all economic activity).

Measurement validity refers to how closely the data match our hypotheses. Measurement reliability refers to the accuracy and consistency of our measure. Describing data is a good way to determine validity and reliability.

Validity

Do the data represent the idea or concept we wish to measure? Suppose we are interested in the relationship between education and democracy. We want to know whether an educated society is likely to be democratic. In this example, we need measures of education and democracy. Education can be measured in a number of different ways: enrollment, years of schooling, literacy, and so forth. Which one of these measures represents the best way to gauge the level of education in society?

Similarly, democracy could be measured in a number of different ways: Are there free and fair elections? Is the rule of law observed? Can anyone run for office? Further, which of these characteristics of democracy best represents the degree to which a government is democratic? A brief look at what is one of the more popular measures of democracy (*polity2*) reveals that Cape Verde, Hungary, Mongolia, Uruguay, and Mauritius all have the same scores as Canada, Sweden, the United Kingdom, and the United States. Does the *polity2* measure provide a valid measure? We could spend the rest of this book and many others debating the issue. The point here is that a quick look at the data may raise questions about whether the variable does, in fact, provide a valid measure.

Let's consider another example. Suppose we want to understand the relationship between education and party affiliation in the United States. We hypothesize the more educated the state, the more liberal its population. In our *states* data set we are limited to using a variable that records the percentage of the population that has completed high school (*hsdiploma*). If we were to blindly throw that variable into a model without looking at it first, we would be unaware that the states with the highest percentages of high school graduates are the sparsely populated midwestern and western states: South Dakota, Washington, Utah, Alaska, Nebraska, Wyoming, and Montana. A brief look at a table that ranks the top 15 states would give pause.

To produce the table, I turn to the 'dplyr' package again. First, I define a data set called *ed* and tell dplyr to use the variable *state* and *hsdiploma* from the *states* data set. I then use the *arrange()* function to list the states in descending order according to the *hsdiploma* variable. Finally, I use the *slice()* function to select only the first 15 states.

Code Chunk 3-11a

```
ed <- states %>%
  dplyr::select(state, hsdiploma) %>%
  arrange(desc(hsdiploma)) %>%
  slice(1:15)
```

Once we have that result, it's easy to present it in tabular form using the *kable()* command. The *pandoc* option is used to produce a simple, nice-looking table in Microsoft Word (Table 3-4).

Code Chunk 3-11b

```
knitr::kable(ed, format = "pandoc", caption = "Table 3-4: Top 15 States")
```

TABLE 3-4 Top 15 States

STATE	HSDIPLOMA
South Dakota	91.8
Washington	91.8
Minnesota	90.8
Utah	90.7
Alaska	90.4
Nebraska	90.4
Vermont	90.0
Wyoming	90.0
Colorado	89.7
Iowa	89.7
Montana	89.6
Maine	89.3
Connecticut	88.2
Kansas	88.1
New Hampshire	88.1

Either educated states tend to be Republican (i.e., our hypothesis is wrong) or we're using a measure that really doesn't capture what we're trying to get at (i.e., the measure is not valid). We might want, instead, to use the percentage of the population with college degrees. Note how a quick look at a simple table led us to question the validity of our measure. Quick descriptions of the data before we start the modeling exercise can pay huge dividends. If we do not have a valid measure, it might be wise to collect more data.

Reliability

The reliability of our measure is also a concern. Reliability is about accuracy. Reliable measures can be repeated (the same result is obtained whether measured today or tomorrow) or replicated by other people (the same result is obtained by two different people). Some measures are poorly constructed, making them difficult to replicate by others. Categorical measures based on poorly defined criteria are unreliable since it would be difficult for others to agree on a score or classification.

Some phenomena are extremely difficult to measure reliably. An individual's intelligence or a country's GDP are two often-used but difficult to reliably measure phenomena. Measuring democracy is difficult as well. Asked to assign a score between 1 and 10 indicating whether a country has an "open" process for candidate selection, two very knowledgeable scholars could come to very different conclusions.

Survey questions that ask about people's attitudes toward candidates or policies may also be unreliable. On Tuesday a survey respondent may give Bernie Sanders a score of 68 out of 100; on Wednesday the same respondent may give Bernie a 50 without any further information. Finally, something as often used as GDP can be unreliable. GDP figures are meant to capture all goods and services produced domestically in a country. In the United States, growth in GDP is regularly corrected months after the initial figures are released. That these figures are regularly revised illustrates the difficulty we have in reliably measuring gross domestic output.

One should go into any data exercise with eyes wide open. Describing your data can help you better understand the limitations of the inferences and conclusions drawn from them.

KNOWLEDGE CHECK: Create valid and reliable measures.

---

13. Which of the following raise concerns of validity?
    a. Large, aggregate measures of an economy
    b. Measuring intelligence by asking someone's classmates
    c. Measuring democracy by asking citizens to rate their country
    d. Using one's level of education or income to measure happiness

14. Which of the following are reliable measures?
    a. Income generated from a lemonade stand
    b. Year-end income generated by Microsoft
    c. Asking a respondent whether they will vote
    d. Crowd size at an inaugural address

15. Which are valid but not reliable?
    a. Measuring climate change from the global temperature for the year
    b. Measuring economic output from GDP for a country
    c. Measuring kindness by how much one donates each year
    d. A film's quality measured by the number of Oscars it wins

16. Indicate whether the following measures are valid, reliable, or both.
    a. Measuring happiness from a person's levels of joy and contentment as observed by others
    b. Measuring government legitimacy by the number of protests
    c. Measuring social justice by the number of people in prison
    d. Measuring one's popularity by the number of Twitter followers

SUMMARY

There is no substitute for being familiar with your data. There are a whole host of problems that can be avoided when we know what the data look like. In addition to avoiding problems, there are discoveries to be made with simple descriptions. In this chapter we saw how descriptions can help us identify interesting puzzles, ask good questions, shape the hypotheses and theories we wish to test, and identify additional clues. A rigorous description can also indicate that the data we have on hand are inadequate to the task, in which case we can reformulate our question or get more data.

COMMON PROBLEMS

- Sometimes it's hard to tell the difference between a continuous and categorical variable, because some continuous variables have so few possible values that they're almost like categorical variables. Some categorical variables have so many categories, they're almost like continuous variables. Take, for instance, the *fhliberties* variable in the *world* data set. It registers how many rights and liberties countries afford individuals (1–7, with 1 being the most). Even though there are only seven possible values, some use it as a continuous variable. In some cases, there's no right answer (continuous vs. categorical). It depends on your question and how you intend to use the measure.

- Determining whether a distribution is skewed or normal is often in the eye of the beholder. While there are more technical ways to measure skewness, if you suspect a variable is skewed, always bear in mind that certain summaries of that variable can be unduly influenced by extreme values.

- Too many times, too many NAs are ignored. The impatient analyst will often drill down, making finer distinctions with the data. Always pay attention to the number of observations you're actually dealing with since it can have a huge impact on how representative the data are for a particular population.

- If pictures are worth a thousand words, then read all of the words! As I've indicated throughout this chapter and will continue to emphasize in others, take your time with each figure you produce. The biggest obstacle I see in data analysis is the temptation to move quickly on to the next figure without realizing what the one you're looking at is trying to tell you.

REVIEW QUESTIONS

1. Why do we describe the data?
2. What is measurement validity?
3. What is measurement reliability?
4. How can a description of the data change our question?
5. How can a description of the data change our hypothesis?
6. Why is it important to know the number of observations?
7. Why is the distinction between questions and puzzles important?
8. What is a data set and what is a variable?
9. What is an ordered categorical variable?
10. What is a continuous variable?

PRACTICE ON ANALYSIS AND VISUALIZATION

1. Indicate which are variables or data sets.

    a. *world*
    b. *inf*
    c. *states*
    d. *nes*

2. Provide the code that shows the first six rows of the *world* data set.

3. Write the code that shows the first six rows of the following variables: *state*, *st*, *murderrate*, and *hsdiploma*.

4. Using the code from Code Chunk 3-3b, reverse the order of the categories in the variable *nes$educ*. Whenever you reverse the order, be sure to create a new variable so that you don't overwrite the original *nes$educ*.

5. Which are categorical variables?
   a. *world$colony*
   b. *world$region*
   c. *states$weed*
   d. *nes$amer_ident*

6. Which are continuous variables?
   a. *world$young*
   b. *world$aclpregion*
   c. *states$trumpwin*
   d. *states$stand*

7. Which are ordered categorical variables?
   a. *nes$pid3*
   b. *nes$educ*
   c. *states$region*
   d. *states$democrat*

8. Which are puzzles and which are questions?
   a. Why are homicide rates higher in southern states?
   b. Are predominantly Republican states wealthier than their Democratic counterparts?
   c. Is spending on education higher in Democratic states?
   d. Do democratic countries grow faster than dictatorships?

9. Which have problems associated with measurement validity, reliability, or both?
   a. Measuring climate change using annual global temperatures
   b. Measuring a country's income using GDP per capita
   c. Measuring democracy using the *polity2* measure
   d. Measuring happiness using income

10. Identify which are the problems associated with measurement reliability.
    a. It's difficult to reproduce results.
    b. The results will not indicate what the analyst wants to infer.
    c. The variable does not fit the question.
    d. It is difficult to tell whether that being measured has increased, decreased, or stayed the same.

ANNOTATED R FUNCTIONS

The following functions appear in this chapter. They are listed in order of their first appearance (with the code chunk number in parentheses) and annotated here to give a very brief description of their use. Some are not stand-alone functions and only work in combination with other commands. As a reminder, the code in every chapter will work properly if executed in the order it appears. Proper execution also depends on typing the author-defined *libraries()* command, which loads the required R packages.

head(): lists the first six rows of a data frame. (3-1a)

select(): used in dplyr to select the variables from the data frame. (3-1a)

kable(): used to display data in table form. (3-1b)

factor(): used to manipulate levels (categories) in a variable. (3-2a)

ggplot(): defines the basic structure of a plot (usually the x and y variables). (3-2b)

aes(): the *aes* (called "aesthetics") function is used in ggplot to define the basic structure of the plot, which often includes the variables you want to use and any shapes or colors. (3-2b)

stat_summary(): calculates various statistics for a data frame to be displayed in ggplot. (3-2b)

ylab(): labels the y axis in ggplot. (3-2b)

xlab(): labels the x axis in ggplot. (3-2b)

theme_minimal(): specifies a minimalist style for ggplot. (3-2b)

theme(): specifies font, size, and so on in a ggplot. (3-2b)

coord_flip(): flips the coordinates of a ggplot. (3-2b)

grid.arrange(): displays two or more different plots together. (3-2c)

levels(): used to list categories (levels) in a variable. (3-3a)

data.frame(): creates a data frame (data set). (3-3a)

melt(): stacks variables on top of each other in a data frame. (3-4a)

geom_density(): draws a density plot in ggplot. (3-4b)

ggtitle(): provides the title for a ggplot. (3-4b)

geom_histogram(): draws a histogram in ggplot. (3-6)

as_tibble(): creates an object resembling a table to use in dplyr. (3-8a)

subset(): chooses the cases based on a logical expression. (3-8a)

group_by(): used in dplyr to carry out calculations based on the groups specified. (3-8b)

summarize(): used in dplyr to calculate statistics specified. (3-8b)

mutate_if(): creates a new variable in dplyr if the logical condition holds. (3-8b)

basicStats(): calculates statistics for the data frame specified. (3-9b)

geom_point(): draws a scatterplot in ggplot. (3-10)

arrange(): used in dplyr to order cases according to the specified variable. (3-11a)

desc(): used in dplyr to order cases in descending order. (3-11a)

slice(): used in dplyr and specifies which rows of the data to use in an operation. (3-11a)

ANSWERS

KNOWLEDGE CHECK

1. b
2. d
3. b, c
4. variable (a), variable (b), data set (c), variable (d)
5. a, b
6. b, c
7. b
8. b, c, d
9. a, d
10. question (a), puzzle (b), question (c), puzzle (d)
11. false (a), false (b), true (c), true (d)
12. b
13. b, c, d
14. a
15. a, b
16. neither (a), reliable (b), reliable (c), reliable (d)

PRACTICE ON ANALYSIS AND VISUALIZATION

1. data set (a), variable (b), data set (c), data set (d)

2. head(world)

3. `head(dplyr::select(states, state, st, murderrate, hsdiploma))` or `df <- states %>% head(dplyr::select(state, st, murderrate, hsdiploma))`

4. `nes$var <- factor(nes$educ, levels(nes$educ)[c(6,5,4,3,2,1)])`

5. a, b, c, d

6. a

7. a, b

8. puzzle (a), question (b), question (c), question (d)

9. reliability (a), reliability (b), validity (c), validity (d)

10. a, d

 Access digital resources, including datasets, at http://edge.sagepub.com/brownstats1e.

4 Central Tendency and Dispersion

CHAPTER OUTLINE

Learning Objectives

Overview

Measures of Central Tendency: The Mode, Mean, and Median

Mean Versus Median

Measures of Dispersion: The Range, Interquartile Range, and Standard Deviation

Interquartile Range Versus Standard Deviation

Summary

Common Problems

Review Questions

Practice on Analysis and Visualization

Annotated R Functions

Answers

LEARNING OBJECTIVES

- Calculate the mode, mean, and median.
- Explain the difference between the mean and median.
- Calculate the range, interquartile range, and standard deviation.
- Explain the difference between the interquartile range and standard deviation.

Overview

Describing data usually starts by answering two questions: (1) What is the typical case? and (2) How typical is the typical case? The first question relates to central tendency and the second concerns dispersion (or spread). This chapter explores these two fundamental characteristics of data. Central tendency is so important to our daily lives that we often use the concept without realizing it. Our daily wardrobe choices are governed, in part, by what we think the temperature will be that day—usually based on the typical or average temperature for the season. Doctors make diagnoses based on central tendency. Medications are prescribed based on whether a patient's heartbeat is below or above average. Is a patient's blood pressure too low or too high? These calls are all based on central tendency.

The same is true with dispersion (also called spread). For example, when calculating the travel time for an important meeting, we may decide not to rely on the average time but the spread. If the travel time varies dramatically (a large spread), it should inform our time of departure. Knowing the dispersion of income helps policymakers design policies to reach those in greatest need. The average household income for a family of four may allow for adequate food, housing, and education. However, are most families close to that average or do conditions vary widely? If some households realize incomes much lower than what is needed to survive, policymakers need to know.

This chapter covers central tendency and dispersion. You'll learn what they are and the kinds of questions they answer. You will also learn different ways to measure them. How to calculate and interpret different representations of central tendency and dispersion form the basis of this chapter.

Measures of Central Tendency: The Mode, Mean, and Median

In our attempt to understand the world around us, we naturally gravitate toward the typical case. We want to know how the typical child does in school, what public services does the typical citizen want, or what is the average amount of time it takes to finish college. Policy decisions at all levels of government, decisions in the medical field, and democracy itself depend on the typical case. Measures of central tendency provide that information.

There are several ways to measure central tendency. Which measure you decide to use is based on two things: the kind of data and the presence of outliers. With respect to the kind of data, is the variable continuous or categorical? When the variable is continuous, we generally ask "What is the average?" When the variable is categorical, we usually want to know what category occurs most frequently.

In terms of outliers, the measure we use depends on whether extreme values exist on the low or high end. As you'll see, one of the measures we use, the mean, is influenced by extreme values. A recent *New York Times* article on the number of guns in the United States provides a nice example (Fisher & Keller, 2017). A vast majority of the world's countries have fewer than 20 guns per 100 people. The United States stands alone at 80! Since the United States represents such an extreme value at the high end, we need to carefully consider how we measure central tendency. If we use the mean to indicate central tendency in this example, it will be influenced (pulled toward a higher value) by the U.S. case. In this example, we might use the median instead of the mean. How to calculate both measures and when to use them also will be discussed.

With categorical data, our question changes. When using categories, we're not really looking for the average, or the middle case. Instead, we want to know which category holds the most observations. The mode, rather than the mean or median, indicates the category with the most cases. Elections are a good example since we want to know which candidate received the most votes. Although press accounts rarely express poll numbers by explicitly mentioning the mode, that is our focus.

To summarize, we have three different ways to measure central tendency: the mode, the mean, and the median. As you'll see, even though each measure is designed to identify the typical case, the differences between them can be substantial. Our choice of measure matters and the stakes can be high. While misleading summaries of central tendency and dispersion might lead to mildly irritating consequences (being late for a meeting or picking the wrong clothes), some are more substantial both for society (forecasting hurricanes) and the individual (admission to college).

The following table serves as a quick guide to the measures of central tendency.

Mode

The **mode** is the value which occurs most frequently in a variable. The mode is especially handy when analyzing categorical data. For example, if we add up all of the Republicans,

Measures of Central Tendency

KIND OF DATA	QUESTION	MEASURE
Continuous (no outliers)	What is the average?	Mean
Continuous (outliers)	What is the value of the middle case?	Median
Categorical	What category is the most popular?	Mode

Democrats, and Independents in the United States, the mode simply refers to the category that holds the most cases. Noted earlier, elections are based on determining the mode: we want to know which candidate received the most votes.

Establishing the mode figures prominently in politics, society, and business. Surveys that ask respondents to describe themselves ultimately depend on establishing the mode. We often want to know how many respondents regard themselves as African American, Asian American, Latinx, Native American, or white. Establishing which of those groups predominate in a state or congressional district holds important implications for politics. Concerns about immigration are based on the mode: what nationalities constitute the largest number of immigrants? Surveys designed to determine the top-selling brand or most popular product—something CEOs obsess about—all involve the mode.

Calculating the Mode

To identify the mode, consider the bar plot for the party identification variable *pid7* in the *NES* data set. In this survey, we see that the number of respondents indicating they are "Strong Democrats" is close to 300, nearly 100 more respondents than the next largest category: Independents. The mode is therefore Strong Democrats. To the degree that the NES survey respondents provide a good representation of the American voter, the histogram in Figure 4-1 will inform political strategies and keep some politicians up late at night.

As with previous chapters, the following code is meant to be copied and used in RStudio. In Code Chunk 4-1, I draw a simple bar plot. Since a bar plot describes one variable, only one variable, *pid7*, is included in the aesthetics (i.e., within the *aes()* function). Note that I tell R to fill the bars with the shade of blue whose code is #0000ff. You can indicate colors by using their name (red, blue, etc.), but to indicate a very specific shade it's helpful to use its code. The other commands should be familiar: *theme_minimal()* indicates a minimalist style for the plot, *coord_flip* changes the orientation of the plot to horizontal, and the *theme()* function specifies the font, style, and size of the labels. It's a lot to type, but note that once you've established this look for a bar plot, to create other bar plots you need only to switch out *pid7* and the data set name (*nes*).

Code Chunk 4-1

```
ggplot(nes, aes(x=pid7)) +
  geom_bar(fill="#0000bf") +
  theme_minimal() +
  ggtitle("Figure 4-1: Strong Democrats Outnumber
```

```
                    Strong Republicans") +
    xlab("Party Identification") +
    ylab("Number of Respondents") +
    coord_flip() +
    theme(plot.title = element_text(size = 8, face = "bold"),
          axis.title = element_text(size = 8, face = "bold"))
```

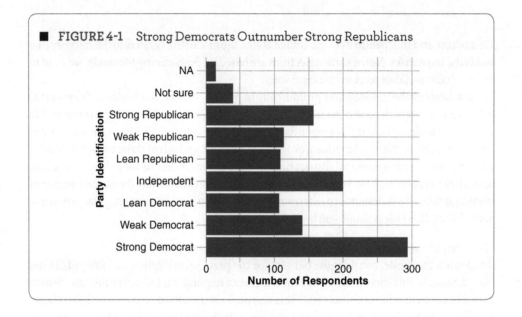

FIGURE 4-1 Strong Democrats Outnumber Strong Republicans

Mean

The **mean** is perhaps the most frequently used numerical summary, stemming from its intuitive appeal and ease of calculation. Everything from a baseball player's success at the plate (batting average) to the academic preparedness of an incoming freshman class (average GPA, average ACT or SAT score) is based on calculating the mean. The mean (or average) is a familiar construct and describes many important aspects of our lives. Consequently, we have a good intuition for what it means and represents.

Calculating the Mean

Suppose we are interested in calculating the average amount of cash the students in your statistics class have on hand. To calculate the mean, we sum all of the cash and divide by the number of students in the class. Let's use the mean to get some terms and mathematical operations under our belt.

First, it is important to note that in this context the words *individuals*, *cases*, and *observations* are synonymous: all three terms can be used interchangeably. To represent the number of individuals, cases, or observations in a data set or variable, we use the letter n. You may be asked, "What is the n?" This question refers to how many observations, individuals, or cases there are in a variable or data set.

Second, even though calculating the mean is fairly simple, it's important to become familiar with its mathematical representation since you'll see it quite often. Parts of the formula are used in many other calculations and will be helpful to know (e.g., the summation sign, Σ). To begin, we start with the lowercase Greek letter μ, which is used to denote the mean.

$$\text{mean} = \mu$$

Here is the mathematical formula for the mean:

$$\mu = \frac{1}{n}\sum_{n=1}^{n} x$$

Let's break this down into its component parts. First, consider the summation sign and the x right next to it. This part of the expression refers to summing—adding up—all of the values: the amount of cash each student has in their pocket. We *sum* the amount of cash on hand (x) for everyone in the class:

$$\sum_{n=1}^{n} x$$

The Greek symbol Σ is the summation sign. In its general form (when we're not referring specifically to our example), we use n to represent the number of individuals, cases, or observations. The summation sign provides a compact and helpful way to indicate we should sum all of the values of x for all of our observations. Without the summation sign, expressing the act of adding up all of the observations would be tedious.

$$x_1 + x_2 + x_3 + \ldots + x_n$$

In calculating the mean cash on hand, x_1 refers to the first student we survey, x_2 represents the second student, x_3 the third student, and so on. The three dots indicate that we continue to survey each individual until we get to the last student in the class, which is denoted by x_n. If there were 10 students in the class, we'd sum over those 10 students. The resulting summation sign would be written as follows:

$$\sum_{n=1}^{10} x$$

This indicates we should repeat the operation 10 times. For completeness, without the summation sign the operation would be written in the following way:

$$x_1 + x_2 + x_3 + \ldots + x_{10}$$

When we write the expression in its general form and are not referring to a specific example, we use the lowercase n instead of 10.

$$\sum_{n=1}^{n} x$$

Now that we've added up all of the cash on hand in the class, we need to divide by the number of students. To divide by the number of students, you'll often see the sum Σ multiplied by $\frac{1}{n}$.

This means take the sum and divide by the number of observations (n). The resulting expression is as follows:

$$\frac{1}{n}\Sigma$$

When we put both expressions together, we're simply summing all of the cash from each individual from 1 to 10, then dividing by 10. In general terms, the summation sign Σ indicates the following: (1) execute the operation and record the result and (2) sum all of the results obtained from each execution of the operation. In this case, the operation refers to finding out how much cash each student has in their pocket (denoted by x). The summation sign performs an important operation in statistics; it is shorthand for "repeat the operation and add up all the values produced" (usually on all of the values $x_1 \ldots x_n$ of a variable). Putting it all together for our specific example, we have the following:

$$\mu = \frac{1}{10}\sum_{n=1}^{10} x$$

Let's make things even more concrete. Suppose that there are 10 students in your class and they have from $1 to $10 in their pockets. So the computer can calculate the mean, we record the number of dollars for each student and put it in a list:

$$x = (1,2,3,4,5,6,7,8,9,10)$$

Given that there are 10 individuals, we can add some information to the formula. One individual in class has $1, another has $2, another has $3, and so on. The first part of the operation involves summing those amounts.

$$1+2+3+4+5+6+7+8+9+10 = 55$$

Now that we've summed the amount of cash on hand for all 10 students in the class ($55), we divide by 10 (multiply the sum by $\frac{1}{10}$).

$$\frac{1}{10}\sum_{n=1}^{10} x$$

The mean amount of cash on hand for the class is $5.50.

So far we've concentrated on an easy example with only a few individuals (n = 10). Let's take advantage of R and calculate the mean number of homicides per 100,000 population for the countries in the world. To visualize the mean in this example, I've drawn a histogram with a red line that indicates the mean.

Consider the following code (Code Chunks 4-2a and 4-2b) that calculates the mean and draws the histogram. The first command *mean(world$homicide, na.rm = TRUE)* will calculate the mean value. Note that included in the parentheses is *na.rm = TRUE*. This tells R to calculate the mean even though some of the countries are missing data. If you want to practice calculating the mean yourself, you can type the command in a code chunk and execute it or simply type it into the console and hit the return key. Also note the two hashtags (##) and the [1] adjacent to the mean 8.459007. The hashtags appear whenever output is generated from a command in R and then knitted into a document. The [1] merely counts the number of elements generated by the command.

Code Chunk 4-2a

```
mean(world$homicide, na.rm = TRUE)

## [1] 8.459007
```

By drawing a line for the mean and overlaying it on the histogram (the *geom_vline()* command below), we can see how the mean represents the middle of the data. One can imagine that the mean is the fulcrum, balancing the histogram as if it were on a teeter-totter.

Code Chunk 4-2b

```
ggplot(world, aes(homicide)) +
  geom_histogram(bins=20, fill = "#0000ff") +
  labs(title="Figure 4-2: Average Homicide Rate for the World") +
  xlab("Number of Homicides per 100,000 Population") +
  geom_vline(xintercept=8.45, col="#bf0000") +
  annotate("text", x = 12, y = -7, label = "8.45", col="#bf0000",
           size = 3) +
  theme_minimal() +
  theme(plot.title = element_text(size = 8, face = "bold"),
        axis.title = element_text(size = 8, face = "bold"))
```

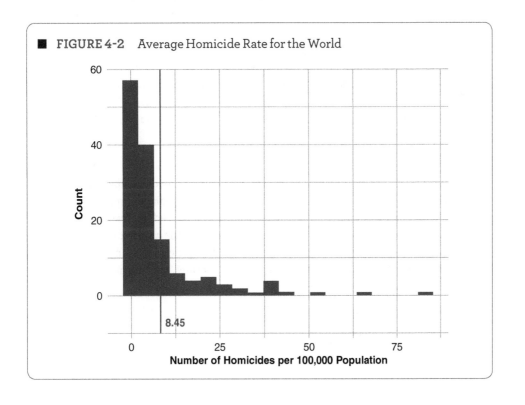

FIGURE 4-2   Average Homicide Rate for the World

The mean homicide rate for the countries in our sample is 8.45. As you can see, most of the countries are below 10 homicides per 100,000. There are a few (three), however, that are above 50! Even though there is a larger blue area in the histogram to the left of the mean, the extreme cases on the right counterbalance the majority of the cases on the left. When variables have extreme values, we may want to use a different measure of central tendency: the median.

> **Art and Practice of Data Visualization**
> **LINES RAISE THE BAR**
>
> Use vertical lines, horizontal lines, and annotation to be exact. Raise the bar when presenting data by identifying and labeling important values in your figures. Figure 4-2 is a good example. The temptation is to state that the mean value for the homicide rate is 8.45 and to let the reader figure out exactly where that might be. Placing a vertical line in the plot leaves nothing to the imagination. Horizontal lines are also useful if you need to point out which cases or groups either surpass or don't quite measure up to an important threshold.

Median

The **median** is not used as often as the mean, but nonetheless serves an important function. The median is used when extreme values exist in the variable. The median is simply the middle case. One way to distinguish it from the mean is that the median asks "What is the typical case?" rather than "What is the average value?" When extreme values exist, the median may give the best representation of what is typical.

The median is useful when we have extreme values. Suppose, for example, we're interested in the average income of a neighborhood that has six families with annual incomes of $50,000 and a seventh household realizing $1,000,000. A more useful measure, the "typical" family in this example, is the middle case ($50,000) rather than the average value (somewhere between $50,000 and $1,000,000).

Calculating the Median

As the histogram in Figure 4-2 suggests, we might want to use the median rather than the mean when describing the central tendency for homicide rates. The median represents the middle or 50th percentile case in a variable (the *middle* case). It is calculated by arranging the data in order from its lowest to highest value, then finding the value of the middle case. If the data set contains an odd number of cases, the median is the middle data point with an equal number of cases on either side. If the data contains an even number of cases, the median is the average of the two middle values in the ordered list.

- Odd number of cases = (1, 2, 3, 4, 5, 6, 7, 8, 9) = 5
- Even number of cases = (1, 2, 3, 4, 5, 6, 7, 8, 9, 10) = 5.5

With the odd number of cases, there are an equal number of cases on each side of the number 5, making 5 the middle case. With an even number of cases, no such middle number exists. Consequently, we take the two middle values in the ordered list and average them:

$$\frac{5+6}{2} = 5.5$$

The median can be calculated by using the *median()* function in R. Note that while R will easily calculate the mean and median with the simple functions *mean()* and *median()*, using the command *mode()* does not calculate the mode. The *mode()* command will tell you instead how a variable was stored in R (not that useful for understanding the central tendency). The easiest way to figure out which category has the most cases is to generate either a bar plot or a frequency table.

Let's go back to the homicide example and calculate the median. Notice in the code that I include the *na.rm=TRUE* option since there might be missing data.

Code Chunk 4-3a

```
median(world$homicide, na.rm = TRUE)

## [1] 3.3
```

Once we have the median calculated, we draw a histogram and place a vertical line at the median: 3.3. Again, use the *geom_vline()* command to produce a vertical line that intersects with the x axis. Use the *annotate()* command to place text within the figure.

Code Chunk 4-3b

```
ggplot(world, aes(homicide)) +
  geom_histogram(bins=20, fill = "#0000bf") +
  ggtitle("Figure 4-3: Median Homicide Rate for the World") +
  xlab("Number of Homicides per 100,000 Population") +
  geom_vline(xintercept=3.3, col="#bf0000") +
  annotate("text", x = 7, y = -7, label = "3.3", col="#bf0000",
           size = 3) +
  theme_minimal() +
  theme(plot.title = element_text(size = 8, face = "bold"),
        axis.title = element_text(size = 8, face = "bold"))
```

As the vertical line in the histogram shows (Figure 4-3), the median number of murders per 100,000 population is 3.3. Note the difference between the mean (8.45) and the median (3.3); the mean is more than double the median. When describing the central tendency of homicide rates, the measure we use for central tendency matters. Let's explore *why* after a few exercises.

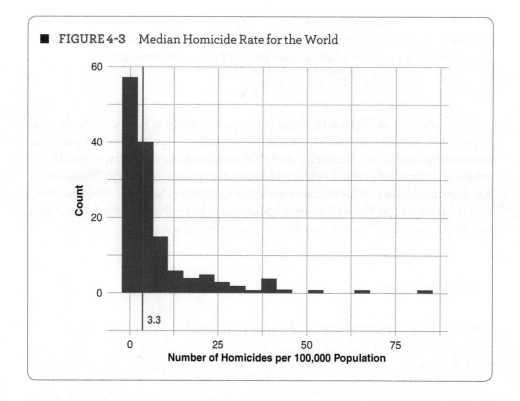

FIGURE 4-3 Median Homicide Rate for the World

KNOWLEDGE CHECK: Calculate the mode, mean, and median.

---

1. Identify the measure of central tendency associated with each question.

    a. Which candidate won the election?

    b. Are housing prices higher in Houston or Dallas?

    c. What's the most popular name for a baby boy this year?

    d. How many students will enroll in your university this year?

2. Calculate the mean and median for the following variables. Which measure provides the best indication of central tendency? You can indicate both if appropriate.

    a. *world$turnout*

    b. *world$gdppc*

    c. *states$hsdiploma*

    d. *states$inc*

3. Determine the mode for the following variables:

    a. *nes$pid7*

    b. *nes$marstat*

    c. *nes$employ*

    d. *nes$race*

Mean Versus Median

Many statistical models and summaries use the mean. As I noted earlier, our decision to use the mean or the median is based on the presence of extreme values. When extreme values are present, the mean can give a very inaccurate sense of the data's central tendency. In the homicide rate example, there are some countries that have an abnormally high number of murders. Other examples abound. Probably the most familiar involves housing prices. Whenever we refer to the price of homes, we use the median rather than the mean because extremely expensive homes can pull the mean way up, providing a distorted sense of the housing market.

To visualize the difference between the mean and the median, consider the variable *durable* from the *world* data. *Durable* records the number of years a country has remained politically stable, as measured by the number of years since a significant change in its democracy score. The histogram of *durable* illustrates how the mean and median are influenced by outliers (Figure 4-4). The histogram indicates there are a few countries that have been stable for more than 100 years. A vast majority of countries, however, have been stable for less than 25 years. In this example, the mean (red line) is located to the right of the median (blue line) because of the extreme cases. Consequently, the two measures differ appreciably: the median is 17 years and the mean is 27.2. To calculate the mean and median for the variable *durable*, simply type *summary(world$durable)*.

Art and Practice of Data Visualization
OPTION *ALPHA*

You've seen everything before in Code Chunk 4-4 except for the inclusion of the option *alpha=.5*. The *alpha* option allows you to lighten the shade of the color being used, which can be extremely helpful. In this instance, it allows us to see the vertical lines drawn with greater clarity.

Code Chunk 4-4

```
ggplot(world, aes(durable)) +
  geom_histogram(bins=20, fill = "#0000bf", alpha=.5) +
  ggtitle("Figure 4-4: The Mean and Median Differ") +
  xlab("Number of Years Since Significant Constitutional Change") +
  geom_vline(xintercept=27.2, col="#bf0000") +
  geom_vline(xintercept=17, col="#0000bf") +
  annotate("text", x = 100, y = 30,
           label = "Mean = 27.2", col="#bf0000", size = 3) +
  annotate("text", x = 100, y = 35,
           label = "Median = 17", col="#0000bf", size = 3) +
  theme_minimal() +
  theme(plot.title = element_text(size = 8, face = "bold"),
        axis.title = element_text(size = 8, face = "bold"))
```

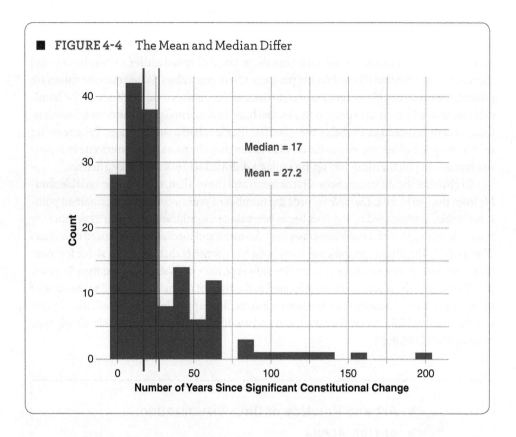

FIGURE 4-4 The Mean and Median Differ

Now suppose when entering the data we accidentally typed an additional "9" for the U.S. case, changing the number of years from 199 to 1999. If we were to calculate the mean and median with that typo, the median would still be 17, but the mean would jump from 27.2 years to 38.6. To illustrate the effect of the typo and how it influences the mean but not the median, I create a new variable called *newdurable* and replace the value 199 with 1999. I then calculate the mean of the *newdurable* variable.

Code Chunk 4-5a

```
world$newdurable <- world$durable

world$newdurable <- replace(world$newdurable,
  world$newdurable==199, 1999)

mean(world$newdurable, na.rm=TRUE)

## [1] 38.61392
```

I redraw the histogram, only this time we include a vertical line (green dashed line) that represents the new mean at 38.6 (Figure 4-5). Note how the typo moves the mean to the right but the median holds fast: even when there is a huge outlier (when the U.S. value was mistakenly 1999), the median does not change, it remains 17.

Code Chunk 4-5b

```
ggplot(world, aes(durable)) +
  geom_histogram(bins=20, fill = "#0000bf", alpha=.5) +
  labs(title="Figure 4-5: Extreme Value Moves the Mean") +
  xlab("Number of Years Since Significant Constitutional Change") +
  geom_vline(xintercept=27.2, col="#bf0000") +
  geom_vline(xintercept=17, col="#0000bf") +
  geom_vline(xintercept=38.6, col="#00ff00",
         linetype = "dashed") +
  annotate("text", x = 125, y = 30,
       label = "Mean With Typo Moves From 27.2 to 38.6",
       col = "#00ff00", size = 3) +
  annotate("text", x = 125, y = 25,
       label = "Median Stays the Same = 17", col = "#0000bf",
       size = 3) +
  theme_minimal() +
  theme(plot.title = element_text(size = 8, face = "bold"),
      axis.title = element_text(size = 8, face = "bold"))
```

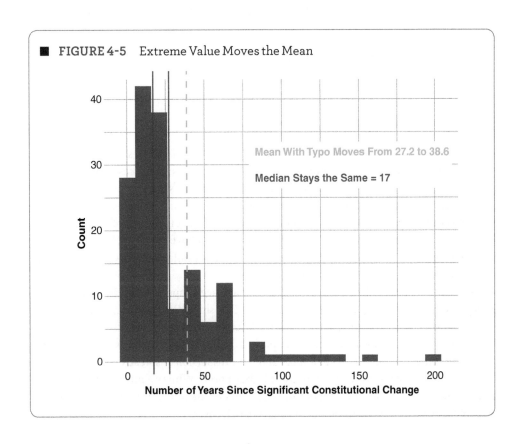

■ **FIGURE 4-5**   Extreme Value Moves the Mean

The degree of sensitivity to outliers is an important distinction between the mean and the median. As we'll discover below, some measures of dispersion are based on the mean. Because some measures of dispersion rely on the mean and outliers could be lurking, we should proceed with caution. The same considerations we make regarding how to measure central tendency also apply to how we measure dispersion.

> **KNOWLEDGE CHECK:** Explain the difference between the mean and median.

4. Indicate which variable is likely to be best described by the mean or median.
   a. Infant mortality
   b. GDP per capita (among the world's countries)
   c. The average percentage of the population having completed high school among the 50 U.S. states
   d. The amount of foreign aid received by each country
5. Describe what happens to the mean and median (whether it increases or decreases) under the following conditions:
   a. An additional case is added that exceeds the variable's maximum value.
   b. An additional case is added that is less than the variable's minimum value.
   c. The maximum is doubled in value.
   d. The minimum value is cut in half.

Measures of Dispersion: The Range, Interquartile Range, and Standard Deviation

Once we know the central tendency of a variable, a second question arises: what is the typical distance from the mean? It's easy enough to calculate the distance of any case from the mean, but we want to know if that distance is typical. For most of us, it's not enough to know if our salary, performance on an exam, or resting heart rate is above or below average; we want to know exactly how our income, grade, or beats per minute relates to the rest of the population. For example, you may know that your grade on an exam is above the mean, but does it place you in the top 10%, 20%, or 40%? The answer could mean the difference between a C+ and an A. Knowing the dispersion provides some perspective.

Another example clarifies. Suppose we know the average temperature in Houston, Texas, is 90 degrees in June. For a trip we might pack shorts and a t-shirt. We know the average temperature in Denver, Colorado, is 75 degrees in July. Although the temperature is somewhat lower in Denver, we'd probably still be tempted to pack shorts and a t-shirt. Here's where knowing dispersion comes in handy. If we knew more about the dispersion or spread of temperatures in both cities, we might radically change our wardrobe options. While we might stay with shorts and a t-shirt for Houston (the temperature does not deviate much from 90), we'd better be prepared for temperatures in the low 60s to high 90s in Denver.

While packing for a trip seems trivial, the dispersion of temperatures could have life or death consequences when planning a trip to Mount Everest or the moon. Knowing the dispersion of temperatures might also influence how we design and store vaccinations. Sometimes knowing the dispersion is more important than knowing the mean or median. If our proposed policy's success depends on accounting for the physical traits of a large population, for example, we had better know its dispersion. A developmental medical device may function properly for the typical patient, but if the dispersions of heart rate, weight, and height are extreme, we may need to alter its design.

Whether planning for hurricanes or financial meltdowns, we need to know the extremes. One cause of the Great Recession (2008) arose from not knowing the extremes: trillions of dollars depended on financial models that did not include data from the 1920s and 1930s (the extremes). Without knowing the full range of financial outcomes (how bad things got during the Great Depression), assumptions of financial risk were highly flawed (Cassidy, 2009). To understand dispersion, we use the range, interquartile range, standard deviation, and variance.

Range

The range of a variable is the distance between its minimum and maximum values. As such, it can be used to help make decisions or develop policies based on the best- or worst-case scenario. For example, if the range of possible outcomes from a new developmental drug includes death, we might decide it is unacceptable for use (especially if its purpose is to cure the common cold). The range of potential outcomes for a particular financial product will influence its price: investors will be asked to pay more for financial products that have a range with few negative returns. Let's get more concrete with another example.

Whether we are getting cable installed, expecting a delivery, or waiting for the plumber, we're often given a range of times for service. Think about how your relative level of patience is affected by the window of time given by the cable technician, delivery person, or plumber. In this instance the window represents the range. If the cable company says there'll be someone at your house between 8 a.m. and 5 p.m., any plans you had away from your house that day will have to be canceled. We often judge the efficiency of these services based on that range. A cable company that gives you a window between 11 a.m. and noon would be considered efficient, if not unheard of.

Consider the range of customers expected at the local Dairy Queen while you're working behind the counter. If the range of customers expected when school lets out is between 10 and 50, your boss may decide to have two to three staff present. If that range is between 10 and 100 and your boss values the mental health of their employees, they would be well advised to have a few extra employees on call in case several school buses show up after an athletic event or band competition. In general, the larger the range, the more flexibility required. As we learned in our discussion about central tendency, extreme values matter. How do extreme values influence the range?

The range of a variable can be sensitive to extreme values, but only in two instances: the extreme value has to be either greater than the maximum or less than the minimum. In other words, adding an extreme case to the variable only changes the range if it becomes the new minimum or maximum. To illustrate, let's examine infant mortality in the United States.

Calculating the Range

To calculate the minimum and maximum of a variable, simply use the *summary()* command. As you can see from Code Chunk 4-6, the *summary()* command is useful because of its simplicity and the amount of information it generates. Not only does it provide the minimum and maximum values of a variable (giving us the range), it provides information on the first quartile, median, mean, and third quartile. The first and third quartiles help establish the boundaries of the interquartile range, which we'll discuss next.

```
Code Chunk 4-6

summary(states$infant)

##    Min. 1st Qu.  Median    Mean 3rd Qu.    Max.
##   4.520   5.938   6.785   6.981   7.878  11.470
```

The output created by R is not always in the most presentable format. The output from the summary command (Code Chunk 4-6) is a good example. While the statistics and numbers are right aligned, they can be a little disorienting. Six different statistics are reported in the output generated by Code Chunk 4-6: Min. = minimum (4.520), 1st Qu. = first quartile (5.938), Median = median (6.785), Mean = mean (6.981), 3rd Qu. = third quartile (7.878), and Max.= maximum (11.470).

While knowing the range of possible outcomes can be useful, we may need more information. For example, let's continue with infant mortality rates (the number of infant deaths per 1,000 live births) in the United States. Infant mortality among the U.S. states ranges from 4.5 to 11.5. Even though we know 11.5 is the maximum, we don't know if most of the states are closer to 4.5 or 11.5. The range may tell us that the degree of inequality in health outcomes is unacceptable, but it doesn't tell us whether the problem is more about inequality (some states need help) or whether we have a health crisis on our hands (49 states could have close to 11.5 deaths per 1,000 live births). We need to know where most of the states lie. Calculating the interquartile range provides that information.

Interquartile Range

The **interquartile range** is the distance between the 25th percentile and the 75th percentile cases: the middle half of the data. The calculation is made as follows. If we have 100 cases (the easiest example), the 25th lowest value is the 25th percentile case and the 75th lowest value is the 75th percentile case. The interquartile range is the distance between those two points. It indicates where the middle half of the data lie. For example, to calculate the interquartile range for infant mortality in the *states* data, the computer orders the data from the lowest to highest number, then determines which states are the 25th and 75th percentile cases. It then reports the infant mortality rates for those cases. It does that all behind the scenes, reporting the numbers when we execute the *summary()* command.

Calculating the Interquartile Range

Like the median, the interquartile range is based on the order in which cases are related to each other. Instead of focusing on the middle case, the interquartile range identifies the 25th and 75th percentile cases. The same command that helps calculate the range (finding the minimum and maximum values of a variable) is used to calculate the interquartile range. I execute that command in Code Chunk 4-7. In the infant mortality case, the interquartile range is between 5.9 and 7.9 deaths per 1,000 live births. (Note that R refers to the 25th percentile as the first quartile and the 75th percentile case as the third quartile.) That means the middle-half of the U.S. states lie within those two values. We now have more information on the extent of the problem. The same code we used to calculate the range is used to calculate the interquartile range.

Code Chunk 4-7

```
summary(states$infant)

##    Min. 1st Qu.  Median    Mean 3rd Qu.    Max.
##   4.520   5.938   6.785   6.981   7.878  11.470
```

Like the median, the interquartile range is not sensitive to outliers. Note that the interquartile range stays the same when we repeat the mistake we made when discussing political stability: we introduced a typo, changing the U.S. case in the variable *durable* from 199 to 1999 years. The summaries of *durable* and *newdurable* in Code Chunk 4-8 provide the minimum, first quartile, median, mean, third quartile, and maximum. Notice that when the maximum is increased to 1999 (*newdurable*), the first and third quartiles do not change.

Code Chunk 4-8

```
summary(world$durable)

##    Min. 1st Qu.  Median    Mean 3rd Qu.    Max.    NA's
##    0.00    8.00   17.00   27.22   38.50  199.00      24
```

Code Chunk 4-9

```
summary(world$newdurable)

##    Min. 1st Qu.  Median    Mean 3rd Qu.    Max.    NA's
##    0.00    8.00   17.00   38.61   38.50 1999.00      24
```

Armed with information on where the middle half of the data lie, we still want additional information. We now know the range and interquartile range of the data, but suppose we want to know how far the typical case is from the mean. While the interquartile range is a step in the right direction, the 25th and 75th percentile cases may not be typical. Calculating the standard deviation will tell us how far the typical case is from the mean.

Standard Deviation

The **standard deviation** is an often-used summary of dispersion (spread). Its popularity stems from two important attributes: its conceptual simplicity and its unit of measurement. Let's clarify what we mean by each.

First, let's talk about units. The standard deviation of a variable is expressed (standardized) in the variable's original units (miles, feet, inches, pounds, years, percentage, etc.). To illustrate I'll use two variables: infant mortality rates and the percentage of a state's legislative seats occupied by women.

Calculating the Standard Deviation

With the standard deviation, we are calculating the average distance (or difference) between all of the cases and the mean. In the following code, I calculate the standard deviation for the two variables. The standard deviation for the percentage of legislative seats held by women is 7.197513 (we'll round it to 7.2). The standard deviation for infant mortality is 1.435324 deaths per 1,000 live births (we'll round that to 1.4). Both of these numbers are reported in their respective units of measurement, making it easy to understand their magnitude.

In Code Chunk 4-10, I calculate the standard deviation using the *sd()* function with the option we are now familiar with: *na.rm = TRUE*. Again, R output is usually rather funky looking (most programming output is). In this case, there is a "[1]" before each standard deviation reported, indicating that there is one number generated by each instance of the *sd()* function.

```
Code Chunk 4-10

sd(states$femleg, na.rm = TRUE)

## [1] 7.197513

sd(states$infant, na.rm = TRUE)

## [1] 1.435324
```

Again, understanding the mechanics of this operation and how it is represented mathematically is important. Calculating the distance (or deviation) of an observation from the mean, squaring those distances, and dividing by *n* are operations that we conduct routinely in statistics. The more comfortable you are with these calculations, the better your intuition for data analysis.

In terms of conceptual simplicity, the standard deviation is the average distance from the mean. As that average grows, so does the dispersion and spread.

Now that we know how to compute the standard deviation using R, let's look underneath the hood to understand the mechanics of the calculation. It is similar to calculating the mean, except that we're interested in the average distance from the mean.

Before proceeding to the math, consider a description of the calculation in English:

Take each case and subtract the mean to calculate each observation's distance from the mean. Square the result to guarantee the distance will be positive. Sum all of the squared distances. Now divide by the number of cases. Then, to express the number in the original units, take the square root.

As you can see, the individual calculations are not difficult, but there are a few steps that need to be taken. Also note that it's a little hard to follow when explained in narrative text. Let's express the calculation by bullet points.

- Calculate the distance of each observation from the mean.
- Square the distance for each observation.
- Sum the squared distances for each observation.
- Divide by the number of observations.
- Take the square root.

Now consider how those same directions are expressed in a mathematical equation. We use the Greek letter σ to denote the standard deviation. Compared to the written narrative or the bullet points, the mathematical formula is compact and exact. While many find this a more elegant and useful representation of the standard deviation, others find the expression impenetrable.

$$\sigma = \sqrt{\frac{\sum(x_i - \mu)^2}{n}}$$

To solidify our understanding of what's going on, let's take each bullet point step by step and show what's happening mathematically.

- Calculate the distance of each observation from the mean.

$$(x_i - \mu)$$

Notice a couple of things here. First, we use the Greek symbol μ to denote the mean of the variable. Also note the subscript i attached to the x. This is notation you will encounter frequently in statistics. The i is simply used to indicate that there are individual observations. The i translates into each observation, which is language we find in three of the bullet points. When we subtract the mean from each observation in this example, we're subtracting the mean from each state's infant mortality rate, which gives us each state's distance from the mean. Once we calculate the distance of a state from the mean, we square the term.

- Square the distance for each observation.

$$(x_i - \mu)^2$$

We square the term so that the value $x_i - \mu$ is always positive. If we didn't square that term, when we sum the distances for each of the 50 states, we would always end up with zero since some states are below and some states are above the mean. Given that the mean represents something close to the middle, we have just as much distance that is positive and negative, canceling each other out when we sum. Squaring the term is an important part of the operation.

Now that we have calculated all the squared distances for each state, we need to add them up, or calculate their sum.

- Sum the squared distances for each observation.

$$\sum(x_i - \mu)^2$$

In our example, we're doing the following:

$$(x_{Alabama} - \mu)^2 + (x_{Alaska} - \mu)^2 + (x_{Arkansas} - \mu)^2 + \ldots + (x_{Wyoming} - \mu)^2$$

As I noted earlier, the total of the distances below the mean will be very close to the total distances that are above the mean. If we didn't square each of those terms, we would always end up with something close to zero. A simple example demonstrates.

Suppose we have two variables with very different spreads: Variable 1 has a range of 4 and a mean of 3. Variable 2 has a range of 40 and a mean of 30.

- Variable 1 = (1, 2, 3, 4, 5)
- Variable 2 = (10, 20, 30, 40, 50)

Let's calculate the distance from the average for each case.

- Variable 1: $(1-3)+(2-3)+(3-3)+(4-3)+(5-3) = 0$
- Variable 2: $(10-30)+(20-30)+(30-30)+(40-30)+(50-30) = 0$

If we don't square each of the terms, we end up with sums of the distances that equal 0 in variables with two very different ranges. Consequently, when calculating the distance from the mean and then summing those distances, we square the distance so that it's always positive. Once those squared distances are summed, we divide by the number of observations much like we did when calculating the mean.

- Divide by the number of observations.

$$\frac{\sum(x_i - \mu)^2}{n}$$

Our final step is to express the number in terms of the original units. Since we squared the distances in an earlier step to avoid negative numbers, we need to convert the number back to its original units by taking the square root.

- Take the square root.

$$\sqrt{\frac{\sum(x_i - \mu)^2}{n}}$$

It's important enough for us to understand the mechanics of this operation to present this with some numbers. Let's go back to the example of calculating the typical amount of cash on hand for students in our class. To make this example more manageable, let's limit the class to five students ($n = 5$). In Code Chunk 4-11, I'll create the following variable called *cash* that records (I made these numbers up) the amount of money in each student's pocket. I calculated the mean, which is $19.

Code Chunk 4-11

```
cash <- c(10, 20, 40, 15, 10)

mean(cash)

## [1] 19
```

Now let's go step by step to calculate the standard deviation for these five students.

- Calculate the distance of each observation from the mean.

$$(x_i - \mu)$$

$$(10-19)+(20-19)+(40-19)+(15-19)+(10-19)$$

- Square the distance for each observation.

$$(x_i - \mu)^2$$

$$(10-19)^2 + (20-19)^2 + (40-19)^2 + (15-19)^2 + (10-19)^2$$

- Sum the squared distances for each observation.

$$\Sigma(x_i - \mu)^2$$

$$(81)+(1)+(441)+(16)+(81)=620$$

- Divide by the number of observations.

$$\frac{\Sigma(x_i - \mu)^2}{n}$$

$$\frac{620}{5} = 124$$

- Take the square root.

$$\sqrt{\frac{\Sigma(x_i - \mu)^2}{n}}$$

$$\sqrt{124} = 12.45$$

We now have the standard deviation for our example: $12.45. To double-check our math, let's have R make the calculation. Note in Code Chunk 4-12 that I don't use the *na.rm = TRUE* option since I know there are no missing data.

Code Chunk 4-12

```
sd(cash)

## [1] 12.4499
```

Let's continue developing our intuition through some visualizations. To better understand what the standard deviation tells us, let's examine how the standard deviation relates to dispersion or spread. To illustrate, I'll create three variables with different standard deviations and plot the corresponding histograms. There's a little preparation involved, so hold on.

First, I'll create a data frame to hold the new variables I'm going to create. I call the new data frame *h* and I specify that it should have 500 observations. Once the data frame is established, I create three new variables (*h$p1*, *h$p2*, and *h$p3*), specifying that each is normally distributed (*rnorm()*), with 500 observations, a mean of 0, and a standard deviation of 1. Note that I set the seed (*set.seed()*) of the random number generator so that you can achieve the same result if you set the seed to the same number (42).

Code Chunk 4-13a
```
h <- data.frame(1:500)

set.seed(42)

h$p1 <- rnorm(500,0,1)
h$p2 <- rnorm(500,0,2)
h$p3 <- rnorm(500,0,3)
```

In Code Chunk 4-13b, I use the function *gather()* from the package 'tidyr' to stack the three variables on top of each other to generate the plot. (In Chapter 3, I used a similar command *melt()* from the 'reshape2' package.) This reshapes the data set *h* so there are two variables: one I'll name *condition*, which records whether the observation is from the p1, p2, or p3 distribution; and the second I'll name *measurement*, which holds the actual number. I then relabel p1, p2, and p3 to indicate which standard deviation they represent.

Code Chunk 4-13b
```
data_long <- gather(h, condition, measurement,
                    p1:p3, factor_key=TRUE)

data_long$condition <- factor(data_long$condition,
                              levels = c("p1","p2","p3"),
                              labels = c("Std=1", "Std=2",
                                         "Std=3"))
```

With the new data set and the new labels, I'm ready to generate the three histograms.

Code Chunk 4-13c
```
ggplot(data_long, aes(measurement)) +
  geom_histogram(aes(measurement), fill = "#0000ff") +
  facet_grid(.~condition) +
  labs(title="Figure 4-6: Standard Deviation of Random Variable
                    Doubled and Tripled") +
  xlab("") +
  theme_minimal() +
  theme(plot.title = element_text(size = 8, face = "bold"),
        axis.title = element_text(size = 8, face = "bold"))
```

FIGURE 4-6 Standard Deviation of Random Variable Doubled and Tripled

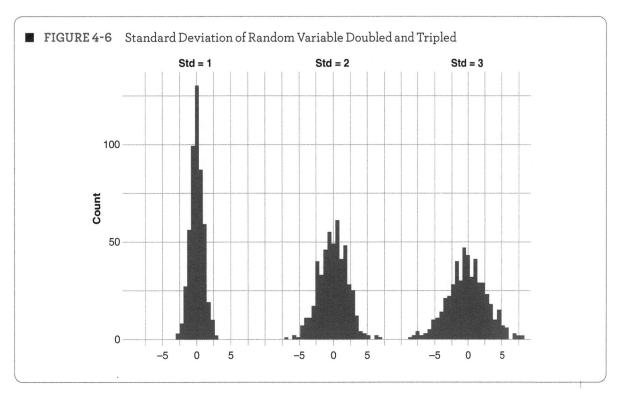

Observe what happens to the spread when we plot the corresponding histograms (Figure 4-6). Notice how the histograms change in each panel, becoming wider and more spread out as the standard deviation increases. In the first panel where the standard deviation is 1, none of the cases are less than −5 or greater than 5. In the second panel where the standard deviation is 2, a few cases are less than −5 and greater than 5. Finally, in panel 3 where the standard deviation is 3, a larger number of cases lie outside the interval −5 and 5.

> **KNOWLEDGE CHECK: Calculate the range, interquartile range, and standard deviation.**

6. Calculate the range, interquartile range, and standard deviation of the following variables:
   a. *world$gdppc*
   b. *states$hsdiploma*
   c. *nes$ftpolice*
   d. *nes$ftobama*

7. Draw a histogram of the percentage of legislators that are women in the world's legislatures (*world$womleg*). Calculate the standard deviation and demarcate it with vertical lines (both above and below the mean) in the figure.

8. Draw a histogram of the percentage of legislators that are women in the world's legislatures (*world$womleg*). Calculate the interquartile range and demarcate it with vertical lines in the figure.

Interquartile Range Versus Standard Deviation

The interquartile range is less sensitive to extreme values than the standard deviation. To get a sense of the distinction between the interquartile range and the standard deviation, let's take the example we used above: political stability as measured by the *durable* variable.

Recall that we introduced a typo to the *durable* variable by accidentally adding a 9 to the United States' 199 years of stability, making it 1999 years. Since the standard deviation is the average distance from the mean, increasing the U.S. entry in *durable* from 199 to 1999 increases the standard deviation from 31 years of regime stability to 159! Again, the output generated without any additional code is difficult to read. The two numbers adjacent to the [1] are the standard deviations calculated from the variables *durable* and *newdurable*. In this example, the 25th and 75th percentile cases remain the same, meaning the interquartile range remains unchanged: the numbers correspond to the two quartiles (8 for the first quartile and 38.5 for the third quartile). Note again that the now familiar *na.rm=TRUE* option is included since there are some missing data in the *world* data set.

Code Chunk 4-14

```
sd(world$durable, na.rm = TRUE)

## [1] 31.1848

summary(world$durable)

##     Min. 1st Qu.  Median    Mean 3rd Qu.    Max.    NA's
##     0.00    8.00   17.00   27.22   38.50  199.00      24

sd(world$newdurable, na.rm = TRUE)

## [1] 159.4293

summary(world$newdurable)

##     Min. 1st Qu.  Median    Mean 3rd Qu.    Max.    NA's
##     0.00    8.00   17.00   38.61   38.50 1999.00      24
```

To summarize, the ability of the interquartile range to withstand extreme values makes it a useful measure when there are outliers. The standard deviation, as the example shows, can be extremely sensitive to outlying cases. Despite its sensitivity to outliers, the standard deviation is the most often-used calculation for dispersion or spread. Knowing exactly how it is calculated will hopefully help us avoid any surprises.

> **KNOWLEDGE CHECK:** Explain the difference between the interquartile range and the standard deviation.

9. Indicate whether you would use the interquartile range or standard deviation to summarize the following variables:
    a. Infant mortality
    b. Income (among the world's countries)
    c. The average level of education in the United States
    d. The average amount of foreign aid received by each country
10. Describe what happens to the interquartile range and standard deviation (whether it increases or decreases) under the following conditions:
    a. An additional case is added that exceeds the variable's maximum value.
    b. An additional case is added that is less than the variable's minimum value.
    c. The maximum is doubled in value.
    d. The minimum value is cut in half.
11. Is the standard deviation or interquartile range a better representation of the variable's spread when examining the variable *world$co2*?
12. Draw a histogram of infant mortality (*world$inf*). Calculate the interquartile range and demarcate it with vertical lines in the figure.
13. Is the standard deviation or interquartile range a better representation of the variable's spread when examining the variable *world$inf*?

A Note on Variance

The **variance** is simply the standard deviation squared. Since the Greek letter σ denotes the standard deviation, we signify variance as follows:

$$Variance = \sigma^2$$

Although the variance is not expressed in a variable's original units, making it less useful as a summary of dispersion, it is important to know since we use it to calculate a number of different statistics. The formula for the variance is as follows:

$$\sigma^2 = \frac{\sum(x_i - \mu)^2}{n}$$

The variance is the average squared distance of each observation from the mean, which is not the most intuitive or helpful statistic by itself. It comes in handy, however, for a number of other calculations. Parts of it—take the numerator $\sum(x_i - \mu)^2$, for example—represent other measures of note. For example, the numerator represents a value called the sum of squared deviations, something you'll see in future endeavors.

$$SS = \sum(x_i - \mu)^2$$

I only mention variance and the sum of squared deviations here so that either can be easily recognized when encountered in other statistics texts.

SUMMARY

Central tendency and dispersion figure prominently in data description. While they are not the only things we want to know, they answer two very important questions: what is the typical case and how typical is the typical case? In this chapter we learned how to calculate and describe a variable's central tendency and dispersion. The most important distinction between the different measures I've presented is their sensitivity to outlying cases. The mean and standard deviation are both sensitive to outlying cases since they are based on calculating the distance from the variable and the variable's mean. The median and interquartile range are calculated differently; they are not as sensitive to the distance from the mean.

Why do we depend on the mean and standard deviation if they are susceptible to outliers? Much of it has to do with the ease of calculation. It is therefore important to understand not only the dispersion of a variable but also what it looks like: its actual shape. The shape of the data is revealed by the descriptive tools provided in the next chapter.

COMMON PROBLEMS

- When calculating the mean, R code requires including the $na.rm = TRUE$ argument when there are missing values or it won't calculate.

- Understand that the summation sign simply means add up the results of the expression following the summation sign for each observation indicated.

- Understand that distances or deviations (usually from the mean) are used interchangeably and describe the same thing.

- Details, details ... pay attention to units of measure. Always know exactly what they are for every variable.

- Understand that the connection between the question being asked and the measure used is important. In the context of this chapter, the main distinction is between determining the central tendency and the spread. For central tendency, we want to know what the average observation looks like. In terms of spread, we want to know if the average (the mean) provides a good indication of the average case.

REVIEW QUESTIONS

1. What is the difference between the mean and median?

2. What is the difference between the standard deviation and the interquartile range?

3. What questions can we answer with the mean?

4. What questions can we answer with the standard deviation?

5. What advantage does the standard deviation have over the variance?

6. Describe what happens to the distribution of a variable as the standard deviation increases.

7. Describe how the summation sign Σ works.

8. What measure would best represent the central tendency when considering life expectancy?

9. What measure would best represent the central tendency for homicide rates?

10. What is the n when considering the *states* data set?

PRACTICE ON ANALYSIS AND VISUALIZATION

1. Generate histograms of the following variables to determine whether the mean or median (or either) is an appropriate measure to indicate central tendency.

a. states$inc
b. world$ethfrac
c. world$co2
d. world$military

2. What category of the *nes$amer_ident* variable is the mode?

3. Generate histograms of the following variables to determine whether the interquartile range or the standard deviation (or either) is an appropriate measure to indicate spread.

a. states$inc
b. world$ethfrac
c. world$co2
d. world$military

4. Calculate the standard deviation for the following variables.

a. world$military
b. nes$ftsci

c. nes$fttrump
d. states$ptratio

5. What calculation is being made by the following formula?

$$\mu = \frac{1}{n}\sum_{n=1}^{n} x$$

a. Mode
b. Mean
c. Standard deviation
d. Interquartile range

6. What's wrong with the following formula for the standard deviation?

$$\sigma^2 = \frac{\sum(x_i - \mu)^2}{n}$$

a. It doesn't sum the squared residuals.
b. It doesn't square the residuals.
c. It doesn't take the square root of the sum of squared residuals divided by n.
d. It doesn't divide by the number of observations.

7. How would you fix the following histogram?

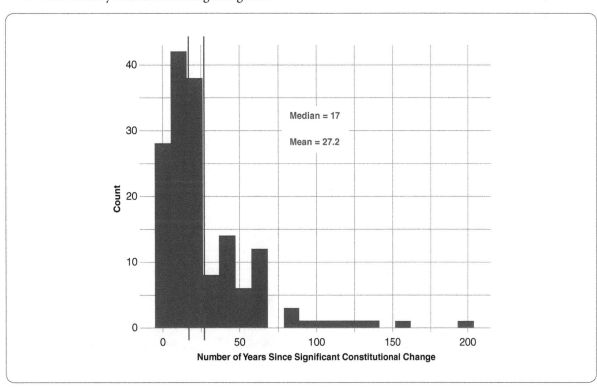

8. Which line represents the mean and median in the following histogram?

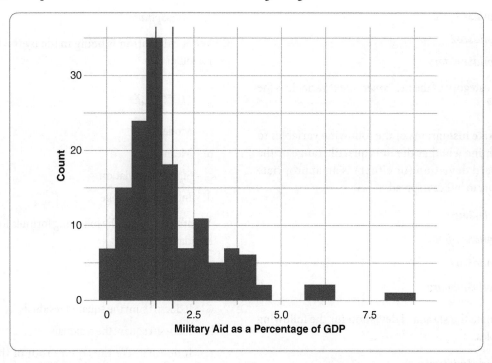

9. Which best describes the distribution depicting military aid as a percentage of GDP in the previous problem?

   a. Normal
   b. Skewed
   c. Skewed right
   d. Skewed left

10. List in order the following distributions according to the size of their standard deviations (ascending order).

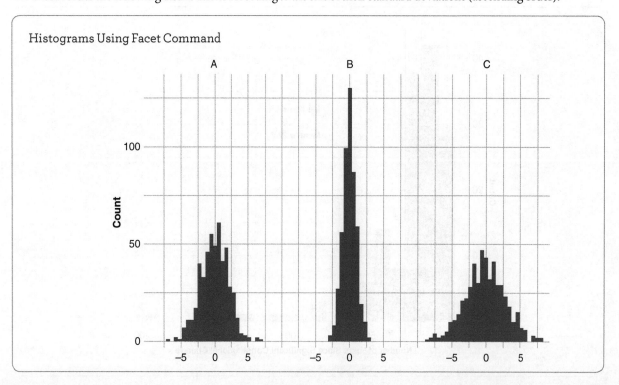

ANNOTATED R FUNCTIONS

The following functions appear in this chapter. They are listed in order of their first appearance (with the code chunk number in parentheses) and annotated here to give a very brief description of their use. Some are not stand-alone functions and only work in combination with other commands. As a reminder, the code in every chapter will work properly if executed in the order it appears. Proper execution also depends on typing the author-defined *libraries()* command, which loads the required R packages.

ggplot(): defines the basic structure of a plot (usually the x and y variables). (4-1)

aes(): the *aes* (called "aesthetics") function is used in ggplot to define the basic structure of the plot, which often includes the variables you want to use and any shapes or colors. (4-1)

theme_minimal(): specifies a minimalist style for ggplot. (4-1)

ggtitle(): provides the title for a ggplot. (4-1)

xlab(): labels the x axis in ggplot. (4-1)

ylab(): labels the y axis in ggplot. (4-1)

coord_flip(): flips the coordinates of a ggplot. (4-1)

theme(): specifies font, size, and so on in a ggplot. (4-1)

mean(): calculates the mean value of a variable. Be sure to include the option *na.rm=TRUE* if your variable has missing values. (4-2a)

geom_histogram(): draws a histogram in ggplot. (4-2b)

geom_vline(): draws a vertical line. (4-2b)

annotate(): places annotation on a figure. (4-2b)

median(): calculates the median of a variable. (4-3a)

replace(): replaces a value in a variable for another. (4-5a)

summary(): generates the mean, median, and interquartile range for a variable. (4-6)

sd(): calculates the standard deviation of a variable. Be sure to use the *na.rm=TRUE* feature if there are missing values. (4-10)

c(): combines elements within the parentheses into a list. (4-11)

data.frame(): creates a data frame (data set). (4-13a)

set.seed(): specifies exactly where to initiate the random number generator. This is useful for reproducing results where a random element is involved. (4-13a)

rnorm(): randomly generates data with a normal distribution. (4-13a)

gather(): converts data from wide form to long form (stacks variables on top of each other much like the *melt()* command used in Chapter 3). The *gather()* command comes from the 'tidyr' package. (4-13b)

factor(): used to manipulate levels (categories) in a variable. (4-13b)

levels(): used to list categories (levels) in a variable. (4-13b)

facet_grid(): used to place figures together using the same dimensions for comparison. (4-13c)

ANSWERS

KNOWLEDGE CHECK

1. mode (a), median (b), mode (c), mean (d)
2. 60.6, 67.5, both (a); 16,326.8, 9,863.2, median (b); 85.5, 86.1, both (c); 40725, 39833, both (d)
3. strong democrat (a), married (b), full-time (c), white (d)
4. median (a), median (b), mean (c), median (d)
5. mean increases, median moves (a); mean decreases, median moves (b); mean increases, median stays (c); mean decreases, median stays (d)
6. 122910, 17934.3, 18456.39 (a); 14.7, 5.64, 3.959 (b); 100, 40, 27.1425 (c); 100, 34, 38.05 (d)

7.  ```
    ggplot(world, aes(womleg)) +
      geom_histogram(bins=20, fill = "#0000ff", alpha=.5) +
      labs(title="Histogram of world$womleg") +
      xlab("Percentage of Seats Occupied by Women") +
      geom_vline(xintercept=(17.5 + 11), col="#bf0000") +
      geom_vline(xintercept=17.5, col="#0000ff") +
      geom_vline(xintercept=(17.5 - 11), col="#00ff00") +
      theme_minimal() +
      theme(plot.title = element_text(size = 8, face = "bold"),
            axis.title = element_text(size = 8, face = "bold"))
    ```

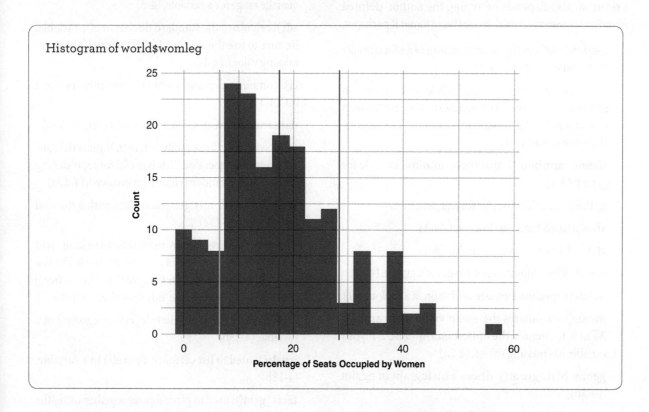

8.  ```
    ggplot(world, aes(womleg)) +
      geom_histogram(bins=20, fill = "#0000ff", alpha=.5) +
      labs(title="Interquartile Range of world$womleg") +
      xlab("Percentage of Seats Occupied by Women") +
      geom_vline(xintercept=9.8, col="#bf0000") +
      geom_vline(xintercept=23.4, col="#0000ff") +
      theme_minimal() +
      theme(plot.title = element_text(size = 8, face = "bold")) +
      theme(axis.title = element_text(size = 8, face = "bold"))
    ```

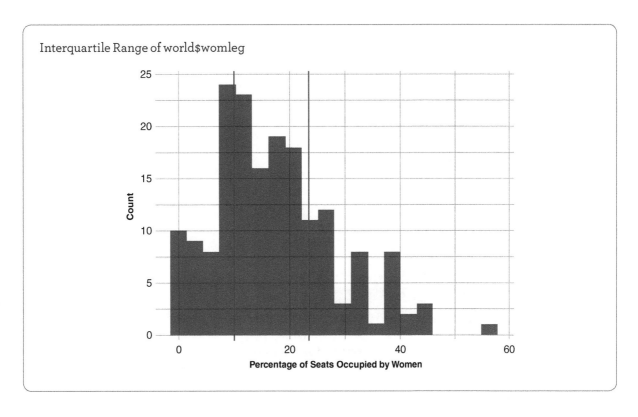

9. interquartile range (a), interquartile range (b), standard deviation (c), interquartile range (d)

10. increases, increases (a); increases, increases (b); no change, increases (c); no change, increases (d)

11. interquartile range

12. 7.7, 28.2

```
ggplot(world, aes(inf)) +
  geom_histogram(bins=20, fill = "#0000ff", alpha=.5) +
  labs(title="Histogram of Infant Morality") +
  xlab("Deaths per 1,000 Live Births") +
  geom_vline(xintercept=7.7, col="#bf0000") +
  geom_vline(xintercept=28.2, col="#0000ff") +
  theme_minimal() +
  theme(plot.title = element_text(size = 8, face = "bold")) +
  theme(axis.title = element_text(size = 8, face = "bold"))
```

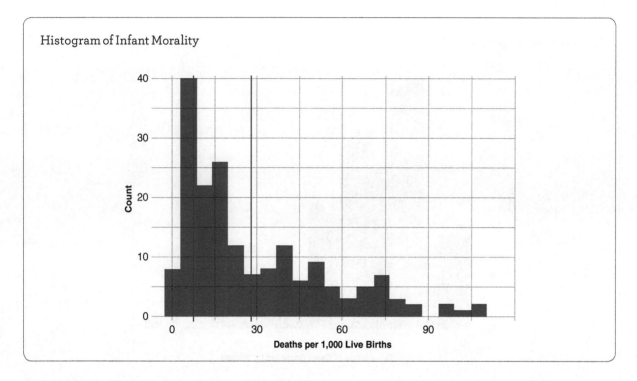

13. interquartile range

PRACTICE ON ANALYSIS AND VISUALIZATION

1. either (a), either (b), median (c), median (d)
2. extremely important
3. either (a), either (b), interquartile range (c), interquartile range (d)
4. 1.49 (a), 23.96 (b), 36.49 (c), 2.55 (d)
5. b
6. c
7. change the shade of histogram's color
8. red = mean, blue = median
9. c
10. b, a, c

 Access digital resources, including datasets, at http://edge.sagepub.com/brownstats1e.

5 Univariate and Bivariate Descriptions of Data

CHAPTER OUTLINE

Learning Objectives
Overview
The Good, the Bad, and the Outlier
Five Views of Univariate Data
Are They in a Relationship?
Summary
Common Problems
Review Questions
Practice on Analysis and Visualization
Annotated R Functions
Answers

LEARNING OBJECTIVES

- Ask the questions univariate views answer.
- Create and analyze univariate views of the data.
- Ask the questions bivariate views answer.
- Create and analyze bivariate views of the data.

Overview

In Chapter 4, we learned how to calculate simple numeric summaries to describe central tendency and dispersion. In this chapter, we learn how to summarize data with pictures. Pictures provide a quick and intuitive way to understand our data. Numerical summaries are helpful, but a picture provides more information. Now that we know *why* data description is so important and *what* it is, we can now proceed to *how*. This chapter is devoted to the mechanics and analysis of univariate and bivariate views of the data.

In terms of univariate data, I'll introduce the frequency table, bar plot, boxplot, histogram, and stem-and-leaf plot. Bivariate views include the scatter plot, mosaic plot, cross-tab, and bubble plot. Boxplots are also helpful when we want to know the relationship between a continuous variable and a categorical variable.

Two important lessons are worth mentioning. First, pictures should be mined extensively for important insights. Before moving on to more sophisticated econometric maneuvers, it is worth giving graphical summaries your full consideration. Second, pictures can be more effective than numerical summaries since they contain more information and reveal empirical patterns. Let's first begin with summaries of single variables: univariate data.

The Good, the Bad, and the Outlier

Univariate views of the data help answer a wide range of questions. Who's the best, who's the worst, and is there someone or something different from everyone or everything else?

Similarly, we can ask what's the best that could happen, what's the worst that could happen, and what's likely to happen? We can also see whether certain characteristics of persons, places, or things are common or are rarely observed. Perhaps an individual case stands out (outliers), uncovering important clues. A surprising amount of information can be gathered by the univariate views of the data introduced in this section.

Important decisions and policies are based on univariate views of the data. For example, public officials and educators want to know about the health risks of their student population. First, they might compile a list of the challenges. Since few schools can afford to adequately address all problems, administrators want to know what kind of illnesses or high-risk behaviors are most prevalent (what is the modal illness or problem behavior?). If the answer is drug use, the school district might decide to employ additional counselors. If the answer is bullying, perhaps deploying tough-looking assistant principals in the hallways is the best way to go.

Univariate pictures of the data can also reveal which people, places, or things are the best, worst, or in the middle. When a class's grades are curved, the professor is looking at a univariate picture of the data for any possible breaks that separate the higher, medium, and lower grades on an exam. Google now tracks human behavior by analyzing the frequency of certain terms typed into their search engine. Internet security is based largely on determining whether there is activity on a company's servers that appears out of the ordinary. Univariate views of the data answer all of these questions.

Examples abound, a quick glance at a histogram can give a financial analyst the likely return on investment for a particular stock or bond. What are the best- and worst-case scenarios for that investment? What's the average household income in the United States? What's the typical voter turnout in the world? What is the most popular stop on the way home from school? Depending on the phenomenon and the question, **univariate descriptions** may be all that are needed to make important discoveries. This chapter shows how to create and analyze views of univariate data.

As waves of data crash in, few take the time to examine in detail the central tendency and dispersion of a single variable. Relying solely on a simple numerical summary of dispersion (the standard deviation) or central tendency (the mean) may be tempting. While any description—even a simple numerical summary—is a step in the right direction, draw a picture instead. Interview each variable before it is employed in your analysis.

With univariate data, we will introduce the frequency table, bar plot, boxplot, and histogram along with the stem-and-leaf plot. You'll learn which kind of summary works with categorical and continuous data. This section also provides examples of their use and the information they convey.

 Art and Practice of Data Visualization
USING DIFFERENT PACKAGES IN R

This book is based on a data environment that is programmed to call up a number of different packages in R. It's helpful to get familiar with installing and loading packages manually since thousands of them exist and can be quite useful. The packages you'll need for this class were installed when you

> typed "installD()" in Chapter 1. All of the packages used in this book are "loaded" every time you type "libraries()". *libraries()* is a function I defined that loads the packages you'll need. If you want to load a package manually, use the R base command *library()*. Most of the descriptions covered in this chapter use the ggplot framework. There are a few commands, however, that come from the 'descr' package: the commands *freq()* and *CrossTab()*.

> **KNOWLEDGE CHECK: Ask the questions univariate views answer.**

1. Which of the following are questions answered by univariate data?
   a. Who is the best, worst, or average?
   b. What is the most likely outcome?
   c. What's the worst thing that could happen?
   d. Am I special?
2. What questions below involve central tendency and which involve dispersion?
   a. Is there income inequality in a country?
   b. What is the most common problem college students face?
   c. Are students' evaluation of their college experience similar?
   d. Is Tom Brady the GOAT? (Greatest of All Time)

Five Views of Univariate Data

To answer the questions posed in the previous section, we have several different tools at our disposal: the frequency table, bar plot, boxplot, histogram, and stem-and-leaf plot. While this is not an exhaustive list, they represent the most common and effective ways to visualize data. Compare the different views presented here to the numeric summaries introduced in the previous chapter. Depending on the question, sometimes a single number is sufficient. Often, a picture is more helpful.

The power of visualizing data is perhaps best illustrated in Edward Tufte's analysis of the *Challenger* disaster when seven astronauts lost their lives only seconds after liftoff (Tufte, 1997, p. 49). At fault were O-rings in the booster rockets that failed because of cold weather. Engineers from the company who built the booster rockets recommended against launch on what was an unseasonably cold morning in Florida. As Tufte shows, the data the engineers used to advise NASA officials were obscured by tables of data and weakly worded text. Of course, we will never know for sure, but had an effective histogram been presented, NASA officials may have been more likely to postpone the launch. Tufte shows how a bar plot depicting the number of booster rocket tests at different temperatures would have put in stark relief the risks involved with launching the space shuttle in temperatures hovering around the low 30s. While

TABLE 5-1 Democrats Are the Mode

```
                  Frequency          Percent
Democrat                449          38.1154
Republican              276          23.4295
Independent             375          31.8336
Other                    74           6.2818
Not sure                  4           0.3396
Total                  1178         100.0000
```

lives are not usually directly at stake when analyzing data, sometimes they are. Pictures aren't always necessary to answer a question, but sometimes they are.

The lessons learned in Chapter 3 regarding continuous and categorical variables will become readily apparent in the next two sections. Our understanding of the distinctions between continuous and categorical variables will guide which views to employ. Below are five different ways to examine univariate data.

Frequency Table

Use a **frequency table** when the variable is categorical. The frequency table indicates how many cases reside in each category, giving the category's relative size. For example, if you want to know how many Democrats, Republicans, and Independents are in the National Election Studies (NES) data, use a frequency table. As the following frequency table indicates, there were 449 respondents who indicated they were Democrats (Table 5-1).

Here are a few words on Code Chunk 5-1. The easiest way to generate a frequency table is to use the *freq()* command included in the R package called 'descr'. The command produces both a frequency table as well as a bar plot of the variable (we'll discuss bar plots later in this section). Since I don't want to generate a bar plot yet, I include the *plot = options(descr.plot = FALSE)* command, which tells R to only generate the frequency table.

Code Chunk 5-1
```
freq(nes$pid3, plot = options(descr.plot = FALSE))
```

Frequency tables reveal the underlying structure of categorical variables, indicating whether to combine or eliminate categories. Suppose we want to know if being employed influences party identification. We have the variable *employ* from the NES data set. A quick look at the frequency table indicates there are nine categories (Table 5-2). We may want to ignore "temporarily laid off" since it only has four respondents. Table 5-2 also suggests that the concept of being employed is more complicated than we may have originally thought. For example, are students unemployed in the same sense as steel workers? Is a homemaker with a dog and a cat at home considered the same as a homemaker with five kids?

TABLE 5-2 Defining Employment Is Complicated

```
                     Frequency            Percent
Full-time                  466            39.5586
Part-time                  141            11.9694
Temporarily laid             4             0.3396
off
Unemployed                 101             8.5739
Retired                    194            16.4686
Permanently                107             9.0832
disabled
Homemaker                   91             7.7250
Student                     60             5.0934
Other                       14             1.1885
Total                     1178           100.0000
```

Code Chunk 5-2

```
freq(nes$employ, plot = options(descr.plot = FALSE))
```

Sometimes, knowing how many people, places, or things belong to a category is sufficient. Firms want to know what percentage of their business originates from abroad, universities want to know how many of their students come from out of state, and political parties want to know what percentage of the voting age population identifies as a member. Frequency tables settle the matter.

Bar Plot

While frequency tables are extremely useful, they may not pack the punch of a visual representation. Fortunately, in R we can draw a **bar plot** to visualize disparities or inequalities between categories. A bar plot of the unemployment variable presents the same information as the frequency table. It helps us visualize the relative size of the different categories. Constructing a bar plot both identifies where most of the cases reside and provides an easy way to make comparisons across categories.

The code for a bar plot is presented in Code Chunk 5-3. Since a bar plot represents one variable, note that the *aes()* function only includes one variable: *employ*. Once we've established those aesthetics, we add a layer with the function *geom_bar()* and we tell it to fill the bars with the color blue (I use the specific shade represented by #0000bf). Everything else in the code chunk should look familiar. If you need a refresher, I provide a glossary of R functions at the end of this chapter.

Code Chunk 5-3

```
ggplot(nes, aes(employ)) +
  geom_bar(fill = "#0000bf") +
```

```
theme_minimal() +
ggtitle("Figure 5-1: Barplot of Employment Variable") +
xlab("") +
ylab("Number of Respondents") +
coord_flip() +
theme(plot.title = element_text(size = 8, face = "bold")) +
theme(axis.title = element_text(size = 8, face = "bold"))
```

Figure 5-1 throws in stark relief the relative frequency of the different categories. A large share of the population, it quickly reveals, is employed full-time. Note the horizontal orientation of the plot using the *coord_flip()* command. Horizontal views of bar plots are helpful when the category names get long (e.g., temporarily laid off).

Boxplot (or Box-and-Whisker Plot)

Boxplots draw the distribution of a continuous variable. They outline the middle of the data (the "box") and identify outlying values (any value that lies outside the "whiskers").[1] They can also be used, as you'll see in the section on bivariate views, to compare the distribution and the median of a variable among different categories. The following boxplot describes the percentage of the voting age population that voted in each country's last national legislative election (Figure 5-2). The median (the thick line in the middle of the box) is slightly above 60% and the interquartile range (the box) is between 55% and 80%.

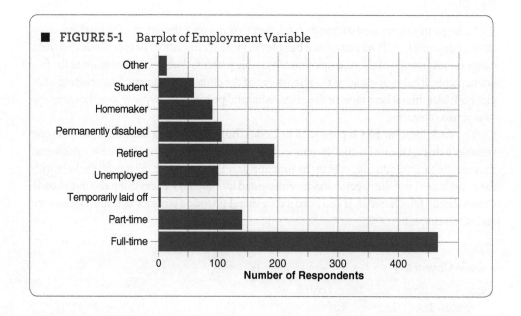

FIGURE 5-1 Barplot of Employment Variable

---

[1] In most statistical packages, the whiskers are drawn to extend 1.5 times the hinge spread (the distance between the median and either the top or bottom of the box) to which they are attached.

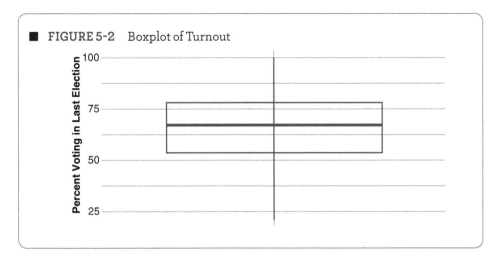

FIGURE 5-2 Boxplot of Turnout

The code in Code Chunk 5-4 generates a boxplot. Note that even though this boxplot shows one variable, we need to specify both an x and a y variable in the aesthetics function. Since we only want to generate a boxplot for one variable in this instance, we simply specify that *x*=" ".

Code Chunk 5-4

```
ggplot(world, aes(x="", turnout)) +
  geom_boxplot(col="#0000bf") +
  theme_minimal() +
  ggtitle("Figure 5-2: Boxplot of Turnout") +
  ylab("Percent Voting in Last Election") +
  xlab("") +
  theme_minimal() +
  theme(plot.title = element_text(size = 8, face = "bold")) +
  theme(axis.title = element_text(size = 8, face = "bold"))
```

This example shows a normal distribution: the box is relatively symmetrical. The distance between the median and the upper and lower hinges (the top and bottom of the box) is roughly the same. The boxplot also shows the median (the thick bar dividing the box; around 65% in this case).

Consider a boxplot of a highly skewed variable, GDP per capita (Figure 5-3).

Code Chunk 5-5

```
ggplot(world, aes(x="", gdppc)) +
  geom_boxplot(color="#0000ff") +
  theme_minimal() +
  ggtitle("Figure 5-3: Skewed Distribution") +
  ylab("GDP in Dollars per Capita") +
  xlab("") +
  theme(plot.title = element_text(size = 8, face = "bold")) +
  theme(axis.title = element_text(size = 8, face = "bold"))
```

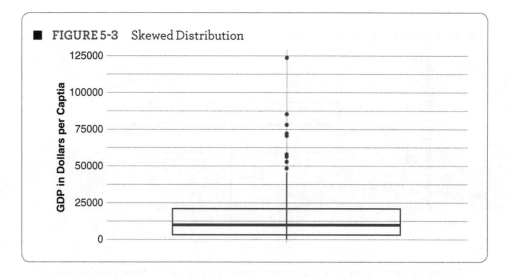

FIGURE 5-3 Skewed Distribution

In Figure 5-3, the distance between the median and the upper hinge is slightly bigger, but notice the outliers that exist beyond the top whisker. In this case, there are at least eight different cases that exist beyond the whisker.

Unfortunately, boxplots don't typically show individual observations. When the number of observations reaches the hundreds or thousands, identifying all of the cases is messy. Fortunately, you can label the points and "jitter" them so that they don't lie directly on top of each other.

Consider the boxplot (Figure 5-4) showing the distribution of states according to the homicide rate (the variable *murderrate*). By labeling and jittering the points, many of the questions we might ask of the data can be answered. Which state has the highest rate? Which state, if any, is typical? How does my state compare to others?

Here are a couple of notes about the code that produces Figure 5-4. I use the jitter feature to spread the points out. Also note that I use *seed = 1* when jittering the points and the labels (so that they match). When using the jitter feature with boxplot, I need to include an x and a y in the *aes()* function. Since I want to draw one boxplot, I specify that the x variable is *factor(0)*, otherwise the code won't run.

Code Chunk 5-6

```
ggplot(states, aes(x=factor(0), y=murderrate)) +
  geom_boxplot(outlier.colour=NA, fill=NA, colour="grey20") +
  geom_point(fill = "#bf0000", size=2,
             shape=21, colour="grey",
             position=position_jitter(seed = 1)) +
  geom_text_repel(aes(label = st), col = "grey",
             position = position_jitter(seed = 1),
             hjust=1, vjust=2) +
  theme_minimal() +
  ggtitle("Figure 5-4: Fancy Boxplot") +
  ylab("Homicide Rate: Murders per 100,000 Population") +
  xlab("") +
```

```
theme(plot.title = element_text(size = 8, face = "bold")) +
theme(axis.title = element_text(size = 8, face = "bold")) +
theme(axis.text.x=element_blank())
```

You can see that by identifying all of the data points with their label, there's a rich amount of information that would otherwise not have been available. First, we see that Louisiana is the state with the highest homicide rate (close to 12 homicides per 100,000 population). Not surprisingly, rural and sparsely populated states have low homicide rates (Vermont, Utah, Minnesota, Iowa, New Hampshire). What may come as a surprise is that New Jersey (NJ) and New York (NY) are close to Kansas (KS).

Histogram

We've already used the histogram to understand the shape of the data. It's such a useful and necessary device that it's been introduced earlier. Below is a formal introduction.

Since the histogram represents a univariate view of the data, I only need to specify one variable in the *aes()* command (Code Chunk 5-7). I then add the layer that includes the histogram: the *geom_histogram()* function. In that function I specify the number of bins to use in the plot (bars) and I fill it with blue (#0000bf). Everything else should look familiar.

Code Chunk 5-7

```
ggplot(world, aes(durable)) +
  geom_histogram(bins=10, fill = "#0000bf") +
  ggtitle("Figure 5-5: Histogram of the durable Variable") +
  xlab("Years Since Last Significant Regime Change") +
  theme_minimal() +
  theme(plot.title = element_text(size = 8, face = "bold")) +
  theme(axis.title = element_text(size = 8, face = "bold"))
```

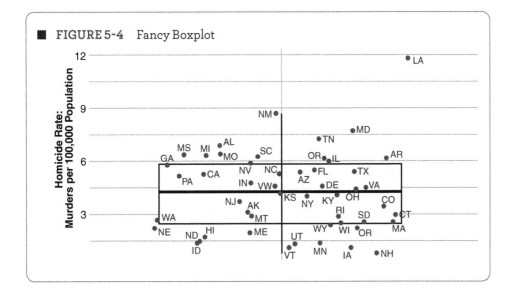

■ **FIGURE 5-4** Fancy Boxplot

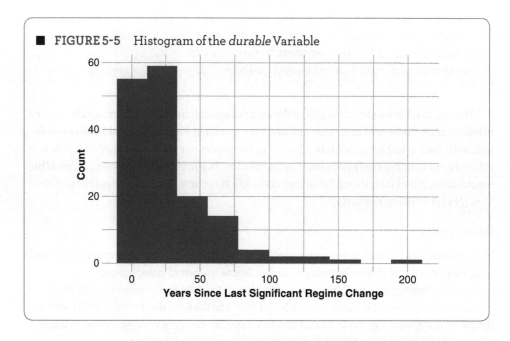

FIGURE 5-5 Histogram of the *durable* Variable

Histograms display the frequency distribution of continuous variables, and they reveal the shape of the variable's distribution. The units on the y axis indicate the number of cases for each range of values. We refer to each range of values as a "bin." The histogram is useful because it provides information about both the center and dispersion of a variable. You can control the number of bins to examine parts of the distribution in more detail. In the example above, I specified 10 bins, which gives a fairly good indication of the variable's overall distribution. Note what happens if I limit the number of bins to an extreme: just 3 (Figure 5-6).

Code Chunk 5-8

```
ggplot(world, aes(durable)) +
  geom_histogram(bins=3, fill = "#0000bf") +
  labs(title = paste("Figure 5-6: Not Enough Bins")) +
  xlab("Years Since Last Significant Regime Change") +
  theme_minimal() +
  theme(plot.title = element_text(size = 8, face = "bold")) +
  theme(axis.title = element_text(size = 8, face = "bold"))
```

In Figure 5-6 there are three bins delineating the following ranges: 0 to 100, 101 to 200, and 201 and beyond. It's hard to learn anything from the histogram since the bins cover such a wide area. Are the cases between 0 and 100 closer to 100 or closer to 0? Be cognizant of bin size since it can influence what you see. The default number of bins is 30, but you shouldn't always take that as a given. Too many or too few bins can hide or obscure important features of the data.

The histogram is well suited to indicate a variable's **skewness**: are the values to the left or right of center more extreme? In Figure 5-7, three histograms are drawn from three different

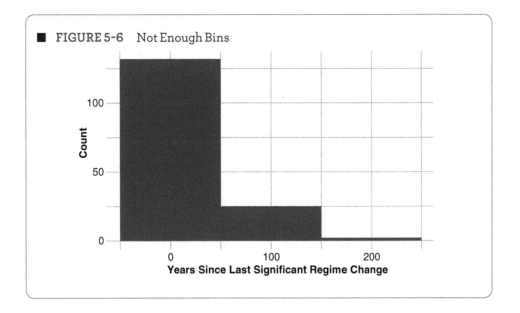

FIGURE 5-6 Not Enough Bins

variables: (1) personal freedom score[2] (Ruger & Sorents, 2009), (2) voting turnout for the 2012 U.S. election, and (3) dollars spent per student in the state. Although more extreme examples of skewness exist, these three variables illustrate a negatively skewed distribution (personal freedom score), a normal distribution (voter turnout), and a positively skewed distribution (dollars spent per student in a state). Understanding these general shapes will prove useful in Chapter 6 when we transform data.

To produce the comparison, I define three different histograms (Code Chunk 5-9): I call them p1, p2, and p3. I then use the *grid.arrange()* function to plot them side by side. The options *nrow=1* and *ncol=3* tell R to place the three objects in one row.

Code Chunk 5-9
```
p1 <- ggplot(states, aes(freedom)) +
  geom_histogram(bins=10, fill = "#0000bf") +
  ggtitle("Freedom") +
  xlab("Personal Freedom Score") +
  theme_minimal() +
  theme(plot.title = element_text(size = 8, face = "bold")) +
  theme(axis.title = element_text(size = 8, face = "bold"))

p2 <- ggplot(states, aes(turnout)) +
  geom_histogram(bins=10, fill = "#0000bf") +
```

---

[2] The personal freedom score constructed by the Mercatus Center (Ruger & Sorents, 2009) records freedom on a number of different dimensions (gun control, tobacco, alcohol, gambling, civil liberties, asset forfeiture, etc.).

```
  ggtitle("Turnout") +
  xlab("Turnout in 2012 Election") +
  theme_minimal() +
  theme(plot.title = element_text(size = 8, face = "bold")) +
  theme(axis.title = element_text(size = 8, face = "bold"))

p3 <- ggplot(states, aes(stuspend)) +
  geom_histogram(bins=10, fill = "#0000bf") +
  ggtitle("Stuspend") +
  xlab("Dollars Spent per Student") +
  theme_minimal() +
  theme(plot.title = element_text(size = 8, face = "bold")) +
  theme(axis.title = element_text(size = 8, face = "bold"))

grid.arrange(p1, p2, p3, nrow=1, ncol=3,
          top = textGrob("Figure 5-7: Examples of Skewness",
                    gp=gpar(fontsize=10,fontface = "bold")))
```

Another example illustrates how histograms highlight skewness. Consider what is probably the most often-used variable with an exaggerated right skew: GDP per capita (the histogram on the left in Figure 5-7). Note how most all of the cases seem to be located to the left of

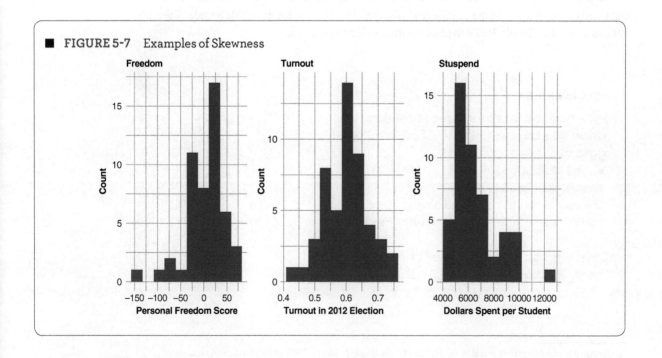

■ **FIGURE 5-7** Examples of Skewness

$50,000, some just to the right of $50,000, and one above $100,000. The distribution is not normal; it is skewed to the right. In the histogram on the right, I logged the values of GDP per capita to generate something closer to a normal distribution for comparison.[3] Histograms reveal skewness at a glance. Note that I use the *options()* command to tell ggplot that I don't want to use scientific notation.

Code Chunk 5-10

```r
options(scipen = 999)

plot1 <- ggplot(world, aes(gdppc)) +
  geom_histogram(bins=20, fill = "#0000bf") +
  ggtitle("Skewed") +
  xlab("GDP per Capita") +
  theme_minimal() +
  theme(plot.title = element_text(size = 8, face = "bold")) +
  theme(axis.title = element_text(size = 8, face = "bold"))

plot2 <- ggplot(world, aes(log(gdppc))) +
  geom_histogram(bins=20, fill = "#0000bf") +
  ggtitle("Normal") +
  xlab("GDP per Capita (logged)") +
  theme_minimal() +
  theme(plot.title = element_text(size = 8, face = "bold")) +
  theme(axis.title = element_text(size = 8, face = "bold"))

grid.arrange(plot1, plot2, nrow=1, ncol=2,
    top = textGrob("Figure 5-8: Skewed and Normal Distribution",
                    gp=gpar(fontsize=10,fontface = "bold")))
```

Stem-and-Leaf Plot

Stem-and-leaf plots provide a handy way to obtain a quick and easy view of a variable's distribution, especially when the number of observations is relatively small. Stem-and-leaf plots are similar to a histogram but have a horizontal orientation. If the values in the stem-and-leaf plot are ordered, it's also an easy way to identify the median case.

The stem-and-leaf plot is constructed by first identifying how best to construct the stem. In many ways, the exercise is similar to calculating how many bins to use in a histogram. With the stem, we usually divide things up by decimal places. What value should each branch represent: ones, tens, hundreds, and so forth? We then place each case (leaf) on the corresponding part of the stem.

---

[3] In Chapter 6 we will discuss log transformations.

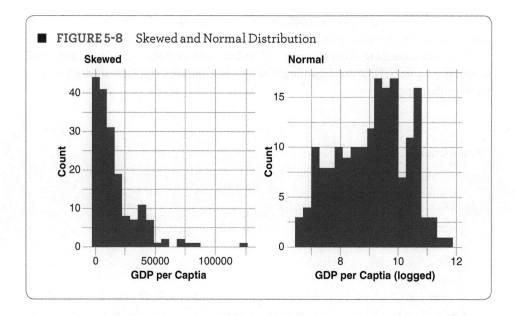

FIGURE 5-8 Skewed and Normal Distribution

To illustrate, the example in Figure 5-9 is a stem-and-leaf plot of women's earnings as a fraction of men's for the 50 states. In this case, the stem is ordered by ones and each branch represents an increment of 2. The number in the leaf represents the first decimal. In the plot, it is easy to spot the two outlying values that exist between 68 and 70. The two 7s to the right of the broken line represent two cases. Since we specified *scale = 2* in the code (Code Chunk 5-11), we don't know if the two 7s represent 68.7 or 69.7. In this plot, the broken vertical line represents the decimal point. Try changing the scale to 3 to see what happens. In any case, we can quickly surmise that the median case is close to 80%.

Code Chunk 5-11
```
stem(states$percwom, scale = 2)
```

As noted, the stem can take on different values depending on the order of magnitude. For example, in the stem-and-leaf plot in Figure 5-9 it made sense to order things by twos: each step in the stem represents 2. In Figure 5-10, when drawing a stem-and-leaf plot of household income for the 50 states, the stem is ordered by 2,000. Each step represents $2,000. In the first row, we see six numbers (six leaves), meaning there are six cases between 32,000 and 34,000. In this case, we don't know exactly what the numbers are; we just know they're between 32,000 and 34,000. If we wanted to know exactly what the numbers were, we could change the *scale = 1* option to *scale = 3* in Code Chunk 5-12.

Code Chunk 5-12
```
stem(states$inc, scale=1)
```

■ **FIGURE 5-9** Stem-and-Leaf Plot for Women's Earnings as a Fraction of Men's Earnings for 50 States

Stem	Leaf
68	7 7
70	
72	6
74	5 7 7 5
76	0 6 7 7 8
78	0 5 6 8 9 0 0 4 6 9
80	0 1 5 6 7 3 4 6
82	1 1 5 5 4 6 6 6 8 9 9
84	0 6 8 0 0
86	0
88	4 5 9

■ **FIGURE 5-10** Stem-and-Leaf Plot of Household Income for 50 States

Stem	Leaf
32	2 3 5 7 7 8
34	0 6 7 6 9
36	1 2 5 5 6 3 8 9
38	0 2 2 6 6 6
40	1 5 5 6 6 6 8
42	5 1
44	0 1 1 3 5 7 5 7 8 9
46	3
48	
50	5 0
52	2 6
54	
56	9

The stem-and-leaf plot is a fairly low-tech way to collect and plot data by hand. It's therefore a quick way to see a distribution if you're without a computer (an admittedly increasingly rare event these days).

The frequency table, bar plot, boxplot, histogram, and stem-and-leaf plot all provide important information we need at the beginning of our analysis. Boxplots and histograms go much further in terms of the amount of information provided relative to numerical summaries.

> **KNOWLEDGE CHECK: Create and analyze univariate views of the data.**

3. Indicate whether the following univariate version of these plots requires a categorical or continuous variable.
   a. Frequency table
   b. Boxplot
   c. Bar plot
   d. Stem-and-leaf plot

4. Draw a picture of the *world$regime* variable and answer the following:
   a. Which type of regime is the most common in these data?
   b. Which type of regime is the rarest?
   c. How many civilian dictatorships are there?
   d. What is the most common type of democracy?

5. Draw a picture of the *nes$terror_worry* variable and answer the following:
   a. Of those asked, are a majority of Americans at least somewhat worried about terrorism?
   b. What is the modal category of those asked?
   c. Are most Americans very worried or extremely worried?
   d. Are Americans polarized on this subject?

6. Draw a picture of the *states$infant* variable, which indicates the number of deaths per 1,000 live births in a state.
   a. What is the range of the variable?
   b. Is the variable slightly skewed to the right?
   c. Is the variable slightly skewed to the left?
   d. How many states have mortality rates above 15?

7. Draw a boxplot of *states$turnout* with labels and jittering.
   a. What state has the highest turnout?
   b. What state has the lowest turnout?
   c. Name three states that are close to the median.
   d. Name the second highest and second lowest states.

Are They in a Relationship?

Earlier we saw that univariate descriptions provide important information about a single variable, allowing us to answer questions about central tendency and dispersion. Now we move on to describing the relationship between two variables. **Bivariate**

descriptions illustrate the relationship between two variables, whether they are continuous or categorical.

We'd like to know whether two different things are related. Climatologists want to know if a rise in temperature of the Earth's atmosphere is associated with the severity of hurricanes, doctors want to know if disease is associated with stress, and sports fans (or university chancellors) want to know whether coaching salaries are associated with wins and losses. Understanding how two things are related represents one of the most basic endeavors we engage in as scholars, analysts, and humans. We visualize data in order to understand those relationships.

Bivariate visualizations reveal the exact nature of that relationship (e.g., is it linear?). Like the univariate context, the visualizations we'll explore in this section can also uncover mistakes made in data collection. For example, a scatter plot will shout if someone entering the data accidentally typed an additional zero. Perhaps most importantly, visualizing relationships generates new and, in many cases, more interesting questions. In this section we will introduce the scatter plot, the mosaic plot, and the bubble plot, all of which help us understand how two or more variables are related. Boxplots can also be used in the bivariate context. They can show how the central tendency and the shape of one variable varies across the categories of another.

Like the views presented for univariate data, this is not an exhaustive list, it merely represents the most common ways to visualize data. The goal is to recognize how visualizing the data, rather than relying solely on a numerical summary, aids our understanding. In many ways, visualizations bring a higher degree of transparency to our enterprise. Single numerical summaries of the data can mask important characteristics of the data. Like sharks swimming beneath a calm ocean's surface, the most important and most relevant features of the data may be lurking behind a calm-looking correlation coefficient, mean, or standard deviation.

> **KNOWLEDGE CHECK: Ask the questions bivariate views answer.**

8. Which of the following questions require two variables?
   a. Are there more Democrats than Republicans in the NES survey?
   b. Are there more presidential regimes than parliamentary regimes in the world?
   c. Are Democrats more worried about terrorism?
   d. Are average incomes higher in presidential regimes than parliamentary regimes?

9. Which are true with bivariate views of the data?
   a. They can help spot mistakes.
   b. They only work if there is a relationship between two variables.
   c. They show if one variable causes another.
   d. They show if there is a relationship between two variables.

10. Indicate whether a scatter plot, boxplot, or mosaic plot is appropriate for showing the relationship described.

    a. Are attitudes toward Obama and scientists related?
    b. Is marital status related to race?
    c. Which region in the United States has the highest infant mortality rates?
    d. Is gender related to worrying about terrorism?

Scatter Plot

The scatter plot is the workhorse of data analysis. Scatter plots are two dimensional, showing the relationship between two continuous variables. By coloring the data or using different symbols by a third variable, scatter plots can provide additional dimensions. In Figure 5-11, voting turnout is plotted against political knowledge—the percentage of a state's population that recognizes the name of the governor. As the scatter plot suggests, there is not much of a relationship between the two variables: where a state is located on the x axis does not help us predict where it is located on the y axis. Knowing the extent of a state's political knowledge does not help us predict how many of its residents will vote.

To be even more specific, consider states where roughly 66% or 67% of the population knows the name of the governor: Massachusetts and Hawaii. This tells us relatively little about a state's turnout. At that level of political knowledge, Hawaii (HI) has roughly 45% turnout, whereas Massachusetts (MA) has just over 65%. There appears to be no relationship between voting turnout and political knowledge.

The structure of the code should look familiar. The only change from the code generating univariate views is that in the *aes()* function we now specify an x variable along with the y variable. In the current example (Code Chunk 5-13), the x variable is *knowgov* and the y variable is *turnout*. The defining layer for scatter plots is *geom_point()*, which draws points on a two-dimensional grid. As we did in previous scatter plots, we only want to label a few interesting cases. We use the *ifelse()* function to say that if the cases are Hawaii (HI) or Massachusetts (MA), then label them with the *st* variable; otherwise leave it blank. We specify "blank" with the two single quotations with nothing in between ('').

Code Chunk 5-13
```
ggplot(states, aes(knowgov, turnout, label = st)) +
  geom_point(col="#bf0000") +
  ggtitle("Figure 5-11: Turnout Not Related to Political Knowledge") +
  ylab("Voting Turnout") +
  xlab("Political Knowledge") +
  theme_minimal() +
  geom_text_repel(size = 2.8,
       aes(label=ifelse(st=="HI" |
                        st=="MA", as.character(st),''),
                        hjust = 0, vjust=-1),
                        show.legend=FALSE) +
  theme(plot.title = element_text(size = 8, face = "bold")) +
  theme(axis.title = element_text(size = 8, face = "bold"))
```

FIGURE 5-11 Turnout Not Related to Political Knowledge

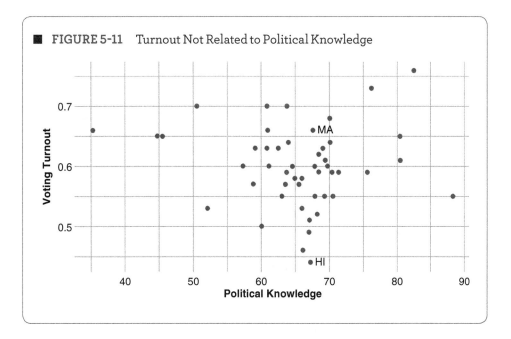

Art and Practice of Data Visualization
DEPENDENT AND INDEPENDENT VARIABLES

When constructing bivariate views of the data, placing the dependent variable on the y axis and the independent variable on the x axis is common practice. The independent variable is cause and the dependent variable is effect. We hypothesize that the independent variable causes the dependent variable. By placing voting turnout on the y axis in the example above, we imply political knowledge causes voting turnout. Although we can remain skeptical about whether political knowledge really does cause turnout, keeping the same x and y axis orientation is key. Switching the independent-dependent variable orientation on the x and y axes disorients both the analyst and the reader.

CONFUSING DEPENDENT AND INDEPENDENT VARIABLES

Students often have a difficult time keeping dependent and independent variables straight. The confusion, I suspect, resides in their definition. Dependent variables are often casually defined as "the thing you're interested in explaining." While this is technically correct, it gets confusing if one declares, "I am interested in democracy's impact on economic growth." When asked to identify the dependent variable in that statement, the novice is tempted to say "democracy" instead of the correct answer: economic growth. If we stick to thinking about dependent and independent variables in terms of causality, there is less confusion.

Code Chunk 5-14

```
ggplot(states, aes(hsdiploma, infant)) +
 geom_point(col="#bf0000") +
 ggtitle("Figure 5-12: Education Reduces Infant Mortality") +
 ylab("Infant Mortality") +
 xlab("High School Diploma") +
 theme_minimal() +
 geom_text_repel(size = 2.8,
    aes(label=ifelse(st=="CA" |
                     st=="LA", as.character(st),''),
                     hjust = 0, vjust=-1),
                     show.legend=FALSE) +
 theme(plot.title = element_text(size = 8, face = "bold")) +
 theme(axis.title = element_text(size = 8, face = "bold"))
```

Figure 5-12 provides an example of a fairly strong relationship between two variables: infant mortality and the percentage of a state's population that has a high school degree. In this plot, knowing the level of education provides some useful information regarding the number of deaths per 1,000 live births.

In the code that generates Figure 5-13, note the addition of the function *geom_smooth()*. This draws a line fit to the data (Code Chunk 5-15). In the *geom_smooth()* function, we include the option *method = "lm"* to specify a straight line.

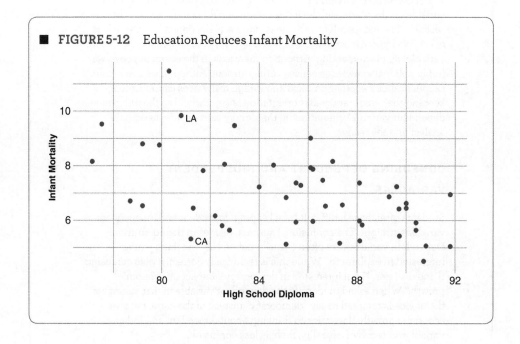

■ **FIGURE 5-12**   Education Reduces Infant Mortality

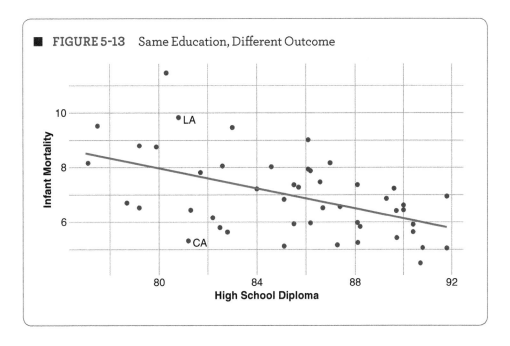

■ FIGURE 5-13   Same Education, Different Outcome

Code Chunk 5-15
```
ggplot(states, aes(hsdiploma, infant)) +
  geom_point(col="#bf0000") +
  geom_smooth(method = "lm", se = FALSE, col="#0000bf") +
  ggtitle("Figure 5-13: Same Education, Different Outcome") +
  ylab("Infant Mortality") +
  xlab("High School Diploma") +
  theme_minimal() +
  geom_text_repel(size = 2.8,
     aes(label=ifelse(st=="CA" |
                         st=="LA", as.character(st),''),
                         hjust = 0, vjust=-1),
                         show.legend=FALSE) +
  theme(plot.title = element_text(size = 8, face = "bold")) +
  theme(axis.title = element_text(size = 8, face = "bold"))
```

Although there does seem to be a relationship between education and infant mortality, there is still considerable variation in infant mortality at each level of education. For example, at roughly 81% of high school completion, note that the difference between Louisiana (LA) and California (CA) is considerable: Louisiana has an infant mortality rate of roughly 11 and California's is slightly over 5.

When looking at real data, patterns are not always recognizable without some help. To the trained eye, the relationship between infant mortality and high school education is negative and fairly linear. To the novice, that relationship may be hard to spot.

The negative relationship becomes more apparent and its linearity becomes obvious with the addition of the line. After fitting the line, we see that the level of education can give us a fairly good indication of where a state might be situated in terms of infant mortality. We would say there is a **negative association** between education and infant mortality: as one variable increases (education), the other decreases (infant mortality). It is also a fairly **linear relationship**. The decline in infant mortality is constant over the range of education.

> **Art and Practice of Data Visualization**
> **CAUTION WITH LINES AND CURVES**
>
> While applying lines and various curves can be helpful in recognizing patterns, also be aware that using them prematurely can make it seem as if there is a pattern when there isn't (a ghost in the graph). Our eyes process information in a certain way and can be tricked. Our natural inclination is to see patterns when they don't exist. Fitting lines to clouds of data may only serve to encourage the eye to find a pattern when it isn't there.

Before continuing, an important point is worth emphasizing: take the time to examine pictures rather than rely solely on numerical summaries. While they are more efficient in some ways, numerical summaries can mask underlying features of the data that deserve attention. Those hidden features can produce misleading estimates down the road. An over-reliance on numerical summaries also limits our ability to make important discoveries. Let's make the trade-offs between simple summaries and visualizing the data more concrete by comparing a scatter plot with a simple numerical summary: the Pearson's R **correlation**. The Pearson's R correlation provides a single numerical summary of the relationship between two continuous variables.

Consider the scatter plot we dissected in Chapter 2 (Figure 2-1). While examining the relationship between ethnolinguistic fractionalization and GDP per capita, we observed three distinct patterns: (1) the industrial countries had homogeneous populations and relatively high incomes, (2) without the industrial countries there was a notable curve to the data, and (3) the curve was probably driven by small island nations. Here I reproduced that scatter plot with all of those features added (Figure 5-14).

There are a couple of things to note in the code (Code Chunk 5-16). First, I put an *ifelse()* command in the color option to designate blue for the industrial countries and red for the rest. I also use an *ifelse()* command to label four countries. Finally, note that in ggplot I can put many layers into the graph. In Figure 5-14 I use two smooths. I use the *subset()* command in one of the smooths so that it fits the countries that are not (!=) the industrial countries.

Code Chunk 5-16
```
ggplot(world, aes(ethfrac, log(gdppc))) +
  geom_point(color=ifelse(world$aclpregion=="Industrial Countries",
                          "#0000bf", "#bf0000")) +
```

```
geom_text_repel(size = 2.8,
          aes(label=ifelse(iso3c=="COM" |
                           iso3c=="HTI" |
                           iso3c=="KIR" |
                           iso3c=="SLB",
                                as.character(iso3c),''))) +
geom_smooth(method="lm", col="grey", se=FALSE,
          linetype = "dashed") +
ggtitle("Figure 5-14: The Relationship Could Be Nonlinear") +
ylab("GDP per Capita (logged)") +
xlab("Ethnolinguistic Fractionalization") +
theme_minimal() +
geom_smooth(data=subset(world,
                        aclpregion != "Industrial Countries"),
                        color="black", se=FALSE) +
annotate("text", x = .15, y = 11.5, size = 3,
          label = "Industrial Countries", col="#0000bf") +
theme(plot.title = element_text(size = 8, face = "bold")) +
theme(axis.title = element_text(size = 8, face = "bold"))
```

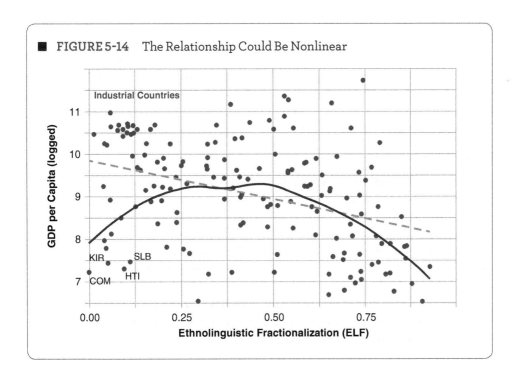

FIGURE 5-14 The Relationship Could Be Nonlinear

The amount of information revealed is considerable, perhaps overwhelming in this case. Nonetheless, there is much to discover and think about by examining the scatter plot.

Now consider the simple numerical summary of the relationship given by the Pearson's R correlation. Pearson's R is calculated by the following formula, generating a number between −1 and 1 depending on whether there is a strong negative correlation (the number approaches −1) or a positive correlation (the number approaches 1):

$$r = \frac{\sum_i (x_i - \bar{x})(y_i - \bar{y})}{\sqrt{\sum_i (x_i - \bar{x})^2} \sqrt{\sum_i (y_i - \bar{y})^2}}$$

Much of the formula depends on the distance between observations and the mean $(x_i - \bar{x})$ and $(y_i - \bar{y})$. As we learned in Chapter 4 when comparing the mean and median, calculations based on the mean are highly susceptible to outliers. Consequently, a high correlation from Pearson's R could result from two very different situations: an outlying case or a strong linear relationship. Visualizing the relationship between two continuous variables will quickly reveal which.

In Code Chunk 5-17, I use the *cor()* command, which produces the correlation between the variables *world$ethfrac* and *world$gdppc*. Below the code chunk, the output produced is [1] −0.3791241, the Pearson's R correlation.

Code Chunk 5-17

```
cor(world$ethfrac, log(world$gdppc), method = "pearson",
        use = "complete.obs")

[1] -0.3791241
```

While Pearson's R indicates that the relationship between the two variables is negative, all of the features gleaned from the scatter plot are unobserved. From the scatter plot we learn about the industrial countries, the possible nonlinear relationship between ethnolinguistic fractionalization and GDP per capita, and the importance of geographic size. From the Pearson's R correlation, we learn that the relationship between the two variables is marginally negative: −.38.

Boxplot (Bivariate)

Boxplots are useful when examining the relationship between continuous and categorical variables. This comes with one caveat: boxplots are less useful if there are more than 30 or so categories. Boxplots are most useful when there are 20 or fewer categories, although this is not a hard or fast rule. The following example illustrates why boxplots are so useful. Their key property resides in the box: they show whether the middle half of the data (the interquartile range) changes over each category. They not only indicate what happens to the median case, but they also show how the distribution changes.

In previous code for boxplots, I only wanted to summarize a single variable. In this case (Code Chunk 5-18), I want to create several boxplots, one for each level of party identification. Consequently, in the aesthetics function, I specify two variables: *pid7* and *ftgay*.

Code Chunk 5-18

```
ggplot(subset(nes, pid7!="NA" & pid7!="Not sure"),
       aes(pid7, ftgay)) +
```

```
geom_boxplot(col="#0000bf") +
theme_minimal() +
theme(axis.text.x = element_text(size=8, angle=45, vjust=.7)) +
ggtitle("Figure 5-15: Partisanship Shapes Attitudes") +
ylab("Gay Thermometer") +
xlab("Party Identification") +
coord_flip() +
theme(plot.title = element_text(size = 8, face = "bold")) +
theme(axis.title = element_text(size = 8, face = "bold"))
```

Figure 5-15 illustrates how the medians and interquartile ranges vary as we look at support for gays and lesbians by party identification (Party ID).[4] Notice how the median value does not change when comparing Independents with Republicans. If we were only considering medians, we would conclude there is no difference. The boxplots, however, tell a very different story: the interquartile range for the strong Republicans includes much lower scores on the thermometer compared to Independents.

Consider another example. An interesting variable in the *world* data set records the ratio of girls' to boys' primary and secondary school enrollment. The larger the ratio, the better educational outcomes for girls relative to boys. When comparing medians, the boxplots in Figure 5-16 indicate that relatively little difference exists between authoritarian regimes (civilian and military) and democratic regimes (parliamentary or presidential).

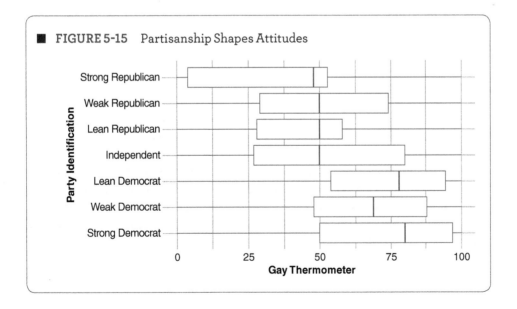

FIGURE 5-15 Partisanship Shapes Attitudes

---

[4] Many variables in the NES data are feeling thermometers—variables ranging from 0 to 100 that indicate how survey respondents feel about a particular kind of person or policy. For example, respondents are asked in the gay feeling thermometer to indicate on a scale of 0–100 how they feel about gays and lesbians.

Code Chunk 5-19

```
ggplot(world, aes(regime, gtbeduc, na.rm=TRUE)) +
  geom_boxplot(col="#0000bf", na.rm=TRUE) +
  theme_minimal() +
  ggtitle("Figure 5-16: Large Variation Under Authoritarian Regimes") +
  ylab("Ratio of Girls-to-Boys Educational Achievement") +
  xlab("") +
  coord_flip() +
  theme(plot.title = element_text(size = 8, face = "bold")) +
  theme(axis.title = element_text(size = 8, face = "bold"))
```

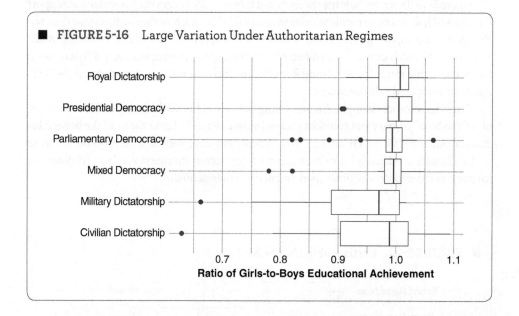

FIGURE 5-16 Large Variation Under Authoritarian Regimes

There is, however, a noticeable difference when we compare distributions. Specifically, the lower quartile dips in authoritarian regimes. Apparently, authoritarian regimes don't guarantee that girls receive an adequate education.

Look inside the boxes to identify the culprits. To construct Figure 5-17, I limited the sample to the two offending regime types (civilian and military dictatorships) by creating a new data set using the *subset()* command in R. I then instructed R to *jitter* their position so that the labels don't overlap. Also note that I include *scale_fill_manual()* to specify exactly which colors I want to use (Code Chunk 5-20).

Code Chunk 5-20

```
regimedata <- subset(world, regime=="Civilian Dictatorship" |
              regime=="Military Dictatorship")
```

```
ggplot(regimedata, aes(regime, gtbeduc, na.rm=TRUE)) +
  geom_boxplot(outlier.colour=NA, fill=NA, colour="grey20") +
  geom_text_repel(size = 2.8, aes(label = iso3c), col = "grey",
           position = position_jitter(seed = 1), hjust=2, vjust=1) +
  geom_point(aes(fill = regime), size=5, shape=21, colour="grey",
           position=position_jitter(seed = 1)) +
  theme_minimal() +
  ggtitle("Figure 5-17: Dictatorships Especially Bad for Girls") +
  ylab("Ratio of Girls-to-Boys Educational Achievement") +
  xlab("") +
  scale_fill_manual(values=c("#0000bf", "#bf0000")) +
  theme(plot.title = element_text(size = 8, face = "bold")) +
  theme(axis.title = element_text(size = 8, face = "bold")) +
  theme(legend.position = "none")
```

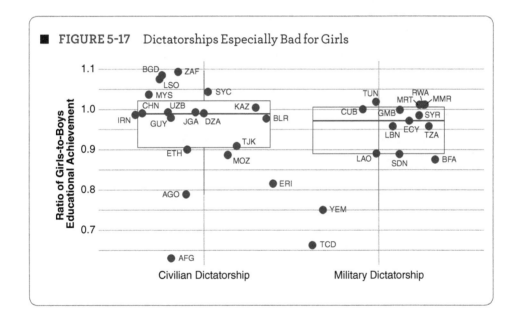

■ **FIGURE 5-17** Dictatorships Especially Bad for Girls

Afghanistan (AFG), Angola (AGO), Eritrea (ERI), and Mozambique (MOZ) represent the poor performing civilian dictatorships. Under military dictatorship, the poor performers are Burkina Faso (BFA), Laos (LAO), Sudan (SDN), Yemen (YEM) and Chad (TCD).

To review, in the bivariate context boxplots provide five different pieces of information for each category of a variable: the 25th, 50th, and 75th percentiles (the box) and the minimums and maximums (the whiskers). Labeling boxplots provides additional information. Consequently, a series of boxplots aligned side by side makes it easier to identify important differences or similarities.

Code Chunk 5-21

```
nes$f.obama <- as.factor(nes$ftobama)

ggplot(nes, aes(f.obama, ftmuslim)) +
  geom_boxplot(col="#0000bf") +
  theme_minimal() +
  theme(axis.text.x=element_blank(),
        axis.ticks.x=element_blank()) +
  ggtitle("Figure 5-18: Too Many Categories!") +
  ylab("Muslim Thermometer") +
  xlab("Obama Thermometer") +
  theme(plot.title = element_text(size = 8, face = "bold")) +
  theme(axis.title = element_text(size = 8, face = "bold"))
```

Boxplots are less useful, however, when taken to an extreme. Suppose I'm interested in the relationship between attitudes or "feelings" respondents have toward Muslims and President Obama. To illustrate, I first converted the *ftobama* variable from a numerical variable to a factor which produces 100 distinct categories. I then plotted a boxplot for every level (100) of the Obama feeling thermometer (Figure 5-18). While we can observe there is a positive trend, the sheer number of boxplots makes it difficult to spot any of their specific features. It is also problematic since at each level of the feeling thermometer for Obama, there may only be a few observations. Given there are five specific features of the boxplot (two whiskers, two hinges, and a median), less than five observations for a category is problematic. In those cases, a full boxplot cannot be formed.

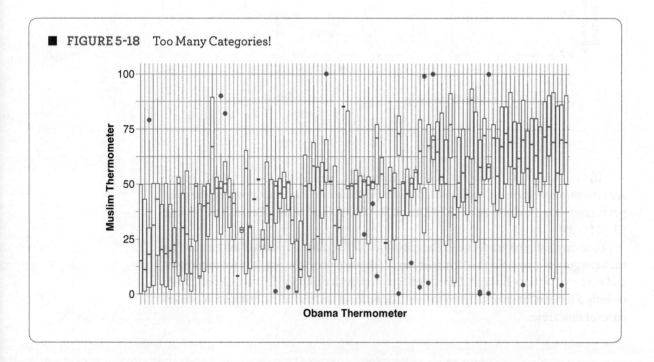

■ **FIGURE 5-18** Too Many Categories!

As noted earlier, boxplots are used when comparing continuous and categorical variables. We saw two examples of its use: (1) the relationship between feelings toward gays and party identification and (2) the ratio of girl's to boy's education and regime type. There are many instances, however, when we want to examine the relationship between two categorical variables. Use mosaic plots and cross-tabs when examining the relationship between two categorical variables.

Mosaic Plot

The **mosaic plot** generates a graphical representation of cross-tabs (tables that show the association between two categorical variables). The advantage of mosaic plots over cross-tabs lies in their ability to provide a visually intuitive way to identify relationships. Mosaic plots not only show how two categorical variables are related, but the width of their columns also indicates the number of observations in each category along the x axis.

Let me begin with an example that examines political partisanship and attitudes toward the police. First, I need to "clean up" the variable recording partisanship. I don't want to include certain categories ("other" and "not sure"), so I designate those cases as "NA." The code in Code Chunk 5-22 creates a new variable called *pid3.new*. The *ifelse()* command reads as follows: if the *pid3* variable registers "other," then assign it NA; if the *pid3* variable registers "not sure," then label that NA; otherwise keep the *pid3* value.

Code Chunk 5-22

```
nes$pid3.new <- ifelse(nes$pid3 == "Other", NA,
                       ifelse(nes$pid3 == "Not sure", NA,
                              nes$pid3))
```

After removing the "other" and "not sure" cases, I label the new variable with the proper names by designating the variable as a factor. Then I assign names to the different categories and change their order (Code Chunk 5-23).

Code Chunk 5-23

```
nes$pid3.new <- as.factor(nes$pid3.new)

levels(nes$pid3.new)=c("Democrat", "Republican",
                       "Independent")

nes$pid3.new = factor(nes$pid3.new, levels(nes$pid3.new)
                      [c(1,3,2)])
```

After that bit of housecleaning, we're ready to draw the mosaic plot (Figure 5-19). Everything in Code Chunk 5-24 should seem familiar. The only new feature is the use of a new layer, *geom_mosaic()*. The aesthetics are a little different in that they ask for *x = product()* where

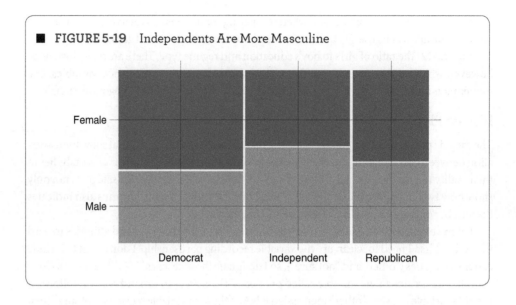

■ FIGURE 5-19 Independents Are More Masculine

Code Chunk 5-24
```
ggplot(data = subset(nes, pid3.new!="NA")) +
  geom_mosaic(aes(x = product(gender, pid3.new),
                                  fill=gender,
                                  na.rm=TRUE)) +
  xlab("") +
  ylab("") +
  ggtitle("Figure 5-19: Independents Are More Masculine") +
  theme_minimal() +
  scale_fill_brewer(palette="Blues") +
  theme(legend.position = "none") +
  theme(plot.title = element_text(size = 8, face = "bold")) +
  theme(axis.title = element_text(size = 8, face = "bold"))
```

you specify the x and y variables. Note that I don't include labels for the x and y axes since the categories are self-explanatory.

Suppose we hypothesize that women are generally more conservative than men. The mosaic plot in Figure 5-19 tests that hypothesis. The mosaic plot indicates males make up a slightly larger percentage of Independents than either Republicans or Democrats. So while gender might be related to party identification, we would conclude that women are not systematically more conservative than men.

We could conclude, however, that women are slightly more decisive in declaring their party allegiance. Note how Figure 5-19 changes the question: we may have originally been interested in gender and conservatism, but now our focus turns to whether women are more likely to establish a partisan identity. Perhaps women more readily relate to political parties. Or perhaps women are simply more ideologically extreme. There are now additional questions worth exploring.

The width of the vertical columns is also worth noting. Figure 5-19 indicates that the largest percentage of the population identifies with the Democratic party, then Independents, followed by Republicans.

Code Chunk 5-25

```
CrossTable(nes$gender, nes$pid3.new,
           main="Cross-Tabulation of Gender and Party ID",
           prop.chisq=FALSE)
```

In Code Chunk 5-25, I provide the code that generates the same information in tabular form (with the more common cross-tab). Observe the ease with which information is conveyed in the mosaic plot compared to the cross-tab in Table 5-3. While the cross-tab provides more exacting information (the actual numbers associated with each category), the relationship between the two categorical variables is more evident in the mosaic plot.

TABLE 5-3 Cross-Tabulation of Gender and Party ID

```
   Cell Contents
|-------------------------|
|                       N |
|           N / Row Total |
|           N / Col Total |
|         N / Table Total |
|-------------------------|

=================================================================
             nes$pid3.new
nes$gender    Democrat   Independent   Republican    Total
-----------------------------------------------------------------
Male               187           208          128      523
                 0.358         0.398        0.245    0.475
                 0.416         0.555        0.464
                 0.170         0.189        0.116
-----------------------------------------------------------------
Female             262           167          148      577
                 0.454         0.289        0.256    0.525
                 0.584         0.445        0.536
                 0.238         0.152        0.135
-----------------------------------------------------------------
Total              449           375          276     1100
                 0.408         0.341        0.251
=================================================================
```

Cross-Tab

Although the **cross-tab** is the bread and butter of descriptive statistics when examining categorical variables, it can be hard to discern patterns from rows and columns of numbers. More than a few otherwise perfectly intelligent students have no doubt decided to abandon the social sciences and never take a math class again after having been exposed to cross-tabs (this author contemplated it). Just to make things clear, the cross-tab represents the numbers that generated the mosaic plot; they are two different representations of the same information.

When examining the cross-tab, note that the table gives both the count and the column percentage. For example, in Table 5-3 there were 577 females in the sample (the count). The number below that (.525) records the percent of females in the sample (52.5%). The percentages should be added *down* the column to sum to 100%.

Another example illustrates the more intuitive nature of the mosaic plot when compared to the cross-tab. Let's ask if there is a relationship between party identification and whether the respondent worries about terrorism. First, in Code Chunk 5-26 I eliminate one of the categories from the *terror_worry* variable by creating a new variable *var2* that drops the level "not asked."

Code Chunk 5-26
```
nes$var2 <- droplevels(nes$terror_worry, "Not asked")
```

Now we're ready to create a mosaic plot. In Code Chunk 5-27 I subset the data by telling R not to include cases in which either variable includes missing values (NAs).

Code Chunk 5-27
```
ggplot(data = subset(nes, pid3.new!="NA" & var2 !="NA")) +
    geom_mosaic(aes(x = product(var2, pid3.new),
                                    fill=var2,
                                    na.rm=TRUE)) +
    xlab("") +
    ylab("") +
    ggtitle("Figure 5-20: Democrats Are Less Worried About Terrorism") +
    theme_minimal() +
    scale_fill_brewer(palette="Blues") +
    theme(legend.position = "none") +
    theme(plot.title = element_text(size = 8, face = "bold")) +
    theme(axis.title = element_text(size = 8, face = "bold"))
```

Recall that NES survey respondents were asked how much they worried about terrorism. They could choose responses that ranged from "not at all worried" to "extremely worried." Are Democrats, Independents, or Republicans more worried about terrorism? Since we're comparing two categorical variables, the mosaic plot fits our needs (Figure 5-20).

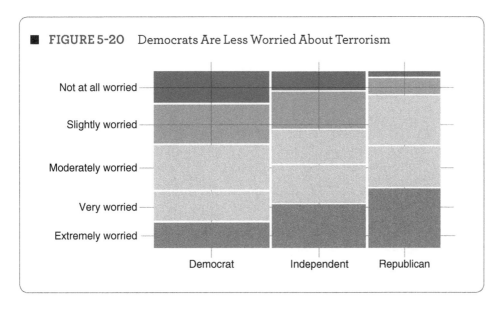

FIGURE 5-20 Democrats Are Less Worried About Terrorism

An interesting pattern emerges in Figure 5-20. As one moves across the ideological spectrum from liberal to conservative, the level of concern rises systematically. More accurately, Democrats are the least worried about a terrorist attack, followed by Independents and then Republicans. The mosaic plot throws this into stark relief. It also quickly shows by the thickness of the columns the relative size (their numbers) of the Democrats, Independents, and Republicans as a share of those who participated in the survey. That Republicans are more worried about the next terrorist attack is a fascinating result. While I won't explore it further, the result does beg the question, why?

Now consider the same information presented in a cross-tab (Table 5-4).

Code Chunk 5-28

```
CrossTable(nes$var2, nes$pid3.new,
           main="Cross-Tabulation of Fear of Terrorism and
           Party ID",
           prop.chisq=FALSE)
```

Although the cross-tab provides the exact numbers for each category, the overall message is more difficult to discern.

Pictures, as opposed to numbers in tabular form, are a more effective way to discover patterns or to make your point. If we only needed to convey a relatively small amount of information, a mosaic plot might be overkill. Pretty shades of blue or green that use up half the page to represent two or three numbers is too much. Avoid using graphs as an excuse to generate a gallery of vivid colors and interesting shapes. Instead, use those colors and shapes to effectively present patterns and relationships.

Bubble Plot

R offers many ways to examine data, which is one of its greatest strengths. The **bubble plot** starts with a scatter plot and adjusts the size of the dots according to a third variable, adding a

TABLE 5-4 Cross-Tabulation of Fear of Terrorism and Party ID

```
   Cell Contents
|-------------------------|
|                       N |
|             N / Row Total |
|             N / Col Total |
|           N / Table Total |
|-------------------------|
```

	nes$pid3.new			
nes$var2	Democrat	Independent	Republican	Total
Extremely worried	43	61	64	168
	0.256	0.363	0.381	0.233
	0.145	0.254	0.346	
	0.060	0.085	0.089	
Very worried	51	53	44	148
	0.345	0.358	0.297	0.205
	0.172	0.221	0.238	
	0.071	0.074	0.061	
Moderately worried	78	47	54	179
	0.436	0.263	0.302	0.248
	0.264	0.196	0.292	
	0.108	0.065	0.075	
Slightly worried	68	52	17	137
	0.496	0.380	0.124	0.190
	0.230	0.217	0.092	
	0.094	0.072	0.024	
Not at all worried	56	27	6	89
	0.629	0.303	0.067	0.123
	0.189	0.112	0.032	
	0.078	0.037	0.008	
Total	296	240	185	721
	0.411	0.333	0.257	

third dimension to a two-dimensional representation of the data. Varying the size of the dots represents just one of many different ways to add a third dimension to a scatter plot. Use shapes or colors to add additional dimensions.

Code Chunk 5-29 draws a bubble plot by specifying *size=evangel* in the aesthetics command for *geom_point()*. Note also that I use the *ifelse()* command within the *geom_text()* command to label seven different states. Finally, I add the *scale_size()* function to specify the size of the dots.

Code Chunk 5-29

```r
ggplot(states, aes(hsdiploma, femleg)) +
  geom_point(aes(size=evangel), col="#0000bf") +
  theme_minimal() +
  ggtitle("Figure 5-21: Women Legislators, Education,
                      and Evangelism") +
  ylab("Percentage of Legislature That Are Women") +
  xlab("Percentage of the State's Population With High School Diploma") +
  scale_size(range = c(.1, 12), name="Evangelism") +
  geom_text(size = 2.8,
      aes(label=ifelse(st=="CA" |
                       st=="LA" |
                       st=="UT" |
                       st=="SC" |
                       st=="NV" |
                       st=="PA" |
                       st=="MD", as.character(st),''),
                       hjust = 2.1, vjust=-1),
                       show.legend=FALSE) +
  theme(plot.title = element_text(size = 8, face = "bold")) +
  theme(axis.title = element_text(size = 8, face = "bold"))
```

In the bubble plot in Figure 5-21, I graphed the percentage of state legislators who are women against the percentage of the population with a high school degree. The bubbles are sized by the percentage of a state's population that self-identifies as evangelicals. What do we learn from adding the third dimension to our graph?

First, Utah stands out among the states with relatively high levels of high school graduates. The large bubble for Utah indicates it has a relatively high level of evangelism. (One might question whether Mormons should be classified as evangelical.) Second, the cases with low levels of education and low percentages of women in the legislature also have relatively large bubbles, indicating sizable evangelical populations. Evangelism seems to be related to both education and to attitudes toward women in politics. Another case to investigate might be South Carolina, which has a relatively low level of female participation in the legislature. Why do the Western states of California, New Mexico, and Nevada all have relatively high levels of female participation? Maryland and Pennsylvania have similar levels of education and similar levels of evangelism, but there is a huge difference in the share of seats held by women. Why?

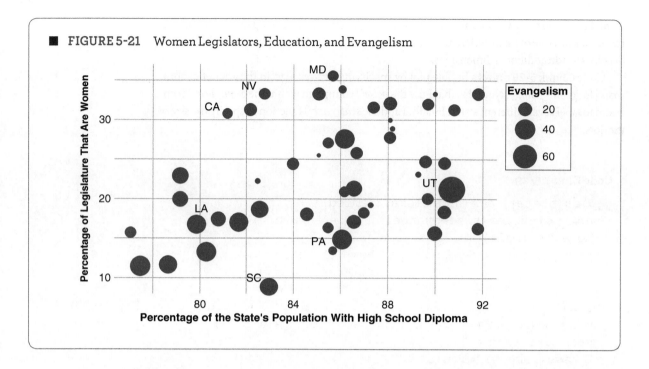

FIGURE 5-21 Women Legislators, Education, and Evangelism

Another example demonstrates how varying the size of the plot points reveals interesting patterns. Figure 5-22 is a standard scatter plot of infant mortality (deaths per 1,000 live births) and democracy (the polity index of democracy). While there may be a negative relationship between democracy and infant mortality, some authoritarian countries with very low infant mortality rates complicate the story. It seems some authoritarian countries perform just as well as their democratic counterparts.

Code Chunk 5-30
```
ggplot(world, aes(polity2, inf)) +
  geom_point(col="#bf0000") +
  geom_smooth(method="lm", se = FALSE) +
  theme_minimal() +
  ggtitle("Figure 5-22: Democracy's Weak Effect on Infant Mortality") +
  ylab("Deaths per 1,000 Live Births") +
  xlab("Polity Score") +
  theme(plot.title = element_text(size = 8, face = "bold")) +
  theme(axis.title = element_text(size = 8, face = "bold"))
```

A possible third dimension to explore is GDP per capita. A country's level of wealth surely has an influence on health outcomes. Let's turn the scatter plot in Figure 5-22 into a bubble plot that sizes the data points by GDP per capita (see Figure 5-23).

Similar to the previous example, all the work is done in the *geom_point()* layer. In the aesthetics we specify *size = gdppc*, which tells R to size the dots by the value of *gdppc* (Code Chunk 5-31).

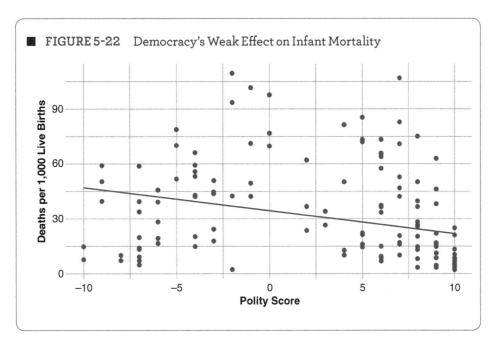

FIGURE 5-22 Democracy's Weak Effect on Infant Mortality

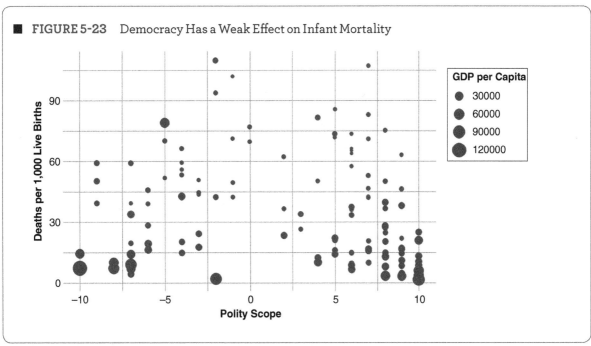

FIGURE 5-23 Democracy Has a Weak Effect on Infant Mortality

Code Chunk 5-31

```
ggplot(world, aes(polity2, inf)) +
    geom_point(aes(size = gdppc), col = "#bf0000") +
    ggtitle("Figure 5-23: Democracy Has a Weak Effect
        on Infant Mortality") +
```

```
ylab("Deaths per 1,000 Live Births") +
xlab("Polity Score") +
scale_size(name="GDP per Capita") +
theme_minimal() +
theme(plot.title = element_text(size = 8, face = "bold")) +
theme(axis.title = element_text(size = 8, face = "bold"))
```

As the bubble plot makes clear, GDP per capita is an important factor to consider in our examination of democracy and infant mortality. There are clearly some wealthy authoritarian regimes that have low infant mortality rates. Of course, the next question is who are these wealthy, authoritarian countries with relatively low infant mortality rates? Figure 5-24 labels a few of the cases we might be interested in.

Code Chunk 5-32
```
ggplot(world, aes(polity2, inf)) +
  geom_point(aes(size = gdppc), col="#bf0000") +
  theme_minimal() +
  geom_text(size = 2.8,
      aes(label=ifelse(iso3c=="QAT" |
                       iso3c=="SAU" |
                       iso3c=="SGP",
                       as.character(iso3c),''),
                       hjust = -.5, vjust=-2),
                       show.legend=FALSE) +
  ggtitle("Figure 5-24: Authoritarian Regimes With
          Low Infant Mortality Rates") +
  ylab("Deaths per 1,000 Live Births") +
  xlab("Polity Score") +
  theme(plot.title = element_text(size = 8, face = "bold")) +
  theme(axis.title = element_text(size = 8, face = "bold"))
```

The three cases identified are Saudi Arabia (SAU), Qatar (QAT), and Singapore (SGP), all of which are countries unlike most others (either for their oil wealth or city-state status). Identifying the actual cases not only indicates wealth is important, but it also explains why we have some authoritarian regimes that are also extremely wealthy.

To summarize, we have a number of different tools to examine our data. The tools we use depend largely on the kind of variables we want to examine. The number of cases or categories is also an important consideration. Figures are preferred over tables when there is a large amount of information to convey. For example, patterns are difficult to discover when looking

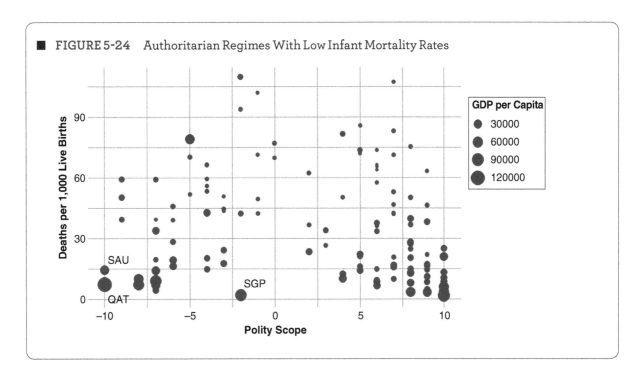

FIGURE 5-24 Authoritarian Regimes With Low Infant Mortality Rates

at the rows and columns of numbers in a cross-tab. The mosaic plot is a more effective means of detecting patterns in the same data. We also learned how to add a third dimension to a scatter plot, giving us additional information.

> **KNOWLEDGE CHECK: Create and analyze bivariate views of the data.**

11. Create and analyze the relationship between the amount of money the state spends on each student (*stuspend*) and homicides (*murderrate*).
    a. What happens to the homicide rate when spending increases?
    b. There are three outliers in the plot. Who are they?
    c. What makes Utah such an interesting case?
    d. What variable should be on the x and y axes of this plot?
12. What is the relationship between GDP per capita and regime type?
    a. Which regime type has the highest median value?
    b. Which regime type has the most spread?
    c. Label the points to identify interesting cases. Which mixed democracies have relatively high incomes?
    d. Which military dictatorship has the highest income?

13. Draw a plot of the relationship between marital status and partisanship and indicate which statements are true.
    a. The majority of respondents are married or single.
    b. The majority of respondents are married.
    c. Republicans favor being single.
    d. Democrats favor marriage.
14. Add income to a scatter plot of *stuspend* and *murderrate*.
    a. Is income related to how much states spend on education?
    b. Is income related to the homicide rate?
    c. What makes Maryland such an interesting case?
    d. Which state in the union has the lowest income?

SUMMARY

Describing data is a powerful exercise. As Minard's antiwar poster demonstrates (Chapter 3), elegant and informative pictures of data make bold and important statements. In this chapter we learned some basics of visualizing data. First, we learned how to describe single variables in different forms (continuous and categorical). We then learned how to describe the relationships between continuous and categorical variables. Different colors, sizes, and shapes provide important information that helps us test hypotheses, make discoveries, and generate new questions.

While visualizing data can be a powerful exercise, the combination of words and numbers can be powerful too. Consider an example from Nobel Laureate Economist Robert Lucas. In an article entitled "On the Mechanics of Economic Development," Lucas describes the differences in economic growth among various countries (Lucas, 1988). Specifically, he notes that over the span of two generations the average income in India will increase by a factor of 2, while South Korea's will increase by a factor of 32! Simply by describing (in a simple sentence) the huge differences that exist between countries, he motivated many to think about the problem. After revealing the differences between India and Korea, Lucas (1988) observed, "Once one starts to think about them, it is hard to think about anything else" (p. 5). Lucas's simple description and statement have served as an orienting theme in the economic growth literature ever since.

Pictures provide important information about the characteristics of the data, how different variables relate to one another, and whether there are some errors lurking in the dark. Good descriptions also generate additional and hopefully more informed questions. It bears repeating that the more you understand your data, what they look like, and how they relate to each other, the better questions asked and the more discoveries made.

COMMON PROBLEMS

- *Categorical versus continuous variables.* One cannot overemphasize the importance of understanding the distinction between categorical and continuous measures. Although it's important to understand the difference before moving on, I find that most will figure it out once confronted with the inability to produce pictures of the data. It's a nice example of learning by doing.

- *When to label, what to label.* R provides the ability to label all points in a scatter plot. It also provides the means to label selected points. When exploring the data, it can be useful to label all of the points so long as they don't obscure important patterns. Once you've explored the data and you want to present your findings, label only the outlying cases or the ones that help make your argument.

- *Dependent versus independent variables.* Probably because of the alliteration or similarity between the words dependent and independent, it's sometimes hard to keep the two straight. The way the two are sometimes defined also makes keeping them straight difficult (see the "Art and Practice of Data Visualization" feature on dependent and independent variables earlier in this chapter). Understanding the difference between dependent and independent variables is extremely important because it provides an important vocabulary we use to communicate when analyzing data.

REVIEW QUESTIONS

1. What is the difference between dependent and independent variables?

2. When will a variable be positively or negatively skewed? Can you think of any likely examples?

3. What views of the data are helpful when our dependent variable is continuous or categorical?

4. Why is a bubble plot useful?

5. What variable should always go on the y axis? Why?

6. In what situation are mosaic plots helpful? What are their advantages over cross-tabs?

7. What features of the data do boxplots provide?

8. What advantage do mosaic plots hold over cross-tabs?

9. What advantage does a bubble plot hold over a scatter plot?

10. Why use a cross-tab instead of a mosaic plot?

PRACTICE ON ANALYSIS AND VISUALIZATION

1. Which of the following are questions that univariate views of the data answer?

    a. What is the most, least, or most common?

    b. What cases are different?

    c. What is the shape of the data?

    d. What is the relationship between x and y?

2. Draw a frequency table of *nes$race* and answer the following questions.

    a. What is the mode?

    b. What category has the least number of cases?

    c. How many different races are recorded in the variable?

    d. How many respondents are white?

3. Draw a boxplot of the variable *world$womleg* and answer the following questions.

    a. What is the median? (approximate)

    b. What is the lower quartile? (approximate)

    c. What is the upper quartile? (approximate)

    d. Does the variable approximate a normal distribution?

4. Draw a histogram of the variable *world$urban* and answer the following questions.

    a. Does the shape of the variable approximate a normal distribution?

    b. What is the minimum value? (approximate)

    c. What is the maximum value? (approximate)

    d. What is the mean? (approximate)

5. Indicate which is the dependent variable in the following questions.
   a. What impact does democracy have on economic growth?
   b. Does income influence happiness?
   c. Are infant mortality rates larger or smaller in African countries relative to the world's other regions?
   d. Are women politicians more likely to enact reforms that benefit education?

6. Indicate which is the independent variable in the following statements.
   a. Good quarterback play is crucial to winning games.
   b. Attendance is always low on Friday mornings.
   c. I never do well in classes that meet at 8:00 a.m.
   d. Her performance on the exam was a testament to her dedication.

7. What does Pearson's R tell us?
   a. Is a variable normally distributed?
   b. Is a variable skewed?
   c. What is the relationship between two variables?
   d. Are there any outliers in a variable?

8. When are boxplots best used when examining the relationship between two variables?
   a. When both variables are continuous
   b. When both variables are categorical
   c. When one variable is continuous and one is categorical
   d. When both variables are normally distributed

9. Plot the relationship between *nes$amer_ident* and *nes$terror_worry*.
   a. Are the two variables related?
   b. What is the modal category for *nes$amer_ident*?
   c. What is the modal category for *nes$terror_worry*?
   d. What category of *nes$amer_ident* seems most worried about terrorism?

10. Which of the following does a bubble plot help accomplish?
    a. If points are clustered together, it does a good job of separating them.
    b. It adds a third dimension to a histogram.
    c. It adds a third dimension to a scatter plot.
    d. It makes identifying outliers easier.

ANNOTATED R FUNCTIONS

The following functions appear in this chapter. They are listed in order of their first appearance (with the code chunk number in parentheses) and annotated here to give a very brief description of their use. Some are not stand-alone functions and only work in combination with other commands. As a reminder, the code in every chapter will work properly if executed in the order it appears. Proper execution also depends on typing the author-defined *libraries()* command, which loads the required R packages.

ggplot(): defines the basic structure of a plot (usually the x and y variables). (5-3)

aes(): the *aes* (called "aesthetics") function is used in ggplot to define the basic structure of the plot, which often includes the variables you want to use and any shapes or colors. (5-3)

theme_minimal(): specifies a minimalist style for ggplot. (5-3)

xlab(): labels the x axis in ggplot. (5-3)

ylab(): labels the y axis in ggplot. (5-3)

theme(): specifies font, size, and so forth in a ggplot. (5-3)

geom_boxplot: draws a boxplot in ggplot. (5-5)

position_jitter(): used in the *aes* feature in ggplot to "jitter" points to be more easily identified if they are clustered. (5-6)

geom_histogram: draws a histogram in ggplot. (5-7)

grid.arrange(): displays two or more different plots together. (5-9)

options(scipen=999): used to indicate to not use scientific notation. (5-10)

stem(): used to draw a stem-and-leaf plot. (5-11)

ifelse(): logical function that allows you to construct an if-then statement. Useful for selecting specific cases to label and also for creating categorical variables from continuous variables. (5-13)

annotate(): places text in the figure. (5-16)

cor(): calculates the Pearson's R correlation. (5-17)

subset(): chooses a specified subset of the data. (5-20)

as.factor(): changes the classification of a variable to being a factor. (5-23)

levels(): lists the levels associated with a categorical variable. (5-23)

geom_mosaic(): draws a mosaic plot using ggplot. (5-24)

scale_filler_brewer(): fills colors using a shading scheme when using ggplot. (5-24)

CrossTable(): creates a cross-tab. (5-25)

droplevels(): drops a level from a categorical variable. (5-26)

scale_size(): function that allows you to adjust the scale of objects in a scatter plot in the 'ggplot' package. (5-29)

ANSWERS

KNOWLEDGE CHECK

1. a, b, c, d
2. dispersion (a), central tendency (b), dispersion (c), dispersion (d)
3. categorical (a), continuous (b), categorical (c), continuous (d)
4. parliamentary democracy (a), royal dictatorship (b), 35 (c), parliamentary democracy (d)
5. yes (a), extremely worried (b), extremely worried (c), no (d)
6. 8 (a), yes (b), no (c), 0 (d)
7. MN (a); HI (b); NE, MS, ID, VT, or LA (c); WI and WV (d)
8. c, d
9. a, d
10. scatter plot (a), mosaic plot (b), boxplot (c), mosaic plot (d)
11. decreases (a); LA, MD, NY (b); low spending and low murder rate (c); y = murder rate and x = student spending (d)
12. royal dictatorship (a); royal dictatorship (b); SVK, PRT, FRA, ISL, FIN, AUT, IRL (c); SGP (d)
13. a, b
14. yes, (a) yes (b), high income and high homicide rate (c), MS (d)

PRACTICE ON ANALYSIS AND VISUALIZATION

1. a, b, c
2. white (a), Middle Eastern (b), 8 (c), 863 (d)
3. 17 (a), 60 (b), 24 (c), no (skewed right) (d)
4. yes (a), 15 (b), 100 (c), 60 (d)

5. growth (a), happiness (b), infant mortality (c), education reform (d)

6. quarterback play (a), day/time (b), time (c), dedication (d)

7. c

8. c

9. yes (a), extremely important (b), extremely worried (c), extremely worried (d)

10. c

 Access digital resources, including datasets, at http://edge.sagepub.com/brownstats1e.

6 Transforming Data

CHAPTER OUTLINE

Learning Objectives
Overview
Theoretical Reasons for Transforming Data
Transforming Data for Practical Reasons
Transforming Data—Continuous to Categorical Variables
Transforming Data—Changing Categories
Box-Cox Transformations
Summary
Common Problems
Review Questions
Practice on Analysis and Visualization
Annotated R Functions
Answers

LEARNING OBJECTIVES

- Connect hypotheses to the data.
- Create views that uncover patterns by transforming data.
- Explain why and how to create categories from continuous data.
- Create categorical variables to connect hypotheses to the data.
- Describe the Box-Cox ladder of transformation.

Overview

Data rarely come in the form we want. They are almost always created by people with different agendas, goals, and hypotheses. Since our hypotheses require appropriate measures, **transforming data** is often necessary. This chapter illustrates the relationship between hypotheses and the data we use to test them. We transform data so that they more closely fit our hypotheses. Data also take forms difficult to analyze, obscuring important patterns and relationships. Consequently, there are practical reasons to transform data. The theoretical and practical reasons for transforming data are presented here.

The relationship between hypotheses and data goes both ways. Sometimes we generate hypotheses given the data we have. Sometimes we transform the data given the hypotheses we formulate. If either is done poorly, the conclusions we draw from the exercise are suspect. For example, if we hypothesize that increases in income generate movement toward democracy, using a categorical variable that classifies people as rich or poor and government as democratic or authoritarian is inappropriate. Let's build an intuition that helps match our hypotheses to our measures.

In terms of *how* we transform data, we'll focus on the most common and effective methods. You will learn the mechanics of transforming categorical and continuous variables. I will introduce the logarithmic transformation, one of the most useful in our toolbox. Finally, the Box-Cox ladder of transformations will be explained in order to provide a conceptual framework for transforming data.

Theoretical Reasons for Transforming Data

Both theory and practice motivate transforming data. Transform data when there is a mismatch between theory and data. In their raw form, the data may simply prevent us from directly answering or testing our hypothesis. From a practical standpoint, the data may come in a form too difficult to examine visually. Put simply, we transform data so that we can properly test our hypotheses and visualize the data more effectively. Let's start by describing the theoretical reasons for transforming data.

Transform the Data to Fit the Theory

Suppose we're interested in the relationship between age and voting behavior. Do voters in their twenties (millennials) hold different political attitudes than retired voters (baby boomers)? The *nes$birthyr* variable records the year a respondent was born. Since the variable is a continuous measure, it is difficult to discern whether millennials behave differently than the boomers. We would be better served with a categorical variable indicating whether a respondent is in their twenties, thirties, forties, fifties, or sixties. Another example will help build your intuition.

An important question economists, psychologists, and philosophers ask is "Does money lead to happiness?" Perhaps with each additional dollar, our happiness grows. Recent findings suggest there is a threshold (Binswanger, 2006). Specifically, once we've reached a certain income, each additional dollar does not buy more happiness. Much like the example earlier concerning age, it might be difficult to spot any patterns in the data if we simply treat income as a continuous variable. Dividing the population into those who make less and those who make more than a certain level of income might be a better strategy.

When our hypothesis involves a natural barrier or threshold, ask if the measure should be continuous. What at first glance seems to generate inconclusive results, at second glance (with a transformation) unearths an important empirical pattern.

Sometimes there's no natural barrier or threshold, but the relationship is nonlinear. Suppose we're interested in understanding the relationship between income and the probability of voting. While a $50,000 increase in the average voter's income might have a substantial impact on voting, a similar increase in Bill Gates's or Warren Buffet's income (both are billionaires) will not. Consequently, income and voting may be related, but not in a linear fashion.

The same logic applies to countries. Consider the following question often asked in political science and economics: what is the relationship between per capita income and democracy? In their raw form, GDP per capita figures range from $300 to $500 (the poorest countries in Africa) to $30,000 to $40,000 (the richest Organisation for Economic Co-operation and Development member countries). Most countries are on the low end. If moving from $300 to $1,000 (a $700 increase) is more important than $30,000 to $30,700 (a $700 increase as well), we might want to try a log transformation so that a $700 increase at the low end means more or is given more weight in the analysis. Using the log transformation not only maps onto a specific theory about the importance of increasing income in rich and poor countries, but it also generates more informative views of the data. Let's consider one more example.

Do levels of violence (homicides) influence spending on the police and public safety? Similar to the earlier examples, an increase in the homicide rate could have a very different effect on policy, attitudes, or election outcomes at different levels. One can imagine that increasing the number of homicides by five in a town that averages one homicide a year is very different than an increase of five in a town that experiences 100. Here again we may be tempted to use a log transformation since an increase at the low end of the distribution should not be treated the same as a similar increase at the high end.

Transform Both Data and Question to Match Each Other

As the earlier examples demonstrate, sometimes we need to change the data so that they map cleanly onto the hypothesis. Sometimes we need to change our hypothesis so that it maps cleanly onto the data. Suppose we want to know if being arrested influences attitudes toward the judicial system (defense attorneys commonly ask potential jurors if they've been arrested). If our survey records how many times respondents have been arrested, we are likely to encounter many individuals with no police records and only a very few with multiple arrests. We may want to change the question and the data: hypothesize that the single event of an arrest (even just once) is important and transform the variable so that it indicates whether someone has been arrested. Another example illustrates how both the hypothesis and data can change in order to gain additional insight.

Does having more education increase job satisfaction? While we could count up the years one has matriculated, perhaps obtaining an MBA or a master's degree in engineering (roughly the same amount of years) produces very different outcomes. For example, one could have either of those two degrees, work in the same industry, yet experience very different career trajectories. Rather than recording how many years an employee went to school, the better question asks what kind of degree helps one rise through the ranks. Here again we've changed both the question and the data.

Is marijuana a gateway drug to heroin use? Does having tried marijuana make taking heroin more likely? The answer holds important implications for drug policy. If we hypothesize that trying marijuana once is sufficient, on a survey we would ask, "Have you ever tried marijuana?" If our hypothesis instead posits that frequent users of marijuana are more likely to experiment with heroin, we'd ask, "How frequently do you use marijuana?" Note that we could collect the same data using the frequency question and then transform the data so they are made of two categories: has the respondent tried marijuana or not? The distinction between the two different hypotheses is important for crafting drug policy. Notice how both the question and the data change as we think about the problem.

To summarize, an important part of data analysis tightly connects the data to our question. Sometimes we transform the data, sometimes we change the hypothesis, and sometimes we do both. In addition to transforming the question, the data, or both for theoretical reasons, we may want to transform our data purely for practical reasons. Theoretical concerns involve how we think one variable impacts another, whereas practical concerns involve generating more informative views of the data. Let's explore the practical reasons for transforming data in the next section.

> **KNOWLEDGE CHECK: Connect hypotheses to the data.**

1. Why do we transform data?
   a. So the data fit with our hypothesis
   b. To help more effectively visualize the data
   c. To help test our hypotheses
   d. To achieve better results
2. The data to fit the theory involved which of the following in the examples?
   a. We changed counts or amounts to categories.
   b. There are anticipated threshold effects in the relationship under examination.
   c. There are anticipated nonlinearities in the relationship under examination.
   d. We changed categories to counts and amounts.
3. Which of the following pertained to transforming the data to fit the theory?
   a. The available data were categorical.
   b. The available data were continuous.
   c. Collecting different data was necessary.
   d. The hypothesized relationship was nonlinear.
4. Which of the following represents a situation when you can change your hypothesis and the data without collecting more data?
   a. The distribution of continuous data conforms to categories.
   b. The hypothesized relationship is qualitative and the data are continuous.
   c. The hypothesized relationship is continuous and the data are categorical.
   d. The hypothesized relationship is nonlinear and the data are continuous.

Transforming Data for Practical Reasons

Although it doesn't have to be an either/or proposition, sometimes we transform data for theoretical reasons and sometimes we transform data for practical reasons. Neither is better than the other. When possible, however, it's best to justify any transformation using both. Let's continue with describing the practical reasons for transforming data.

Perhaps we are interested in the relationship between the size of a state's population and attitudes toward immigrants. Because there are so many states with relatively small populations (e.g., North Dakota, South Dakota, Montana, etc.), it might be hard for us to visually examine relationships between population size and views toward immigration. Consequently, it might be helpful to transform the data.

Population, income, and homicide rates feature distributions with many observations at the low end with a few outliers at the high end, requiring that we transform the variable. Another variable with similar characteristics records the amount of CO_2 emissions (in kilograms per $2,000 USD of GDP) in a country, state, or city. While we may not have a hypothesis that recommends there is a nonlinear relationship, the distribution of the variable makes data analysis difficult. Scatter plots would choke on the CO_2 emissions variable in its raw form.

Consider the scatter plot in Figure 6-1. I've compounded the difficulty of analyzing CO_2 emissions by plotting them against an unlogged form of GDP per capita. Both variables are in their raw form and are characterized by distributions with very large outliers. Rendered in this form, it is extremely difficult to determine whether there is any relationship.

There are a couple of things to note in Code Chunk 6-1. First, the basic structure of the *ggplot()* command should look familiar. In this case, since I want to draw a scatter plot, I use the *geom_point()* command. I also employ an *ifelse()* command as an easy way to label all points above a GDP per capita of $50,000 and a CO_2 value above 2 kg per $2,000 USD of GDP. As in the previous chapter, I indicate no label for the rest of the points by using two single quotes (' ').

Now consider the plot in Figure 6-2, the same plot with both variables transformed: CO_2 emissions and GDP per capita are now expressed in their logged form (a complete treatment of using the log transformation comes later in this chapter). Notice how the outlying countries are now closer to the rest of the data. Also note that the data are no longer clustered at the low end of either scale. With the transformation, it is possible to see there is a positive relationship between GDP per capita and CO_2 emissions in the lower half of the data. Once a certain level of GDP per capita is reached (around 9 in logged form; $8,095), as countries become richer, their CO_2 emissions begin to decline. They start to decline rapidly at the high end of GDP per capita, with Norway (NOR), Switzerland (CHE), and Singapore (SGP) leading the pack. In fact, without Kuwait (KWT), United Arab Emirates (ARE), Brunei (BRN), and Qatar (QAT), the improvement (drop) in CO_2 emissions would be even more

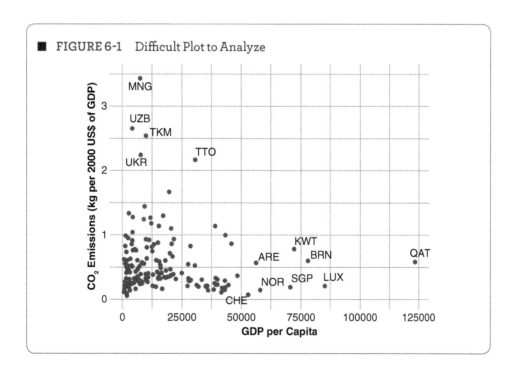

FIGURE 6-1 Difficult Plot to Analyze

Code Chunk 6-1

```
ggplot(world, aes(gdppc, co2)) +
  geom_point(col="#bf0000") +
  geom_text_repel(size = 2.8,
      aes(label=ifelse(gdppc > 50000 | co2 > 2,
                        as.character(iso3c),''),
                        hjust = 0, vjust=-1),
      show.legend=FALSE) +
theme_minimal() +
theme(plot.title = element_text(size = 8, face = "bold")) +
theme(axis.title = element_text(size = 8, face = "bold")) +
ggtitle("Figure 6-1: Difficult Plot to Analyze") +
ylab("CO2 Emissions (kg per 2000 US$ of GDP)") +
xlab("GDP per Capita")
```

pronounced as GDP increases. To illustrate, I drew a second smooth (green) for the sample without those cases.

To draw the two curves fitting logged values of GDP per capita and CO_2 emissions, I can simply specify in the aesthetics function that the two variables should be logged: *log()*. In one of the curves, I indicate that the curve should not (!=) consider Kuwait (KWT), Qatar (QAT), Brunei (BRN), or United Arab Emirates (ARE).

To summarize, there are both theoretical and analytical reasons to transform data. Get in the habit of transforming data to fit your hypotheses to the data. Data come to us in forms

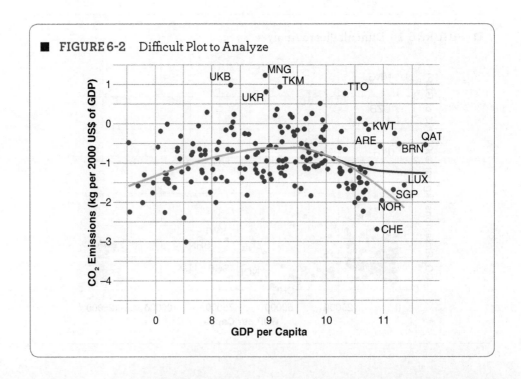

■ FIGURE 6-2   Difficult Plot to Analyze

Code Chunk 6-2

```
ggplot(world, aes(log(gdppc), log(co2))) +
  geom_point(col="#bf0000") +
  geom_smooth(se=FALSE) +
  geom_text_repel(size = 2.8,
      aes(label=ifelse(gdppc > 50000 | co2 > 2,
                       as.character(iso3c),''),
                       hjust = 1, vjust= -1),
                       show.legend=FALSE) +
theme_minimal() +
theme(plot.title = element_text(size = 8, face = "bold")) +
theme(axis.title = element_text(size = 8, face = "bold")) +
geom_smooth(data=subset(world,
                        iso3c != "KWT" &
                        iso3c != "QAT" &
                        iso3c != "BRN" &
                        iso3c != "ARE"),
                        color="green", se=FALSE) +
ggtitle("Figure 6-2: Difficult Plot to Analyze") +
ylab("CO2 Emissions (kg per $2,000 USD of GDP)") +
xlab("GDP per Capita")
```

that were useful to others. As I hope to have shown, the forms data take alter our hypotheses, our analyses, and the conclusions we draw from them. Good analysis does not take the data as given but recognizes that important information can be generated by transforming our questions, hypotheses, and the data. Now that we know *why* we transform data, let me explain *how*.

> **KNOWLEDGE CHECK: Create views that uncover patterns by transforming data.**

5. Which of the following are analytical reasons for transforming data?
   a. The relationship between two variables is nonlinear.
   b. There are extremely large values in the data.
   c. One of the variables in a relationship is skewed.
   d. There are extremely small values in the data.

6. Which of the following are practical reasons to transform data?
   a. There are clumps of data that need to be spread out.
   b. There are outlying cases that need to be brought closer to the rest of the data.
   c. The data are highly skewed.
   d. The data are normally distributed.

Transforming Data—Continuous to Categorical Variables

Certain situations call for taking a continuous variable and assigning each value to a category (e.g., low, high, medium). From a purely data-analytic perspective, note that when we categorize a continuous measure we are throwing data away. Consider the age variable and voting behavior. If we transform a continuous age variable to one that simply indicates whether a respondent is in the boomer generation (1) or not (0), we're throwing away information. With that transformation, all ages below 55 would be 0 and all ages above 55 would be 1. We can no longer distinguish, for example, between respondents in their twenties, thirties, or forties. Discarding data is never a good practice, but if our theory posits an important difference exists between millennials and baby boomers, there's a good theoretical reason to examine the data as a categorical variable (is the respondent older than 55 years or not?).

Art and Practice of Data Visualization
DEFINING NEW VARIABLES

One thing to remember is to never create a new categorical variable with the same name as the continuous variable. In other words, don't write over the old variable with a new one. A typical error made by many is to use the same variable name to create the categorical variable. Once the original continuous variable has been overwritten with categories, it is gone and the data have truly been thrown away!

To illustrate, let's consider age and its relationship with attitudes toward then-candidate Donald Trump. The survey used here was conducted during the 2016 campaign. As can be observed in the simple scatter plot of the Trump feeling thermometer and year of birth (Figure 6-3), it is hard to discern any pattern given the basic characteristics of the data.

Code Chunk 6-3

```
ggplot(nes, aes(birthyr, fttrump)) +
  geom_point(col="#bf0000") +
  ggtitle("Figure 6-3: Feelings Toward Trump and Year of Birth") +
  ylab("Trump Thermometer") +
  xlab("Year of Birth") +
  theme_minimal() +
  theme(plot.title = element_text(size = 8, face = "bold")) +
  theme(axis.title = element_text(size = 8, face = "bold"))
```

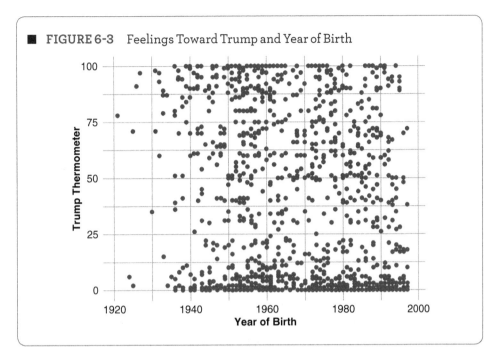

FIGURE 6-3 Feelings Toward Trump and Year of Birth

A more useful view is generated by dividing the age variable into categories. Toward that end, let's first convert the *birthyr* variable to the age of an individual by subtracting *nes$birthyr* by 2016 (the year the survey was conducted), then divide it into different categories (Code Chunk 6-4). The *cut()* command takes the variable *nes$age* and delineates five different categories.

Code Chunk 6-4

```
nes$age <- 2016 - nes$birthyr

nes$age5 = cut(nes$age, c(19,31,41,61,99,120))
```

Now that the continuous *birthyr* variable has been transformed to a categorical variable with the *cut()* command, let's plot the relationship using boxplots (Figure 6-4).

Code Chunk 6-5

```
ggplot(nes, aes(age5, fttrump)) +
  geom_boxplot(col="#0000bf") +
  theme_minimal() +
  theme(plot.title = element_text(size = 8, face = "bold")) +
  theme(axis.title = element_text(size = 8, face = "bold")) +
  ggtitle("Figure 6-4: Trump Thermometer by Age") +
  ylab("Trump Thermometer") +
  xlab("Age of Respondent")
```

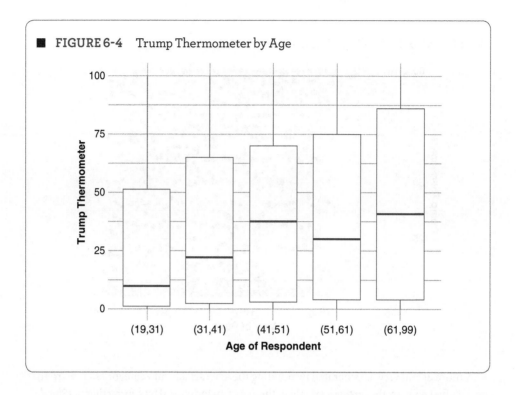

FIGURE 6-4 Trump Thermometer by Age

The resulting set of boxplots are more informative (Figure 6-4). We can see that the millennials (in this case, respondents 18–31 years old) typically viewed Trump negatively (a median thermometer score of around 12) relative to respondents between 61 and 99 years old (the median is roughly 37).

Another example, perhaps more controversial, is the democracy measure used by many scholars. The controversy stems from how we conceptualize democracy. Is it a distinct phenomenon or is it best thought of as a point on a scale? Those who argue for a continuous variable make the argument that democracy is a continuum upon which different regimes could be placed. Small differences may characterize different regimes on that continuum, making them more or less democratic. Suppose two regimes are exactly the same except one has more freedom of the press. Rather than calling one a dictatorship and the other a democracy, some scholars prefer to use a measure that delineates that small but important difference.

Others might argue that without a completely free press, there is no democracy. In a noted study of democracy and development, Przeworski and colleagues (2000) argue that democracy is like being pregnant—you either are a democracy or you are not. Others building on that argument claim that rather than two kinds of regime types (democracy or dictatorship), there is a third kind in the middle: mixed (Goldstone et al. 2010).

While the theoretical debate is extremely important, the practical consequences are significant as well. While categorizing a continuous variable to fit the hypotheses is important, it can come at an important cost. Let's see what happens to the analysis when we take a continuous measure of democracy (the *polity2* variable) and impose the categorical approach on it by specifying that cases scoring 5 and above are democratic and the rest are authoritarian. Note that in Code Chunk 6-6 we use the logged version of GDP per capita. Again, in the aesthetics command, we need only place *gdppc* within the *log()* function to log the variable.

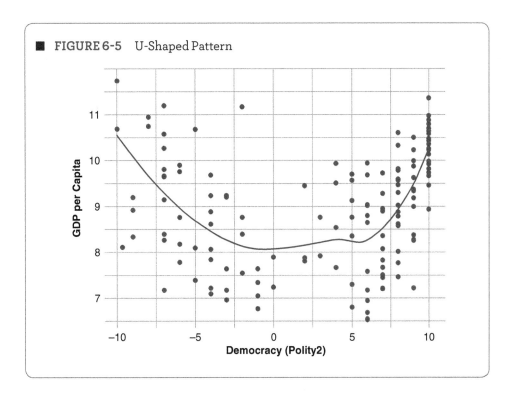

FIGURE 6-5 U-Shaped Pattern

Code Chunk 6-6

```
ggplot(world, aes(polity2, log(gdppc))) +
  geom_point(col="#bf0000") +
  geom_smooth(se=FALSE, col="#0000bf") +
  theme_minimal() +
  theme(plot.title = element_text(size = 8, face = "bold")) +
  theme(axis.title = element_text(size = 8, face = "bold")) +
  ggtitle("Figure 6-5: U-Shaped Pattern") +
  ylab("GDP per Capita") +
  xlab("Democracy (Polity2)")
```

To illustrate the consequences of categorizing a continuous measure, I plotted the *polity2* variable against GDP per capita (logged). There is a distinct U-shaped pattern that indicates very authoritarian countries have relatively high incomes. Then, as countries become more democratic, the associated levels of income drop. After democracy scores above 5 are reached, there is an abrupt increase in wealth as we move along the x axis. This pattern suggests one possible explanation: perhaps the kind of regime is less important than the severity of its form. Now observe what happens when we categorize the continuous measure.

To categorize the continuous measure of democracy, I transformed the variable to record 1 when the *polity2* score is above 5 and to record 0 for all cases below (Code Chunk 6-7).

Code Chunk 6-7

```
world$regimetype <- ifelse(world$polity2 > 5, 1, 0)

world$regimetype <- factor(world$regimetype,
  levels = c(0, 1),
  labels = c("Dictatorship", "Democracy"))
```

Now that we've created a categorical variable from a continuous measure, let's make boxplots of each regime type (Code Chunk 6-8).

Code Chunk 6-8

```
ggplot(subset(world, regimetype !="NA"), aes(regimetype,
log(gdppc))) +
  geom_boxplot(col="#0000bf") +
  geom_smooth(method="lm", se=FALSE) +
  theme_minimal() +
  theme(plot.title = element_text(size = 8, face = "bold")) +
  theme(axis.title = element_text(size = 8, face = "bold")) +
  ggtitle("Figure 6-6: Previous Patterns Go Unobserved") +
  ylab("GDP per Capita") +
  xlab("Democracy (Polity2)")
```

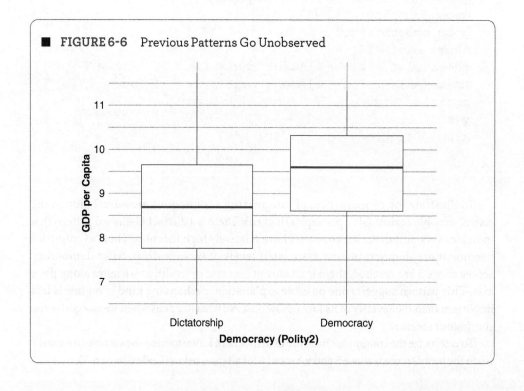

■ **FIGURE 6-6** Previous Patterns Go Unobserved

Noted earlier, information is thrown away when we categorize a continuous variable,. In this case, we're not making any distinctions between dictatorships scoring between −10 and 4 or democracies scoring between 6 and 10. Consequently, the steep decline in wealth when moving from the most authoritarian regimes to the least is not observed. Nor, for that matter, is the steep increase in wealth when we move along the x axis from 6 to 10. The conclusion we draw from the boxplots in Figure 6-6 is that democracies outperform their authoritarian counterparts. While it does answer the direct question "Does democracy help or hinder economic development?" some of the information we gained from Figure 6-5 is lost. Clearly, *how* we look at the data determines what we see.

> **KNOWLEDGE CHECK: Explain why and how to create categories from continuous data.**

7. Which are some considerations when transforming a continuous variable into a categorical variable?

    a. Data will be discarded, or thrown away.

    b. Patterns of the data will become hidden.

    c. Viewing the data with transformed data will change what you see.

    d. Changing what you see should be avoided since it leads to bias.

8. Which of the following are involved in transforming continuous data to categorical data?

    a. The *cut()* command

    b. The *geom_boxplot()* function

    c. Subsetting the data

    d. The *ifelse()* function

9. Which of the following statements best describe transforming data from continuous variables to categorical measures?

    a. Transforming data ties data more closely to the hypothesis.

    b. Transforming data provides an alternate view of the data.

    c. Transforming data focuses our attention on a specific part of the data.

    d. Concerns regarding discarding data always outweigh the gains.

Transforming Data—Changing Categories

In the previous examples, we transformed continuous variables into categorical variables. In some contexts, it might be worthwhile to reduce the number of categories. A political scientist or policymaker, for example, may be interested in the age citizens can vote, purchase alcohol, or retire. In these cases, we require two categories. Yet at the same time, additional categories might be helpful. Combining so many age groups into two categories might obscure important patterns. One doesn't always know how adding or subtracting categories will alter the view in advance. Theory can guide in this instance, but it's also important to recognize that our theory could change as we explore.

First let's consider an example with too many categories. Some data in the National Election Studies (NES) come with more than 20 categories (e.g., the income variable *nes$faminc*). The *faminc* variable records whether a respondent's income lies between $0 and $10,000, $10,000 and $20,000, and so forth. The variable contains over 15 categories. The resulting series of boxplots appears in Figure 6-7.

There is a little preparation involved to view these data properly (Code Chunk 6-9). I create two different variables for comparison. First, I turn the *faminc* variable into a factor by creating a new variable *nes$faminc.f*, which is used to draw boxplots of the raw data. I create another variable—a copy of the *nes$faminc* variable—so that I don't write over the original data. The new variable is called *nes$nfam*. Since the new *nfam* variable has missing values and other categories labeled 31, 97, and 98, I replace all categories above 16 with "NAs" since I don't intend to use them.

Code Chunk 6-9

```
nes$faminc.f <- as.factor(nes$faminc)

nes$nfam <- nes$faminc

nes$nfam[nes$nfam > 16] <- NA
```

The next step involves cutting the variable using the *cut()* command (Code Chunk 6-10). I decided to cut it into a four parts: low (Low), medium low (MedLow), medium high (MedHigh), and high (High).

Code Chunk 6-10

```
nes$faminc4=cut(nes$nfam, breaks=c(0,3,7,10,16))

levels(nes$faminc4)=c("Low", "MedLow", "MedHigh", "High")
```

Now we're ready to plot, comparing a view of the data when there are too many categories with a view that has a more manageable number (Code Chunk 6-11). In Code Chunk 6-11, I plot all of the categories.

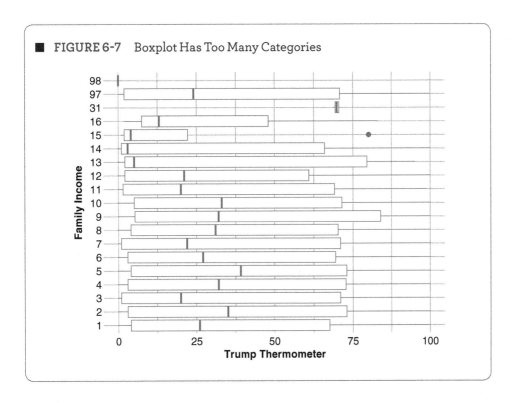

FIGURE 6-7 Boxplot Has Too Many Categories

Code Chunk 6-11
```
ggplot(nes, aes(faminc.f, fttrump)) +
  geom_boxplot(col="#0000bf") +
  theme_minimal() +
  theme(plot.title = element_text(size = 8, face = "bold")) +
  theme(axis.title = element_text(size = 8, face = "bold")) +
  ggtitle("Figure 6-7: Boxplot Has Too Many Categories") +
  ylab("Trump Thermometer") +
  xlab("Family Income") +
  coord_flip()
```

While there are only 19 different categories, they make no intuitive sense and therefore it's hard to understand the relationship between income and support for Trump. It's also difficult to discern any pattern when there's too many categories. Finally, observe how the medians in the boxplots jump up and down from category to category. This could result from having too few respondents in each category, causing the median to move up and down in no particular pattern. Now consider the same kind of plot, with the 19 categories reduced to four (Figure 6-8)

Code Chunk 6-12
```
ggplot(subset(nes, faminc4!="NA"), aes(faminc4, fttrump)) +
  geom_boxplot(col="#0000bf") +
  theme_minimal() +
  theme(plot.title = element_text(size = 8, face = "bold")) +
  theme(axis.title = element_text(size = 8, face = "bold")) +
  ggtitle("Figure 6-8: Trump Is a Middle-Class Phenomenon") +
  ylab("\n Trump Thermometer") +
  xlab("Family Income\n") +
  coord_flip()
```

Although we can see a pattern in Figure 6-7 (the wealthier respondents have lower thermometer scores for Trump), reducing the categories to four (Figure 6-8) might be more helpful. In Figure 6-8, we can detect an interesting pattern that may not have appeared otherwise: there seems to be an uptick between the lowest income and the medium-low categories. While the pattern observed is dependent on how the categories were constructed, it suggests Trump's supporters come not from the poorest of society but from those just above. We could investigate further.

Reducing categories is one approach, and adding categories (if they are available) is another. Consider the different questions, hypotheses, and conclusions generated by expanding the number of categories. To illustrate, let's examine a variable in the *nes* data that records whether the respondent feels better or worse-off financially than 20 years ago (Figure 6-9). Using the same dependent variable (the Trump thermometer), I generated two figures (Figures 6-9 and 6-10) that vary the number of categories.

Art and Practice of Data Visualization
COLLECT MORE DATA THAN YOU NEED

Always collect more data than you think you need. In the example with financial well-being, we compare views of the data when the categories are too few. Note that adding categories to a view can only be accomplished if the more detailed data were collected in the first place. When collecting data or deciding what to collect, it's always good to err on the side of too much information since it can always be pared down. For example, asking someone's age gives more information than asking someone if they are old enough to vote (yes/no).

To illustrate, I took the variable *nes$finwell* and created a new variable called *nes$newfin*, indicating whether a respondent feels better or worse-off financially than they did 20 years ago (Code Chunk 6-13). The *ifelse()* command is used to indicate that if *nes$finwell* registers "the same as 20 years ago," "a little worse," "moderately worse," or "a great deal worse," then assign it a 0; otherwise assign it a 1. I then labeled those categories "Worse" and "Better."

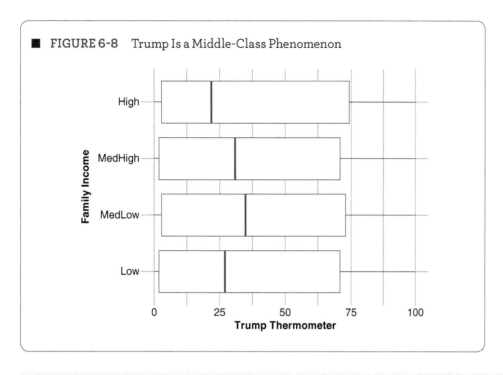

■ **FIGURE 6-8**   Trump Is a Middle-Class Phenomenon

Code Chunk 6-13

```
nes$newfin <- ifelse(nes$finwell=="The same as 20 years ago" |
                     nes$finwell=="A little worse" |
                     nes$finwell=="Moderately worse" |
                     nes$finwell=="A great deal worse", 0, 1)

nes$newfin <- factor(nes$newfin, levels = c(0,1),
                     labels = c("Worse", "Better"))
```

I then plotted the *nes$newfin* variable using boxplots. In Figure 6-9 the boxplots represent the respondents that felt either worse-off or better-off financially.

Code Chunk 6-14

```
ggplot(subset(nes, finwell!="Skipped"), aes(newfin, fttrump)) +
  geom_boxplot(col="#0000bf") +
  theme_minimal() +
  ylab("\n Trump Thermometer") +
  xlab("") +
  theme(plot.title = element_text(size = 8, face = "bold")) +
  theme(axis.title = element_text(size = 8, face = "bold")) +
  ggtitle("Figure 6-9: Do You Feel Financially Better
          or Worse-Off Than 20 Years Ago?") +
  coord_flip()
```

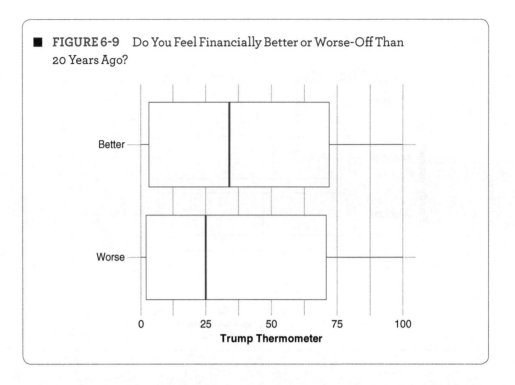

FIGURE 6-9 Do You Feel Financially Better or Worse-Off Than 20 Years Ago?

According to Figure 6-9, respondents who felt they were financially better-off over the last 20 years support Trump. Since we only have two categories, it's hard to figure out what, exactly, is going on. Too many different levels of attitudes are grouped together. The two categories mask important differences that may exist. Now consider Figure 6-10, which draws boxplots for all of the available categories.

Code Chunk 6-15
```
ggplot(subset(nes, finwell!="Skipped"), aes(finwell, fttrump)) +
  geom_boxplot(col="blue") +
  theme_minimal() +
  theme(plot.title = element_text(size = 8, face = "bold")) +
  theme(axis.title = element_text(size = 8, face = "bold")) +
  ggtitle("Figure 6-10: Do You Feel Financially Better
           Off Than 20 Years Ago? ") +
  ylab("Trump Thermometer") +
  coord_flip() +
  ylab("\n Trump Thermometer") +
  xlab("")
```

As Figure 6-9 indicates, there does seem to be an important change in the median level of support. This indicates that respondents' views toward Trump were higher for people who consider themselves better-off. If, instead, we examine the same measure but with more refined

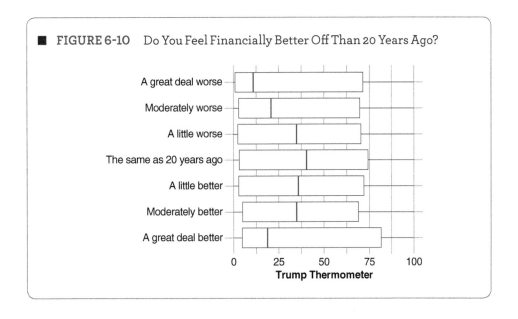

■ **FIGURE 6-10** Do You Feel Financially Better Off Than 20 Years Ago?

categories, we see something different (Figure 6-10). Figure 6-10 indicates that respondents with more moderate evaluations of their financial well-being supported Trump. Put differently, respondents who experienced only moderate change were more likely to vote for Trump. Those at the high end of financial success may have been the wealthiest and most educated, while those at the bottom were probably the least educated and the poorest.

The example shows that the question we ask can affect how we use the data, which can, in turn, influence our conclusions. When we take whatever data are given and blindly proceed with the analysis, someone else's decisions, goals, and motivations could have an important impact on what we find.

> **KNOWLEDGE CHECK: Create categorical variables to connect hypotheses to the data.**

10. Indicate which statements more accurately describe fitting categories to hypotheses.
    a. Having more categories is always better since they help refine the hypothesis.
    b. Having fewer categories is always better since they help simplify complexity.
    c. Theory should drive how we categorize data.
    d. Exploring different categories can change the theory.
11. Why is having too many categories a problem?
    a. There are too few observations in each category.
    b. Too much information should be avoided.
    c. Too many categories indicates a lack of theory.
    d. It can make visualization difficult.

Box-Cox Transformations

Noted statistician John Tukey constructed what is called the **Box-Cox ladder of transformations**: an ordered list of transformations to use on continuous variables (Tukey, 1977). The list contains a number of mathematical expressions that reshape the distribution of the data:

$$\frac{1}{x^2}, \frac{1}{x}, \frac{1}{\sqrt{x}}, \log(x), \sqrt{x}, x, x^2$$

Let's start with what is by far the most common transformation of continuous data, the log transformation. Although there are different kinds of **logarithmic transformations**, the **natural log** and the **log to the base 10** chief among them, let's examine taking logs to the base 10 or $\log_{10}$ to help build an intuition:

$$Log_{10}(100) = 2$$
$$Log_{10}(1,000) = 3$$
$$Log_{10}(10,000) = 4$$

When we "take logs" using log to the base 10, we calculate what exponent of 10 would generate that number. For example, when we take the $\log_{10}$ of 1,000, we calculate the exponent of 10 that produces 1,000: 3. When we take the natural log, we're simply replacing the number 10 with *e* or 2.718. To convert our data back to its raw form, we take the antilog. The antilog of 3 takes 10 ($\log_{10}$) and exponentiates it by 3:

$$\text{The antilog of 3 when using base}_{10} = 1,000$$

When we transform our data, every value can be converted back to the original number. Another nice property of transformations is that we keep the order of our cases the same. In other words, if we lined up the cases from lowest to highest before we transformed the measure, transforming them would not alter their place in line.

Taking logs might not always be the right solution. In some cases, taking the square root of a variable may produce the desired result (a normal distribution). Still, in other cases, taking the inverse $\frac{1}{x}$ might be the best alternative. Tukey's Box-Cox ladder of transformation orders these different transformations along a spectrum that helps us choose the best.

Art and Practice of Data Visualization
TRANSFORMING VARIABLES CHANGES THEIR UNITS

Remember that when data are transformed, the scale has changed. For example, when a number is logged to the base 10, an increase in one unit is a 10-fold increase. When a number is logged to the base 2, a one-unit increase translates into doubling the number.

Art and Practice of Data Visualization
A TIP FOR TAKING THE LOG OF VARIABLES WITH ZEROS

Taking the log of any variable containing zeros is problematic since the log(0) is undefined. Consequently, the transformation can't be made. Add a constant to the variable (usually an arbitrarily small number like 1 or smaller), then take the log.

Each transformation does something different to the data. A log transformation, for example, "pulls" in large outliers and helps spread out small values. Transforming a variable using the square of each observation spreads the distribution at the higher end. To visualize, consider a histogram of GDP per capita from the *world* data set (Figure 6-11).

The code in Code Chunk 6-16 should look familiar. First, I define two objects (both histograms), then I plot them using the *grid.arrange()* function. Note how I control the number of bins in each histogram with the option *bins=10* in the two *geom_histogram()* functions.

Code Chunk 6-16

```
options(scipen = 999)
p1 <- ggplot(world, aes(gdppc)) +
  geom_histogram(bins=10, fill = "#0000bf") +
  labs(title = paste("Figure 6-11: Panel A")) +
  xlab("GDP per Capita") +
  theme_minimal() +
  theme(plot.title = element_text(size = 8, face = "bold")) +
  theme(axis.title = element_text(size = 8, face = "bold"))

p2 <- ggplot(world, aes(log(gdppc))) +
  geom_histogram(bins=10, fill = "#0000bf") +
  labs(title = paste("Panel B")) +
  xlab("GDP per Capita (logged)") +
  theme_minimal() +
  theme(plot.title = element_text(size = 8, face = "bold")) +
  theme(axis.title = element_text(size = 8, face = "bold"))

grid.arrange(p1, p2, ncol=2)
```

In Panel A of Figure 6-11, the majority of the data are concentrated at low income levels, with a few observations located at the opposite extreme ($120,000). This kind of distribution lends itself to taking logs. Taking logs pulls large outlying values closer to the rest of the distribution and generates separation among the smaller values. In raw form,

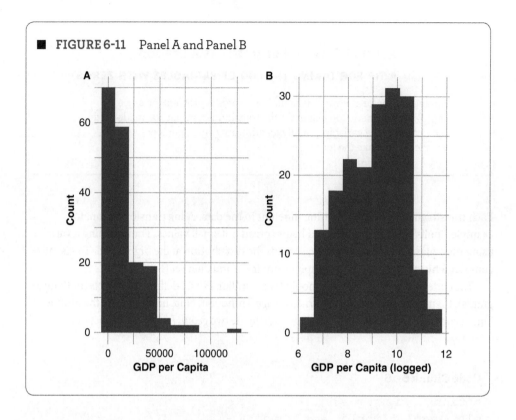

FIGURE 6-11   Panel A and Panel B

we would say the income variable is positively skewed (Panel A). The histogram in Panel B is the logged form of GDP per capita. It approximates what we'd like to see, a normal distribution.

Another view of the data is helpful. The log transformation's effect is particularly evident when we look at scatter plots. Suppose we're interested in the relationship between the percentage of women in the lower house of the legislature (*world$women05*) and GDP per capita income (*world$gdppc*). When we use the raw form of the income data, patterns are difficult to discern. The logged form of the GDP per capita variable allows a better view. Note how the cases at the low end in Panel B of Figure 6-12 are now more spread out, and we have a better view of the data. In this case, the better view reveals there is no relationship between the two variables.

Code Chunk 6-17

```
p3 <- ggplot(world, aes(gdppc/1000, womleg)) + theme_minimal() +
  geom_point(col="#bf0000") +
  labs(title = paste("Figure 6-12: Panel A")) +
  ylab("Percent of Legislators That Are Women") +
  xlab("GDP per Capita (in thousands)") +
  theme(plot.title = element_text(size = 8, face = "bold")) +
  theme(axis.title = element_text(size = 8, face = "bold"))
```

```
p4 <- ggplot(world, aes(log(gdppc), womleg)) +
  geom_point(col="#bf0000") + theme_minimal() +
  labs(title = paste("Panel B")) +
  xlab("GDP per Capita (logged)") +
  ylab("") +
  theme(plot.title = element_text(size = 8, face = "bold")) +
  theme(axis.title = element_text(size = 8, face = "bold"))


grid.arrange(p3, p4, ncol=2)
```

To understand how moving up and down the ladder of transformation affects the distribution of the data, consider the four histograms in Figure 6-13. The upper-left histogram shows the data in their raw form. Again, as noted earlier, there are a lot of countries with relatively low per capita incomes. The next histogram, which takes the square root of GDP per capita, starts to pull those large values ($120,000) down, approximating a normal distribution. The next rung on the ladder of transformation is the log. The log transformation, compared to the rest, seems to produce a histogram that approximates the normal distribution. Finally, we have a histogram that shows the inverse transformation, illustrating that we've probably gone too far down the ladder of transformation since the data appear to have a positive skew.

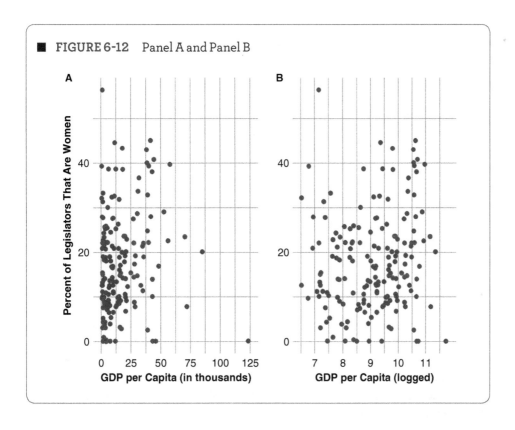

FIGURE 6-12   Panel A and Panel B

Code Chunk 6-18

```
p1 <- ggplot(world, aes(gdppc)) +
  geom_histogram(bins=10, fill = "#0000bf") +
  labs(title = paste("Figure 6-13: Panel A")) +
  xlab("GDP per Capita") +
  theme_minimal() +
  theme(plot.title = element_text(size = 8, face = "bold")) +
  theme(axis.title = element_text(size = 8, face = "bold"))

p2 <- ggplot(world, aes(sqrt(gdppc))) +
  geom_histogram(bins=10, fill = "#0000bf") +
  labs(title = paste("Panel B")) +
  xlab("GDP per Capita (square root)") +
  theme_minimal() +
  theme(plot.title = element_text(size = 8, face = "bold")) +
  theme(axis.title = element_text(size = 8, face = "bold"))

p3 <- ggplot(world, aes(log(gdppc))) +
  geom_histogram(bins=10, fill = "#0000bf") +
  labs(title = paste("Panel C")) +
  xlab("GDP per Capita (logged)") +
  theme_minimal() +
  theme(plot.title = element_text(size = 8, face = "bold")) +
  theme(axis.title = element_text(size = 8, face = "bold"))

p4 <- ggplot(world, aes(1/gdppc)) +
  geom_histogram(bins=10, fill = "#0000bf") +
  labs(title = paste("Panel D")) +
  xlab("GDP per Capita (inverse)") +
  theme_minimal() +
  theme(plot.title = element_text(size = 8, face = "bold")) +
  theme(axis.title = element_text(size = 8, face = "bold"))

grid.arrange(p1, p2, p3, p4, ncol=2, nrow=2)
```

To build our intuition even further, consider the four scatter plots in Figure 6-14. The scatter plots illustrate the same transformations as Figure 6-13. Each scatter plot has different forms of GDP per capita on the x axis plotted against the human capital index (*world$pwthc*).

The code in Code Chunk 6-19 is rather voluminous since I'm defining four different scatter plots. I then arrange them in a grid with the function *grid.arrange()*.

Code Chunk 6-19

```r
p1 <- ggplot(world, aes(gdppc, pwthc, label=iso3c), pos=3) +
  geom_point(col="#bf0000", size=0.5) +
  theme_minimal() +
  theme(plot.title = element_text(size = 8, face = "bold")) +
  theme(axis.title = element_text(size = 8, face = "bold")) +
  ggtitle("Figure 6-14: Four Transformations") +
  ylab("") +
  xlab("\n Raw GDP per Capita") +
  geom_text_repel(size=3, data=subset(world, iso3c=="QAT"),
                  aes(label=iso3c), hjust=-.5, vjust=2,
                        show.legend=FALSE)

p2 <- ggplot(world, aes(sqrt(gdppc), pwthc)) +
  geom_point(col="#bf0000", size=0.5) +
  theme_minimal() +
  theme(plot.title = element_text(size = 8, face = "bold")) +
  theme(axis.title = element_text(size = 8, face = "bold")) +
  ylab("") +
  xlab("\n Square Root") +
  geom_text_repel(size=3, data=subset(world, iso3c=="QAT"),
                  aes(label=iso3c), hjust=-.5, vjust=2,
                        show.legend=FALSE)
```

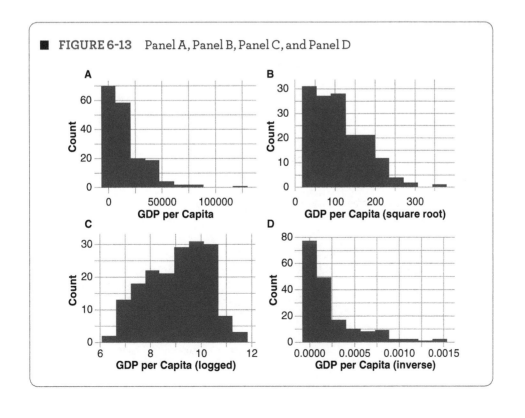

FIGURE 6-13 Panel A, Panel B, Panel C, and Panel D

```r
p3 <- ggplot(world, aes(log(gdppc), pwthc)) +
  geom_point(col="#bf0000", size=0.5) +
  theme_minimal() +
  theme(plot.title = element_text(size = 8, face = "bold")) +
  theme(axis.title = element_text(size = 8, face = "bold")) +
  ylab("") +
  xlab("\n Logged (base 10)") +
  geom_text_repel(size=3, data=subset(world, iso3c=="QAT"),
            aes(label=iso3c), hjust=-.5, vjust=2,
                 show.legend=FALSE)

p4 <- ggplot(world, aes(1/gdppc, pwthc)) +
  geom_point(col="#bf0000", size=0.5) +
  theme_minimal() +
  theme(plot.title = element_text(size = 8, face = "bold")) +
  theme(axis.title = element_text(size = 8, face = "bold")) +
  ylab("") +
  xlab("\n Inverse") +
  geom_text_repel(size=3, data=subset(world, iso3c=="QAT"),
            aes(label=iso3c), hjust=4, vjust=4,
                 show.legend=FALSE)

gridExtra::grid.arrange(p1,p2,p3,p4)
```

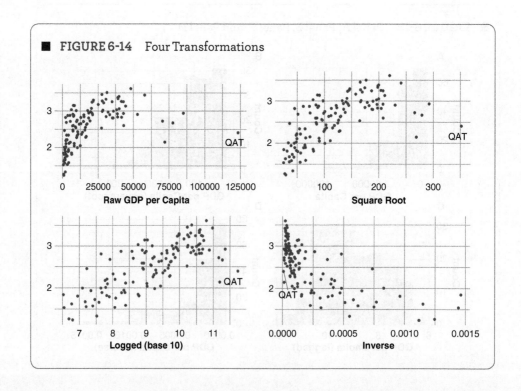

FIGURE 6-14 Four Transformations

Notice the distance between Qatar (QAT) and the rest of the data when GDP per capita takes its raw form. Qatar records a GDP per capita income greater than $125,000 (the United States records just over $48,000). As we travel down the Box-Cox ladder, taking the square root of GDP per capita brings Qatar closer to the other observations. Though closer, it still remains removed from the rest of the data. In the lower-left quadrant of Figure 6-14, the log of GDP per capita is plotted against the human capital index. Qatar appears even closer to the rest of the data and the relationship between income and human capital is linear. The log transformation, therefore, seems to be a good solution. If we travel even further down the ladder and take the inverse of GDP per capita, Qatar no longer stands out, but now we have other outliers and are moving away from a linear relationship.

> **KNOWLEDGE CHECK:** Describe the Box-Cox ladder of transformation.

12. What is the Box-Cox ladder of transformation?
    a. A list of transformations to use with categorical data
    b. An ordered list of transformations to use on continuous data
    c. A set of transformations all based on taking logs
    d. Transformations designed by John Tukey

13. Which statements accurately describe logarithmic transformations?
    a. They are helpful when data are normally distributed.
    b. They are helpful when cases are clustered at the low end of the distribution with a few outliers at the high end.
    c. They are helpful when cases are clustered at the high end of the distribution with a few outliers at the low end.
    d. They spread values out at the low end of a distribution and pull outlying values at the high end closer to the rest of the data.

14. Answer the following questions.
    a. What is the $\log_{10}$ of 1,000?
    b. What is the $\log_{10}$ of 10,000?
    c. What is the antilog (base $_{10}$) of 100?
    d. What is the antilog (base $_{10}$) of 10?

SUMMARY

Transforming data is done for two primary reasons. First, there may be important theoretical reasons to expand or contract the number of categories in a variable, change a continuous measure to a categorical one, or use a logarithmic transformation of a continuous measure. Second, transforming data improves visualization. The goal here was to present the most often-used methods of transformation since data rarely come in the form that we'd like. An important rule of thumb is to always record how the variable was transformed so that the original data can be recovered. Finally, never take the data as given. Always question its substance and form. More than likely, the data were collected and assembled by somebody else for different purposes and with different motivations. Consequently, the data are less likely to fit your needs. By transforming data, you can draw a tighter connection between your hypotheses and the data you use to test them.

COMMON PROBLEMS

- Rather than take data as given, transform data freely and often. Most data you'll use were not collected specifically for you. Transform the data in order to better fit your hypotheses or to help obtain good visualizations of the data. Considerations regarding transforming the data should be made while describing data. Beginning analysts all too often take the data as given without questioning whether they match the question, hypothesis, or goal of visualizing patterns.

- Don't write over existing variables: when transforming variables, create a new variable with a new name so that you've preserved the original variable.

- When taking logs of a variable with zeros, add a constant to the variable first since taking the log of 0 is undefined and will not compute. When dealing with small numbers, adding .001 to the variable is perfectly fine. When the variable you want to transform has bigger values (including the zeros), simply transform the variable by adding 1 before taking logs.

- Before removing outliers from the data to perform analysis, see if transforming the variable (usually by taking logs) first eliminates the outliers by pulling them in closer to the data.

REVIEW QUESTIONS

1. Why do we transform data?

2. What are the most common forms of transformation?

3. What does taking logs do to a variable's distribution?

4. Why does GDP per capita typically need to be transformed?

5. What is Tukey's Box-Cox ladder of transformation?

6. When does it make sense to divide a continuous variable into categories?

7. What two log expressions are normally used in transforming data?

8. In what part of the data analysis process should we think about transformation?

9. How can transforming the data throw away information?

10. Why do data rarely come in the form we want?

PRACTICE ON ANALYSIS AND VISUALIZATION

1. Provide the code to cut the age of respondents from the *nes* data into three parts: people younger than age 30, respondents in their thirties, and those aged 40 and older.

2. Provide the code that transforms the *nes$race* variable into one that records whether a respondent self-identifies as White or not. Label the new categories "White" and "Non-White."

3. What kind of independent variable do the following hypotheses require: continuous or categorical?

   a. Do tastes in comedy differ tremendously between those in their twenties and those aged 50 years and older?

   b. The older you get, the wiser you become.

   c. Older voters are more conservative because they own homes.

   d. More educated people make more money because they have a diploma.

4. What kind of dependent variable do the following hypotheses require: continuous or categorical?

   a. Voters with more education are more likely to vote Democrat.

   b. African Americans will have more contact with the police.

   c. Race is associated with ever having been arrested.

   d. Race is associated with income.

5. Compare histograms of *states$infant* in raw and logged versions. What version of the data more closely approximates a normal distribution?

6. What is the $\log_{10}$ of the following?

   a. 10

   b. 1

   c. 1,000,000

   d. 1,000

7. What is the antilog ($base_{10}$) of the following?

   a. 3

   b. 4

   c. 1

   d. 0

8. Which of the following variables should be logged?

   a. GDP per capita (*world$gdppc*)

   b. Ethnolinguistic heterogeneity (*world$ethfrac*)

   c. Percentage of a state's population with a high school diploma (*states$hsdiploma*)

   d. Income per household (*states$poptotal*)

9. Which of the following are benefits of transforming data?

   a. The data more closely match the hypothesis.

   b. Underlying patterns can be better observed.

   c. Outliers no longer exist.

   d. It's not a good idea to analyze raw data.

10. What are the disadvantages to transforming data?

    a. Analysts transform data to get the answer they want.

    b. Transforming data is simply a way to change your answer to a question.

    c. Transforming the data changes the order of cases.

    d. Once transformed, the original raw form cannot be recovered.

ANNOTATED R FUNCTIONS

The following functions appear in this chapter. They are listed in order of their first appearance (with the code chunk number in parentheses) and annotated here to give a very brief description of their use. Some are not stand-alone functions and only work in combination with other commands. As a reminder, the code in every chapter will work properly if executed in the order it appears. Proper execution also depends on typing the author-defined *libraries()* command, which loads the required R packages.

ggplot(): defines the basic structure of a plot (usually the x and y variables). (6-1)

aes(): the aes (called 'aesthetics') function is used in ggplot to define the basic structure of the plot, which often includes the variables you want to use and any shapes or colors. (6-1)

geom_text_repel(): adds text to a plot so that labels don't overlap. (6-1)

ifelse(): logical function that allows you to construct an if-then statement. Useful for selecting specific cases to label and also for creating categorical variables from continuous variables. (6-1)

theme_minimal(): specifies a minimalist style for ggplot. (6-1)

theme(): specifies font, size, and so forth in a ggplot. (6-1)

ylab(): labels the y axis in ggplot. (6-1)

xlab(): labels the x axis in ggplot. (6-1)

geom_smooth(): fits a curve or straight line to the data. (6-2)

subset(): chooses a specified subset of the data. (6-2)

cut(): divides continuous data into different levels or categories. (6-4)

geom_boxplot: draws a boxplot in ggplot. (6-5)

as.factor(): changes the classification of a variable to being a factor. (6-9)

levels(): lists the levels associated with a categorical variable. (6-10)

coord_flip(): flips the coordinates on a figure so that it changes to either a vertical or horizontal orientation. (6-11)

geom_histogram: draws a histogram in ggplot. (6-16)

grid.arrange(): displays two or more different plots together. (6-16)

options(scipen=999): used to indicate to not use scientific notation. (6-16)

ANSWERS

KNOWLEDGE CHECK

1. a, b, c
2. a, b, c
3. b, d
4. a, b, d
5. a, c
6. a, b, c
7. a, b, c
8. a, d
9. a, b, c
10. c, d
11. a, d
12. b, d
13. b, d
14. 3 (a), 4 (b), 2 (c), 1 (d)

PRACTICE ON ANALYSIS AND VISUALIZATION

1. nes$agevar <- cut(nes$age, c(30,40)). The variable *nes$age* is one you create and the variable *nes$agevar* is the name you give the new variable that groups the ages.

2. First, recategorize the *nes$race* variable:

   nes$whitevar <- ifelse(nes$race=="White", 1, 0).

   Then label the new variable *nes$whitevar*:

   nes$whitevar <- factor(nes$whitevar, levels = c(0,1), labels = c("Non-White", "White"))

3. categorical (a), continuous (b), categorical (c), categorical (d)

4. categorical (a), continuous (b), categorical (c), continuous (d)

5. logged

6. 1 (a), 0 (b), 6 (c), 3 (d)

7. 1,000 (a), 10,000 (b), 10 (c), 1 (d)

8. a, d

9. a, b, c

10. a

 Access digital resources, including datasets, at http://edge.sagepub.com/brownstats1e.

7 Some Principles of Displaying Data

CHAPTER OUTLINE

Learning Objectives
Overview
Some Elements of Style
The Basic Elements of a Story
Documentation (Establishing Credibility as a Storyteller)
Build an Intuition (Setting the Context)
Show Causation (The Journey)
From Causation to Action (The Resolution)
Summary
Common Problems
Review Questions
Practice on Analysis and Visualization
Annotated R Functions
Answers

LEARNING OBJECTIVES

- List some elements of style.
- Tell your story.
- Describe the elements and importance of documentation.
- Explain how to set the context.
- Establish causation.
- Identify the actions implied by the analysis.

Overview

Data are often better described by pictures than presented in tables with numbers. Poorly designed representations of data can mislead, obscure, and neglect important information that should otherwise figure prominently. In this chapter we will explore some basic rudiments of presenting data.

Due in part to the evolution of statistical programs and increasing attention to the design and aesthetics of data visualization, a growing emphasis in statistics concerns its graphical representation. For those wanting a deep dive, Edward Tufte's four books on displaying data are a good place to start. Tufte's work touches on graphical integrity, aesthetics, principles of analytic designs, and the intellectual tasks upon which they are based. I borrow liberally from Tufte, focusing on what aspects of data visualization I think are the most important: (1) documentation, (2) causation, and (3) integration. These three aspects of data visualization will carry us a long way in understanding our problem and presenting its solution.

I also borrow liberally from Cole Nussbaumer Knaflic (2015) and Nancy Duarte (2010) who emphasize both the clarity and structure that underpin effective presentations of data. For Knaflic and Duarte, presenting data is storytelling. Aspects of Tufte's principles overlap and reinforce the points emphasized by Duarte and Knaflic. For example, Tufte's emphasis on documentation serves to establish the credibility of the storyteller, a key component of Knaflic and Duarte's approach.

Presenting data is an ethical, moral, and emotional act. Just as Strunk and White (1979) remind with writing, data analysis and the presentation of it reflect traits that lie deep within. Impatience is the enemy. Impatience manifests itself in many ways: poorly labeled graphs, half-baked theories, and the tendency to stop the conversation when the data aren't done talking. The impatient don't have time for truth and beauty, settling instead for whatever appears in front of them. Consider the following quote from William Strunk Jr. and E. B. White's famous pamphlet, *The Elements of Style*:

> Style takes its final shape more from attitudes of mind than from principles of composition, for, as an elderly practitioner once remarked, "Writing is an act of faith, not a trick of grammar." This moral observation would have no place in a rule book were it not that style is the writer, and therefore what a man is, rather than what he knows, will at last determine his style. (Strunk & White, 1979, p. 84)

Analyzing data and presenting the results is a moral exercise, just like writing. Part of the enjoyment we have in the enterprise is not only what we discover about our world, but what we discover about ourselves.

Some Elements of Style

In the spirit of Strunk and White's famous collection of dos and don'ts for writing in the English language, I offer a much smaller, less ambitious list of things to consider when visualizing data. The list, I hope, provides a set of directions that encourages clarity, transparency, and truth, enabling the author to make their point and allowing the reader to quickly understand. The first two points help with clarity: (1) eliminate the clutter and (2) construct pictures that sharpen our focus. The third point, integrate words and pictures, invites using the available white space in a figure or table to provide as much information as possible in as little space available. Too often we ask the reader to unnecessarily shift their gaze between a figure and the written text on the page. Going back and forth between text and graphics disorients. Often the reader abandons one or the other, ignoring half of the information. Place the text right next to the data.

The next instruction, pictures should be worth a thousand words, is derived from Tufte's order to show restraint. Too often, when it is time to relay information in numerical form, we turn to a figure when a table or a simple sentence will suffice.

The final set of instructions is less technical, but no less important. Know your audience and know your purpose are two directives that should weigh heavily. Figures adorned with cartoon characters are probably not appropriate for the boardroom or a meeting of the faculty senate. While regression tables should be used sparingly, we can probably all agree that they have no place in a presentation at the Elks Lodge, the Rotary Club, or the community senior center.

Knowing your purpose is just as important: to explore, to explain, and to inform. Visualizing data helps in each one of these, but we should assemble views of the data accordingly. When exploring the data, arrange visuals with objective neutrality; you don't want to be led down the wrong path. Once your conclusions have been drawn, your job is to persuade. Present pictures that have focus and are easily digested. Finally, if the goal is to inform, more ink per square inch can be spilled. In other words, provide as much information in as small a space as possible.

These are a set of principles, guides, or suggestions that I've collected over time. In each instance the idea was expressed by someone else in greater detail with a series of examples. In each case, I'll point to the source so that you can explore it more fully.

Eliminate the Clutter

In many ways, this is the direct parallel to Professor Strunk's never-ending quest to eliminate words. Unnecessary words confuse, and in so many ways dilute the message lying underneath. So too do unnecessary embellishments or ornamentation. Tufte's books provide entertaining examples of charts and graphs gone wrong in this respect. Look there for more artistic and historical examples of **chart clutter**. In a more recent book, Knaflic (2015) provides excellent examples of chart clutter in the context of graphics produced in Excel.

There are many ways clutter creeps in. Unfortunately, early advances in computer graphics got too far out in front of good practice, providing us with a large collection of unnecessary features and embellishments. Fortunately, more recent versions of data graphics packages are more attuned to good presentational practice.

Let's start with a standard figure in ggplot and illustrate how to eliminate chart clutter. In Code Chunk 7-1 is the by now familiar *ggplot* command for a scatter plot. In Code Chunk 7-1, I plot the relationship between ethnolinguistic fractionalization and GDP per capita. Notice that it doesn't include the command *theme_minimal()*, which would have removed the gray background.

Code Chunk 7-1

```
ggplot(world, aes(ethfrac, log(gdppc), col=aclpregion)) +
  geom_point() +
  ggtitle("Figure 7-1: Ethnolinguistic Fractionalization Influences
      GDP per Capita") +
  ylab("GDP per Capita (logged)") +
  xlab("Ethnolinguistic Fractionalization") +
  theme(plot.title = element_text(size = 8, face = "bold")) +
  theme(axis.title = element_text(size = 8, face = "bold"))
```

While the 'ggplot' graphics package in R obeys many of the best practices that have evolved with the growth of big data and data analysis more generally, there are a few things to note.[1] Consider the scatter plot in Figure 7-1. Suppose my goal is to show how the "industrial countries" cluster together at high incomes and relatively low levels of ethnolinguistic heterogeneity.

---
[1] There is a proliferation of books about presenting data and I will not run through the countless number of ways one can declutter. For what I think is one of the most elegant and useful texts in this regard, I would suggest *Storytelling With Data* by Knaflic (2015).

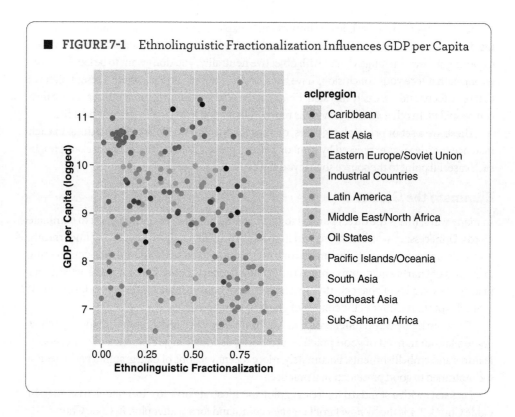

FIGURE 7-1 Ethnolinguistic Fractionalization Influences GDP per Capita

Most would color the data points by region, generating a legend on the right. While this is better than most graphics in terms of clutter, there's still work to be done.

Code Chunk 7-2
```
ggplot(world, aes(ethfrac, log(gdppc))) +
  geom_point(color=ifelse(world$aclpregion=="Industrial Countries",
             "#0000bf", "grey")) +
  ggtitle("Figure 7-2: Cleaned-Up Version of Figure 7-1") +
  ylab("GDP per Capita (logged)") +
  xlab("Ethnolinguistic Fractionalization") +
  annotate("text", x=.15, y = 11.5, label = "Industrial Countries",
           col="#0000bf") +
  theme_minimal() +
  theme(plot.title = element_text(size = 8, face = "bold")) +
  theme(axis.title = element_text(size = 8, face = "bold"))
```

Consider Figure 7-2, a cleaned-up version of Figure 7-1. First note the removal of the gray background by including the command *theme_minimal()*. Removing it lets us gray out features of the figure that we'd like to de-emphasize. The command *theme_minimal()* also replaces a traditional representation of the x and y axes with a light grid. I've also removed the legend,

■ **FIGURE 7-2** Cleaned-Up Version of Figure 7-1

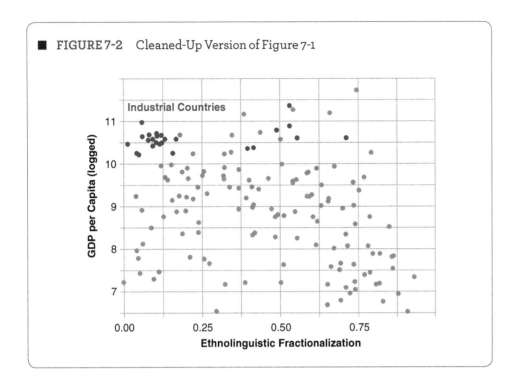

which occupied one-third of the available space in the previous figure. As already noted, I use this theme throughout the book as it removes unnecessary features of the standard ggplot figure.

Since identifying a cluster of cases was the goal, coloring those cases dark blue and labeling them provides focus.

Focus

When we remove clutter, we naturally sharpen our focus; Figure 7-2 was a good example. Now consider Figure 7-3. First, if these were curves or line plots that could change direction, it would be difficult to follow as they crossed each other. Second, the number of regions and the different shades of colors makes distinguishing between them difficult: Southeast Asia and Sub-Saharan Africa are indistinguishable as are Latin America and the Middle East.

Code Chunk 7-3
```
ggplot(world, aes(ethfrac, log(gdppc), color = aclpregion)) +
  geom_point() +
  theme_minimal() +
  theme(plot.title = element_text(size = 8, face = "bold")) +
  theme(axis.title = element_text(size = 8, face = "bold")) +
  geom_smooth(method="lm", se=FALSE) +
  ggtitle("Figure 7-3: Colorfully Confusing") +
  ylab("GDP per Capita (logged)") +
  xlab("Ethnolinguistic Fractionalization")
```

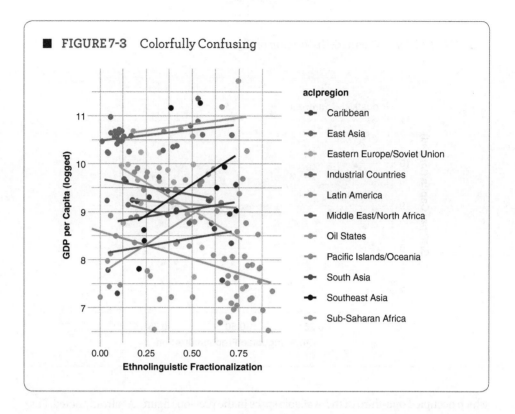

■ FIGURE 7-3   Colorfully Confusing

Because of the kaleidoscopic nature of Figure 7-3, it's hard to notice that two of the regions are very different from the others: Southeast Asia and the Pacific Islands/Oceania. The code in Code Chunk 7-4 focuses our attention on those two regions by restricting the colors to three: blue, black, and gray. Note how I've included three *smooth()* commands in the code. I also color the regions with an *ifelse()* command in the *geom_point()* layer and include plenty of annotation with the *annotate()* command.

Code Chunk 7-4
```
ggplot(world, aes(ethfrac, log(gdppc), by=aclpregion)) +
 geom_point(color=ifelse(world$aclpregion=="Southeast Asia", "black",
                ifelse(world$aclpregion=="Pacific Islands/Oceania",
                       "#0000bf", "grey"))) +
 theme_minimal() +
 theme(plot.title = element_text(size = 8, face = "bold")) +
 theme(axis.title = element_text(size = 8, face = "bold")) +
 geom_smooth(data=subset(world,
                      aclpregion == "Southeast Asia"),
                      color="black", method="lm", se=FALSE) +
 geom_smooth(data=subset(world,
                      aclpregion == "Pacific Islands/Oceania"),
```

```
                              color="blue", method="lm", se=FALSE) +
   geom_smooth(data=subset(world,
                           aclpregion != "Southeast Asia" &
                             aclpregion != "Pacific Islands/Oceania"),
                           color="grey", method="lm", se=FALSE) +
   ggtitle("Figure 7-4: These Two Are Not Like the Others") +
   ylab("GDP per Capita (logged)") +
   xlab("Ethnolinguistic Fractionalization") +
   annotate("text", x = .85, y = 10.5,
            label = "Southeast Asia", col="black") +
   annotate("text", x = .12, y = 7.1,
            label = "Pacific Islands/Oceania", col="#0000bf")
```

When you are dealing with multiple lines, especially if they are not straight, highlight the one or two lines you want to emphasize and gray out the rest. In this case, attention is drawn to Southeast Asia and Oceania, two regions that buck the trend: as ethnic heterogeneity increases, so does income. Although the result is not as colorful, drawing the stark contrast strengthens the point. Again, removing the legend and placing the labels in proximity to the points increases the actual size of the graph by about one-third without taking up any additional space.

Integrate Words and Pictures

Integrating words with pictures is an effective way to communicate. More than anything, it places the words where they need to be, in the data. Although there have been significant developments in data visualization software, we're still often caught in a limited medium that makes

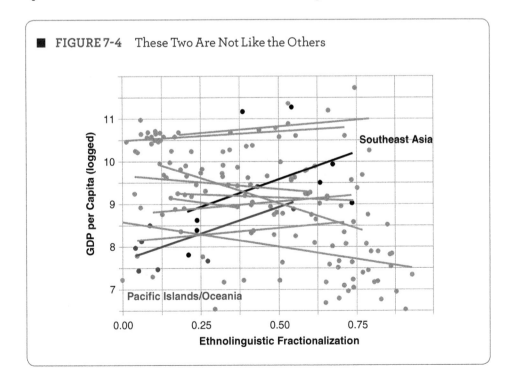

■ **FIGURE 7-4**   These Two Are Not Like the Others

it difficult to combine the two. While labeling points and providing annotations can be difficult, some software packages make it easier than others. Adobe Illustrator is particularly useful in combining words with pictures. Consider several different versions of the following figure (Figure 7-5).

Remember, we noted that while there does seem to be a negative relationship between GDP per capita and ethnolinguistic fractionalization, the relationship resembles more of an inverted U-shape than a straight line. To illustrate the power of placing text in and around a figure, let's start with a stripped-down version of the scatter plot (Figure 7-5). The code in Code Chunk 7-5 generates a very basic scatter plot with no labeling in the graph and does not call attention to any specific cases or groups of cases. Its only refinements are the *theme_minimal()*

Code Chunk 7-5

```
ggplot(world, aes(ethfrac, log(gdppc))) +
 geom_point(color="red") +
 ggtitle("Figure 7-5: GDP per Capita Versus Ethnolinguistic
                   Fractionalization") +
 ylab("GDP per Capita (logged)") +
 xlab("Ethnolinguistic Fractionalization (ELF)") +
 theme_minimal() +
 theme(plot.title = element_text(size = 8, face = "bold")) +
 theme(axis.title = element_text(size = 8, face = "bold"))
```

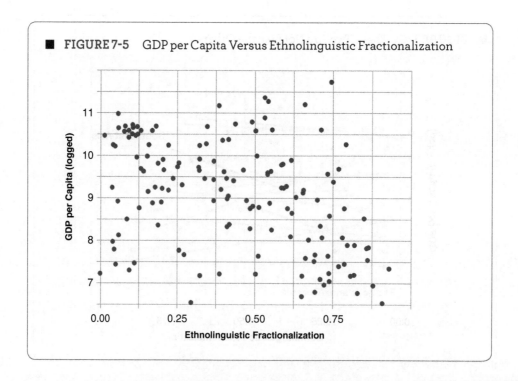

■ **FIGURE 7-5** GDP per Capita Versus Ethnolinguistic Fractionalization

command and additional *theme()* commands that set the font style and size for the figure. While we can discern a negative relationship between income and ethnolinguistic fractionalization, many of the important features of the plot remain undetected by the lack of labeling.

Now consider a more informative scatter plot that includes labeling and a line. To draw what you see in Figure 7-6, I used the code in Code Chunk 7-6 to draw the basic elements of the scatter plot. I then imported the figure into Adobe Illustrator to place the comments and lines exactly where I want them. Most all of the work to generate the figure was done by overlaying three different lines using *geom_smooth()*, labeling several small island nations, and adding text to highlight the industrial and sub-Saharan African nations.

Code Chunk 7-6
```
ggplot(world, aes(ethfrac, log(gdppc), label = iso3c)) +
 geom_point(color=ifelse(world$aclpregion=="Industrial Countries",
                         "blue",
                 ifelse(world$aclpregion=="Sub-Saharan
                                 Africa", "black", "grey"))) +
 geom_smooth(method="lm", se=FALSE, col="grey", linetype = "dashed") +
 geom_smooth(data=subset(world,
                         aclpregion != "Industrial Countries"),
                 color="black", se=FALSE) +
ggtitle("Figure 7-6: ELF Lowers Income") +
ylab("GDP per Capita (logged)") +
xlab("Ethnolinguistic Fractionalization (ELF)") +
annotate("text", x = .12, y = 11.2,
         label = "Industrial Countries", col="blue") +
annotate("text", x = .75, y = 6.5,
         label = "Sub-Saharan Africa", col="black") +
theme_minimal() +
geom_text_repel(size = 3, col="blue",
    aes(label=ifelse(iso3c=="COM" |
                iso3c=="HTI" |
                iso3c=="SLB" |
                iso3c=="KIR",
                as.character(iso3c),''),
        hjust = 0, vjust=-1), show.legend=FALSE)
```

While at first glance Figure 7-6 violates the principle of too much information, every piece of extra information helps explain the relationship's nonlinear form and the cases responsible for it. There are several aspects to the relationship that bear noting. If we discard, for example, the cluster of European democracies, the curve becomes more evident. Having said that, an important part of the curve involves states with low GDP per capita and low levels of ethnic heterogeneity. Upon closer inspection, these cases are very small island nations: Kiribati (KIR), Comoros (COM), Haiti (HTI), and the Solomon Islands (SLB).

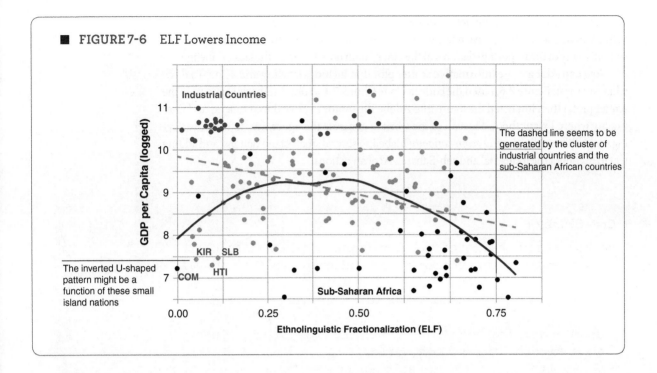

■ FIGURE 7-6   ELF Lowers Income

Standard practice presents a scatter plot with labels of the different points. Usually, the real description occurs in a paragraph placed either above or below the figure or on an adjacent page. A more effective way to make the argument is to include the words and additional features (arrows, lines drawn, etc.) in the figure. Figure 7-6 provides an illustration of how additional information can be included.

With a more informative title, different colors for interesting cases, a visual aid (the curve), and some extra text placed in the figure, the argument is stronger. It's much easier for our eyes to gather information with objects that are adjacent to each other than situated either above, below, or on another page. While R makes annotation within graphics possible, software specifically designed to combine all manner of graphic devices and effects is preferred. Figure 7-6 was generated by R with some help by Adobe Illustrator.

 Art and Practice of Data Visualization
FIGURE AND TABLE PROXIMITY

While integrating text within a figure can be extremely effective, positioning figures and tables as close to the text that describes them is crucial. Whenever possible, place your tables and figures as close to the text that describes them. Word processing programs are sophisticated enough these days to keep most words and the pictures they describe on the same page without creating large swaths of blank space.

Pictures Should Be Worth a Thousand Words

Integrating words and pictures increases the amount of information conveyed. According to Tufte, maximize the amount of information per square inch (Tufte, 2006). There is a direct parallel between a picture that informs and writing that is clear, direct, and succinct. Much like the loosely written paragraph, a series of empty figures and tables dilutes the message. Avoid using figures instead of tables or tables instead of words when a simple sentence will do.

Consider a bar plot of party identification (Figure 7-7). Note how I use the *subset()* command in Code Chunk 7-7 to remove two categories from the plot: "Other" and "Not sure."

Code Chunk 7-7
```
ggplot(subset(nes, pid7!="Other" & pid7!="Not sure"), aes(pid7)) +
  geom_bar(fill="#0000bf") +
  theme_minimal() +
  theme(plot.title = element_text(size = 8, face = "bold")) +
  theme(axis.title = element_text(size = 8, face = "bold")) +
  coord_flip() +
  ggtitle("Figure 7-7: Bar Plot of Party ID") +
  ylab("Number of Respondents")
```

The bar plot in Figure 7-7, while aesthetically pleasing if you like the color blue, only provides seven points of information. Given the size of the graph, the information conveyed per square inch is fairly small. In this case, use a table.

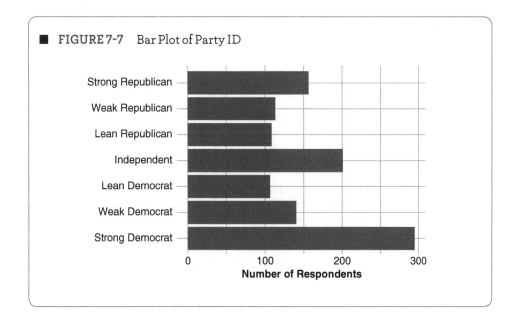

FIGURE 7-7   Bar Plot of Party ID

Code Chunk 7-8

```
count(nes, pid7)

# A tibble: 9 x 2
  pid7              n
  <fct>           <int>
1 Strong Democrat   295
2 Weak Democrat     141
3 Lean Democrat     107
4 Independent       201
5 Lean Republican   109
6 Weak Republican   114
7 Strong Republican 157
8 Not sure           40
9 <NA>               14
```

In the raw R output in Code Chunk 7-8 (a tibble), the exact number of respondents is presented in compact fashion within a fairly small area. Take the exercise one step further. If your point is that there are more Democrats than Republicans, consider using a sentence.

> In the NES survey, respondents identifying as Strong Democrats outnumbered Strong Republicans by a ratio of 2 to 1.

We can get carried away with graphics when a simple sentence suffices. Only use pictures if they are worth a thousand words.

Art and Practice of Data Visualization
BLINDED BY THE LIGHT

Using R to generate figures opens up an entire world of fonts, shapes, symbols, and colors. When they are confronted for the first time by so many choices, I've seen students make color choices designed to express their personalities rather than convey information effectively. The tendency for many is to use bright pinks, blues, yellows, and florescent greens. While certainly colorful and eye-catching, they can also blind. Colors and patterns can vibrate and be hard to look at. Using lighter shades, at the risk of stunting personal expression, makes figures easier to view and understand. When viewing data containing different colors, varying the shade of the same color can give a more professional look to your work.

■ **FIGURE 7-8**

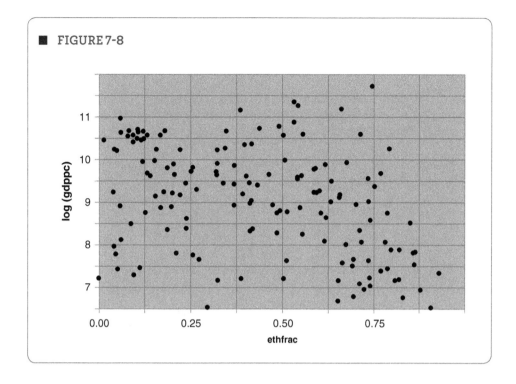

As a final set of examples, the next four scatter plots illustrate how labels, shapes, and colors provide additional information with elegance. The scatter plot in Figure 7-8, while emphasizing the relationship between GDP per capita (logged) and ethnolinguistic fractionalization, doesn't say much. It also has the clutter of the gray background that many software programs incorporate.

Notice in Code Chunk 7-9 the small amount of code. Not much code is required to generate the basics of a scatter plot, histogram, or bar plot. To add titles, specify fonts, label cases, and fit lines, more code is required. To explore data without details does not require much typing.

Code Chunk 7-9

```
ggplot(world, aes(ethfrac, log(gdppc))) +
  geom_point()
```

An important tradeoff exists between cleanliness and information. The tradeoff is magnified when there are over 100 different categories (or individual cases). When confronted with the choice between cleanliness and information, some practitioners choose cleanliness. For example, a "clean" plot may contain no labels. At the opposite extreme, consider what happens when every observation is labeled. So that country abbreviations don't overlap, I use the *geom_text_repel()* command, which connects the abbreviations to the points (Code Chunk 7-10).

Code Chunk 7-10

```
ggplot(world, aes(ethfrac, log(gdppc))) + geom_point() +
  theme_minimal() +
  theme(plot.title = element_text(size = 8, face = "bold")) +
  theme(axis.title = element_text(size = 8, face = "bold")) +
ggtitle("Figure 7-9: Ethnolinguistic Fractionalization
                     Influences GDP per Capita") +
  ylab("GDP per Capita (logged)") +
  xlab("Ethnolinguistic Fractionalization") +
  geom_text_repel(size=3, aes(label = iso3c, size = 1,
           hjust = 0, vjust=-1))
```

Although Figure 7-9 avoids the gray background, the labels and their pointers distract from understanding the relationship. Fortunately, we can reduce the noise by incorporating color. Note that by coloring the data points red and the remaining information gray, attention is drawn back to the relationship while preserving the information provided by the labels (Figure 7-10). Also note that I reduced the labels' size (Code Chunk 7-11).

Code Chunk 7-11

```
ggplot(world, aes(ethfrac, log(gdppc))) +
  geom_point(col="#bf0000") +
  theme_minimal() +
  ggtitle("Figure 7-10: Ethnolinguistic Fractionalization
           Influences GDP per Capita") +
  ylab("GDP per Capita (logged)") +
  xlab("Ethnolinguistic Fractionalization") +
  theme_minimal() +
  theme(plot.title = element_text(size = 8, face = "bold")) +
  theme(axis.title = element_text(size = 8, face = "bold")) +
  geom_text_repel(size=2,
      aes(label = iso3c, size = 1, hjust = 0, vjust=-1),
             col="grey", show.legend=FALSE)
```

Fortunately, by changing the color and size of the points and labels, we can provide more information without obscuring the relationship between the two variables. In Figure 7-10, we now have much more information, we still retain our focus on the two variables, and we know where each country is situated. Notice that most of the Western European countries are located in the upper-left quadrant (the industrialized countries identified earlier). The negative relationship seems to be driven by Western Europe and sub-Saharan Africa.

If we want to add more information to Figure 7-10, we can color the points according to region. The colors confirm that the big difference is between Western Europe and Africa. We can go further: let's check to see whether democratic regimes figure in this story. Perhaps the

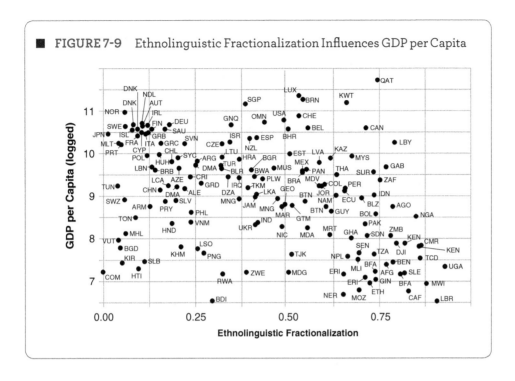

FIGURE 7-9 Ethnolinguistic Fractionalization Influences GDP per Capita

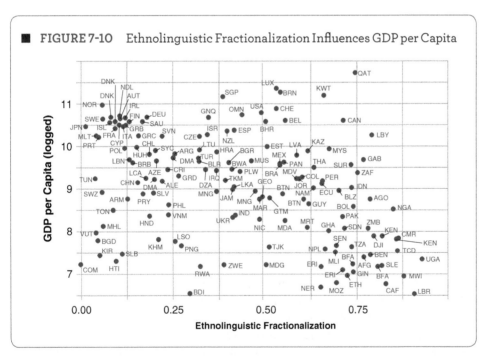

FIGURE 7-10 Ethnolinguistic Fractionalization Influences GDP per Capita

relationship between ethnolinguistic fractionalization and GDP per capita changes when ethnic groups enjoy the protections afforded by the rule of law and human rights—two regime characteristics associated with democracy. We've already used color, so let's use different shapes to distinguish between democratic and authoritarian regimes.

Using the *ifelse()* command (Code Chunk 7-12), I took the *regime* variable in the *world* data set and assigned the parliamentary and presidential democracies the score of 1, while everything else received a 0. In Figure 7-11, triangles were assigned to the democratic cases. I used the option *shape=reg* in the aesthetics of the *ggplot()* command.

While there do seem to be some important regional differences between Western Europe and sub-Saharan Africa, assigning triangles to the democratic cases does not reveal any obvious patterns. Admittedly, Figure 7-11 may provide too much information. If we want to focus specifically on regime type, it might be helpful to drop color as a distinguishing feature. While Figure 7-11 might strain the eyes (and overtax the brain), there are ways we could shade the democratic countries differently to see if a pattern emerges.

```
Code Chunk 7-12
world$reg <- ifelse(world$regime=="Parliamentary Democracy", 1,
          ifelse(world$regime=="Presidential Democracy", 1, 0))

world$reg <- as.factor(world$reg)


ggplot(world, aes(ethfrac, log(gdppc), col=region, shape=reg)) +
  geom_point() +
  ggtitle("Figure 7-11: Ethnolinguistic Fractionalization Influences GDP per Capita") +
  ylab("GDP per Capita (logged)") +
  xlab("Ethnolinguistic Fractionalization") +
  theme_minimal() +
  theme(plot.title = element_text(size = 8, face = "bold")) +
  theme(axis.title = element_text(size = 8, face = "bold"),
        legend.position="bottom") +
  labs(color = "Region", shape = "Regime") +
  geom_text_repel(size=2,
        aes(label = iso3c, size = 1, hjust = 0, vjust=-1),
        col=ifelse(world$iso3c=="MDG", "#0000bf",
              ifelse(world$iso3c=="LUX", "#0000bf", "grey")))
```

Let's review. Pictures should be worth a thousand words, but they can be counterproductive if they're much more than that. There is a tradeoff between how streamlined a figure looks and how much information it conveys. That stark tradeoff can, however, be softened by elegance. The careful and judicious use of colors and shapes (along with inserting text in the figure) can help maintain an aesthetically pleasing picture that makes an important point. There are no hard and fast rules here and your choices depend on what, exactly, you want to argue, display, or illustrate.

Know Your Audience

Your audience should dictate not only what you present, but *how* you present. There are two dimensions upon which to judge your audience: their experience with data and their time. Let's talk about experience with data first.

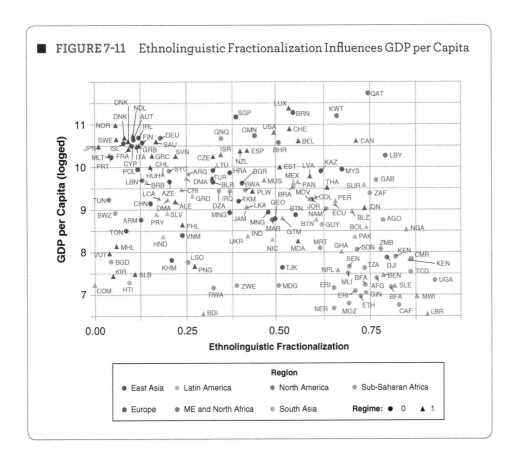

■ **FIGURE 7-11** Ethnolinguistic Fractionalization Influences GDP per Capita

Here is a simple rule to keep in mind: if you have to spend more time explaining how the graph works, or what it represents, rather than what it actually says, reconsider the graph. Most everything you need to communicate can be achieved with a bar plot, line plot, histogram, or scatter plot. Of course, the highlighting, the text, and the colors make all of the difference. If the audience has to decode the evidence you're presenting, you're fighting an uphill battle.

In terms of their time, presidents and CEOs of large corporations have little or no time to navigate the intricacies of a complex plot. Not only is it best to stick with traditional forms with this crowd, pay particular attention to decluttering your visuals and focusing on the main point. This group has little time or bandwidth to come away with anything other than your main point.

Finally, realize there's a difference between presenting in front of students you're trying to get excited about using R and department chairs or deans who are contemplating budget cuts to their departments or colleges. With the former group, a multicolor cat plot (Figure 7-12) might just change some cat lover's attitude toward learning statistics and data analysis. The latter group might just as easily become offended at your lack of recognizing the gravity of the situation, especially since it involves figures on tuition and possible layoffs.

The cat plot is drawn here to excite the cat lover. There are a couple of things to note when drawing the cat plot (Code Chunk 7-13). First, you can see how to install a package and load it with the *library()* command and the *install_github()* command since I'm grabbing this package from a GitHub page (normally the *install.packages()* command is used). I also use made-up data. I define x as ranging from 0 to 20 and create a y that is an expression of x. Once those variables are created, the *multicat()* command follows.

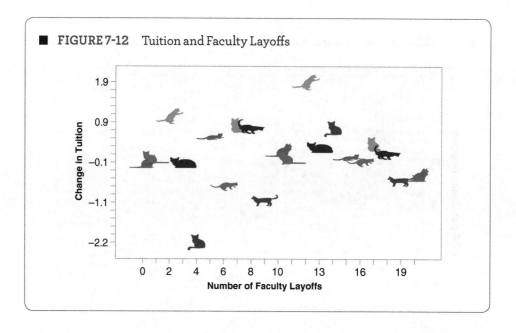

FIGURE 7-12 Tuition and Faculty Layoffs

Code Chunk 7-13

```
install_github("Gibbsdavidl/CatterPlots")

library(CatterPlots)

x <- 0:20
y <- -x^2 + 10
set.seed(33)

multicat(xs=x, ys=rnorm(21),
         cat=c(1,2,3,4,5,6,7,8,9,10),
         catcolor=c('blue', 'green', 'red', 'purple', 'brown'),
         canvas=c(-.1,1.1, -0.1, 1.1),
         xlab="Number of Faculty Layoffs",
         ylab="Change in Tuition",
           main="Figure 7-12: Tuition and Faculty Layoffs")
```

Know Your Purpose: Explanatory, Exploratory, or Informational

Inextricably tied to knowing your audience is knowing your purpose. Here I borrow from work by Nancy Duarte (2010). Duarte distinguishes between two purposes to which I'll add a third. For Duarte, there are **explanatory** and **exploratory** visualizations of data. When in explanatory mode, the main goal is to focus laser-like on the main point. Hit the reader right between the eyes with the most compelling visuals. Simplicity is the rule.

The code in Code Chunk 7-14 grays out the data and fits a line that prominently illustrates there is a positive and linear relationship between income and abortions in the 50 U.S. states. The actual code is as bare bones as you can get. The resulting figure (7-13) is an example of a visualization you'd use to explain something or to make a very specific point.

When in exploratory mode, neutrality is the goal. Don't stack the deck for or against any one particular explanation. Rather, construct an experiment that gives alternative hypotheses an equal chance. Highlighting specific features should be avoided until after you've explored.

Code Chunk 7-14

```
ggplot(states, aes(inc, abort)) +
  geom_point(col="grey") +
  geom_smooth(method="lm", se=FALSE, col="#0000bf") +
  theme_minimal() +
  theme(plot.title = element_text(size = 8, face = "bold")) +
  theme(axis.title = element_text(size = 8, face = "bold")) +
  ggtitle("Figure 7-13: Abortion Is More Available to the Wealthy") +
  ylab("Abortions per 1,000 Women") +
  xlab("Income in Dollars per capita")
```

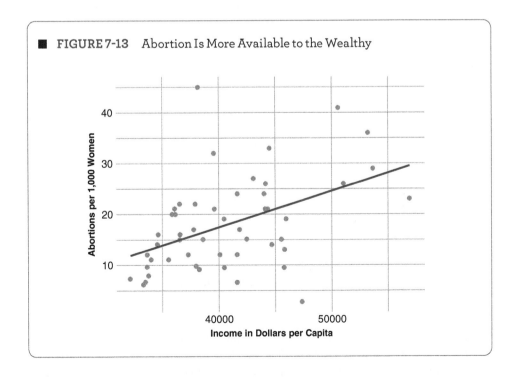

■ **FIGURE 7-13** Abortion Is More Available to the Wealthy

Code Chunk 7-15

```
ggplot(states, aes(inc, abort, col = region)) +
  scale_color_manual(values=c("#bf0000", "#00bf00",
          "#00ffff", "#263333")) +
  geom_point(size=2) +
  theme_minimal() +
  theme(plot.title = element_text(size = 8, face = "bold")) +
  theme(axis.title = element_text(size = 8, face = "bold")) +
  ggtitle("Figure 7-14: Exploring Abortion and Income by Region") +
  ylab("Abortions per 1,000 Women") +
  xlab("Income in Dollars per Capita") +
  annotate("text", x = 45000, y = 30,
        label = "West", col="#00bf00") +
  annotate("text", x = 32000, y = 12,
        label = "South", col="#bf0000") +
  annotate("text", x = 50000, y = 35,
        label = "Northeast", col="#263333") +
  annotate("text", x = 39000, y = 5,
        label = "Midwest", col="#00ffff") +
  guides(col=FALSE, shape=FALSE)
```

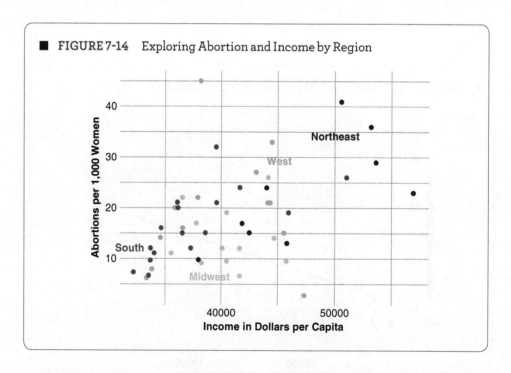

FIGURE 7-14 Exploring Abortion and Income by Region

Figure 7-14 represents a visual designed to explore whether geographic region influences the relationship between abortion rates and income in the United States. In this scatter plot we're not shading one region darker than the other to make an argument. Here we're just trying to determine whether there is a pattern; we're not trying to convince or persuade anyone.

Admittedly, the difference between a graph designed for explanation and one designed for exploration can be subtle. If our purpose is explanation, we might show one graph with the South and Northeast colored differently and everything else grayed out to highlight regional differences. We could also highlight one region at a time (see Figure 7-15).

In Code Chunk 7-16, notice how I've used the *size = 2.5* option in the *geom_point()* function to make all of the points slightly bigger. Also note how I've used the *ifelse()* command to color the Southern states blue and everything else gray.

Code Chunk 7-16
```
ggplot(states, aes(inc, abort, label = "state")) +
  geom_point(size = 2.5, col=ifelse(states$region=="South", "#0000bf",
                                    "grey")) +
  theme_minimal() +
  theme(plot.title = element_text(size = 8, face = "bold")) +
  theme(axis.title = element_text(size = 8, face = "bold")) +
  ggtitle("Figure 7-15: Explaining Abortion and Income") +
  ylab("Abortions per 1,000 Women") +
  xlab("Income in Dollars per Capita") +
  annotate("text", x = 32000, y = 12,
           label = "South", col="blue") +
  guides(col=FALSE, shape=FALSE)
```

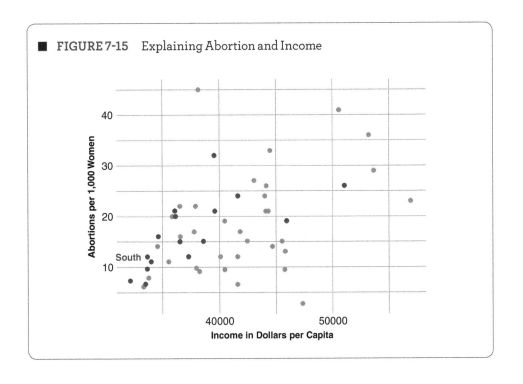

■ **FIGURE 7-15**   Explaining Abortion and Income

To the explanatory and exploratory, I add **informational**. Sometimes the goal is not only to explain or explore, it is also to inform or to provide a reference. In this mode, communicate as much information as possible. In Tufte's words, more ink per square inch. While avoiding clutter, assume the audience can refer back to the visual as it needs. Visuals that can be studied over time can be extremely useful. All too often the informational role of data visualization is ignored, replaced by visuals that help explain and explore but do not inform.

Figure 7-16 graphs the same relationship depicted in the last two figures, only this time it provides as much information as possible. Because there is a lot of information shown, it is difficult to make out any one particular relationship or theme, but that's not the point. Rather than explaining or exploring the relationship, the goal is to provide as much information as possible. From Figure 7-16 we understand the relationship between abortions and income, the values for each state, and the mean values for abortions and income, and we add a third variable (the log of population density) to the mix. While it is difficult to discern a single explanation or to make a quick discovery since there is a lot of information displayed, such figures can be useful.

Code Chunk 7-17
```
ggplot(states, aes(inc, abort, col = region)) +
  geom_point(size=log(states$density)) +
  scale_color_manual(values=c("#bf0000", "#00bf00",
        "#00ffff", "#263333")) +
  theme_minimal() +
  theme(plot.title = element_text(size = 8, face = "bold")) +
  theme(axis.title = element_text(size = 8, face = "bold")) +
  geom_text_repel(size=2.5,label=states$st, col="grey", vjust=-2) +
  ggtitle("Figure 7-16: Abortion in the United States") +
  ylab("Abortions per 1,000 Women") +
  xlab("Income in Dollars per Capita") +
  annotate("text", x = 45000, y = 30,
        label = "West", col="#00bf00") +
  annotate("text", x = 32000, y = 12,
        label = "South", col="#bf0000") +
  annotate("text", x = 50000, y = 35,
        label = "Northeast", col="#263333") +
  annotate("text", x = 39000, y = 5,
        label = "Midwest", col="#00ffff") +
  guides(col=FALSE, shape=FALSE) +
  annotate("text", x = 52000, y = 17,
        label = "Mean of Y", col="grey") +
  annotate("text", x = 42200, y = 41,
        label = "Mean of X", col="grey") +
  annotate("text", x = 52000, y = 12,
        label = "point size = log of population density",
                col="grey") +
  guides(col=FALSE, shape=FALSE) +
```

```
geom_hline(aes(yintercept = mean(states$abort)), col="grey",
           linetype = "dashed") +
geom_vline(aes(xintercept = mean(states$inc)), col="grey",
           linetype = "dashed")
```

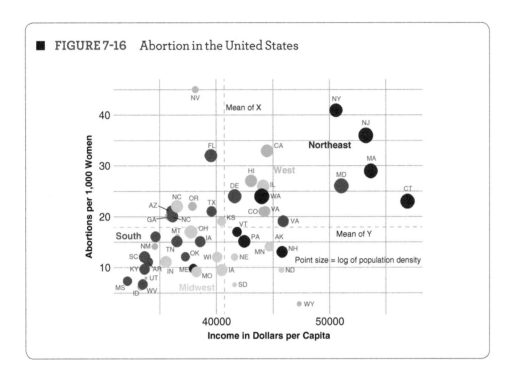

FIGURE 7-16 Abortion in the United States

KNOWLEDGE CHECK: List some elements of style.

---

1. Which of the following represent chart junk or chart clutter?
   a. Unnecessary legends or axes
   b. Gray background
   c. Labeling points in a scatter plot
   d. Too many colors in a plot
2. Which of the following are useful ways to focus attention on certain features of a figure?
   a. Label the points.
   b. Color the points according to groups.
   c. Highlight the feature with a distinct color.
   d. Highlight the feature by changing its size.

3. Which of the following are effective ways to focus attention?
    a. Color
    b. Shape
    c. Size
    d. Integrate words with pictures
4. When should you use a picture?
    a. When you really want to emphasize something
    b. When you have more than three pieces of information
    c. When the page could use a little color
    d. When you're confronted with a table of multiple numbers
5. When should you use an informational approach?
    a. When you want to provide a reference
    b. When you want to provide a lot of information
    c. When you want to understand relationships and provide information
    d. When you want to make a single point

The Basic Elements of a Story

While we may not be engaged in writing the next great American novel, the latest HBO crime drama, or the next iteration of the Marvel Comics story line, present your data analysis with care. Current work in psychology suggests that telling stories is the most effective way to communicate information.[2] Toward that end, here are the basics.

Every story has a hero and every hero has a mentor: a trusted individual who serves as an adviser, confident, or guide. Famous mentors abound in the most popular stories: Yoda in the *Star Wars* films, Dumbledore in the *Harry Potter* books, or "Doc" Brown in the *Back to the Future* trilogy. You are the mentor. Guide and advise the hero through the journey.

The audience is the hero. Too often authors assume that role, losing the audience's trust. When the audience is treated like the hero, presentations are respectful and cater to its needs, providing the advice and tools necessary to complete the journey.

The framework we'll use is a device known as the hero's journey, a structure found in many tales. First articulated in 1949 by Joseph Campbell as a simple three-part story, screenwriter Christopher Vogler (2007) expanded on it by identifying 12 different components. We'll keep it simple by focusing on the three acts of the hero's journey. In act I, the hero is introduced, the context is established, and there is an event or development that encourages the hero to embark on the journey. In act II, the hero begins the journey, enters another world, and is met with a series of obstacles and challenges that require the hero to grow, develop, or mature. In act III, the conflict is resolved or the obstacle is overcome and the hero returns victorious. While we're not writing a screenplay for a major studio, paying attention to the basic components of storytelling persuades an audience to act.

---

[2] For a complete treatment of the reasons why storytelling is an effective method of communication, see Nancy Duarte's book, *Resonate* (Duarte, 2010). Her book also does a nice job of building contrast and tension in the story to make for more effective presentations.

With these different roles in mind and the basic structure of a story (context, journey, resolution), you're ready to start constructing effective presentations. It bears mentioning again, the more we pay attention to communicating our results, we increase the odds that our point will find its mark.

> **KNOWLEDGE CHECK: Tell your story.**

6. Setting the context in a story is synonymous with which of the following?
   a. Model estimation
   b. Diagnostics
   c. Describing data
   d. Generating hypotheses
7. The journey is analogous to which of the following?
   a. Making your argument
   b. Identifying the puzzle
   c. Solving the puzzle
   d. Generating new questions
8. Act III, resolution of the conflict, is analogous to which of the following?
   a. Offering evidence in support of your hypothesis
   b. Identifying the hypothesis
   c. Your conclusion
   d. Generating the next question

Documentation (Establishing Credibility as a Storyteller)

Now that we've established some principles of displaying data and the basic outlines of telling a story, the next three sections show how they connect. Let's start with documentation and establishing credibility as a mentor. For those lucky enough to see Tufte's presentations in person, one is struck by his focus on documentation. Good mentors document everything. Put simply, properly documenting your data establishes you as a credible storyteller. Below are aspects of documentation that build credibility.

Where do the data come from? Science depends on the ability to reproduce results. Analysis can't be reproduced unless others know its source. At the most basic level, documentation refers to indicating where the data come from. Scientific progress is accelerated when the origins of the data are identified *and* the data are made publicly available.

The first rule in documenting data analysis is to identify the data source. Reading a document or hearing a presentation should directly connect the reader or audience member to the data. Most scholars will provide a citation somewhere in the body of the text, but including the information in the notes of a table or figure can be helpful. There is no substitute for making your data public.

What do the data say? Be very clear about the data with accurate labels. Impatience is the enemy. Anxious to proceed, the analyst pays little attention to the data's units. Not knowing whether the number represents an amount or a rate, for example, causes confusion. The three examples in our data that create the most chaos are measures of homicide (*world* data set), infant mortality (*states* data set), and abortions (*states* data set): the murder rate represents the number of murders per 100,000 population; the infant mortality rate is the number of deaths per 1,000 live births; the abortion figure records the number of abortions in a state per 1,000 women.

How were the data made? Continuing with the homicide rate example, how were those numbers collected? What agencies compiled the data and how were they constructed? What organization entered them into the computer and is providing them for public consumption? In the case of homicide rates, one might expect to see very different numbers if they were collected by a national health ministry, a national police agency, or a lobby group funded by a private corporation involved in prison construction.

Some of the most important debates of our time hinge on where the data originate and how they are constructed. A recent cover story in *The Economist*, as an example, reported on the current debate on measuring income inequality ("Economists Are Rethinking the Numbers on Inequality," 2019). Over the last few years, a number of books and articles have been written on the growing income inequality in Western Europe and the United States (Piketty, 2014; Piketty & Saez, 2003; Piketty & Zucman, 2014; Piketty et al., 2018). Some argue we now live in a new gilded age where a small percentage of the population controls a growing percentage of national wealth. Much of those data are based on tax returns. While an improvement over previous work, data based on tax returns are not perfect. Without getting into the details, a debate now rages among economists about whether previous figures based on tax returns accurately portray current trends. Electoral campaigns and policy hinge on whether income inequality is increasing or decreasing. Establishing where the data come from and how they were made is fundamental.

How were the data altered? In addition to documenting where the data were obtained, data analysis almost always involves analyzing a subset of the data. For example, when examining GDP per capita, the oil states of the Gulf region stand out because of their vast oil wealth and small populations. Does it make sense to include those countries in the analysis? Always provide a very clear treatment of how the subset was constructed. What cases were left out and why? Place that discussion front and center in the presentation. If you feel the need to bury them in an obscure place that will not catch the reader's attention, there's probably a reason why.

> **KNOWLEDGE CHECK: Describe the elements and importance of documentation.**

9. Why is documentation so important?
    a. It establishes your credibility.
    b. The hero is more likely to accept you as a mentor.
    c. Science depends on replication.
    d. It establishes the data's credibility.

10. Which of the following are the key elements of documentation?
    a. Accurately describe the units.
    b. Indicate who produced the data.
    c. Outline the coverage or boundaries of the data.
    d. Demonstrate that the measure is valid.

Build an Intuition (Setting the Context)

The first part of any story involves setting the **context**. Familiarize your audience with the data. Give them a firm grasp of the problem and an accurate picture of the dependent variable and the main independent variable. The more novel your measures, the more time you should spend explaining their construction and main characteristics.

Sometimes it's helpful to show all of the data, clearly labeling all of the cases in a scatter plot or bar plot. With big data—lots of observations—it often helps to pick a few cases that will resonate with the audience.

Focusing attention on the problem, gap, struggle, or anomaly is another way to set the context for your audience. We would all be better served if all PowerPoint presentations started with text scrolling upward, floating into the galaxy explaining the current state of affairs between the Empire and the Rebels. Try to achieve the same purpose with helpful views of the data, setting the stage for the hero's journey.

> **KNOWLEDGE CHECK: Explain how to set the context.**

11. Setting the context rests on which of the following?
    a. Describing data
    b. Building an intuition
    c. Showing your data
    d. Connecting the audience to the data

Show Causation (The Journey)

Once the context is set, explain why the problem, gap, struggle, or anomaly exists. Visuals conveying a sense of exploration and information are not as useful in this endeavor as ones that hit the audience right between the eyes. Below are some hints that help tell the causal story.

Correlation is not causation, but your writing should reflect one or the other. Tables and figures should make clear statements that directly relate to the issue at hand. Extraneous words and sentences muddle our writing; extraneous tables and figures muddle our analysis and argumentation. The best way to avoid extraneous material is to tell a very clear and decisive causal story.

Careful analysts know that their descriptions and comparisons don't prove **causation**. Consequently, the language they use either stays strictly in the realm of correlation or it is explicitly causal. We run into trouble when our language goes back and forth between correlation and causation, attempting to push the boundaries of what the data can actually say. It's better to stick with one approach or the other. Perhaps the best way to proceed is to acknowledge the limitations of the data, then tell your causal story.

Breaking the chains of causation. Some variables are correlated because of a long chain of events. When we focus on the two endpoints of this chain, this is the **reduced form** of the causal relationship. Take, for example, Robert Putnam's work on social capital and good governance (Putnam, 1993). Putnam finds a strong empirical relationship between a population's level of social capital—participation in civic life—and the effectiveness of government. Linking civic activity to policy outcomes involves a fairly extensive causal chain of events. Individuals who participate in social and athletic clubs, for example, develop the requisite skills for effective political participation. Governments faced with populations characterized by high levels of effective political participation will be compelled to provide effective policy.

The reduced form of Putnam's argument is the correlation between the number of social organizations in society and measures of institutional effectiveness: number of family clinics, daycare facilities, cabinet stability, and so forth. Connecting these variables involves a fairly long chain of events: participation in social organizations → higher levels of social capital → effective participation in government → good government outcomes. While it is important to establish whether the two endpoints of the process are correlated, always endeavor to show the connections between them. Tables and figures that reveal the intermediate steps are not always possible, but the analyst should leave no stone unturned in efforts to verify they exist.

A table or figure can be judged by its title. Tables and figures make causality explicit with strong, declarative titles. While there's a tradition of being overly cautious, vacuous titles can be less than helpful. While an approach that avoids making causal claims is laudable, a title that explicitly states what the figure or table shows seems more effective if not more honest. A title for a scatter plot will usually read "Scatter Plot of Y Against X." Instead, try "X Causes Y." Analysts will differ. The advice here is to stick with one approach or the other. Avoid the pretense of correlation while slipping in causal language when the reader least suspects it.

> **KNOWLEDGE CHECK: Establish causation.**

12. Act II, the journey, is synonymous with which of the following?
    a. Accurate descriptions of the data.
    b. Figure and table titles that make strong, declarative statements.
    c. A strong causal argument.
    d. Describing all links that connect the causal chain.

From Causation to Action (The Resolution)

What are the implications of your findings? What do you hope your audience will do after they are presented with the evidence (the story)? Often we conduct the analysis, generate the findings, and then present them without thinking about what they imply. Presenting data without an end goal in mind is tantamount to an abrupt ending to a book, movie, or song. It also leaves the audience with the question, so what?

Work backward. The story you tell and the evidence you marshal should be a call to action, whether it's organizing a protest, persuading a board to vote for your proposal, or convincing policymakers to adopt a policy. Focusing on the action you want to encourage helps eliminate extraneous information. This is particularly useful since the temptation to share too much is particularly strong given the increasing ease technology affords us in generating colorful, exotic, and even cute (think "catterplots") representations of the data.

> **KNOWLEDGE CHECK: Identify the actions implied by the analysis.**

13. How does calling for action help the story?
    a. It identifies extraneous material.
    b. It provides a sense of purpose.
    c. It gives the storyteller credibility.
    d. It provides a logical end to the story.

SUMMARY

Generating pictures of quantitative data is an art. While there are no hard and fast rules, there are some best practices for effective presentation. Good documentation plays an essential role, resulting in a more persuasive argument. Being explicit about causation raises the level of transparency and clarity. Finally, using all modes of information available—text and pictures—enriches, refines, and strengthens our arguments. Our motivations, intentions, and level of patience become readily evident, ultimately affecting our ability to persuade.

When the author's words come both from the head and the heart, it's much easier to establish credibility. It's also clear when the audience is treated with respect. These aspects of storytelling are important since they build a bond of trust between the author (mentor) and the audience (hero).

Storytelling is increasingly recognized as a necessary form of communication when presenting data. Many of the most important elements of storytelling fit nicely with the best practices associated with visualizing data. As argued in the beginning of this chapter, many of the more practical rules for presenting data—respect for the audience, clarity, and decisiveness—are also foundational aspects of how we express ourselves with writing. In that sense, how we present data is a lot like writing; it provides a window into who we are.

COMMON PROBLEMS

- *Graphs are used when sentences will do.* Once you learn how to produce a number of different kinds of visualizations, there is a tendency to "visualize" everything when simple sentences will work. The easiest way to determine whether a figure or table is actually needed is to do a quick inventory on how many actual pieces of information are conveyed in the figure. If fewer than 10, use a table. If fewer than five, stick with a sentence.

- *Know your purpose.* Before constructing a visualization, determine whether you're trying to explain, persuade, or inform. More than anything, each of these three purposes has the greatest bearing on what your visualization will look like. If these purposes are confused (your intent does not match the actual visualization), the audience will be confused.

- *Know your audience.* Is the room filled with experts on the subject, students eager to learn, or community members looking for a call to action? When your audience isn't considered, your ability to appear as a credible source of information is diminished. No single presentation is appropriate for all audiences.

- *Don't confuse clutter with information.* Too much information displayed poorly is inelegant, not clutter. Clutter is simply stuff that appears in a figure or table that conveys no information whatsoever. Figures that contain a lot of information may seem busy and dense, but they may not contain any clutter. In some cases, the dense and information-packed figure may be appropriate. If the purpose is informational, the large amount of densely-packed information might just fit the bill.

- *Spend the time necessary to learn your craft.* Miles Davis, the famous jazz musician, is quoted as saying, "Sometimes you have to play a long time to be able to play like yourself." Since the goal is expression and the most effective form of communication comes from inside, being able to effectively manipulate colors, shapes, and words in a figure is crucial. Facility with R is analogous to knowing your way around a trumpet.

- *Devote enough time to the enterprise.* Introductory statistics books rarely spend any time with considerations on presenting data. That is unfortunate. Nothing can be done if the data are poor, uninformative, or uninteresting. Having said that, all too often genuinely useful and interesting data are poorly presented, obscuring important truths.

REVIEW QUESTIONS

1. What are the components of documentation?
2. How can we make a causal argument with figures and tables?
3. Why does integrating words and pictures together matter?
4. How does impatience manifest itself in communicating data analysis?
5. How is writing like presenting pictures of data?
6. How is presenting arguments based on quantitative data a moral exercise?
7. What colors should one use when presenting data?
8. When should you use a bar plot, table, or sentence?
9. How should tables and figures be titled?
10. What is the reduced form version of an argument?

PRACTICE ON ANALYSIS AND VISUALIZATION

1. Write an *ifelse()* command that identifies Kansas and Nebraska and colors their points blue and all other points gray.
2. Write an *ifelse()* command that identifies Brazil and Russia and colors their points red and all other points gray.
3. Which are the defining features of an exploratory visualization?
   a. An emphasis on maximizing the amount of data per square inch
   b. Highlighting the main point of the figure

c. Giving all aspects of the visualization equal weight

d. A title that makes a statement

4. Which are the defining features of an explanatory visualization?

   a. An emphasis on maximizing the amount of data per square inch
   b. Highlighting the main point of the figure
   c. Giving all aspects of the visualization equal weight
   d. A title that makes a statement

5. Which are the defining features of an informational visualization?

   a. An emphasis on maximizing the amount of data per square inch
   b. Highlighting the main point of the figure
   c. Giving all aspects of the visualization equal weight
   d. A title that makes a statement

6. Which of the following is one of the three acts of a story?

   a. The journey
   b. The end
   c. Setting the context
   d. The mentor

7. Which of the following are good figure or table titles?

   a. Scatter plot of *murderrate* and *inc*
   b. Scatter plot of homicides per 100,000 and average household income
   c. Income and murder are related
   d. Conditions of poverty fuel homicide

8. When presenting data, who of the following is the hero of the story?

   a. The author
   b. The reader
   c. The audience
   d. The data

9. What command is helpful when drawing two separate curves in a figure?

   a. The *ifelse()* command
   b. The *subset()* command
   c. The *aes()* command
   d. The *theme_minimal()* command

10. Which of the following are important principles to follow when presenting data?

    a. Focus the audience's attention.
    b. Know your audience.
    c. Eliminate the clutter.
    d. Integrate words and pictures.

ANNOTATED R FUNCTIONS

The following functions appear in this chapter. They are listed in order of their first appearance (with the code chunk number in parentheses) and annotated here to give a very brief description of their use. Some are not stand-alone functions and only work in combination with other commands. As a reminder, the code in every chapter will work properly if executed in the order it appears. Proper execution also depends on typing the author-defined *libraries()* command, which loads the required R packages.

ggplot(): defines the basic structure of a plot (usually the x and y variables). (7-1)

aes(): the aes (called "aesthetics") function is used in ggplot to define the basic structure of the plot, which often includes the variables you want to use and any shapes or colors. (7-1)

ggtitle(): main title of the ggplot. (7-1)

ylab(): labels the y axis in ggplot. (7-1)

xlab(): labels the x axis in ggplot. (7-1)

theme(): specifies font, size, and so forth in a ggplot. (7-1)

ifelse(): logical function that allows you to construct an if-then statement. Useful for selecting specific cases to label and also for creating categorical variables from continuous variables. (7-2)

annotate(): places text in the figure. (7-2)

theme_minimal(): specifies a minimalist style for ggplot. (7-2)

geom_smooth(): draws a curve or a straight line in a scatter plot. (7-3)

subset(): chooses a specified subset of the data. (7-4)

geom_text_repel(): labels points and draws them so that the labels don't overlap. (7-6)

geom_bar(): draws a bar plot in ggplot. (7-7)

count(): returns the number of cases or observations in a category or a variable. (7-8)

labs(): specifies the legend titles in a ggplot. (7-12)

geom_hline(): draws a horizontal line. (7-17)

geom_vline(): draws a vertical line. (7-17)

set.seed(): tells R on what number to start the random number generator. Setting the seed allows you to duplicate results based on random numbers if it matches the number in the example. (7-13)

multicat(): a fun function that draws a "catterplot." (7-13)

guides(): allows you to make changes to the legends and guides that are automatically produced in ggplot. (7-15)

scale_colour_manual(): allows you to specify exactly what colors to use in a plot. (7-17)

ANSWERS

KNOWLEDGE CHECK

1. a, b
2. c, d
3. a, b, c, d
4. d
5. a, b, c
6. c
7. b
8. a
9. a, b, c, d
10. a, b, c
11. a, b, c, d
12. b, c, d
13. a, b, d

PRACTICE ON ANALYSIS AND VISUALIZATION

1. col=ifelse(states=="KS" | states=="NE", "blue", "grey")
2. col=ifelse(iso3c=="BRA" | iso3c=="RUS", "red", "grey")
3. c
4. b, d
5. a
6. a, c
7. d
8. c
9. b
10. a, b, c, d

 Access digital resources, including datasets, at http://edge.sagepub.com/brownstats1e.

8 The Essentials of Probability Theory

CHAPTER OUTLINE

Learning Objectives
Overview
Populations and Samples
Sample Bias and Random Samples
The Law of Large Numbers
The Central Limit Theorem
The Standard Normal Distribution
Summary
Common Problems
Review Questions
Practice on Analysis and Visualization
Annotated R Functions
Answers

LEARNING OBJECTIVES

- Explain the distinction between populations and samples.
- Recognize potential sources of bias and why randomization helps.
- Identify uses of the law of large numbers.
- Explain why the central limit theorem is so important.
- Connect sampling distributions to the standard normal distribution.

Overview

We often want to know things about a population but don't have the time or money to interview each individual, measure every object, or observe every event. Politicians want to know if their constituents support a proposed piece of legislation. Health officials want to know what percentage of the population has been vaccinated. Politicians question whether students from different ethnic backgrounds, sexual orientations, and political ideologies feel free to express their views on campus. Society depends on this information and getting it can be extremely time-consuming and expensive. Consequently, a significant aspect of data analysis involves drawing a random sample from a population, calculating statistics from that sample, and then determining how closely those statistics match reality.

This chapter lays the theoretical groundwork for the rest of the book. Since we can't count and observe everything, we rely on examining parts of a population: samples. Some important foundational concepts in probability theory allow us to determine whether those samples provide good representations of the entire population. For example, in political polling, we want to know if the information we collected from a random sample of 1,500 likely voters provides a good representation of the entire nation. An admissions officer will want to know if a college's incoming first-year class (the sample) accurately represents the pool of qualified students (the population). Finally, epidemiologists at the Centers for Disease Control and Prevention (CDC) want to know if the number of measles cases in a community is relatively large, indicating a potential outbreak. All of these situations, some with life or death consequences, depend on knowing how closely a sample statistic (whether the mean, median, standard deviation, etc.) matches the population. This chapter presents the essential concepts of probability theory that undergird statistical inference.

We will proceed as follows. First, the problem: we need to know something about a population but can't afford (either because of time or money) to count or observe everything or everyone. The solution is to take a random sample, calculate a statistic from the sample (perhaps a mean, median, or standard deviation), then use probability theory to calculate precisely how closely that sample statistic approximates the true mean, median, or standard deviation. Underpinning this exercise are two fundamental concepts of probability theory: the *law of large numbers* and the *central limit theorem*. Both concepts rely on repetition. It need not get any more complicated than understanding how flipping a coin 10 times (an experiment) could produce eight heads and two tails but if we repeat the experiment 100 times, the mean of all those experiments will converge toward five heads and five tails. We learn this from the law of large numbers.

If we increase the number of times we flip the coin in each experiment (e.g., 10 to 50), the resulting distribution of those means will begin to resemble the shape of a normal curve. We learn this from the central limit theorem. This holds important implications for using samples to understand a population. We use the normal curve to calculate how close our sample estimate is to what's going on in a population. Why is that important? In many circumstances, a rough guess will do. For example, what percentage of the U.S. population prefers Coca-Cola versus Pepsi? In life and death situations (e.g., the percentage of the population that has developed antibodies for COVID-19), we need more certainty.

Populations and Samples

A **population** is a defined set of individuals, events, or objects. For example, we could define the set of individuals residing in the United States as a population. Whether it's the U.S. population or the students at a university, a population is a defined set of individuals, events, or objects that we want to know something about. Every 10 years the U.S. government conducts a census to count every individual living in the country, an expensive and cumbersome process. Since the census only occurs every decade, if we want to know something about the U.S. population during the intervening years, we're forced to examine a subset of the population.

Samples are a subset of the population and are used to determine population characteristics without surveying every member of the population. When populations number in the millions, taking samples is the only way to proceed. We also take samples when the population is much smaller. When universities want to know about their student population (e.g., roughly 30,000 students in a large public institution), it's more cost-effective to take a random sample of the population, calculate a statistic from that sample, and make inferences about the student body. To determine the characteristics of a population, we take **random samples**: small subsets of the population in which each object or individual has an equal chance of being selected. Random selection undergirds the entire enterprise.

When referring to populations, we use a specific vocabulary and notation: we use the capital N for the number of individuals, events, or objects in a population, and we use the term *population parameters* to describe the characteristics of the population (means, standard deviations, etc.). By convention we use Greek letters for population parameters such as the mean μ and variance σ. When referring to samples, we use a small n for the number of individuals, events, or objects and we use the term *sample statistic* to describe the characteristics of the sample. By convention we use English letters when referring to sample statistics such as the sample mean $\bar{x}$ and sample variance s.

Art and Practice of Data Visualization
BE PRECISE WITH LANGUAGE

Statistics, like any language, has users or speakers who are less precise than others. When we talk about calculating characteristics of our sample, we use the term "statistics." When we are referring to the population, we use the term "parameters." While these are seemingly overly fine distinctions, not using the proper terms causes confusion. Distinguishing between statistics and parameters may seem unnecessary. My experience indicates it is necessary. If you don't keep the difference in mind, the clear distinction between samples and populations will forever be muddled, clouding your understanding as we march forward through the chapters to come.

KNOWLEDGE CHECK: Explain the distinction between populations and samples.

---

1. In each instance, indicate whether the statistics class described is a sample or a population.
    a. The class is surveyed to find out whether the professor performed well.
    b. The class is surveyed to find out whether more liberal-leaning students were treated fairly in the class.
    c. The class is surveyed to get a sense of whether college students plan to vote in the next election.
    d. The class is surveyed by student government to decide on whether to expand the recreation center.
2. Which are the defining characteristics of a sample?
    a. It is a subset of a population.
    b. It includes all of the people, places, or things of a population.
    c. We use parameters to define the characteristics of a sample.
    d. We use statistics to define the characteristics of a sample.

Sample Bias and Random Samples

Let's use the university example to introduce the notion of **sample bias** and explain why taking a random sample is so important to this enterprise. Sample bias occurs when the sample is not drawn randomly from the entire population. Every member of the population does not have an equal chance of being drawn.

Suppose a university's board of regents wants to learn things about the climate on campus. Having every single individual on campus fill out a survey is not only expensive, it's time-consuming and would be very difficult to implement. Instead, the administration decides it must draw a sample from the population to find out, for example, if students on either end of the political spectrum feel free to express their opinions in class or if different minority groups feel welcome on campus. Could the administration simply have the students from this statistics class take the survey and draw inferences from that sample? Are the students in this class representative of the population as a whole?

In order to obtain unbiased estimates, the sample we draw from the population must be random. Using the statistics class as the sample might introduce bias in a number of different ways: (1) students in statistics may be more conservative or liberal relative to their colleagues, (2) the class meets during a time only available to students who don't have work study, or (3) perhaps the class is primarily made up of first-years and their opinions about the university's climate are likely to change over time. The list could go on.

When we sample randomly, each and every individual in the population must have an equal probability of being drawn. From a practical point of view, we'd like to put every student's name in a hat and draw our sample randomly. Since 30,000 names won't fit in a hat, we can assign each student a number and use a random number generator to choose a subset from the population. Drawing randomly eliminates the possibility of sample bias. Since all confounding factors (e.g., partisanship, work-study status, or year in school) that could bias the results are equally likely to be present in our random sample, the statistics we calculate from the sample will be unbiased.[1]

Sampling randomly is not a trivial exercise. In 1969, there was a famous incident of random sampling gone wrong. Before the United States went to an all-volunteer army, it relied on drafting individuals to serve. At the onset of the draft, the U.S. government planned to randomly select individuals (young males) to serve in Vietnam. Each day of the year was printed on a small piece of paper and then inserted into a small plastic capsule. On live television, the capsules were placed in a large bin where they were mixed. A blind-folded head of the selective service reached in the bin and chose a capsule. The first capsule chosen was assigned draft day 1. The process was repeated 366 times since leap days were included. The numbers assigned to each day of the year would then determine the order young men were drafted. Those drafted eventually noticed that days toward the end of the year were higher up in the draft order. Apparently, the capsules were not mixed adequately and thus they were not completely randomized. Consequently, in the beginning of the draft, young American males born in the later part of the year were more likely to fight in Vietnam.

---

[1] There is a variant of random sampling that some scholars insist on: **sampling with replacement**. In this example, when a student is selected, his or her name would be placed back into the pool of possible candidates to be drawn again. This is called sampling with replacement and ensures that each individual has the same chance of being drawn and that every draw is independent of the other: whoever is drawn first has no impact on the probability that others will be drawn. While sampling with replacement does guarantee independence (the object selected does not influence the probability of others being selected in subsequent draws), random selection without replacement is usually adequate to achieve accurate estimates, especially if being selected multiple times would affect the object or individual (they might have to change their answer if selected more than once).

Random digit dialing, a common technique used to conduct surveys, represents another example of the challenge associated with randomization. With the advent of cell phone numbers that have effectively been blocked and because many low-income Americans don't have land lines or cell phones, the practice of randomly dialing telephone numbers across the country increasingly fails to provide a representative sample of all people living in the United States. Caller ID and the ability to block callers presents the survey industry with a particularly difficult challenge.

The two previous examples illustrate the rigor random sampling entails. Failure to account for small details (like mixing the capsules up completely) can have an important impact, introducing bias into the study. With a good understanding of samples, populations, and random sampling, we're ready to discuss two concepts that serve as the foundation of statistical inference.

> **KNOWLEDGE CHECK: Recognize potential sources of bias and why randomization helps.**

3. What are some reasons why our statistics class might be a biased sample of university students?
    a. It's an introductory course and most students will be first-year students or sophomores.
    b. Most of the students are from the social sciences.
    c. It's not a randomly drawn sample.
    d. Perhaps social science students are more liberal than the rest of the population.
4. Which are characteristics of randomly drawn samples?
    a. Everyone or everything in the population has an equal chance of being drawn.
    b. Each draw is independent of the previous draws.
    c. All confounding factors have an equal chance to be present in the sample.
    d. None of the confounding factors will be present in the sample.
5. Why was there bias in the U.S. draft example?
    a. The capsules were not properly mixed.
    b. Each capsule did not have an equal chance of being chosen.
    c. The random draw did not account for leap years.
    d. There was no sampling with replacement.

The Law of Large Numbers

Established centuries ago and lying at the heart of all probability theory is the **law of large numbers**. Insights associated with the law of large numbers were first recorded by the Swiss mathematician Jakob Bernoulli in a posthumously published book in 1713 (Tijms, 2004, p. 21). The law states the following:

> If a certain chance experiment is repeated an unlimited number of times under exactly the same conditions, and if the repetitions are independent of each other, then the fraction of times that a given event A occurs will converge with probability 1 to a number that is equal to the probability that A occurs in a single repetition of the experiment. (Tijms, 2004, p. 20)

Let's place this statement in the context of political polls. In this case, a chance experiment refers to a random sample of n eligible voters in a U.S. senator's state. If these random samples are repeated and are independent of each other, the law of large numbers implies that the percentage of our respondents who indicate they support Senator A will converge with probability 1 to a number that is equal to the probability that A occurs in a single repetition of our experiment. If 30% of the population supports Senator A, the probability that support for Senator A will occur in one replication of our experiment will be 30%. Put more concretely, as we keep repeating the experiment and calculate the percentage supporting Senator A from each sample, the percentage of voters who say they support Senator A will converge to 30%.

To further develop the concept, we need to introduce an extremely important construct: the **sampling distribution**. A sampling distribution is the distribution of sample statistics collected from each sample. When we draw a sample, we calculate a statistic from it and then we repeat by drawing another sample and calculating a statistic from that sample. We repeat this many times, recording the statistic from each sample. We then take all of those sample statistics we recorded from our samples and plot them using a histogram or a density plot. The resulting distribution is called the sampling distribution. To illustrate, consider the following example.

Visualizing the Law of Large Numbers

Visualizing the law of large numbers involves several steps. First, we need to generate some numbers to create a population. Second, we need to draw random samples from that population. Third, we calculate the sample statistic from each sample. Finally, we plot those statistics using a histogram or density plot.

Generating random data, taking a sample, calculating the sample mean, and plotting that mean take a little more code than usual. Fortunately, R provides the tools necessary to specify these steps and repeat them.

First, we set the random number generator in R to a specific number with the *set.seed()* command so that you can repeat this on your computer and get the same exact random numbers I generated here (Code Chunk 8-1). We can set that number to be any number; I set it to 3,376. After setting the random number generator to 3,376, we arbitrarily define the population to have a mean of 100 and a standard deviation of 10. We define the mean *mu* and we define the standard deviation *sigma*.

Code Chunk 8-1

```
set.seed(3376)
mu=100; sigma=10; n=10
```

After defining the basic characteristics of the population we're going to draw from and the number of observations in each sample (10), we let R know that we want to create a variable where we can put the mean from each sample we draw (Code Chunk 8-2). We tell R that we want a variable that will hold five observations: *xbar1=rep(0,5)*. We then instruct R that we want to conduct an operation five times, *for (i in 1:5)*, and we want that operation to calculate the mean of a sample we draw from a population whose mean = *mu* and whose standard deviation = *sigma*, which we've just defined (Code Chunk 8-1) as 100 and 10.

Code Chunk 8-2

```
xbar1=rep(0,5)
for (i in 1:5) {xbar1[i]= mean(rnorm(n, mean=mu, sd=sigma))}
```

Code Chunk 8-3 simply repeats that step for the sampling distributions we want to create when we draw 10 and 100 samples. This time, each mean calculated is stored in a vector (a variable) called *xbar2* and *xbar3*.

Code Chunk 8-3

```
mu=100; sigma = 10; n=10
xbar2=rep(0,10)
for (i in 1:10) {xbar2[i]=mean(rnorm(n, mean=mu, sd=sigma))}

mu=100; sigma = 10; n=10
xbar3=rep(0,100)
for (i in 1:100) {xbar3[i]=mean(rnorm(n, mean=mu, sd=sigma))}
```

Now that we've calculated the means for 5, 10, and 100 samples and we've placed the results in three vectors (xbar1, xbar2, and xbar3). We need to put the data in a list and then stack each list on top of each other so that we can graph it. The two commands in Code Chunk 8-4 reorder the structure of the data so that we can place density plots over each other to illustrate. Figure 8-1 provides a schematic diagram of what we're doing to the data. In the diagram, *V1* represents the numbers we collected when we sampled 5 times, *V2* represents the numbers we collected when we sampled 10 times, and *V3* represents the numbers we collected when we sampled 100 times. Once they are put together in a list, we "reshape" that list with the *melt()* command. We're treading into some deep waters with respect to data structure here. It's important to understand, however, the steps necessary to create some of the figures. This is a useful exercise since other kinds of visualizations require this kind of manipulation.

Code Chunk 8-4

```
x <- list(v1=xbar1,v2=xbar2,v3=xbar3)
data <- melt(x)
```

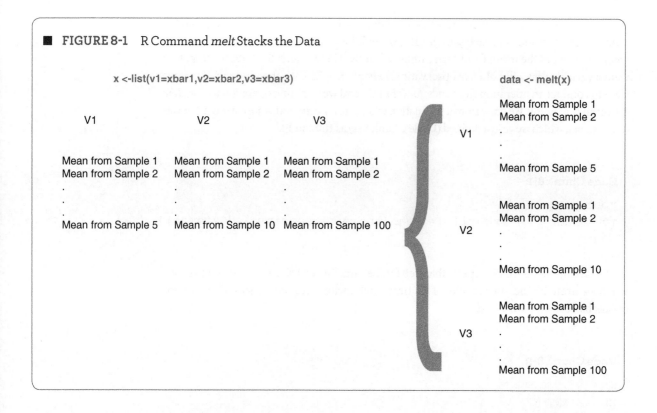

FIGURE 8-1 R Command *melt* Stacks the Data

Finally, with all of the means calculated and all of them arranged in the correct way (stacked into one variable), we can plot the three density plots and overlay them to illustrate how the mean of each sampling distribution approaches the population parameter as we repeat the number of times we run the experiment. This is the principle concept of the law of large numbers. As we increase the number of times we repeat an experiment—flip 10 coins or take a sample of 10 observations—the mean of the resulting sampling distribution will approach the population parameter.

As the density plots of the three sampling distributions illustrate (Figure 8-2), when we increase the number of experiments—from 5 to 10, then 100—the mean of the sampling distribution gets closer to 100. The colored vertical lines show where the mean is for each density plot. When we repeat the experiment five times, plot the means, and calculate the mean of that sampling distribution, we get 101.811. When we increase the number of experiments to 10, the mean of that sampling distribution is 100.480. Finally, when we increase the number of experiments to 100, the resulting mean is 100.071.

There are a couple of things to note about the code (Code Chunk 8-5) that draws Figure 8-2. First, we calculate the means for each sampling distribution xb1, xb2, and xb3. We're then ready to draw the distributions. We use a density plot, which we've used before. It's similar to a histogram in that it shows the picture of a variable's distribution. Also note that within the *geom_density()* function I use the *alpha=.50* option. This changes the opacity of the colors so that when we overlay the density plots, the first one isn't completely masked.

Code Chunk 8-4

```
xb1 <- mean(xbar1)
xb2 <- mean(xbar2)
xb3 <- mean(xbar3)

ggplot(data,aes(x=value, fill=L1)) +
  geom_density(alpha=.50) +
  ggtitle("Figure 8-2: Increased Draws, Increased Accuracy") +
  theme_minimal() +
  theme(plot.title = element_text(size = 8, face = "bold")) +
  theme(axis.title = element_text(size = 8, face = "bold")) +
  geom_vline(xintercept=xb1, col="#bf0000") +
  geom_vline(xintercept=xb2, col="#008b00") +
  geom_vline(xintercept=xb3, col="#0000ff") +
  annotate("text", x = 95, y = .15, label =
        "Mean with 5 draws = 98.895", col="#bf0000") +
  annotate("text", x = 95, y = .175, label =
        "Mean with 10 draws = 100.481", col="#008b00") +
  annotate("text", x = 95, y = .20, label =
        "Mean with 100 draws = 100.071", col="#0000ff") +
  theme(legend.position = "none")
```

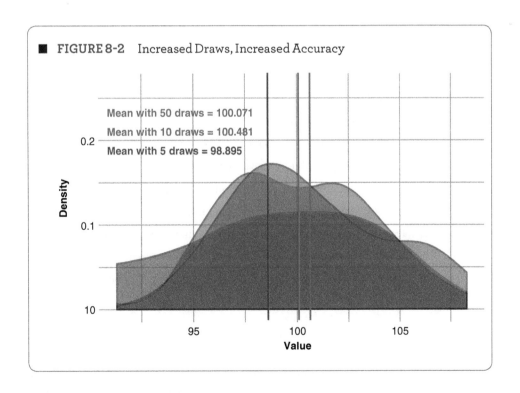

FIGURE 8-2 Increased Draws, Increased Accuracy

The law of large numbers states that the more we repeat the experiment, the mean of the resulting sampling distribution will converge with probability 1 to the number that is equal to the population mean. Note how the sampling mean is 100.811 when we take 5 samples, 100.481 when we take 10 samples, and 100.071 when we take 50 samples. As we increase the number of samples, the mean of our sampling distribution is getting closer to true population mean that we specified: 100.

The law of large numbers allows us to determine certain characteristics of a population so long as we conduct enough experiments (i.e., take enough samples). As we increase the number of experiments, we know we're getting closer to the population parameter. We're heading in the right direction.

Unfortunately, given our time and resources we rarely have time to take more than one sample. How do we know how close we are to the population parameter and how confident can we be about that estimate with just one sample? The central limit theorem provides the answer.

KNOWLEDGE CHECK: Identify uses of the law of large numbers.

---

6. Which of the following statements are true about the law of large numbers?
   a. As the number of possible outcomes increases, the probability of outcome A decreases.
   b. The fraction of times that a given event A occurs will converge with probability one to a number that is equal to the probability that A occurs in a single repetition of the experiment.
   c. Each experiment has to be independent of the previous ones.
   d. It only applies to random draws with more than 30 observations.
7. Why is the law of large numbers helpful?
   a. We rarely have time to repeat experiments.
   b. It allows us to determine certain characteristics of a population.
   c. We can use sampling distributions instead of surveying the entire population.
   d. It explains the logic behind the business model of modern-day casinos.
8. The visualization of the law of large numbers demonstrated which of the following.
   a. Repeated experiments or draws bring us closer to the population parameter.
   b. The accuracy of our estimate increases as the number of experiments increase.
   c. Our accuracy increases with each additional sampling distribution.
   d. The sampling distribution is considered "the experiment."

The Central Limit Theorem

As noted earlier, the law of large numbers implies that the *long-run* behavior of randomly choosing people for a survey or flipping a coin can be predicted with 100% accuracy. In other words, repetition—surveying more and more people or flipping coins repeatedly—allows us to achieve better accuracy. If we keep taking random samples in our experiment above, the mean of the sampling distribution would get closer and closer to 100. Suppose we drew random samples of voters from a population and asked about support for a U.S. senator. The *long-run* behavior of that process (asking survey respondents about political preferences) can be predicted with 100% accuracy. But time and money are not infinite (we don't have the time or money to keep repeating the experiment), so we rely on the **central limit theorem** to calculate the level of accuracy.

Consider the work of pollsters. They (1) draw a random sample from a state or congressional district, (2) calculate the amount of support for Politician A, and (3) establish a margin of error, claiming that they are 95% confident the *real* percentage point advantage Politician A has over her competitors lies within that range. The central limit theorem provides the probabilistic and mathematical basis for making such claims. How did the central limit theorem come about and what does it say?

The central limit theorem began to take form shortly after Bernoulli's work. In 1730, the French-born mathematician Abraham de Moivre found that increasing the number of trials (flipping coins, in this case) generated a distribution of outcomes that approximated what is now known as the normal curve (Stigler, 2016). The compelling part of de Moivre's discovery was later recognized by the Belgian astronomer Adolphe Quetlet (1796–1894) who found that the normal curve approximated the distribution of many naturally occurring and human-made phenomena (Tijms, 2004, p. 145). If we draw a histogram or density plot of the outcomes—heights of women, wingspans of butterflies, scores on an exam—the distribution is symmetrical around a mean, approximating a curve with specific characteristics. It's hard to overstate the importance of the central limit theorem. The theorem, in conjunction with the law of large numbers, forms the foundation of probability theory. Consider the following quote from Sir Francis Galton (Galton, 1889, p. 66) found in Tijms (2004):

> I know of scarcely anything so apt to impress the imagination as the wonderful form of cosmic order expressed by the "Law of Frequency of Error". The law would have been personified by the Greeks and deified, if they had known of it. It reigns with serenity and in complete self-effacement, amidst the wildest confusion. The huger the mob, and the greater the apparent anarchy, the more perfect is its sway. It is the supreme law of unreason. Whenever a large sample of chaotic elements are taken in hand and marshalled in the order of their magnitude, an unsuspected and most beautiful form of regularity proves to have been latent all along. (p .169)

The central limit theorem says three things:

1. The sampling distribution of means approaches a normal distribution as the number of observations from each sample increases.
2. The sampling distribution of sums $\sum X_i$ is also approximately normally distributed with mean $n\mu$ and standard deviation $\sqrt{n}\sigma$.
3. If the population (each X_i) is normally distributed, then the sample distribution of the mean ($\bar{X}$) and the sum ($\sum X_i$) are normally distributed, no matter the sample size.

The central limit theorem has an important payoff: the three key implications listed here allow us to calculate the accuracy of our estimates as we gather more information. We know that if we repeatedly sample from a population or repeatedly flip a coin (as the number of times we sample or flip approaches infinity), our estimate will converge on the population parameter. Since we don't have the time or money to repeat experiments that many times, it's extremely helpful to know how accurate our estimates are given the time and money we do have. In some contexts we only need to be accurate within several miles (e.g., will we make it to the next city on our last gallon of gas?). In other contexts the level of accuracy must be extremely high (e.g., calculating the tolerances of a crucial material used to manufacture aircraft). Let's take each of the three points listed above and elaborate.

Sampling Distribution of Means Approaches Normality as n Increases

The first point states that if we increase the number of observations in each sample (say from 50 to 500), the sampling distribution of means will approach a **normal distribution**. Remember, the law of large numbers says that if we increase the number of samples, then the mean of the sampling distribution will approach the population parameter (the true population mean). Here, the central limit theorem says that if we increase the number of observations in each sample, then the shape of the sampling distribution will converge to a normal distribution. Recall that a normal distribution is simply a distribution that resembles a bell-shaped curve that is symmetrical and is centered over the sample or population mean. Let's use R to see if it's true.

To illustrate, we'll use the code from our demonstration of the law of large numbers. In this example, however, we'll increase the number of observations rather than the number of samples (Code Chunk 8-6). We'll draw from the same population we did before—one that has a mean of 100 and a standard deviation of 10. To construct each sampling distribution, we'll perform the experiment 10 times by taking 10 samples. To construct the first sampling distribution, we'll draw 5 observations in each sample. In the second, we'll draw 10 observations in each sample. In the last, we'll draw 100 observations.

Code Chunk 8-6

```
set.seed(3376)
mu=100; sigma=10; n=5
xbar1=rep(0,10)
for (i in 1:10) {xbar1[i]=mean(rnorm(n, mean=mu, sd=sigma))}

mu=100; sigma = 10; n=10
xbar2=rep(0,10)
for (i in 1:10) {xbar2[i]=mean(rnorm(n, mean=mu, sd=sigma))}

mu=100; sigma = 10; n=100
xbar3=rep(0,10)
for (i in 1:10) {xbar3[i]=mean(rnorm(n, mean=mu, sd=sigma))}
```

Now that we've created the data, we need to manipulate them so that we can overlay the density plots as we did in Figure 8-2. Again we use the *melt()* function to stack the data on top of each other to draw the plot. The mean for each sampling distribution is calculated, and the ggplot code for the density curves is executed (Code Chunk 8-7).

Code Chunk 8-7
```
x <- list(v1=xbar1,v2=xbar2,v3=xbar3)
data <- melt(x)

ggplot(data,aes(x=value, fill=L1)) +
  geom_density(alpha=.70) +
  theme_minimal() +
  theme(plot.title = element_text(size = 8, face = "bold")) +
  theme(axis.title = element_text(size = 8, face = "bold")) +
  ggtitle("Figure 8-3: Increasing Observations Increases
Normality") +
  annotate("text", x = 95, y = .15, label =
           "5 Observations per Draw", col="#bf0000") +
  annotate("text", x = 95, y = .175, label =
           "10 Observations per Draw", col="#008b00") +
  annotate("text", x = 95, y = .20, label =
           "100 Observations per Draw", col="#0000ff") +
  theme(legend.position = "none") +
  xlab("Mean of Sample Means")
```

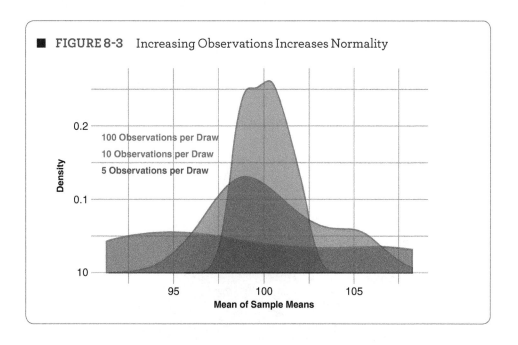

■ **FIGURE 8-3** Increasing Observations Increases Normality

As evident in Figure 8-3, the sampling distribution takes on a normal shape as the number of observations in each draw increases: the sampling distribution colored blue takes on a bell-shaped curve with increased symmetry over the mean of 100. Note how our accuracy improves as we increase the number of observations in each sample. For example, note that when there are five observations per draw (the red sampling distribution), we end up with sample means less than 95 and greater than 105. A little more accuracy is achieved when the number of observations is 10 (the green sampling distribution). When we increase the number of observations per sample to 100 (the blue sampling distribution), the accuracy increases dramatically: all of the sample means are well within the boundaries of 95 and 105. Increasing the number of observations in each sample increases the accuracy of our estimate.

Art and Practice of Data Visualization
STATISTIC'S GOLDEN RULE

Like so many randomly drawn samples, this point is worth making again and again: increasing the number of observations in each sample increases the accuracy of our estimates. Increasing the number of observations is simply getting more information. The more information we have, the more accurate our predictions. What is true with any hypothesis or theory (i.e., getting more information generates better hypotheses and theories) is true in random sampling (i.e., increasing the number of observations increases the accuracy of our estimates).

Sampling Distribution of Sums Is Normal

The second point of the central limit theorem is fairly straightforward given the first. Suppose we take 50 samples of 1,000 observations but instead calculate the sum of each sample rather than the mean. The distribution of those 50 sums will be normally distributed. To demonstrate, all that is needed here is to change our mean calculation to a sum in the code (Code Chunk 8-8).

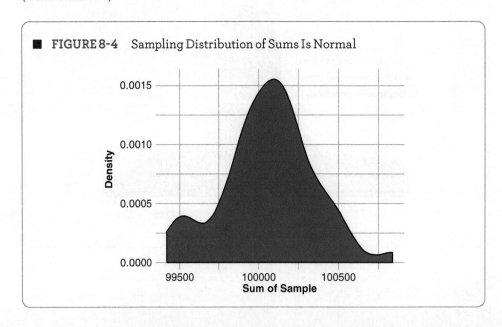

■ **FIGURE 8-4** Sampling Distribution of Sums Is Normal

Code Chunk 8-8

```
set.seed(3376)
mu=100; sigma=10; n=1000
xbar1=rep(0,50)
for (i in 1:50) {xbar1[i]=sum(rnorm(n, mean=mu, sd=sigma))}

x <- data.frame(v1=xbar1)

ggplot(x,aes(x=v1)) +
  geom_density(fill="#0000ff", alpha=.7) +
  theme_minimal() +
  theme(plot.title = element_text(size = 8, face = "bold")) +
  theme(axis.title = element_text(size = 8, face = "bold")) +
  ggtitle("Figure 8-4: Sampling Distribution of Sums Is Normal") +
  xlab("Sum of Sample") +
  theme(legend.position = "none")
```

As the density plot in Figure 8-4 demonstrates, the shape of the sampling distribution of sums approximates normality. It's not perfect since it represents data that we generated randomly. It does show, however, that other sample statistics (other than the mean) exhibit the same behavior when we construct a sampling distribution.

Number of Observations When Drawing From a Normal Distribution

The third point of the central limit theorem states if the population from which we are drawing is normally distributed (something we don't always know beforehand), the sampling distribution will approach normality even when the number of observations in our sample is small. This result implies that if we know we're sampling from a normal distribution, learning about the population will be less time-consuming and less expensive because we don't need as many observations.

To demonstrate, we draw randomly from a population 100 times, but in each random sample we only take five observations (Code Chunk 8-9).

Code Chunk 8-9

```
set.seed(3376)
mu=100; sigma=10; n=5
xbar1=rep(0,100)
for (i in 1:100) {xbar1[i]=mean(rnorm(n, mean=mu, sd=sigma))}

x <- data.frame(v1=xbar1)

xb1 <- mean(x$v1)
```

```
ggplot(x,aes(x=v1)) +
  geom_density(fill="#0000bf", alpha=.7) +
  theme_minimal() +
  theme(plot.title = element_text(size = 8, face = "bold")) +
  theme(axis.title = element_text(size = 8, face = "bold")) +
  ggtitle("Figure 8-5: Sampling Distribution From Normal
          Population") +
  xlab("Mean of Sample") +
  theme(legend.position = "none")
```

As the density plot in Figure 8-5 indicates, we don't need a large number of observations in each sample for the sampling distribution to approach normality. While not perfectly symmetrical (again, we're using randomly generated data), we see that with very few observations per sample, a fairly symmetrical sampling distribution results, centered over the population parameter: the population mean in this case is 100.

Let's summarize the central limit theorem's main points. First, as the number of observations increases in each sample, the shape of a sampling distribution will approximate a normal curve. Second, a sampling distribution of sums will also approximate a normal curve. Finally, when drawing samples from a normal distribution, the resulting sampling distribution of means or sums will approximate a normal distribution regardless of how many observations are drawn in each sample. These represent the main points of the central limit theorem. But wait, there's more: there remains an extremely useful property of the central limit theorem that deserves exploring.

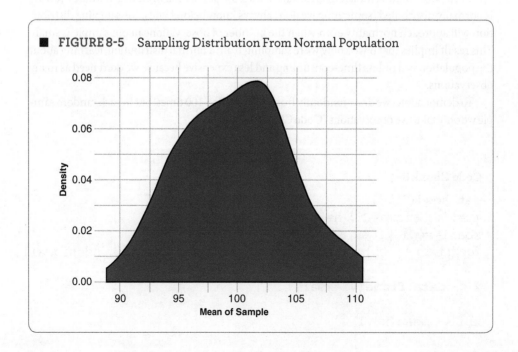

■ FIGURE 8-5 Sampling Distribution From Normal Population

A Useful Property of the Central Limit Theorem

The law of large numbers stipulates that when we repeat an experiment (i.e., draw a sample) and begin to construct a sampling distribution, the mean of the sampling distribution will converge to the population parameter. By increasing the number of trials, we're getting closer to the truth. The central limit theorem implies something slightly different: as we increase the number of observations in each sample, the *shape* of the sampling distribution will approach normality.

This is helpful because if we know that the sampling distribution will begin to look like a standard normal curve, we can calculate exactly how far we are from the population parameter (e.g., the true mean of the population). If we're short on time and money, being in the same neighborhood as the truth might suffice. If knowing the population parameter is a life or death consideration, we'd like to be much closer to the truth than "somewhere in the neighborhood." Given that the shape of our sampling distribution approaches the standard normal curve, we can make precise calculations about how far our sample statistic might be from the population parameter. We can also obtain a sense of how confident we should be in that estimate.

Perhaps the most miraculous aspect of the central limit theorem is that it holds true even when the distribution of the population is not normal. In other words, we can assume that our sampling distribution will approximate normality regardless of how the population is distributed. As we demonstrated in Figure 8-5, sampling from a population with a normal distribution can save us time and money since we can still produce a normally distributed sampling distribution with just a few observations in each sample (it will be relatively cheap). Fortunately, we can use the same machinery to determine the accuracy of our estimates regardless of the population's distribution, so long as we have enough observations. The next section demonstrates this extremely useful phenomenon.

Sampling From Different Distributions

When we draw random samples and calculate a statistic from each, we want to know how close we are to reality. How close does our sample statistic match the population parameter? Sampling distributions are the link that connects random samples to the probabilistic statements we make about our estimates. For example, the margin of error we report in a political poll is based on what we know about sampling distributions and how they approximate the standard normal distribution.

As noted in the previous section, it turns out that no matter the shape of our population's distribution, the sampling distribution we construct from it will approximate a symmetrical, normal-looking distribution. The tendency for sampling distributions to approximate normal distributions links the statistics we generate from our samples to the probabilistic statements we make about them.

To illustrate, let's start with a uniform distribution. A **uniform distribution** resembles the form of a square block since the frequency of cases does not change over the range of possible values. The number of cases at the low, middle, and high end of the range are all equal. The density plot in Figure 8-6a represents a uniform distribution. There are a few bumps along the top since it was generated by a random number function in R, *runif()*.

To produce the illustration, I create a population with 100,000 observations that is uniformly distributed between 50 and 150. In Code Chunk 8-10, I create a variable with a uniform distribution called *xunif* using the command *runif()*, which produces a population of 100,000 observations with a minimum at 50 and a maximum of 150.

Code Chunk 8-10
```
xunif <- runif(100000, min = 50, max = 150)

x <- data.frame(xunif)

p <- ggplot(x, aes(xunif)) +
  geom_density(alpha=.7, fill="#0000ff") +
  theme_minimal() +
  theme(plot.title = element_text(size = 8, face = "bold")) +
  theme(axis.title = element_text(size = 8, face = "bold")) +
  annotate("text", x = 80, y = .012, label =
            "Mean = 100", col="#bf0000") +
  geom_vline(xintercept=100, col="#bf0000") +
  ggtitle("Figure 8-6a: Uniform Distribution") +
  xlab("Uniform Distribution")
```

If we take random samples from that uniformly distributed population and produce a sampling distribution of their means (Figure 8-6b), our sampling distribution will approach normality as n increases. In this particular example (Code Chunk 8-11), I take 500 samples from the uniform distribution. The size of each sample is 50. I then calculate the mean for each sample, constructing the sampling distribution and then using a density plot to graph it. To align it on the page correctly, I use the *grid.arrange()* function.

Code Chunk 8-11
```
set.seed(3376)
min=50; max=150; n=50
xbar1=rep(0,500)
for (i in 1:500) {xbar1[i]=mean(runif(n, min, max))}

x <- data.frame(v1=xbar1)

xb1 <- mean(x$v1)

p1 <- ggplot(x, aes(x=v1)) +
  geom_density(fill="#0000ff", alpha=.7) +
  geom_vline(xintercept=xb1, col="#bf0000") +
  theme_minimal() +
  theme(plot.title = element_text(size = 8, face = "bold")) +
```

```
  theme(axis.title = element_text(size = 8, face = "bold")) +
  annotate("text", x = 107, y = .10, label =
        "Mean = 99.689", col="#bf0000") +
  ggtitle("Figure 8-6b: Sampling Distribution
        From Uniform Distribution") +
  xlab("Mean of Sample") +
  theme(legend.position = "none")

grid.arrange(p, p1, ncol=2)
```

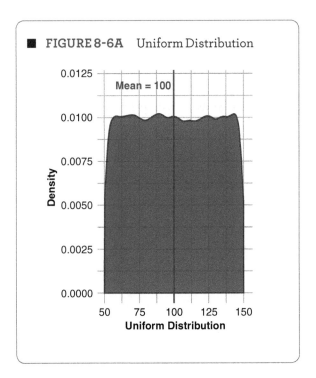

■ FIGURE 8-6A   Uniform Distribution

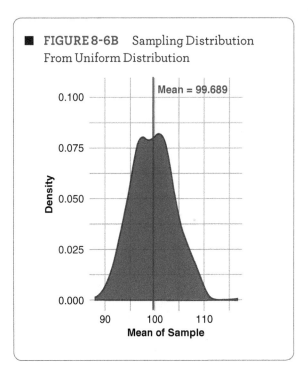

■ FIGURE 8-6B   Sampling Distribution From Uniform Distribution

As the density plots in Figure 8-6a and 8-6b make clear, with a sample of n = 50, the sampling distribution approximates a normal distribution. So even though with each observation we're just as likely to draw a number close to 50 as we are 150, the resulting shape approaches the standard normal curve when we take the mean of each sample and plot it (the sampling distribution).

Let's go further. Consider an **exponential distribution**—a distribution that is badly skewed in the positive direction (Figure 8-7a). In the exponential distribution, a large majority of cases are clumped together at the low end, with gradual declining frequency occurring at values exponentially higher than those at the low end. First, in Code Chunk 8-12 we construct the exponential distribution with the *rexp()* command and define a density plot called p.

Code Chunk 8-12
```
expd <- rexp(100000, rate = .01)

xexp <- data.frame(expd)

p <- ggplot(xexp, aes(expd)) +
  geom_density(alpha=.7, fill="#0000ff") +
  theme_minimal() +
  theme(plot.title = element_text(size = 8, face = "bold")) +
  theme(axis.title = element_text(size = 8, face = "bold")) +
  annotate("text", x = 300, y = .008, label =
           "Mean = 100", col="#bf0000") +
  geom_vline(xintercept=100, col="#bf0000") +
  ggtitle("Figure 8-7a: Exponential Distribution") +
  xlab("Exponential Distribution")
```

Then, we draw randomly (Code Chunk 8-13) as we did with the previous distribution and define a density plot of the means (called p1).

Code Chunk 8-13
```
set.seed(3376)
n=50
xbar1=rep(0,500)
for (i in 1:500) {xbar1[i]=mean(rexp(n, rate = .01))}

x <- data.frame(xbar1)

xb1 <- mean(x$xbar1)

p1 <- ggplot(x,aes(xbar1)) +
  geom_density(fill="#0000bf", alpha=.7) +
  geom_vline(xintercept=xb1, col="red") +
  theme_minimal() +
  theme(plot.title = element_text(size = 8, face = "bold")) +
  theme(axis.title = element_text(size = 8, face = "bold")) +
  annotate("text", x = 125, y = .027, label =
           "Mean = 100.774", col="#bf0000") +
  ggtitle("Figure 8-7b: Sampling Distribution
          From Exponential Distribution") +
  xlab("Mean of Sample") +
  theme(legend.position = "none")

grid.arrange(p, p1, ncol=2)
```

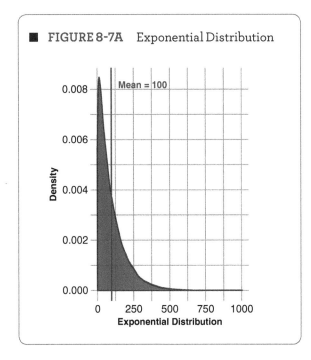

FIGURE 8-7A Exponential Distribution

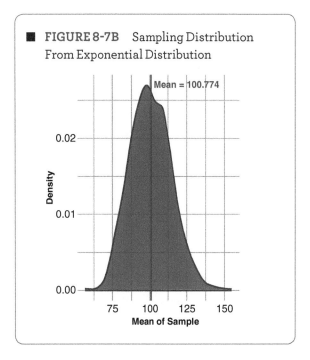

FIGURE 8-7B Sampling Distribution From Exponential Distribution

Figure 8-7b represents the sampling distribution from taking 500 samples with 50 observations each. Again, as with the case in Figure 8-6 which featured a uniform distribution, the sampling distribution generated by taking draws from the exponential distribution produces a normal-looking bell-shaped curve.

The same story holds for a bimodal distribution (Figures 8-8a and 8-8b). A **bimodal distribution** is characterized by a shape with two humps: rather than having one mode, there are two. A little extra work is needed to generate a population characterized by a bimodal distribution (Code Chunk 8-14). To produce a bimodal distribution in R, we must combine or "mix" two normal distributions together. Once we've done that, we can plot the density plot. For the two normal distributions, I specified that one has a mean of 75 and the other has a mean of 125. Both were given the same spread (a standard deviation of 10). These two distributions were then combined and graphed (Figure 8-8a).

Code Chunk 8-14

```
set.seed(3376)
n=500

y1 = rnorm(n, 75, 10); y2 = rnorm(n, 125, 10)
w = rbinom(n, 1, .5)

x2 = w*y1 + (1-w)*y2

bm <- data.frame(x2)
p <- ggplot(bm, aes(x2)) +
```

```
geom_density(alpha=.7, fill="#0000ff") +
theme_minimal() +
theme(plot.title = element_text(size = 8, face = "bold")) +
theme(axis.title = element_text(size = 8, face = "bold")) +
annotate("text", x = 80, y = .017, label =
         "Mean = 100", col="#bf0000") +
geom_vline(xintercept=100, col="#bf0000") +
ggtitle("Figure 8-8a: Bimodal Distribution") +
xlab("Bimodal Distribution")
```

Now that the bimodal distribution has been created, it's time to draw samples from it and plot it (Code Chunk 8-15).

Code Chunk 8-15
```
set.seed(3376)
n=50
xbar1=rep(0,500)
for (i in 1:500) {xbar1[i]= mean(sample(bm$x2, n, replace = TRUE))}

x <- data.frame(xbar1)

xb1 <- mean(x$xbar1)

p1 <- ggplot(x,aes(xbar1)) +
  geom_density(fill="#0000bf", alpha=.7) +
  geom_vline(xintercept=xb1, col="#bf0000") +
  theme_minimal() +
  theme(plot.title = element_text(size = 8, face = "bold")) +
  theme(axis.title = element_text(size = 8, face = "bold")) +
  annotate("text", x = 110, y = .10, label =
           "Mean = 101.811", col="#bf0000") +
  ggtitle("Figure 8-8b: Sampling Distribution
          From Bimodal Distribution") +
  xlab("Means From Each Sample") +
  theme(legend.position = "none")

grid.arrange(p, p1, ncol=2)
```

Drawing 500 samples with a size of n = 50 produces a sampling distribution of the sample means that approximates the normal curve. The density plot in Figure 8-8b is the sampling distribution from the bimodal distribution. Even though the samples are drawn from an oddly shaped bimodal distribution, the sampling distribution approximates a normal curve.

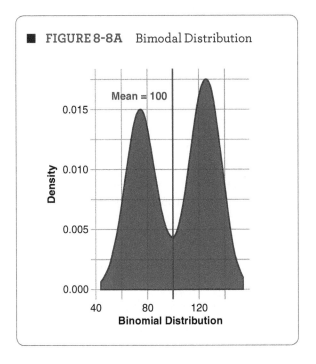

FIGURE 8-8A Bimodal Distribution

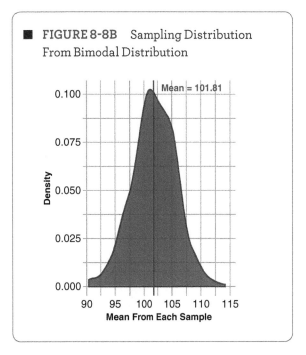

FIGURE 8-8B Sampling Distribution From Bimodal Distribution

Since the sampling distributions in Figures 8-6 through 8-8 approximate the normal distribution, we've met the conditions necessary to use the normal curve, also referred to as the *standard normal distribution*, a device to evaluate the accuracy of the statistics we draw from each sample. We not only know how sure we can be of our estimates, as we'll see in the next chapter we can calculate the sample size required to achieve specific levels of accuracy. Now that we've seen how sampling distributions are related to the normal curve, we can discuss why the normal curve figures so prominently and how it helps with the construction of confidence intervals, hypothesis testing, and regression coefficients in the chapters that follow.

> **KNOWLEDGE CHECK:** Explain why the central limit theorem is so important.

9. How was the central limit theorem discovered?

   a. Abraham de Moivre discovered it.

   b. Sir Francis Galton discovered it.

   c. Adolphe Quetlet drew on work from Abraham de Moivre.

   d. An astronomer observed that the patterns noticed by de Moivre were present in many astronomical phenomenon.

10. What does the central limit theorem say?

    a. The fraction of times that a given event A occurs will converge with probability one to a number that is equal to the probability that A occurs in a single repetition of the experiment.

    b. The sampling distribution of means approaches a normal distribution as the number of observations from each sample increases.

    c. The sampling distribution of sums $\sum X_i$ is approximately normally distributed with mean $n\mu$ and standard deviation $\sqrt{n}\sigma$.

    d. If the population (each X_i) is normally distributed, then the sample distribution of the mean ($\bar{X}$) and the sum ($\sum X_i$) are normally distributed, no matter the sample size.

11. Which of the following explains why the central limit theorem is so important?

    a. We can calculate how far away from the population parameter we are given the size of our sample.

    b. So many naturally occurring phenomenon have a normal distribution.

    c. As the number of observations increases, our sample statistic will approach the population parameter.

    d. To use its machinery, we do not have to assume that the population we are studying is normally distributed.

12. Which of the following statements are true?

    a. Repeated draws from a uniform distribution will produce a sampling distribution that is normally distributed.

    b. Repeated draws from an exponential distribution will produce a sampling distribution that is normally distributed.

    c. Repeated draws from a bimodal distribution will produce a sampling distribution that is normally distributed.

    d. If drawing from a normal distribution, the sample size does not need to be large to produce a normally distributed sampling distribution.

The Standard Normal Distribution

The **standard normal distribution** is a construct that provides the scaffolding to make precise statistical statements about the accuracy of our estimates. This is made possible because the standard normal curve has some very specific characteristics.

For example, we stipulate that the standard normal distribution has a mean of 0 and a standard deviation of 1. This allows us to easily "standardize" units. Say, for example, we want to know the average household income of a population measured in dollars. To use the standard normal curve, we convert dollars to standard deviations from the mean. Once expressed in standard deviations, we can make some useful statements with the help of the standard distribution. If, for example, the standard deviation of our sample is $10,000 and the mean is $50,000, we can infer that roughly 68% of households have incomes between $40,000 and $60,000. The next two visualizations will help illustrate.

Let's build a visualization that demonstrates how we use the normal distribution. Although the code is somewhat involved, the payoff is substantial. For completeness, the code in Code Chunk 8-16 uses base R commands. The code in Code Chunk 8-17 generates a similar visualization using ggplot. I use both to show that there are many different ways in R to generate the same plot.

First, I define the mean (*mean=0*) and the standard deviation (*sd=1*) for the curve. To specify the boundaries for the area I want to shade, I define the lower bound (*lb=−1*) and the upper bound (*ub=1*). I then create two variables to be plotted. One I call *x*, which goes from −4 to 4. The other uses *x* by taking each value of *x* and converting it to a density (the height of the curve) using the *dnorm()* function. With these two variables I have a number associated with the x axis (*x*) and the corresponding number associated with the y axis (*hx*).

After *x* and *hx* are defined (my x and y), I plot them using the *plot()* command. Only after the *plot()* command is given can I define the area underneath the curve I want to shade. To shade the area under the curve, I create a variable *i* that helps define the area of a polygon. That is followed by the *lines()* command and the *polygon()* command that actually draw the area underneath the curve and color it blue. As an aside, you'll see in the next example (Code Chunk 8-17) that this part is accomplished using the *geom_ribbon()* command when using ggplot.

The final commands help calculate the area underneath the curve that lies between 1 standard deviation above and below the mean. I define the area underneath the curve by creating a variable called *area*. In that expression, I use the *pnorm()* function, which calculates the area underneath the curve to the left of the value specified. To calculate the area underneath the curve between two values, I subtract the area underneath the curve left of the lower bound from the area underneath the curve left of the upper bound. That sum gives us the area between the lower and upper bounds. Once calculated, I place that value beneath the plot title using the *paste()* command and the *mtext()* command. The final line of the code draws the x axis of the plot.

The code in Code Chunk 8-16 will produce the standard normal distribution where the area 1 standard deviation above and below the mean will be shaded and the area calculated. All of the commands in Code Chunk 8-16 have to be run in sequence to draw the plot and calculate the area underneath the curve. Manipulate the parameters (the upper and lower bounds of the curve; −1, 1) of the code and observe how things change. (Just change the upper and lower bound in the second line of the code.) It's a good way to develop an intuition regarding standard deviations and the area underneath the standard normal curve. You'll also need it for a KNOWLEDGE CHECK question at the end of this section.

Code Chunk 8-16

```
mean=0; sd=1
lb= -1; ub=1

x <- seq(-4,4,length=10000)*sd + mean
hx <- dnorm(x,mean,sd)

plot(x, hx, type="n", xlab="", ylab="",
    main="", axes=FALSE)
```

```
i <- x >= lb & x <= ub
lines(x, hx)
polygon(c(lb,x[i],ub), c(0,hx[i],0), col="#0000bf")

area <- pnorm(ub, mean, sd) - pnorm(lb, mean, sd)

result <- paste("P(",lb,"< 0 <",ub,") =", signif(area, digits=3))
mtext(result, 3)
axis(1, at=seq(-5, 5, 1), pos=0)
```

We use the standard normal distribution since it indicates the area underneath the curve at specific distances from the mean. The total area underneath the curve sums to 100%. As Figure 8-9 illustrates, 68% of the area underneath the curve lies between 1 standard deviation above and below the mean. Roughly 95% of the area underneath the curve lies within 2 standard deviations of the mean. Finally, the area underneath the curve within 3 standard deviations on either side of the mean is 99.7%.[2] Change the lower and upper bounds in Code Chunk 8-16 from −1 and 1 to −2 and 2 and observe what happens to the calculated area underneath the curve (it should read .954). To make that change in the code, you only need to change the definition of the lower and upper bounds in the second line of Code Chunk 8-16.

Since sampling distributions approximate the normal distribution, we can use that information to determine how close our sample statistic is to the true population parameter. All we need to do is convert (standardize) our estimates into what are called z-scores, then use what we know about the area underneath the standard normal curve to calculate how far we could be from the true population parameter (confidence interval) and how confident we are of that calculation (our level of confidence). In the next chapter we will construct confidence intervals and conduct hypothesis testing, both of which use the normal distribution. For now, let's

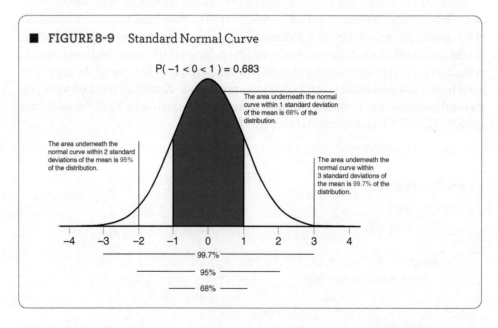

■ **FIGURE 8-9** Standard Normal Curve

---

[2] The actual figure in Figure 8-9 was drawn by first executing the code in Code Chunk 8-16. I then exported the figure to Adobe Illustrator to add the labels.

understand how to link the area under the normal curve to values of z-scores, an important process in understanding the accuracy of our estimates.

The Standard Normal Distribution and Critical z-Scores

In days past, students would spend hours working with a gargantuan table of numbers (the standard normal table) to calculate how much of the distribution lies beneath the normal curve at a given distance from the mean. When confronted with the table—usually printed on the inside cover of their statistics textbook—more than a few students no doubt walked straight to the registrar's office and dropped their statistics class. Simple commands in R now easily tell us what we need to know about the standard normal distribution. The standard normal distribution indicates where our sample statistic lies in the distribution, allowing us to make some probabilistic statements about its accuracy.

In the previous section, we learned that 68% of the distribution lies between 1 standard deviation on either side of the mean. We use the standard normal table (or R) to calculate the percentage of the distribution that lies underneath the curve for all sorts of other values.

Let me demonstrate how a specific example relates to the standard normal curve. To illustrate, I calculated the mean and standard deviation of IQ scores for 2,287 Norwegian eighth graders (the data are found in the 'MASS' data package and are called *nlschools*). Say we draw one student from the distribution who scored a 15 on the IQ exam. Since we know the mean of the population (11.88) and the standard deviation (2.06), we can calculate the corresponding z-score. To calculate the corresponding **z-score**, we use the following formula:

$$Z = \frac{X - \mu}{\sigma}$$

where Z is the z-score, μ is the mean, X is the number we're interested in converting (in this case, 15), and σ is the standard deviation (in this case, 2.06). In this example, we have

$$Z = \frac{15 - 11.88}{2.06} = 1.51$$

The corresponding z-score is 1.51. The back or front cover of many statistics books include the standard normal distribution table, which indicates the corresponding area underneath the curve that lies above or below the z-score. With R, we simply insert 1.51 into the function *pnorm()* and R will provide the area underneath the curve to the left of that number.

Typing *pnorm(1.51)* produces .935, which indicates that approximately 94% of the distribution lies to the left of an IQ score of 15, leaving roughly 6% in the blue-shaded area. Calculating the area underneath the curve for any value in a distribution tells us its position relative to the rest of population. Is it exceptional? In the next chapter, we'll see how using that information allows us to make inferences about the population and about the accuracy of our estimate.

To visualize what we're doing, I use code from the 'ggplot' package. (See if you can figure out which commands correspond with each other in Code Chunks 8-16 and 8-17.) I drew the corresponding standard normal distribution and placed the units of the IQ test on top to show how those scores correspond to the standardized score on the x axis. Again, although the code is somewhat involved, playing around with it is one way to gain a deeper understand of how we use the standard normal curve to make probabilistic statements.

As in the previous example, I start by defining two variables. Variable *v1* ranges from −4 to 4 with 10,000 observations (Code Chunk 8-17). Variable *v2* simply converts each value of *v1* to a density (the height of the curve). The two variables are then combined into a data frame using the *data.frame()* command.

To shade the desired area under the curve, I use the *geom_ribbon()* command, telling R to shade only the values of *v1* between 1.51 and 4 using the *subset()* command. In the aesthetics function *aes()*, the y maximum is specified by the variable we created earlier (*v2*). With these two functions, I've specified the dimensions of the shaded area. Finally, note how I create the second axis that is a linear function of the bottom axis (*2.06+11.88) with the *scale_x_continuous()* command.

As you can see in Figure 8-10, the Swedish student with a score of 15 on the IQ exam is fairly exceptional. The blue-shaded part of the curve corresponds to the 6% of the population that scored higher than 15. Again, as in the previous example, I encourage you to play with the code to see how it changes what is drawn. It's a great way to gain a deeper understanding.

Code Chunk 8-17

```
v1 <- seq(-4, 4, length = 10000)
v2 <- dnorm(v1, 0, 1)

df <- data.frame(v1, v2)

ggplot(df, aes(v1, v2)) +
  geom_line() +
  theme_minimal() +
  theme(plot.title = element_text(size = 8, face = "bold")) +
  theme(axis.title = element_text(size = 8, face = "bold")) +
  geom_ribbon(data=subset(df ,v1>1.51 & v1<4),aes(ymax=v2),ymin=0,
              fill="#0000bf",colour=NA) +
  scale_fill_brewer(guide="none") +
  xlab("Standard Deviations From Mean") +
  ylab("Density") +
  geom_vline(xintercept = 1.51, linetype=2) +
  scale_x_continuous(sec.axis=sec_axis(~.*2.06+11.88))
```

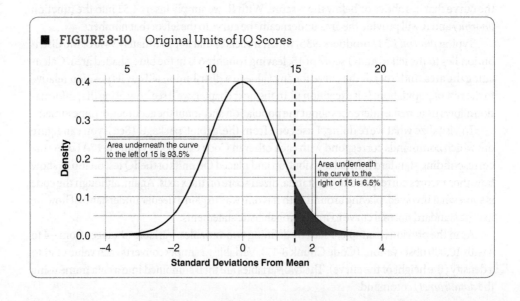

FIGURE 8-10 Original Units of IQ Scores

> **KNOWLEDGE CHECK: Connect sampling distributions to the standard normal distribution.**

13. Which of following describe features of the standard normal curve?

    a. It has a standard deviation of 1.

    b. It has a mean of 0.

    c. It is a uniform distribution.

    d. It represents the distribution of many different phenomena.

14. What percentage of the area lies underneath the standard normal curve 1 standard deviation below and above the mean?

    a. 95%

    b. 68%

    c. 90%

    d. 99%

15. What percentage of the area lies underneath the standard normal curve approximately 2 standard deviations below and above the mean?

    a. 95%

    b. 68%

    c. 90%

    d. 99%

16. What is a z-score?

    a. It is the area lying underneath the standard normal curve.

    b. It represents the units of a variable (height, age, weight, etc.) expressed in standard deviations.

    c. It is calculated by taking an observation, subtracting the mean, and dividing by the variable's standard deviation.

    d. It is calculated by taking an observation, subtracting the mean, and dividing by the standard deviation of the standard normal distribution.

17. What is the z-score helpful for?

    a. Calculating the standard deviation of the variable

    b. Understanding the spread of the variable

    c. Understanding how far our sample statistic is from the true population parameter

    d. Understanding the spread of the sample

18. Using Code Chunk 8-16, calculate the following:

    a. What is the area underneath the standard normal curve that lies between −1.5 and 1.5 standard deviations?

    b. What is the area underneath the standard normal curve that lies between −1 and 2 standard deviations?

    c. What is the area underneath the standard normal curve that lies between 1 and 2 standard deviations?

    d. What is the area underneath the standard normal curve that lies between 2 and 1.3 standard deviations?

SUMMARY

In this chapter we defined populations and samples, explaining how samples help us make inferences about a population in cases when we don't have the time or money to survey or record data for each individual in the population. Fortunately, the law of large numbers and the central limit theorem come to the rescue. With the law of large numbers and the central limit theorem, we can obtain unbiased estimates of the population parameters and we can make some statements about how well the estimates we generate from our sample fit reality. This is made possible since we can invoke the standard normal distribution when taking samples. Since the sampling distributions we generate from our populations approximate a normal distribution no matter the distribution of the population, we can use the architecture of the standard normal distribution to ascertain where our estimates lie in the distribution. This is an extremely powerful feature of sampling since we can gauge how far our estimate is from reality.

COMMON PROBLEMS

- *Determining the population.* In some contexts it's easy to lose track of the population. For example, if we want to know whether most students at a university attended public high school, we might take a sample from a large 200-person statistics class and calculate a sample statistic. In that case, although we're drawing a subset of the class, the class is not the population. The class itself is a sample from the population in this case, and not a random one at that. The population is defined by all of the students that attend the university.

- *Noting the difference between a sample statistic and a population parameter.* While the concept of a sample statistic and population parameter may be clear and distinct, we sometimes get too casual in our use of the terms, using one when we mean the other. Recall that we calculate a statistic from a sample (mean, median, or standard deviation) in order to understand a population parameter (mean, median, or standard deviation). If we were to survey everyone in the population, there'd be no need to take a random sample, and we'd calculate the population parameter (we wouldn't have to guess). Keeping the two straight helps us keep straight whether we've calculated something from a sample or we've surveyed or counted all individuals or objects in a population. When the two are mixed, so is the distinction between a sample and a population. When that happens, we confuse our best guess with reality.

- *Not having a clear understanding of sampling distributions.* Sampling distributions get confusing since they represent summaries of repeated samples. A sampling distribution does not, for example, directly represent the individuals or objects drawn from a population. Each observation for a sampling distribution is a summary itself. We take a sample, calculate a sample statistic, record it, then repeat. A sampling distribution represents the aggregation of the summaries we calculate from each sample. In that sense, it is once removed. If we take 100 samples with 500 observations each, our sampling distribution is made up of 100 observations: one summary statistic for each of the 100 samples.

- *Increasing observations versus increasing samples.* Again, the distinction between an observation and a sample is clear, but if we confuse the two in our language, keeping things straight is difficult. The distinction here is important since the law of large numbers (increasing samples) and the central limit theorem (increasing observations) focus on one or the other. When we increase observations, we're increasing the size of our sample. When we increase the number of samples, we're repeating the exercise of taking a sample over and over.

- *Attention to detail when taking random samples.* Assigning each individual or object in a population a number and then using a random number generator on a computer to select a sample is the best way to take random samples. Once selected, putting the objects or individuals back in the population for possible selection again is the best practice unless doing so influences the outcome. Anything that involves a different process should be scrutinized carefully to make sure it is truly random: every individual or object has an equal chance of being drawn.

REVIEW QUESTIONS

1. What is a population?
2. What is a random sample?
3. What is a sampling distribution?
4. What is the standard normal distribution?
5. Why do we care about the standard normal distribution?
6. What are the units of the standard normal distribution?
7. What area lies beneath the standard normal curve 1 standard deviation on either side of the mean?
8. How do we convert our units of measurement to z-scores?
9. What is a z-score?
10. Why is the z-score so helpful?

PRACTICE ON ANALYSIS AND VISUALIZATION

1. Which of the following are true about populations?
    a. Statistics describe the features of populations.
    b. Parameters describe the features of populations.
    c. We use Greek symbols and notation when describing populations.
    d. We use lowercase notation when describing populations.

2. Which is a population and which is a sample?
    a. We survey the entire state of Nebraska to determine the percentage of liberals in the state.
    b. We survey the entire state of Nebraska to determine the percentage of liberals in the country.
    c. Nebraska in option a
    d. The United States in option b

3. Why do we take random samples to determine the population parameter?

   a. We don't have enough time or money to count everything or everyone.
   b. We want to eliminate bias from our sample.
   c. We want to eliminate bias from the population.
   d. We want all possible confounding factors to have an equal chance of being represented in our sample.

4. Why do some use sampling with replacement?

   a. It builds a truly random sample.
   b. Sampling with new cases is too expensive.
   c. Each draw is independent of all others.
   d. It keeps us from running out of observations.

5. Which is true about the law of large numbers?

   a. Each experiment has to be independent of the others.
   b. As we increase the number of experiments, the mean of our sampling distribution will approach the population parameter with 100% certainty.
   c. It is based on the number of repeated experiments.
   d. It only works if the sampling distribution is normally distributed.

6. Which of the following are true of the central limit theorem?

   a. The fraction of times that a given event A occurs will converge with probability one to a number that is equal to the probability that A occurs in a single repetition of the experiment.
   b. The sampling distribution of means approaches a normal distribution as the number of observations from each sample increases.
   c. The sampling distribution of means approaches a normal distribution as the number of samples increases even if we're drawing from a population that is not normally distributed.
   d. It allows us to calculate how close our sample estimate is from the true population parameter.

7. Which of the following are true about the standard normal distribution?

   a. It has a mean of 0 and a standard deviation of 1.
   b. Roughly 68% of the area underneath the curve lies between 1 standard deviation above and below the mean.
   c. Roughly 99% of the area underneath the curve lies between 1 standard deviation above and below the mean.
   d. It represents the units of the variable we're examining.

8. What are z-scores?

   a. They are derived by subtracting the mean from our observation and dividing by the mean.
   b. They are derived by subtracting the standard deviation from our observation and dividing by the mean.
   c. They are derived by subtracting the sample mean and dividing by the sample standard deviation.
   d. They are derived by subtracting the population parameter and dividing by the population's standard deviation.

9. What are z-scores used for?

   a. They help us convert from the sample's units to units that accord with the standard normal distribution.
   b. They allow us to use the standard normal distribution to make inferences about the accuracy of our sample statistics.
   c. They allow us to determine whether our sample is normally distributed.
   d. They allow us to determine whether our sample was randomized properly.

10. Which are important implications of the central limit theorem?

    a. The population is normally distributed, so the number of observations in a sample drawn is less important.
    b. The sampling distribution of sample statistics will approach normality regardless of the shape or distribution of the population.

c. As the number of experiments increases, the mean of the sampling distribution from your samples will approach the population mean.

d. Most natural phenomena have populations that approximate a normal distribution.

ANNOTATED R FUNCTIONS

The following functions appear in this chapter. They are listed in order of their first appearance (with the code chunk number in parentheses) and annotated here to give a very brief description of their use. Some are not stand-alone functions and only work in combination with other commands. As a reminder, the code in every chapter will work properly if executed in the order it appears. Proper execution also depends on typing the author-defined *libraries()* command, which loads the required R packages.

set.seed(): tells R on what number to start the random number generator. Setting the seed allows you to duplicate results based on random numbers if it matches the number in the example. (8-1)

rep(): where you specify the number of repetitions to repeat a command in a for-loop. (8-2)

for(): specifies the number of times a loop should occur. It is followed in the same line by the command line that is to be repeated. (8-2)

mean(): calculates the mean of a variable. (8-2)

rnorm(): randomly generates numbers to create a normally distributed population. (8-2)

list(): combines specified objects into a list. (8-4)

melt(): takes variables aligned in columns and "stacks" them into one column, one on top of the other. (8-4)

ggplot(): defines the basic structure of a plot (usually the x and y variables). (8-5)

aes(): the aes (called "aesthetics") function is used in ggplot to define the basic structure of the plot, which often includes the variables you want to use and any shapes or colors. (8-5)

ggtitle(): main title of the ggplot. (8-5)

theme_minimal(): specifies a minimalist style for ggplot. (8-5)

theme(): specifies font, size, and so forth in a ggplot. (8-5)

geom_vline(): draws a vertical line. (8-5)

geom_density(): draws a density plot in the ggplot framework. (8-5)

annotate(): places text in the figure. (8-5)

xlab(): labels the x axis in ggplot. (8-7)

data.frame(): combines columns to form a data set. (8-8)

runif(): generates numbers to create a uniformly distributed population. (8-10)

grid.arrange(): arranges plots on a page according to the pattern specified by rows and columns. (8-11)

rexp(): randomly generates numbers to create an exponentially distributed population. (8-12)

sample(): takes random samples as directed from a specified population. (8-15)

dnorm(): calculates the probability density function. (8-16)

plot(): base R plotting function. (8-16)

ylab(): labels the y axis in ggplot. (8-16)

lines(): base R plotting functions for lines. (8-16)

polygon(): base R plotting function to generate shapes (polygons). (8-16)

paste(): base R function for pasting text together. (8-16)

signif(): indicates what decimal place to round to. (8-16)

mtext(): used to write text in the margin of a plot. (8-16)

axis(): used to specify axis characteristics in a plot using base R. (8-16)

geom_line(): draws a line in ggplot. (8-17)

geom_ribbon(): helps draw an area with a minimum and maximum value in ggplot. (8-17)

scale_fill_brewer(): allows you to specify a specific palette of colors to use in a plot. (8-17)

scale_x_continuous(): helps draw an x axis according to specifications given. (8-17)

ANSWERS

KNOWLEDGE CHECK

1. population (a), population (b), sample (c), sample (d)
2. a, d
3. a, b, c, d
4. a, b, c
5. a, b
6. b, c
7. b
8. a
9. c, d
10. b, c, d
11. a, b, c, d
12. a, b, c, d
13. a, b, d
14. b
15. a
16. c
17. c
18. .87 (a), .82 (b), .14 (c), .07 (d)

PRACTICE ON ANALYSIS AND VISUALIZATION

1. b, c
2. population (a), sample (b), population (c), population (d)
3. a, d
4. a, c
5. a, b, c
6. b, d
7. a, b
8. c
9. a, b
10. a, b

Access digital resources, including datasets, at
http://edge.sagepub.com/brownstats1e.

9 Confidence Intervals and Testing Hypotheses

CHAPTER OUTLINE

Learning Objectives
Overview
Confidence Intervals With Large Samples
Small Samples and the t-Distribution
Comparing Two Sample Means
Confidence Levels
A Brief Note on Statistical Inference and Causation
Summary
Common Problems
Review Questions
Practice on Analysis and Visualization
Annotated R Functions
Answers

LEARNING OBJECTIVES

- Calculate confidence intervals around proportions and means.
- Use the t-distribution to calculate confidence intervals.
- Determine if the means of two populations are different.
- Explain what confidence means in statistics.
- Connect confidence intervals to statistical estimates.

Overview

This chapter takes the theoretical concepts from Chapter 8 and demonstrates their application. We often want to know something about a larger population but have limited resources: counting or measuring every individual is either impractical or too expensive. Statistical inference aids our endeavor by providing the mathematical and probabilistic machinery to make important claims.

The central limit theorem instructs that we can learn about a population by taking a small sample, calculating a sample statistic, and then constructing a confidence interval to indicate how close the sample statistic is to the population parameter. Calculating confidence intervals to test hypotheses will be introduced with several examples. You will learn how to construct a confidence interval in several different contexts. Finally, I will use simulations to explain what we mean by levels of confidence. By chapter's end you'll have the basis to understand the probabilistic statements that accompany findings in surveys, hypothesis tests, and regression analysis.

This chapter proceeds as follows. First, we'll construct confidence intervals in different contexts. Through several examples, you'll learn how to establish confidence intervals around proportions and means when the sample size is greater than 100. Second, I introduce the t-distribution, the theoretical distribution used when calculating confidence intervals for samples with less than 100 observations. Third, we'll explore two-sample t-tests, a useful tool to compare the means from two groups. Fourth, through simulations, we'll deepen our understanding of confidence levels. This gets to the heart of what we mean by statistical significance. Finally, I'll discuss how the last few chapters relate to the next few: the same statements of statistical significance that apply to means and proportions apply to the estimates we generate when running regressions.

Confidence Intervals With Large Samples

Important court cases regarding discrimination, quality control in industrial manufacturing, public opinion polls, and the outcomes of elections themselves are based on constructing confidence intervals. With the law of large numbers and the central limit theorem in place, we now have a way to take a random sample, calculate a statistic from it, and figure out how close we might be to the population parameter. We can also establish how much confidence to place in our estimate. This is not likely the first time you've run across confidence intervals. Most every election poll makes reference to them. When poll results are reported, they are usually accompanied by the phrase "within ±3 percentage points" or "within a 3 percentage point margin of error." **Confidence intervals** establish the range of values on either side of the sample statistic that are likely to include the population parameter. In an election survey, the ±3 percentage point margin of error establishes the range within which we think the true support for the candidate lies.

There are three distinct contexts under which we draw samples to make inferences about the population. First, we may want to know the proportion of the population intending to vote for a politician. This question often takes the form of "How many?" Second, we might want to know the disposable income of the average citizen in our country. This question often takes the form of "How much?" Finally, we want answers to these questions when our sample size is relatively small (perhaps we can't afford a large survey).

Stated differently, we calculate confidence intervals for the following reasons: (1) when we want to know the proportion of a population (e.g., the percentage of the electorate that supports a particular candidate), (2) when we want to know the population mean (e.g., what is a country's average level of income), and (3) we want to know how many or how much when the sample is small.

Finding the Population Proportion

Epidemiologists want to know the proportion of a population that has received the vaccination for a disease. School officials want to know what proportion of a school district's students have experimented with drugs. Particularly during political campaigns, we want to know the proportion of the population that supports a candidate or a particular policy. Democratic systems depend on making calculations about what the majority (51%) of society wants. Do voters support armed action against ISIS, gay marriage, gun control, or fracking? These are all important questions whose answers should inform our politicians if not influence their policies. Since we don't have the money to survey every individual or measure every object, we need to make a best guess about the **population proportion** and understand how confident we can be about our guess.

Example: Winning Elections
Let's take a simple example from politics. The election is a few weeks away and in order to convince donors to contribute, the candidate needs to know their chances. Before we take the sample and make the calculation, we need to establish a level of confidence. In other words, we need to establish how sure we want to be that our estimate approximates the population parameter.

If we're making a life and death decision, we may want to calculate a confidence interval at the 99.99% level of confidence. If the stakes aren't as high, we may be happy to calculate a confidence interval at 90%. Levels of confidence indicate something very specific. For example, if we were to perform our experiment (draw a sample) 100 times, the level of confidence indicates how many times out of 100 we would expect the confidence interval we construct from our sample to contain the population parameter. I will perform several simulations toward the end of the chapter to help illustrate the concept more fully.

Most political polling is executed at the 95% level of confidence. We need to establish the level first since we'll use that as part of our calculation. That is, we want to be 95% confident that the interval we calculate includes the true population mean. Later in the chapter we'll develop a better intuition for levels of confidence. But for now, let's focus on calculating the interval. Let's start with the step-by-step formulation:

1. We represent the confidence interval with the following formula:

$$.95 \approx Pr\left(\hat{p} - 1.96 SE(\hat{p}) \leq p \leq \hat{p} + 1.96 SE(\hat{p})\right)$$

Since the formalization of the interval can be intimidating, let's break it down. The .95 at the beginning of the equation represents the level of confidence we've chosen (95%). The $\hat{p}$ represents the sample statistic we calculate. In this case it would be the proportion of the sample that supports a particular candidate. The p represents the true population proportion: the actual proportion of the population that supports the candidate. The *SE* in the equation stands for the standard error of the sample. Finally, the 1.96 is the **critical z-value** that is associated with being 95% confident.

We find the 1.96 from the standard normal distribution table usually found in the appendices of statistics textbooks. Or, we can obtain it in R by typing *qnorm(.025)*. The .025 represents the area in each tail of the distribution that adds to .05%. With the standard normal distribution, 95% of the area underneath the curve lies 1.96 standard deviations above and below the mean. That is, the formula merely states that we're 95% confident that the true population proportion (p) lies between ±1.96 multiplied by the standard error or our sample.

Art and Practice of Data Visualization
DECIMALS AND PERCENTAGES

Remember your decimals when dealing with percentages. Percentages expressed as decimals are always between 0 and 1. For example, .025 would be 2.5% and .05 is 5%.

2. Calculate $\hat{p} = \dfrac{x}{n} = .55$. Calculating $\hat{p}$ is easy in the case of proportions. The symbol $\hat{p}$ simply represents the number of respondents in our sample who said they support our candidate divided by n.

3. Calculate the standard deviation of the sample: $s = \sqrt{\hat{p} * (1 - \hat{p})} = .497$.

4. Calculate the standard error of the sample: Standard error of $\hat{p} = \dfrac{s}{\sqrt{n}} = \dfrac{(.497)}{\sqrt{1,000}} = .0157$.

5. Plug values into the expression for the 95% CI: 95% CI = .550 ± (1.96)(.0157) = .550 ± .031. We then need to convert .031 to a percentage, which is 3.1%. This is what polls refer to as the margin of error (usually 3 percentage points, as in this case).

Surveys are expensive. How much will it cost (how many observations are required) to produce an estimate at the 95% confidence level with a margin of error of ±3 percentage points? To determine the number of observations required, we use the following formula:

$$n = \dfrac{z_{a/2}^2 p^*(1-p^*)}{E^2} = \dfrac{(1.96^2)(.50(1-.50))}{.03^2} = 1,068$$

In the equation above, $z_{a/2}^2$ is the corresponding critical z-value for the level of confidence chosen. Since we're not favoring one choice (candidate) over the other, we set p^* at .50, which represents the most conservative guess of the population parameter. Finally, E^2 represents our preferred margin of error. With all of those choices made, we make the calculation. The result is 1,067.11 and standard practice is to round observations up, which yields 1,068. This means that at the 95% level of confidence and a margin of error of ±3 percentage points, we need to survey 1,068 people.

Suppose there is a billionaire candidate in the race and they want to be 99% confident of the result. Moreover, they are not comfortable with a 3 percentage point margin of error; they want to be more exact. Let's suppose they want to be 99% confident with only a margin of error at ±1 percentage point. In that instance, we plug in the new numbers and calculate the number of people we will have to survey. In this case, instead of using 1.96, which is the z-critical value for 95%, we use 2.576, which is the z-critical value for the 99% level of confidence. The only other difference is that we substitute .01 for .03 since we're interested in establishing a confidence interval that is ±1 percentage point rather than 3. The resulting equation is as follows:

$$n = \dfrac{z_{a/2}^2 p^*(1-p^*)}{E^2} = \dfrac{(2.576^2)(.50(1-.50))}{.01^2} = 16,590$$

As you can see, our anxious candidate *needs* to be a billionaire in order to afford the poll. Simply by increasing the level of confidence from 95% to 99% and decreasing our margin of error from 3 percentage points to 1, the number of people we need to poll has increased by a factor of 16! Next time you see poll numbers reported in the media, you'll have a better appreciation for the precision of ±3 percentage points.

Example: The Karl Rove Meltdown

Let's go back in time to a crisp November eve, the night of the 2012 U.S. presidential election. With results just in from the state of Ohio, some of the major networks began to call President Obama the victor.

On the Fox News Election Night Studio set, the famous Republican strategist Karl Rove wasn't so sure. With a number of Ohio counties still left to report their results, Rove thought it was premature to declare a victor. With 77% of the vote reported, the smallest of margins separated Obama from Republican challenger Mitt Romney. With 23% of the vote left to come in, Obama had 50.3% of the vote and Romney had 49.7%. In raw votes, Obama had 2,151,158 votes and Romney had 2,122,402. Only 28,756 separated the two candidates, with roughly 1,276,518 left to come in. Surely there were enough Republican votes left in the remaining counties to throw Ohio into the Romney win column.

As much as it can on a live, nationally televised news broadcast, pandemonium ensued. Rove's reluctance to accept Obama's victory clearly wasn't in the script. A visible level of unease could be read on the faces of Megyn Kelly and Chris Wallace. Far from the glitzy set from where Kelly, Wallace, and Rove debated the outcome, Fox News had banished its statisticians (its "decision desk") to a small, florescent-lit room deep in the bowels of Fox Studios.

The decision desk did not agree with Rove. Apparently too boring to warrant an on-air camera, there was no way to talk with the decision desk. Determined to get answers, Kelly set off on a mission. After several awkward minutes with cameras following as she walked by staffers, down hallways, and past a picture of General Lee, Kelly arrived at the decision desk and posed the question, "How certain are you that Obama has won Ohio?" The less telegenic decision desk expert answered, "99.95%." Armed with the equation above, let's see why Rove was wrong.

We know that the election was close, so instead of a 3 percentage point margin of error, let's go down to 0.1 percentage point. Let's also use a z-value of 3.89, which is associated with a 99.99% level of confidence (this is a nationally televised broadcast, after all). If we plug those values into our equation, we get the following:

$$n = \frac{z_{a/2}^2 p^*(1-p^*)}{E^2} = \frac{(3.89^2)(.50(1-.50))}{.001^2} = 3,783,025$$

In other words, we need 3,783,025 votes for a margin of error at .1 percentage point and to be 99.99% confident of our result. Note that by this time, there were 4,273,560 votes counted (77% of the total vote), more than enough votes to call the race. None of the remaining precincts yet to report their vote were in Republican strongholds.

Kelly eventually returned to the set where several more experts attempted to convince Rove that the race was over. It took at least one additional commercial break for Rove to surrender. To view the incident, just google "Karl Rove meltdown."

The networks are more sophisticated now than they were in 2012 in terms of monitoring and calling elections. Granted, much of the difficulty with calling elections centers on the way voting returns are reported: they are not reported as random draws from a population. There is bias in how electoral returns come in. To be fair, this is what was giving Rove some pause. The first returns coming in might represent urban areas, ones that generally favor democratic candidates. In any case, we can make some calculated guesses as to the victor much earlier than Rove was willing to accept.

Example: Is the Federal Government Biased?

Let's look at another example for practice. In the NES survey, which represents a sample of 1,178 respondents randomly drawn from the U.S. population older than 18 years, a question was asked whether the respondents felt the federal government favored Blacks or whites.

Before we get started, let's take a look at the variable with a frequency table (Code Chunk 9-1). Since we're only interested in the frequency table and not the accompanying bar plot, in the *freq()* command we include the option *plot = options(descr.plot = FALSE)*.

Code Chunk 9-1

```
freq(nes$disc_fed, plot = options(descr.plot = FALSE),
     main="Table 9-1: Most Americans See No Bias")
```

How the responses break down by category is shown in Table 9-1. Suppose we want to know what percentage of the population felt that the federal government was either race neutral or favored Blacks. To match that question with our data, we need to transform the data. Specifically, we need to collapse some categories. The code in Code Chunk 9-2 provides a series of *ifelse()* statements. That is, the code directs R accordingly: if the variable is *Treats whites much better*, *Treats whites moderately better*, or *Treats whites a little better*, then assign the value 0. Otherwise, assign the variable a 1. This recodes the variable accordingly, allowing us to find the percentage of respondents that either thought there was no favoritism or felt the federal government favored Blacks.

$$.95 \approx Pr\left(\hat{p} - 1.96 SE(\hat{p}) \leq p \leq \hat{p} + 1.96 SE(\hat{p})\right)$$

We can now take the new variable *nes$fed* and make the calculations. The standard error is obtained by first calculating the standard deviation of the variable, which is .495. The resulting standard error is .014. We then plug .576 ($\hat{p}$), .014 (SE), and our critical value (1.96) into the equation. We find that 57% of the population that answered the question feel that there is no favoritism or Blacks are favored with a margin of ±2.7 percentage points.

$$.95 \approx Pr\left(\hat{p} - 1.96 * .014 \leq p \leq \hat{p} + 1.96 * .014\right)$$

TABLE 9-1 Bias in the Federal Government

	Frequency	Percent
Treats whites much better	199	16.89304
Treats whites moderately better	138	11.71477
Treats whites a little better	162	13.75212
Treats both the same	353	29.96604
Treats Blacks a little better	152	12.90323
Treats Blacks moderately better	87	7.38540
Treats Blacks much better	86	7.30051
Skipped	1	0.08489
Total	1178	100.00000

Code Chunk 9-2

```
nes$fed <-
  ifelse(nes$disc_fed=="Treats whites much better", 0,
    ifelse(nes$disc_fed=="Treats whites moderately better", 0,
      ifelse(nes$disc_fed=="Treats whites a little better", 0, 1)))
```

Making the calculations, we get the following:

$$.95 \approx Pr(57 - 2.7 \leq p \leq 57 + 2.7)$$

That is, we're 95% confident that the true proportion of the population that feels there is either no favoritism or that the federal government favors Blacks is 57% ± 2.7 percentage points.

Finding the Population Mean

In addition to knowing the population proportion (i.e., how many), we want to know the population mean (i.e., how much). Instead of asking how many people smoke, attend college, get vaccinated, or support the President, we want to know how much the average income is, the average distance to work, or the average calories consumed. When the outcome is continuous rather than categorical, the basic mechanics are the same—we take a random sample and calculate the sample statistic—but the formula is slightly different.

Examples abound. When proposing tax increases, politicians need to know the mean income of their constituency. When deciding to open a new daycare facility, potential owners want to know how many kids live in the surrounding areas. When relocating a business, managers want to know the average commute time of their employees. Epidemiologists, hospitals, and the police want to know the average amount of alcohol consumed to assess the effort required to curb alcoholism, to evaluate the need for internists, and to determine how many sobriety checkpoints need to be deployed on New Year's Eve. When interested in means rather than proportions, we use a slightly different formula to calculate the confidence interval.

Example: Hours Studied per Week

Let's say we want to know the average amount of time students from your college or university study a week. We can't afford to survey all students so we take a random sample of 100 students to find out. Suppose from the start that the chancellor wants to be 90% confident that our estimate is within a certain distance of the real population mean. First, consult a table with z-critical values and find the critical value for 90%. The critical value tells us how many standard deviations from the mean of the standard normal distribution correspond to 90% of the area underneath the curve. We see that 90% is 1.65, 95% is 1.96, and 99% is 2.58. A 90% confidence interval for the population mean is computed with the following formula:

$$\bar{X} \pm 1.65 \left(\frac{s}{\sqrt{n}} \right)$$

To plug in the numbers and calculate, we first need the sample mean for the 100 students drawn. Let's make it easy and say that the sample mean is 50. Students in the sample, on average, study 50 hours every week. Suppose the sample standard deviation is 10. This is calculated by the following formula:

$$s = \sqrt{\frac{1}{n-1} \sum_{i=1}^{100} (x_i - \bar{x})^2}$$

Now that we have the sample mean and the sample standard deviation, we're set. This leads to the following calculations:

$$CI_{90\%} = 50 \pm 1.65\left(\frac{10}{\sqrt{100}}\right)$$
$$CI_{90\%} = 50 \pm 1.65$$
$$CI_{90\%} = (48.35, 51.65)$$

With these calculations, we would say that we're 90% confident that the population mean (how much students actually study) lies somewhere between 48.35 and 51.65 hours. The lesson here is as follows: when you don't know the population parameter, you use the mean and standard deviation of the sample to help calculate the confidence interval for the estimate. In this case our estimate is the mean number of hours students study in a week.

Example: Attitudes Toward Scientists

Suppose the federal government is contemplating a significant investment in NASA or in an expensive Large Hadron Collider. In order to gauge the country's appetite for such an investment, the administration wants to know the average attitude toward scientists in the country. We're interested in the population mean. The mean, in this case, represents the center of a continuous variable—a feeling thermometer on scientists. The thermometer ranges from 0 to 100, with 100 being the highest (the highest level of respect for scientists). We're also in the context of a large sample size. We want to say some things about the average attitude toward scientists among citizens in the United States. To do so we draw a sample of 1,178 citizens (the number of respondents in the NES survey). The average feeling thermometer score of that sample is 72.9. Let's calculate the 95% confidence interval for this sample.

1. Formally, we represent the confidence interval with the following formula:

$$.95 \approx Pr(\bar{x} - 1.96 SE(\bar{x}) \leq \mu \leq \bar{x} + 1.96 SE(\bar{x}))$$

2. Calculate $\bar{x} = \frac{1}{n}\Sigma x_i = 72.9$.
3. Calculate the sample standard deviation: $s = \sqrt{\frac{1}{n-1}\sum_{i=1}^{1178}(x_i - \bar{x})^2} = 23.96$ (given from the example).
4. Calculate the sample standard error: $SE(\bar{x}) = \frac{s}{\sqrt{n}} = \frac{23.96}{\sqrt{1178}} = .70$.
5. The 95% CI is $72.9 \pm (1.96)(.69) = 72.9 \pm 1.35$ points on the thermometer.

In this particular example, we had a sample mean of 72.9 points on the thermometer, and the confidence interval is 72.9 ± 1.35 points. This means that we're 95% confident that the true mean lies between 71.55 and 74.25 points.

Example: Age of the Population

Public officials need to monitor demographic trends for a whole host of reasons. Whether the concern is the viability of the social security system, the number of elementary schools required to educate our children, or the number of hospitals required for an aging population, policymakers need to have an accurate view of demographic trends.

Although knowing demographic trends over time would be helpful, sometimes it's necessary simply to know the mean age of the population at a given time. In the NES survey, we have data on the year a respondent was born. In this example, we don't know the average age of the population, but we can calculate the mean age—at least the mean age of those over 18—from the sample and use that to estimate the population mean.

Let's calculate the mean and a confidence interval at the 95% level of confidence from our NES sample of 1,178 respondents. Remember, we don't know what the true population mean is but we want to find out. The NES survey provides a random sample of 1,178 individuals from which we can calculate our confidence interval. The mean and standard deviation of our sample is calculated simply by calculating the mean and standard deviation of the variable nes$birthyr (Code Chunk 9-3).

Code Chunk 9-3

```
mean(nes$birthyr)
```
[1] 1967.978

```
sd(nes$birthyr)
```
[1] 16.98526

With our sample of 1,178 respondents, we calculate the mean, which is approximately 1968. The average respondent in the survey was born in 1968, which means the average age of the respondent is 48 years old since the survey was taken in 2016. To calculate the confidence interval, we also need the standard deviation of the sample (16.985).

With those two numbers, we go through the same five steps that we did in the previous example.

1. Formally state the interval: .

$$.95 \approx Pr(\bar{x} - 1.96 SE(\bar{x}) \leq \mu \leq \bar{x} + 1.96 SE(\bar{x}))$$

2. Calculate the sample mean: $\bar{X} = \frac{1}{n}\sum x_i = 48$.

3. Calculate the sample standard deviation: $s = \sqrt{\frac{1}{n-1}\sum_{i=1}^{1178}(x_i - \bar{x})^2} = 16.985$ (given from the example).

4. Calculate the sample standard error: $SE(\bar{x}) = \frac{s}{\sqrt{n}} = \frac{16.985}{\sqrt{1178}} = .46$.

5. The 95% CI = $48 \pm (1.96)(.46)$ y = $48 \pm .9$ years.

As the calculations indicate, we can be fairly accurate when we're dealing with a large number of respondents. We are 95% confident that the true mean age of the adult population is 48 years ± 1 year. Unfortunately, we're not always flush with data. In many cases, samples of over a thousand are not possible. In cases when the number of observations is smaller, we can still calculate a confidence interval, but there is a slight adjustment as we'll see in the next section.

> **KNOWLEDGE CHECK: Calculate confidence intervals around proportions and means.**

1. Using the NES variable *nes$amer_ident* as your sample, calculate what percentage of Americans regard their identity as an American to be "extremely important." Provide your best guess and the confidence interval at the 95% level.

2. Using the NES variable *nes$healthspend* as your sample, calculate what percentage of Americans favor "increasing a great deal" government spending to pay for health. Provide your best guess and the confidence interval at the 99% level. Be sure to account for those who didn't answer the question.

3. Suppose you took a random sample of 100 countries from the *world* data set and found that the sample mean of military spending as a percentage of GDP was 1.96% with a standard deviation of .5 percentage points. Construct a confidence interval at the 95% level. Using the countries in the *world* data set as your population, is the true population mean within the confidence interval? Construct a confidence interval for the 90% level. Is the true population mean now within the confidence interval?

4. In the previous problem, what happened to the size of the confidence interval when moving from a 95% level of confidence to the 90% level?

    a. The confidence interval increased.

    b. The confidence interval decreased.

    c. The confidence interval stayed the same.

5. Suppose you took a random sample of 100 countries from the *world* data set and found that the sample mean of homicides was 6.5 per 100,000 population with a standard deviation of 5. Construct a confidence interval at the 99% level of confidence. Using the countries in the *world* data set as the population, is the true population mean within the interval?

Small Samples and the t-Distribution

In most cases, we don't know the standard deviation of the population we're sampling from: we can't assume it resembles the standard normal distribution with 95% of the area beneath the curve lying 1.96 standard deviations on either side of the mean. We know that many distributions have a normal shape, but we don't know beforehand if the population resembles the standard normal distribution. We're also not always blessed with the time or the money to draw samples of more than 100 individuals or objects. When our sample has a relatively small number of observations (say less than 100) and we don't know the population standard deviation beforehand, we rely on a different theoretical distribution: the **t-distribution** instead of the z-distribution (the standard normal distribution).

To account for the relative uncertainty, the t-distribution has a larger spread—more dispersion. Put differently, the t-distribution's height is shorter and its tails are wider. With less of the distribution centered over the mean and a larger portion of the distribution residing in the tails, the t-distribution accounts for estimates more likely to be found further away from the mean.

To compare the two different distributions, let's overlay them in a ggplot. First we create the three variables we need: a variable for the x axis (*x1*) and two variables that contain the corresponding density values for a t-distribution (*x2*) and a normal distribution (*x3*). For *x1*, we simply say we want a variable with 10,000 observations between −4 and 4. For *x2*, we use the *dt()* function, which takes *x1* and supplies the corresponding density (height of the curve) for the t-distribution with 1 degree of freedom (df = 1). The third variable takes *x1* and supplies the corresponding density for the normal distribution. Code Chunk 9-4 then combines the three variables into a data frame called *mydata*.

Code Chunk 9-4

```
x1 <- seq(-4, 4, length = 10000)
x2 <- dt(x1, df = 1)
x3 <- dnorm(x1)

mydata <- data.frame(x1,x2,x3)
```

Now that we have all of the data, we simply draw a figure with ggplot that has two *geom_line()* functions included, one colored red (normal distribution) and one colored blue (t-distribution) (Figure 9-1). To note, I added labels to this figure in Adobe Illustrator.

Code Chunk 9-5

```
ggplot(mydata, aes(x1)) +
  geom_line(aes(y = x2), col="#0000bf") +
  geom_line(aes(y = x3), col="#bf0000") +
  theme_minimal() +
  theme(plot.title = element_text(size = 8, face = "bold"),
        axis.title = element_text(size = 8, face = "bold")) +
  xlab("Standard Deviations") +
  ylab("Density") +
  ggtitle("Figure 9-1: The t-Distribution Has Fatter Tails")
```

As you can see, the t-distribution (blue curve) has a lower peak and wider tails, a function of the distribution's greater variance. Although the difference between the two distributions does not seem dramatic, the small differences in the tails can result in large differences between the two as we travel further from the mean.

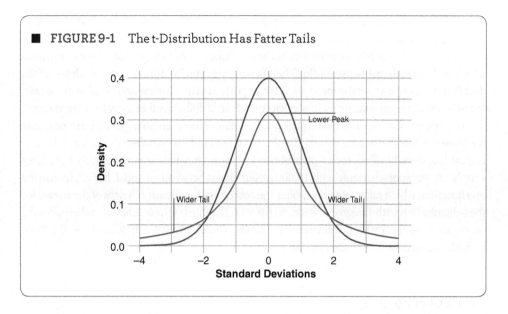

FIGURE 9-1 The t-Distribution Has Fatter Tails

Perhaps the most important difference between the standard normal distribution and the t-distribution, however, is that there are many t-distributions, each based on the number of observations of the sample. We use the number of observations to determine which t-distribution to use with a slight adjustment: we use the number of observations minus 1 or n − 1. We also use a different term instead of observations—we say the **degrees of freedom**.

To illustrate what I mean by many t-distributions, let's draw several different curves all with different degrees of freedom. Using ggplot, the exercise is fairly easy (Code Chunk 9-6). First, note that I'm creating three different curves with the *geom_line()* command and I'm specifying that each curve has a t-distribution with a different degree of freedom (1, 2, and 6). The only other thing worth noting is that with the *scale_color_discrete()* command, I'm changing the title and labels on the legend.

Figure 9-2 shows the different t-distributions. Note that as the degrees of freedom gets larger, the shape of the curve changes: the peak gets higher and the tails narrow. Since we're

Code Chunk 9-6
```
ggplot(mydata, aes(x1)) +
  geom_line(aes(y = dt(x1, df = 1), col = "#bf0000")) +
  geom_line(aes(y = dt(x1, df = 2), col = "#008b00")) +
  geom_line(aes(y = dt(x1, df = 6), col = "#0000bf")) +
  xlab("Standard Deviations") +
  ylab("Density") +
  ggtitle("Figure 9-2: The t-Distribution Approaches Normality") +
  scale_color_discrete(name = "Degree of Freedom",
                       labels = c("df=6", "df=2", "df=1")) +
  theme_minimal() +
  theme(plot.title = element_text(size = 8, face = "bold"),
        axis.title = element_text(size = 8, face = "bold"),
        legend.title = element_text(size = 8),
        legend.position = c(0.75, 0.7))
```

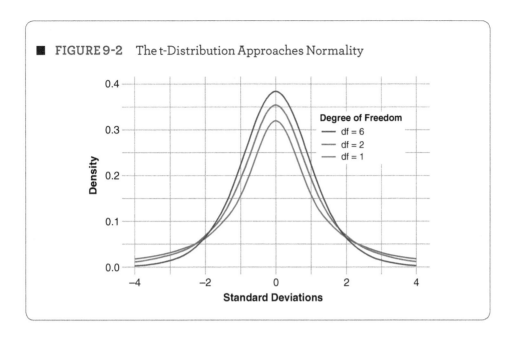

FIGURE 9-2 The t-Distribution Approaches Normality

dealing with very small samples, the difference between a sample size of 2 and 6 can be important. Figure 9-2 illustrates how t-distributions with more degrees of freedom begin to resemble a normal distribution.

Degrees of Freedom

While learning how to make some rudimentary calculations without fully comprehending the underlying theory is sometimes necessary, understanding the degrees of freedom before we continue will come in handy down the line. So before we continue, let's take a short detour to learn some things about the degrees of freedom.

It's important to note that the degrees of freedom comes up in a number of different contexts. We don't need to cover them all here. For our purposes, let's concentrate on one. Fundamental to our detour is the definition of degrees of freedom: the degrees of freedom is the number of *independent* observations needed to make a calculation. Rather than using the term *independent*, we can also define the degrees of freedom as the number of *critical* observations needed to make a calculation.

Since we don't know the population standard deviation (we're not assuming it's a standard normal distribution), we need to estimate it by taking a sample. And, when calculating the sample standard deviation, we have to make one calculation first: we need to calculate the sample mean. Since we need the sample mean first to calculate the standard deviation, we use up one of those critical observations. We therefore have only n − 1 critical inputs left to make the calculation. Let's consider a simple example.

Suppose we sample from a larger population, randomly drawing the following numbers (1, 2, 3, 4, 5). The sample mean is 3. Now consider an important component of our calculation for the sample standard deviation: the distance of each point from the mean. We know mathematically that the distance of each point from the mean has to sum to 0:

$$(x_1 - \bar{x}) + (x_2 - \bar{x}) + (x_3 - \bar{x}) + (x_4 - \bar{x}) + (x_5 - \bar{x}) = 0$$

$$(1-3) + (2-3) + (3-3) + (4-3) + (5-3) = 0$$

Given that, if we have four of the five numbers, we can calculate the fifth. For example, it's clear that if we know the first four numbers and we know they all have to sum to 0, the fifth number (represented by the question mark) has to be 5:

$$(1-3)+(2-3)+(3-3)+(4-3)+(?-3)=0$$

In this example, we have five observations and the degrees of freedom is 5 – 1 since we only need four independent observations to calculate the standard deviation. This is why the degrees of freedom plays a role when we use t-distributions. It stems from the fact that we don't know the population standard deviation beforehand and we have to rely on a sample to calculate it. So, when choosing which t-distribution to use, we take the number of observations we have (n) and subtract 1 (n – 1). When we have 10 observations, for example, we use the t-distribution associated with 10 – 1 or 9 degrees of freedom.

Sample Standard Deviation With Small Samples

The quantity n – 1 plays another important role when calculating confidence intervals. Specifically, you may have noticed we use (n – 1) rather than (n) when calculating the sample standard deviation. Why use (n – 1)?

It turns out, when our sample sizes get smaller and smaller, our calculations of the sample standard deviation do an increasingly poor job of estimating the population standard deviation. In statistical language, our estimate of the standard deviation is biased. Specifically, the estimate becomes biased downward unless we correct it by using n – 1 in the denominator rather than simply n. Let's do an experiment to illustrate the bias.

To set up the experiment, let's produce several sampling distributions of the standard deviation. Five different sampling distributions where n = 100, n = 50, n = 10, n = 5, and n = 3 should illustrate the point. To begin, I use the command *set.seed()* and set it at 34 so that you can reproduce this experiment. I also set the mean to 50 and the standard deviation to 10. The next series of commands creates a new variable (*xbar1*) with room to hold 50 repetitions: *rep(0,50)*. I then use a for-loop, which is helpful in repeating commands. The for-loop in this instance tells R to take 50 draws (i in 1:50) of 100 observations from a normal distribution with a mean of 50 and a standard deviation of 10. I then calculate the sample standard deviation using the *sd()* command. Since R already corrects for the bias when calculating the sample standard deviation by dividing with the square root of n – 1, I multiply the expression by the square root of (n/n – 1) since we want to see what happens when we divide by n and not the degrees of freedom n – 1. The standard deviation we're calculating in Code Chunk 9-7 is as follows:

$$s = \sqrt{\frac{1}{n}\sum_{i=1}^{n}(x_i - \bar{x})^2}$$

The final two commands in Code Chunk 9-7 collect each variable in a list and then stack them using the *melt()* command so that we can draw the graph.

Code Chunk 9-7

```
set.seed(34)
mu=50; sigma=10

xbar1=rep(0,50)
for (i in 1:50) {
xbar1[i]=sd(rnorm(100, mean=mu, sd=sigma))*sqrt((100-1)/100)}

xbar2=rep(0,50)
for (i in 1:50) {
xbar2[i]=sd(rnorm(50, mean=mu, sd=sigma))*sqrt((50-1)/50)}

xbar3=rep(0,50)
for (i in 1:50) {
xbar3[i]=sd(rnorm(10, mean=mu, sd=sigma))*sqrt((10-1)/10)}

xbar4=rep(0,50)
for (i in 1:50) {
xbar4[i]=sd(rnorm(5, mean=mu, sd=sigma))*sqrt((5-1)/5)}

xbar5=rep(0,50)
for (i in 1:50) {
xbar5[i]=sd(rnorm(3, mean=mu, sd=sigma))*sqrt((3-1)/3)}

x <- list(v1=xbar1,v2=xbar2,v3=xbar3,v4=xbar4,v5=xbar5)
data <- melt(x)
```

Now that we've created the data (five different variables that each represent 50 standard deviations calculated from 50 draws), we're ready to plot. I first transform and label the variables using the *factor()* command. I then use the standard *ggplot()* command to produce a scatterplot with one wrinkle: I include the *stat_summary()* command, which calculates the mean of each variable and connects them with a red line.

Code Chunk 9-8

```
data$L1 <- factor(data$L1,
levels = c("v1","v2","v3","v4","v5"),
labels = c("n=100", "n=50", "n=10", "n=5", "n=3"))

ggplot(data,aes(y=value, x=L1)) +
  geom_point(col="#0000bf") +
```

```
stat_summary(aes(y = value, group=1), fun.y="mean", colour="#bf0000",
            geom="line", group=1) +
ggtitle("Figure 9-3: Sample Standard Deviation Decreases
        With Smaller n") +
xlab("Number of Observations in Each Sample") +
ylab("Sample Standard Deviation") +
theme_minimal() +
theme(plot.title = element_text(size = 8, face = "bold"),
      axis.title = element_text(size = 8, face = "bold")) +
geom_hline(yintercept = 10, linetype = "dashed")
```

As the number of observations used in each sample decreases, the mean value of the sample standard deviations decreases. This is because as our sample size gets smaller, the sample standard deviation calculated with n in the denominator becomes increasingly underestimated. Again, we say that the sample standard deviation estimator is **biased** since it does a poor job of estimating the population parameter. Why does our estimate of the sample standard deviation get worse as our n decreases?

The estimate misses the target because the bias in the estimator is related to n. To demonstrate, let's consider the estimate of the sample standard deviation using n in the denominator:

$$\sqrt{\frac{1}{n}\sum_{n=1}^{n}(x_i - \overline{x})^2}$$

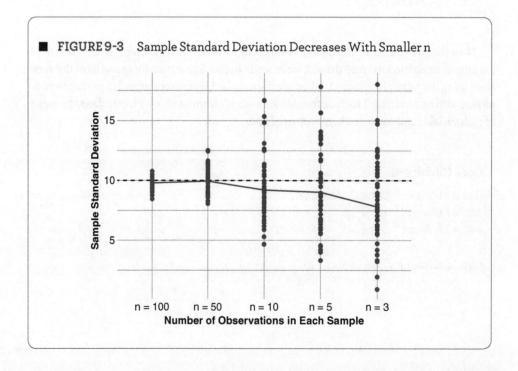

■ FIGURE 9-3   Sample Standard Deviation Decreases With Smaller n

The following equation shows the relationship between the biased estimator for the sample standard deviation (the left side of the equation) and the true population standard deviation σ:

$$\sqrt{\frac{1}{n}\sum_{n=1}^{n}(x_i - \bar{x})^2} \approx \frac{(n-1)}{n}\sigma$$

This means our estimate of the standard deviation gets worse as n decreases. We know this because the quantity $\sqrt{\frac{n-1}{n}}$ gets smaller and smaller as n decreases. As n decreases, the sample standard deviation calculated with n becomes a smaller and smaller fraction of the true population standard deviation.

Let's get concrete with an example. When n = 100, the bias of estimating the standard deviation is equal to $\frac{(100-1)}{100}$. In other words, we're multiplying σ by .99. If our n is much smaller, say 2, the bias is equal to $\frac{(2-1)}{2}$, which is equivalent to multiplying σ by .5. So, in order to obtain a sample estimate that is not biased (to remove $\frac{n-1}{n}$ from the right side of the equation), we simply multiply both sides of the equation by $\frac{n-1}{n}$. This gives us the following:

$$\sqrt{\frac{1}{n-1}\sum_{n=1}^{n}(x_i - \bar{x})^2} = \sigma$$

The resulting equation is what we use to calculate the sample standard deviation. You'll notice that it includes n − 1 in the denominator. When we adjust the estimator to account for the bias, we use n − 1 rather than n.

Now we repeat the previous experiment by calculating the standard deviation of each sample by n − 1. In Code Chunk 9-9 we produce the same data, except that we do not multiply the expression by the square root of (n − 1/n). This leaves R's calculation of the standard deviation in its regular form, in which it divides by n − 1.

Code Chunk 9-9

```
set.seed(34)
mu=50; sigma=10

xbar1=rep(0,50)
for (i in 1:50) {xbar1[i]=sd(rnorm(100, mean=mu, sd=sigma))}

xbar2=rep(0,50)
for (i in 1:50) {xbar2[i]=sd(rnorm(50, mean=mu, sd=sigma))}

xbar3=rep(0,50)
for (i in 1:50) {xbar3[i]=sd(rnorm(10, mean=mu, sd=sigma))}
```

```
xbar4=rep(0,50)
for (i in 1:50) {xbar4[i]=sd(rnorm(5, mean=mu, sd=sigma))}

xbar5=rep(0,50)
for (i in 1:50) {xbar5[i]=sd(rnorm(3, mean=mu, sd=sigma))}

x <- list(v1=xbar1,v2=xbar2,v3=xbar3,v4=xbar4,v5=xbar5)
data <- melt(x)

data$L1 <- factor(data$L1,
levels = c("v1","v2","v3","v4","v5"),
labels = c("n=100", "n=50", "n=10", "n=5", "n=3"))

ggplot(data,aes(y=value, x=L1)) +
  stat_summary(aes(y = value,group=1), fun.y="mean",
               colour="#bf0000", geom="line",group=1) +
  geom_point(col="#0000bf") +
  ggtitle("Figure 9-4: Bias Is Eliminated With n – 1") +
  xlab("Number of Observations in Each Sample") +
  ylab("Sample Standard Deviation") +
  theme_minimal() +
  theme(plot.title = element_text(size = 8, face = "bold"),
        axis.title = element_text(size = 8, face = "bold")) +
  geom_hline(yintercept = 10, linetype = "dashed")
```

Using n – 1 in the denominator is called **Bessel's correction**. When we correct for the downward bias (Figure 9-4), you'll see that as the size of the sample decreases to n = 3, the mean of the standard deviations computed still approximates the population standard deviation, which is 10.[1]

Before continuing on, let's take stock of where we've been. When we don't know the population standard deviation (which is usually the case) and we are dealing with a sample size less than 100, we use the t-distribution to make inferences. There are many t-distributions and we choose the one associated with the degrees of freedom in the sample (n – 1). I defined the degrees of freedom and its relevance to the t-distribution. The quantity n – 1 also figures prominently in another context: the formula used to calculate the sample standard deviation. When using n, the formula for the sample standard deviation does a poor job of estimating the population's standard deviation: it is biased. To eliminate some of the bias, we make a small correction by using n – 1 rather than n.

---

[1] Technically, the calculation for the sample standard deviation using n – 1 is still biased. While the n – 1 adjustment helps, it still remains a biased estimator.

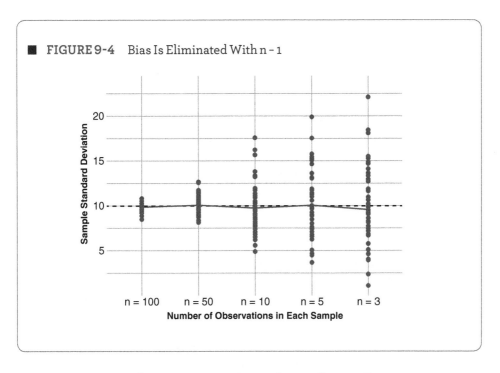

FIGURE 9-4 Bias Is Eliminated With n - 1

Constructing Confidence Intervals With Small Samples

Now we're ready to construct confidence intervals when our sample size is less than 100. How do we go about making inferences about the population mean μ when we have a sample of n = 100 or less? Fortunately, the steps we took earlier for larger samples are the same. The only difference is that we use a **critical t-value** rather than a critical z-value in our calculations. Let's use the previous example to illustrate the problem we have when our sample has an n = 20 rather than n = 1,178. Let's address the same question we posed earlier with respect to people's attitudes toward scientists. Suppose we take a random sample of 20 individuals from the NES survey and again find that the sample mean $\bar{x}$ is 72.9.

> **Art and Practice of Data Visualization**
> **THE MAGIC NUMBER**
>
> Although four out of five statisticians recommend 30 as the threshold to determine using the normal distribution (z-critical score) or the t-distribution (t-critical score), the analyst is advised to take a more conservative approach and use the numbers 50 or even 100.

1. Formally, we calculate the confidence interval with the following formula:

$$.95 \approx Pr\left(\bar{x} - 2.093 SE(\bar{x}) \leq \mu \leq \bar{x} + 2.093 SE(\bar{x})\right)$$

2. Calculate $\bar{X} = \frac{1}{n}\sum x_i = 72.9$.

3. Calculate the sample standard deviation: $s = \sqrt{\dfrac{1}{20-1}\sum_{i=1}^{20}(x_i - \bar{x})^2}$.

4. Calculate the sample standard error: $SE(\bar{X}) = \dfrac{s}{\sqrt{n}}$.

5. The 95% CI = $72.69 \pm 2.23(SE(\bar{X}))$.

Note that the only difference between this example and the previous one lies in two numbers. First, to determine which t-distribution to use, we take the number of observations (20) and subtract 1 (20 – 1), which gives us the degrees of freedom. After we've identified the correct t-distribution, we use a different critical value in this case (2.23, the t-critical value), which gives us the 95% level of confidence.

Example: Women's Pay Versus Men's

Let's try another example to illustrate calculating a confidence interval when the number of observations is relatively small. Suppose we are embroiled in a court case involving unequal pay and we want to know how much women make relative to men nationally. Unfortunately, we've spent so much on lawyers, we only have limited funds to determine the national average; we only have enough money to contact 30 out of the 50 U.S. state departments of labor for our data-gathering exercise. I draw 30 observations from the state data randomly to create a new 30-state data set (called *rstates* in Code Chunk 9-10).

There are a couple of things to note in Code Chunk 9-10. First, to replicate the results I set the seed to 567. I then convert the *states* data set from a data frame to a tibble of class 'tbl_df' so that I can properly use the *sample_n()* function, allowing me to take the new tbl_df *nstates* and randomly draw (with replacement) 30 cases. Once the new tibble *rstates* with 30 randomly drawn states has been created, I can calculate the mean and standard deviation of the variable *rstates$percwom*. Finally, note that I use the command *qt()* to find the critical value for t at the 90% level of confidence when n is equal to 30 (1.697).

Code Chunk 9-10

```
set.seed(567)

nstates <- tbl_df(states)

rstates <- sample_n(nstates, 30, replace = TRUE)

mean(rstates$percwom)
sd(rstates$percwom)
sd(rstates$percwom)/sqrt(30)
qt(.05, 30)


mean(states$percwom)
sd(states$percwom)
```

If we want to be 90% confident, the argument we use in the *qt()* function is .05, which is the percentage of the distribution we want in each tail.

1. Formally, we calculate the confidence interval with the following formula:

$$.90 \approx Pr(\bar{x} - 1.7 SE(\bar{x}) \leq \mu \leq \bar{x} + 1.7 SE(\bar{x}))$$

2. Calculate $\bar{x} = \frac{1}{n}\sum x_i = 79.93$.

3. Calculate the sample standard deviation: $s = \sqrt{\frac{1}{20-1}\sum_{i=1}^{20}(x_i - \bar{x})^2} = 5.11$.

4. Calculate the sample standard error: $SE(\bar{x}) = \frac{s}{\sqrt{n}} = .93$.

$$.90 CI = 79.93 \pm 1.7 SE(\bar{x}) = 79.93 \pm 1.58$$

These calculations indicate that we can be 90% confident that the true population mean lies in the interval 78.35 and 81.51. Since we have data from all 50 states, we can check to see if the mean lies somewhere within that interval. The mean women's salary relative to men's over the entire 50 states is, in fact, 80.44.

While constructing a confidence interval around our estimate is extremely helpful when we have limited time and resources (and therefore limited data), there is another important use of this machinery: comparing the means between two groups.

KNOWLEDGE CHECK: Use the t-distribution to calculate confidence intervals.

---

6. How is the t-distribution different than the standard normal distribution?
    a. It has wider tails and a higher peak.
    b. It has smaller tails and a higher peak.
    c. It has wider tails and a lower peak.
    d. It has smaller tails and a lower peak.

7. Why is the shape of the t-distribution different than the standard normal distribution?
    a. We are dealing with fewer observations when we use the t-distribution.
    b. We are dealing with more observations when we use the standard normal distribution.
    c. We are dealing with less information when we use the t-distribution.
    d. We are dealing with more information when we use the standard normal distribution.

8. What are the degrees of freedom?
    a. The number of observations in a sample.
    b. The number of critical observations in a sample.

c. The number of independent observations in a sample.

   d. The number of observations needed for a random sample.

9. What did we learn in the example of Bessel's correction?

   a. Our estimates of the mean are increasingly biased as n decreases.

   b. Uncorrected estimates of the standard deviation are increasingly biased as n increases.

   c. Our estimates of the mean are increasingly biased as n increases.

   d. Uncorrected estimates of the standard deviation are increasingly biased as n decreases.

10. Set the seed to 38 and randomly draw 25 cases from the *states* data and calculate the following:

    a. What is the sample mean of the variable *murderrate*?

    b. What is the sample standard deviation of the variable *murderrate*?

    c. What is the confidence interval for your estimate at the 95% level?

    d. Does the confidence interval include the population mean?

11. Set the seed to 38 and randomly draw 25 cases from the *states* data and calculate the following:

    a. What is the sample mean of the variable *inc*?

    b. What is the sample standard deviation of the variable *inc*?

    c. What is the confidence interval for your estimate at the 99% level?

    d. Does the confidence interval include the population mean?

12. Set the seed to 38 and randomly draw 25 cases from the *states* data and calculate the following:

    a. What is the sample mean of the variable *turnout*?

    b. What is the sample standard deviation of the variable *turnout*?

    c. What is the confidence interval for your estimate at the 90% level?

    d. What is the confidence interval for your estimate at the 99% level?

Comparing Two Sample Means

One of the more common tests we conduct in the social sciences involves comparing means between two different groups. We often want to know if different ethnic or racial groups achieve the same success as a way to evaluate how society allocates public goods and services or provides opportunities. Fortunately, we simply perform an extension of what we've done already.

When comparing means, we want to know whether the difference between them is significant. When we calculate the means for two different groups, we want to know whether the difference we observe is real or if it is simply due to the samples we drew. Remember, random samples produce different means just due to luck of the draw, what we also refer to as **sampling variance**. It is the case, for example, that had we drawn a different sample from the U.S. population for the

NES survey, we would have arrived at different mean values for all of our variables. Given that, when we calculate the means for different groups from that sample, how likely is that difference due to sampling variance or an important difference between the two groups? If we drew another sample, would we get a similar result? Making inferences about two sample means is simply a way to determine whether the differences we see between two groups is the result of something substantive or the result of random chance (sampling variance). Specifically, we want to know whether the difference we calculate between the means in two samples is different than zero.

We need to be very deliberate and clear about our hypothesis. When comparing means, hypothesize there is a difference between the means. What we call the **null hypothesis** states there is no difference between the means. We say the null hypothesis *is rejected* if the difference in means is not zero.

To make inferences about two sample means, we use the **two-sample t-test**, which is based on the t-distribution introduced above. The two-sample t-test we conduct produces a **t-ratio (for a two-sample t-test)**, which tells us how far the calculated difference is from 0. The t-ratio is calculated with the following formula:

$$\text{t-Ratio for a two-sample t-test} = \frac{\overline{x}_1 - \overline{x}_2}{\sqrt{\frac{s_1^2}{n} + \frac{s_2^2}{n}}}$$

Example: Two Groups and Two Incomes

Suppose we want to know whether to locate a Krispy Kreme franchise in your hometown and we know from careful market research that the amount of cash customers keep on hand is an important determinant of donut purchases. We have a choice of locating the store close to one of two local schools, Hometown High or Ridgeview Academy. Is there a difference between the two student populations in terms of cash on hand? Suppose we perform surveys and find that there is a $10 difference between Hometown High and Ridgeview students. If the t-ratio generated is greater than the absolute value of 2.042, we know that the probability of calculating that $10 difference by random chance is pretty small: it's only going to happen 5% of the time we take a random sample. The larger the t-ratio, the less likely the difference in means is due solely to chance.

To illustrate the mechanics of what's going on, consider Figure 9-5, which shows where the difference of $10 would lie in the t-distribution. To draw the figure, we first need to create two variables. We first create *v1*, which ranges between −4 and 4 and contains 10,000 observations (Code Chunk 9-11). As in previous illustrations of the normal curve or t-distribution, we re-express *v1* as densities of the t-distribution with an n of 30. Once combined into a data frame, we're ready to plot. To shade the area underneath the curve, I use the *geom_ribbon()* command.

Code Chunk 9-11

```
v1 <- seq(-4, 4, length = 10000)
v2<- dt(v1, 30)

df <- data.frame(v1, v2)
```

```
ggplot(df, aes(v1, v2)) +
  geom_line() +
  theme_minimal() +
  theme(plot.title = element_text(size = 8, face = "bold"),
        axis.title = element_text(size = 8, face = "bold")) +
  geom_ribbon(data=subset(df ,v1>2.042 & v1<4),aes(ymax=v2),ymin=0,
              fill="#0000bf",colour=NA) +
  geom_ribbon(data=subset(df ,v1< -2.042 & v1> -4),aes(ymax=v2),ymin=0,
              fill="#0000bf",colour=NA) +
  scale_fill_brewer(guide="none") +
  ggtitle("Figure 9-5: A $10 Difference Is Significant") +
  xlab("Standard Deviations") +
  ylab("Density") +
  geom_vline(xintercept = 2.5, linetype=2) +
  geom_vline(xintercept = -2.5, linetype=2) +
  scale_x_continuous(sec.axis=sec_axis(~.*4.00+0,
                     name = "Difference in Dollars"))
```

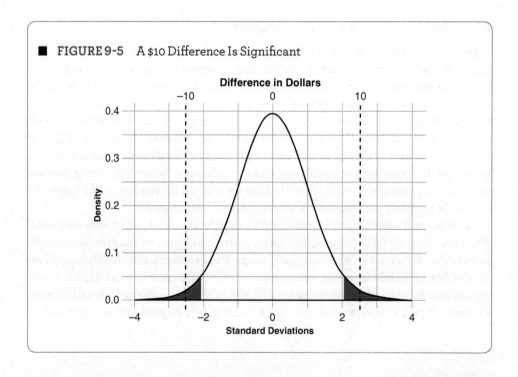

FIGURE 9-5 A $10 Difference Is Significant

The blue-shaded areas represent 5% of the area underneath the curve. One thing to note is that there are two regions shaded: one above 2.042 and one below −2.042. Since one group could have a lower average income than the other, the lower half of the distribution has negative values. The graph in Figure 9-5 reveals that if a group has either $10 less or $10 more than another group, we're pretty sure this is not simply due to sampling variance since the probability is that the difference in the two sample means (the sample means between the two groups) is large.

Since we are now comparing means, we express the confidence interval differently. The confidence interval for large sample sizes is as follows:

$$\mu_1 - \mu_2 = \overline{x_1} - \overline{x_2} \pm z_{\frac{a}{2}} SE\left(\overline{x_1} - \overline{x_2}\right)$$

Example: Race and Attitudes Toward the Police

Fortunately, R provides an easy way to make inferences about two sample means. This is a very common test when we want to know whether the means of a dependent variable differ according to (or what we say is conditional on) an independent variable. For example, we might want to know whether the mean attitude toward police differs between whites and Blacks. For ease of exposition, we draw a distinction between whites and nonwhites in the race variable we have from the NES.

To draw that distinction, we create a new variable called *nes$white*, which uses the *ifelse()* command. If the *nes$race* variable indicates "white," then assign the value 1; otherwise assign the value 0. Once we have that variable, we use the *t.test()* command in R. The *pander()* function produces a serviceable table of our results.

As Table 9-2 indicates, Whites have a more favorable view toward the police (a 11.3-point difference on the feeling thermometer with a t-ratio of 6.25). According to the two-sample t-test, the difference is statistically significant at the 99.99% level of confidence. The null hypothesis (i.e., there is no difference between whites and Blacks) is rejected.

Code Chunk 9-12

```
nes$white <- ifelse(nes$race=="White", 1, 0)

p <- t.test(ftpolice ~ white, data=nes)

pander(p)
```

TABLE 9-2   Large Difference by Race

Welch Two-Sample t-Test: ftpolice **by** white

TEST STATISTIC	DF	P-VALUE	ALTERNATIVE HYPOTHESIS
−6.253	523.6	8.366e-10 * * *	Two-sided
Mean in Group 0		Mean in Group 1	
57.89		69.26	

As noted earlier, when we test whether the means of two groups differ, we're testing whether the null hypothesis (i.e., there is no difference between means) holds. If the t-value indicates we should reject the null hypothesis, we can say with a certain level of confidence that the differences in means we observe are not due to chance but due to something else.

Example: Income and Support for Trump

Let's look at another example to build our intuition when comparing two means. An important exercise political parties undergo after every election—especially after a loss—involves figuring out what happened. The 2016 presidential election is no different than any others in that respect. Many polls leading up to the race turned out to be wrong, predicting a fairly comfortable win for Hillary Clinton. In the aftermath, analysts want to know what groups favored Trump.

Per capita income is a good place to start, specifically the median household income of a state. Were the states Trump won on average richer, poorer, or the same? Traditionally, Republican support comes from wealthier voters. We might expect, therefore, that the states Trump won have higher incomes on average. There are reasons to expect the opposite, however. More urban and educated voters were more likely to support Clinton in 2016. There are good reasons, therefore, to expect that incomes could be higher or lower among the two different groups of states. The null hypothesis is that there is no difference in median incomes between states won by Trump and states won by Clinton. The calculation is fairly easy in this case (Code Chunk 9-13) since we already have a variable that cleanly distinguishes between states Trump won and lost: *states$trumpwin*.

Code Chunk 9-13

```
p <- t.test(medinc ~ trumpwin, data = states)

pander(p)
```

TABLE 9-3 Poor States Favored Trump

Welch Two-Sample t-Test: `Medinc` **by** `Trumpwin`

TEST STATISTIC	DF	P-VALUE	ALTERNATIVE HYPOTHESIS
4.875	32.28	2.8e-05 * * *	Two-sided
Mean in Group 0		Mean in Group 1	
56,061		46,951	

Using the two-sample t-test, we find that the income in states that President Trump won was $46,951 and $56,061 in the states Hillary Clinton won, a difference of $9,110 (Table 9-3). From the t-test, we find the t-ratio is 4.875, well above the critical t-value at the 99.99% level of confidence. The test, therefore, indicates that the likelihood of finding that much of a difference would be very small (less than .001%) if the true difference between those two groups of states were zero. We say, therefore, that the null hypothesis (i.e., that there is no difference between the means) is rejected.

Whether taking samples to determine the proportion of a population, the mean, or to compare means from two different samples, establishing the confidence we have in the result is an important part of the exercise. But what does it mean to be 95% or 99% confident? What do we actually mean when we say that we're 99% confident that the difference between two means is not zero? The next section provides the answer.

> **KNOWLEDGE CHECK: Determine if the means of two populations are different.**

13. What is the difference between states Trump won and lost in terms of state spending on education per student? Perform a two-sample t-test and report whether the null hypothesis is rejected.

14. What is the difference between Northeastern states and the rest of the country in terms of homicides per 100,000 population? Perform a two-sample t-test and report whether the null hypothesis is rejected.

15. What is the difference between countries that were former British colonies and the rest in terms of GDP per capita? Perform a two-sample t-test and report whether the null hypothesis is rejected.

Confidence Levels

To understand levels of confidence, consider the earlier example that examined the average number of hours studied at a university. In that example, we knew beforehand that the true population mean was 50 hours studied a week. If we want to construct intervals at the 90% level of confidence, we're open to being wrong 10% of the time. What does that mean? It means that if we were to take 10 random samples from a population and calculate their means, approximately 1 of the 10 intervals that we calculate from each sample would not include the true population mean. To illustrate, let's take a normally distributed variable that has a mean of 50 and a standard deviation of 1. Let's then draw 20 random samples from the distribution (each sample has an n = 100). We calculate from each sample the mean, the standard deviation, and the associated 90% confidence interval. In 2 out of the 20 cases (10%), we should see that the 90% confidence interval does not include the real population mean = 50. This implies that we'd be wrong 10% of the time.

In order to run this simulation, I use a function written by Randall E. Schumacker (2015) to which I made just a few cosmetic changes (Code Chunk 9-14). I defined the function as *conf()*, which takes as arguments the number of observations per sample (100), the mean of the population (50), the standard deviation of the population (1), the number of repetitions (20), and the critical z-score (1.65). So you can duplicate the output, be sure to use the *set.seed()* command and set it at 556.

TABLE 9-4 Two Simulations Don't Include the Mean

SAMPLE MEAN	CI (LOW TO HIGH)	POPULATION MEAN	WITHIN CI
50.02	49.85–50.18	50	Yes
49.9	49.73–50.06	50	Yes
49.97	49.8–50.13	50	Yes
49.94	49.78–50.11	50	Yes
50.07	49.91–50.24	50	Yes
50.08	49.91–50.24	50	Yes
50	49.84–50.17	50	Yes
49.81	49.64–49.97	50	No
50	49.83–50.16	50	Yes
49.82	49.66–49.99	50	No
49.94	49.78–50.11	50	Yes
49.9	49.73–50.06	50	Yes
50.06	49.89–50.22	50	Yes
49.94	49.78–50.11	50	Yes
50.08	49.91–50.24	50	Yes
49.9	49.74–50.07	50	Yes
49.95	49.79–50.12	50	Yes
49.98	49.82–50.15	50	Yes
49.94	49.78–50.11	50	Yes
50.07	49.91–50.24	50	Yes

Code Chunk 9-14

```
set.seed(556)

conf(100, 50, 1, 20, 1.65)
```

As Table 9-4 indicates, there are indeed two instances where our sample estimate did not have a confidence interval that included 50. That is what we mean by confidence. If our experiment (our sampling procedure) were to be repeated 100 times, we would expect that our confidence interval would contain the true population parameter 90 times.

To keep building our intuition, if we were to perform the same exercise and stipulate we wanted to be confident at the 99.95% level, we would expect that out of 20 draws, the resulting confidence interval would include the true population mean in at least 19 out of the 20 samples—there's just the small probability that one of them would not include the true population mean. In effect, all we're changing from the previous example is that we're raising the level of confidence from 90% to 99% and we do that by changing the critical z-score from 1.65 to 2.576.

Code Chunk 9-15

```
set.seed(556)

conf(100, 50, 1, 20, 2.576)
```

As Table 9-5 indicates, since the level of confidence is so high, in every sample the resulting confidence interval included the true population mean (50). Note also, however, that increasing the level of confidence comes at a cost. As we increase the level of confidence, the size of our confidence interval grows: we can't be as precise with our estimate. In a political poll, as we increase the level of confidence, our margin of error would grow.

Now that we've established the usefulness of the standard normal distribution and the t-distribution for indicating how close our estimates are to the population parameters, let's put what we've learned in Chapters 8 and 9 in the context of where we're going next.

> **KNOWLEDGE CHECK: Explain what confidence means in statistics.**

16. Which of the following is an accurate statement about statistical significance?
    a. The level of statistical significance reflects the number of times out of 100 random samples that we would expect the population parameter to fall within the calculated confidence interval for each sample.
    b. The level of statistical significance reflects the number of times out of 100 random samples that we would expect the population parameter to fall outside of the calculated confidence interval for each sample.
    c. It reflects the level of confidence we have that our sample mean is within 1 standard deviation of the population parameter.
    d. The level of statistical significance indicates the confidence we have that the population parameter accurately describes our sample statistic.

17. Which of the following describes the relationship between the level of confidence and the size of the confidence interval.
    a. As the level of significance rises, the size of the confidence interval decreases.
    b. As the level of significance rises, the size of the confidence interval increases.
    c. As the level of significance decreases, the number of times the population parameter will lie inside the confidence interval for a given number of random draws will decrease.

d. As the level of significance increases, the number of times the population parameter will lie inside the confidence interval for a given number of random draws will increase.

A Brief Note on Statistical Inference and Causation

With the central limit theorem under our belt, we can ask what proportion of a population supports a politician or policy, we can determine what the typical case looks like, and we can compare population parameters between two populations. We also know what being confident at the 95% level actually means. We now have the conceptual, analytical, and probabilistic frameworks for understanding the statistical significance of our regression estimates. This is the machinery of statistical inference.

TABLE 9-5 All Simulations Contain the Population Mean

SAMPLE MEAN	CI (LOW TO HIGH)	POPULATION MEAN	WITHIN CI
50.02	49.76–50.28	50	Yes
49.9	49.64–50.15	50	Yes
49.97	49.71–50.23	50	Yes
49.94	49.68–50.2	50	Yes
50.07	49.82–50.33	50	Yes
50.08	49.82–50.34	50	Yes
50	49.74–50.26	50	Yes
49.81	49.55–50.06	50	Yes
50	49.74–50.26	50	Yes
49.82	49.56–50.08	50	Yes
49.94	49.68–50.2	50	Yes
49.9	49.64–50.15	50	Yes
50.06	49.8–50.32	50	Yes
49.94	49.69–50.2	50	Yes
50.08	49.82–50.33	50	Yes
49.9	49.64–50.16	50	Yes
49.95	49.69–50.21	50	Yes
49.98	49.72–50.24	50	Yes
49.94	49.69–50.2	50	Yes
50.07	49.82–50.33	50	Yes

Regression estimates—the slopes and intercepts of lines we will learn how to estimate—are merely sample statistics used to describe what's going on in the population. As we'll see, the accompanying t-ratios in a regression table provide the information we need to gauge how close we are to the target. With regression analysis, t-ratios indicate how likely we'd arrive at that estimate if the true population parameter—the slope or intercept—was actually zero.

This chapter comes with an important caveat. With most data analysis, especially the exploratory kind, cases are removed from the population and new variables are added to and subtracted from the model. As a result, the idea that all observations have an equal chance of being selected in a random sample is usually violated. The foundation upon which the law of large numbers and the central limit theorem was built—the random sample—no longer holds. Consequently, we need to maintain a very healthy dose of skepticism about the inferences we draw from our estimates. Fortunately, there are steps we can take to mitigate our injurious ways. Diagnostics, introduced later in the book, are an important aid in that endeavor.

> **KNOWLEDGE CHECK: Connect confidence intervals to statistical estimates.**

18. Which of the following accurately describe the relationship between sample statistics and regression estimates?

    a. Regression estimates (the slope and intercept) are similar to sample statistics.

    b. Regression estimates often differ from sample statistics since regression estimates are not always based on random sampling.

    c. Because of a, we can use the machinery of statistical inference when estimating the intercepts and slopes associated with regression analysis.

    d. Because of b, we have to approach the statistical significance of regression estimates with caution.

SUMMARY

Since much of what we do involves analyzing samples of data in order to make inferences about a population, statistics provides some useful tools that allow us to know the accuracy of our inferences. The machinery introduced in this chapter helps with hypothesis testing, sampling, and regression analysis. Understanding the basic statistical properties of samples and populations forms the basis of claims made in industry, the media, and academia.

The difficulty many have with these concepts stems from the disconnect between contexts where randomization is central to the enterprise and contexts where it is not. While the principles described in this chapter do apply directly to surveys or field experiments where randomization is a central feature of the enterprise, a majority of data analysis is based on data not randomly collected or on treatments not randomly assigned. Self-selection by individuals into groups (labor unions, for example) or political self-determination by countries into democratic or authoritarian regimes limits what we can learn from our analysis. In a nutshell, this is why we have to say *correlation is not causation*. Consequently, we have to remain cautious about drawing inferences that assume the treatment has been assigned randomly when it has not. While there are steps that can be taken to minimize our uncertainty, I recommend building a case on as many clues as you can find rather than building a case on as many assumptions as you can make.

COMMON PROBLEMS

- *The distinction between population proportion and population mean.* When we calculate confidence intervals for election polling, think population proportion; when calculating the confidence interval for the average characteristics of a person, place, or thing, think population mean. Again, as with the distinction between sample statistic and population parameter, keep things straight between population proportion and population mean since the two involve different calculations for establishing the confidence interval. Once you know whether it's a population proportion or a population mean, calculating the confidence interval is fairly straightforward.

- *The distinction between small and large samples.* This determines whether you'll be using critical z-values or critical t-values. The distinction is really all about the level of uncertainty. When we have small samples (less information), our guesses will be less accurate. As a result, we calculate our confidence intervals based on the t-distribution, which has a larger variance than the standard normal distribution. Also remember that the larger variance decreases or declines depending on the degrees of freedom. The more observations, the higher the degrees of freedom, the closer our results are to what we'd obtain by using the standard normal distribution. When dealing with samples whose number of observations is less than 100 (some say 30), it's a good rule of thumb to use the t-distribution.

- *Numerous t-distributions.* When making calculations using the standard normal distribution, we need not worry about the number of observations since we have enough to make the same calculations. When making calculations using the t-distribution, whether we have 3 degrees of freedom or 30 can make an important difference in terms of which critical t-values we use. For example, the critical t-value for a 90% level of confidence for 30 degrees of freedom is 1.697. The critical t-value for a 90% level of confidence for 3 degrees of freedom is 2.35.

- *Determining degrees of freedom.* The degrees of freedom refer to the number of independent data points used to compute a statistic. When calculating the mean of a sample, for example, we need all (n) of the observations to make the calculation. To calculate the standard deviation of a sample, we need to first calculate the mean of the sample, which in effect uses one of our observations up, decreasing the number of independent observations to n – 1. If we know the mean of the sample and n – 1 of the n observations, we can figure out what the last observation is given that we know the sample mean. A more intuitive example involves the choice a soccer coach makes in choosing a goalie. Imagine the 11 players on a team are lined up and the coach goes down the line, inspecting each player for consideration. If the coach has inspected the first 10 players and a goalie has not been chosen, we know the 11th has to be the goalie. In this example if we know the 10 players who are not the goalie (10 degrees of freedom), we know with certainty who will keep the goal (the 11th).

- *Difference between one-tailed and two-tailed t-tests.* In some contexts we are interested in the area underneath the curve that lies above a certain critical z-value. In those cases we can use a one-tailed test. In other contexts, we don't know whether our sample statistic will lie above or beneath a value of interest. For example, when examining the difference in sample means, the difference between one sample or the other could be positive or negative. In those cases, it makes sense to use the two-tailed test. In general, or in practice, most use the t-critical or z-critical values associated with the two-tailed tests since they provide a stronger test of significance. They represent more conservative estimates of how different the sample statistic is from zero.

- *Understanding statistical significance.* It's helpful to understand exactly what different levels of confidence represent since they are ubiquitous—we see them in many different contexts. Our level of confidence is based on how many times out of 100 the confidence interval we construct includes the true population parameter. For example, when we construct a confidence interval at the 90% level of confidence, we are saying that out of 100 different samples, the true population parameter will rest inside of the confidence interval we construct 90 times.

REVIEW QUESTIONS

1. What question is associated with population proportions?
2. What question is associated with population means?
3. When should we use the t-distribution?
4. What is bias?
5. Why do we use Bessel's correction?
6. What is the degrees of freedom?
7. What does it mean to be 95% confident?
8. How is calculating population proportions different than population means?
9. What steps need to be taken when dealing with a small sample?
10. What is the biggest challenge with regression analysis?

PRACTICE ON ANALYSIS AND VISUALIZATION

1. Using the NES variable *nes$follow* as a sample of the U.S. population, calculate what percentage of Americans follow politics "most of the time." Provide your best guess and the confidence interval at the 95% level.

2. Using the NES variable *nes$pid3* as a sample of the U.S. population, calculate what percentage of Americans consider themselves to be Independents. Provide your best guess and the confidence interval at the 99% level.

3. Set the seed to 40 and randomly draw 25 cases from the *states* data, then calculate the following:
   a. What is the sample mean of the variable *abort*?
   b. What is the sample standard deviation of the variable *abort*?
   c. What is the confidence interval for your estimate at the 90% level?
   d. What is the confidence interval for your estimate at the 99% level?

4. Set the seed to 40 and randomly draw 25 cases from the *states* data, then calculate the following:
   a. What is the sample mean of the variable *infant*?
   b. What is the sample standard deviation of the variable *infant*?
   c. What is the confidence interval for your estimate at the 90% level?
   d. What is the confidence interval for your estimate at the 99% level?

5. Set the seed to 40 and randomly draw 70 cases from the *world* data, then calculate the following:
   a. What is the sample mean of the variable *womleg*?
   b. What is the sample standard deviation of the variable *womleg*?
   c. What is the confidence interval for your estimate at the 90% level?
   d. What is the confidence interval for your estimate at the 99% level?

6. What is the difference between states Trump won and lost in terms of the state's population that is evangelical? Perform a two-sample t-test and report whether the null hypothesis is rejected.

7. What is the difference between Southern states and the rest of the country in terms of women's earnings as a fraction of men's (*states$percwom*)? Perform a two-sample t-test and report whether the null hypothesis is rejected.

8. What is the difference between whites and nonwhites in terms of their attitudes toward gays and lesbians (*nes$ftgay*)? Perform a two-sample t-test and report whether the null hypothesis is rejected.

9. What is the difference between those divorced in the United States and the rest of the population in terms of their attitudes toward Hillary Clinton (*nes$fthrc*)? Perform a two-sample t-test and report whether the null hypothesis is rejected.

10. Which of the following is an accurate statement about confidence levels?

    a. As the level of confidence grows, the confidence interval gets smaller.

    b. As the level of confidence grows, the confidence interval gets larger.

    c. Statistical significance rests heavily on statements based on taking random samples.

    d. Since regression analysis is often based on samples that are not random, statements about statistical significance in the regression context should be made with great care.

ANNOTATED R FUNCTIONS

The following functions appear in this chapter. They are listed in order of their first appearance (with the code chunk number in parentheses) and annotated here to give a very brief description of their use. Some are not stand-alone functions and only work in combination with other commands. As a reminder, the code in every chapter will work properly if executed in the order it appears. Proper execution also depends on typing the author-defined *libraries()* command, which loads the required R packages.

freq(): produces a frequency table for the specified variable. (9-1)

ifelse(): logical function that allows you to construct an if-then statement. Useful for selecting specific cases to label and also for creating categorical variables from continuous variables. (9-2)

mean(): calculates the mean value of a variable. Be sure to include the option 'na.rm=TRUE' if your variable has missing values. (9-3)

sd(): calculates the standard deviation of a variable. Be sure to use the *na.rm=TRUE* feature if there are missing values. (9-3)

seq(): generates a vector of numbers between specified values and at specified increments. (9-4)

dt(): takes a significance level and the degrees of freedom as arguments and returns the critical t-value. (9-4)

dnorm(): converts number in a variable to its density. In other words, it gives the height of the normal curve at the value of x. (9-4)

data.frame(): converts a collection of columns into a data frame. (9-4)

ggplot(): defines the basic structure of a plot (usually the x and y variables). (9-5)

aes(): the aes (called "aesthetics") function is used in ggplot to define the basic structure of the plot, which often includes the variables you want to use and any shapes or colors. (9-5)

geom_line(): ggplot command that draws a line. (9-5)

theme_minimal(): specifies a minimalist style for ggplot. (9-5)

theme(): specifies font, size, and so forth in a ggplot. (9-5)

xlab(): labels the x axis in ggplot. (9-5)

ylab(): labels the y axis in ggplot. (9-5)

ggtitle(): provides the title for a ggplot. (9-5)

scale_color_discrete(): allows you to specify exactly what colors you want to use in a graph. (9-6)

set.seed(): specifies exactly where to initiate the random number generator. Useful for reproducing results where a random element is involved. (9-7)

sqrt(): calculates the square root. (9-7)

rep(): command that specifies how many times an action should be repeated. As used here, it specifies to create a variable with 50 zeros. It is used to create a variable that will hold the values of the following *for()* command (a for-loop). (9-7)

melt(): takes a data frame and stacks the columns on top of each other. (9-7)

factor(): used to manipulate levels (categories) in a variable. (9-8)

levels(): used to list categories (levels) in a variable. (9-8)

geom_point(): draws points on a ggplot grid. (9-8)

stat_summary(): calculates various statistics for a data frame to be displayed in ggplot. (9-8)

geom_hline(): draws a horizontal line in a ggplot. (9-8)

tbl_df(): produces a tibble which allows the easy manipulation of a table in the 'tidyr' package. (9-10)

geom_ribbon(): specifies the boundaries of a polygon in ggplot. (9-11)

scale_fill_brewer(): colors a ggplot with a specified palette in cases where you need multiple colors. (9-11)

geom_vline(): draws a vertical line in a ggplot. (9-11)

scale_x_continuous(): draws an x axis for a ggplot. (9-11)

t.test(): calculates a two-sample t-test. (9-12)

conf(): a function (written by Randall Schumacker and edited by author) that samples from a normal distribution, calculates the sample mean, standard deviation, specifies a confidence interval, and indicates whether the given population parameter lies within the confidence interval. (9-14)

ANSWERS

KNOWLEDGE CHECK

1. (.4015, .4575), .4295
2. (.208, .232), .22
3. (1.86, 2.058), yes
4. b
5. (4.71, 7.29), yes
6. c
7. b, c, d
8. b, c
9. d
10. 4.36 (a), 1.90 (b), (3.41, 5.17) (c), yes (d)
11. 39,954 (a), 6,418 (b), (3,988, 4,223) (c), yes (d)
12. .5992 (a), .076 (b), (.599, .621) (c), (.579, .641) (d)
13. 1,784, rejected
14. 1.786, rejected
15. 3,045, not rejected
16. a
17. b, c, d
18. a, b, c, d

PRACTICE ON ANALYSIS AND VISUALIZATION

1. (.489, .547), .5178
2. (.82, .354), .318
3. 17.9 (a), 9.58 (b), (14.64, 21.16) (c), (12.54, 23.26) (d)
4. 7.34 (a), 1.87 (b), (6.7, 7.98) (c), (6.3, 8.3) (d)

5. 17.3 (a), 9.7 (b), (14.08, 20.56) (c), (12.18, 22.46) (d)

6. 15.3, rejected

7. 1.26, not rejected

8. .66, not rejected

9. 1.6, not rejected

10. b, c, d

 Access digital resources, including datasets, at http://edge.sagepub.com/brownstats1e.

10 Making Comparisons

CHAPTER OUTLINE

Learning Objectives
Overview
Why Do We Make Comparisons?
Questions That Beg Comparisons
Comparing Two Categorical Variables
Comparing Continuous and Categorical Variables
Comparing Two Continuous Variables
Exploratory Data Analysis: Investigating Abortion Rates in the United States
Good Analysis Generates Additional Questions
Summary
Common Problems
Review Questions
Practice on Analysis and Visualization
Annotated R Functions
Answers

LEARNING OBJECTIVES

- Discuss the reasons why we compare things.
- Identify questions that imply comparison.
- Construct and interpret figures when both variables are categorical.
- Construct and interpret figures with continuous and categorical variables.
- Construct and interpret figures with continuous variables.
- Connect exploring data with generating hypotheses to make discoveries.
- Compose the next question after your analysis.

Overview

Whether interested in governments, societies, or markets, making comparisons is central to our enterprise. Political scientists interested in explaining political polarization want to know how Republican attitudes *compare* to attitudes held by Democrats. Sociologists want to know how suicide rates in one group (defined by race, gender, class, etc.) *compare* to another. Economists *compare* unemployment rates between states with and without minimum wage laws. In this chapter we focus on comparisons. Specifically, we will concentrate on bivariate views of the data to understand the relationship between two variables. Comparisons also offer context; they answer the question "compared to what?" Consequently, comparisons reveal the importance of what we observe.

Since we've already studied the mechanics of *how* to describe bivariate views of the data (Chapter 5), our focus turns to *why*. This chapter links the questions we ask directly to visualizing data. The ability to construct a scatter plot that answers a specific question is an important skill.

This chapter also formally introduces **exploratory data analysis (EDA)**: an orientation to data analysis promoted by the late statistician John Tukey. Tukey argued for an iterative process, starting with exploration (Tukey, 1977). Through exploration and gaining familiarity with the data, better hypotheses are formed. The process does not stop there: once the initial hypotheses are tested, we generate additional hypotheses and return to the data. EDA is analogous to having a discussion. Tukey argued for a conversation between data and theory.

Exploration leads to discovery. Despite the open-ended enterprise that exploration and discovery imply, there are still rules to follow and appropriate techniques to employ. After introducing the mechanics of comparison, this chapter provides an example of how the process of description, hypothesis generation, testing, and making inferences works with a real-world example: abortion rates in the United States. A key point of the exercise will be to illustrate how EDA solves some mysteries but identifies others.

Why Do We Make Comparisons?

We make comparisons for two main reasons. First, we make comparisons to understand the relationship between two variables. Second, to fully understand a description, characteristic, or statistic, placing it in context is extremely important. To understand a fact or figure, we need to ask, "Compared to what?" Without placing data in its proper context, relatively little information is conveyed. Without comparisons, data can mislead. For example, a governor can brag about spending $500 per student on elementary education, but that could be the lowest level in the entire country.

In Chapters 3–6, we described data. What do the variables look like? Are there any obvious mistakes in the data? Do some of the variables need to be transformed? Once we've described the data and understand their basic features, we move to asking questions that require making comparisons. In describing data we want to know "how much." We now move to the next question, "Is that a lot?"

Without some sense of context, we can't answer that question. Is your dorm room nice? Is it hot today? Are interest rates high? Are we ready for the next hurricane? All of these questions require context, which implies we make comparisons. Note that each question contains an adjective: nice, hot, high, and ready. We can't answer those kinds of questions without comparing. Our dorm room may seem *nice* if it's the largest in the dorm and has air conditioning when others don't. A day in the 90s may seem *cool* if the temperature for the last 5 days has been above 100 degrees. Unless we know the context—unless we make comparisons—we are lost. Views of the data not placed in context can be disorienting, vacuous, and at their worst, highly misleading. Ask of any data or statistic, "Compared to what?"

Art and Practice of Data Visualization
IS IT A LOT AND WHY?

Whether you're the producer or consumer of the analysis, insist on comparisons. They help gauge the magnitude of the puzzle or problem. Comparisons also identify potential causes. If we observe that Americans with college degrees are healthier than those without, we would be drawn to examine how and why those two groups differ. Income, diet, exercise, stress, and doctor visits might all be things that differentiate those with degrees and those without. We might find stress levels are the biggest difference between the two groups. Making comparisons provides more clues.

> **KNOWLEDGE CHECK: Discuss the reasons why we compare things.**

1. Why do we compare?
    a. Little information is conveyed without comparison.
    b. Without comparisons, descriptions can be misleading.
    c. It is hard to judge the importance of a statistic without comparison.
    d. We compare to understand the relationship between two variables.

Questions That Beg Comparisons

After we know the central tendency, dispersion, and shape of the variables in our data set, we are often drawn to knowing how two variables relate to each other. When they are both categorical (e.g., gender and party identification), we want to know whether being in one category (being a male) is related to being in another (being a Republican).[1] When one variable is categorical and one is continuous, we want to know if the central tendency or dispersion of the continuous variable (e.g., the percentage of the population with a high school diploma) is related to the categorical variable (e.g., the different regions in the United States). Finally, when both are continuous we examine whether changes in one (e.g., GDP per capita) are associated with changes in another (e.g., homicide rates). While each of the three examples above seeks to *understand the relationship between two variables*, sometimes the questions posed are slightly different and can tip us off to what kind of variables are used.

How are two categorical variables related? Are males more likely to self-identify as Republican? Are first-generation college students more likely to live at home during their studies? Are women more likely to major in psychology? Are individuals under stress more likely to develop diabetes? Notice the phrase *more likely* appears in all of these questions. This is a good tip-off that the comparison involves two categorical variables. At base, we want to know whether having one condition (being male, for example) is related to another (being Republican). The best way to answer these questions is through examining cross-tabs, mosaic plots, and frequency tables.

How is a continuous variable associated with a categorical variable? Are males more conservative than females? Do first-generation college students live further from campus than other students? Do women major in subjects that pay less? Do individuals in stressful jobs have lower life expectancy? Here we want to know if being in a category is associated with *more* or *less* of something else. We could also ask whether there is *more* variation in political ideology, distance from campus, salary, or life expectancy for the different kinds of individuals in these questions. When we are interested in how a categorical variable and a continuous variable relate, we use boxplots and jitter plots.

How are two continuous variables related? Is age related to conservatism? Is income related to distance from campus? Are scores on the SAT related to starting salaries out of college? Is a person's level of stress related to life expectancy? Notice here how the question has changed from "more likely" and "more or less" to "related." Does the amount of one variable help us

---

[1] Some survey questions lose their relevance over time. Given recent societal trends with respect to gender norms, it may be only a matter of time when gender becomes a continuous variable.

determine the amount of another? If so, they are related. Similarly, is one variable related to the variation (spread) of another? In either case, a scatter plot is the best way to answer the question.

While we could simply memorize which plots go with continuous and categorical variables, understanding how specific kinds of questions are associated with each kind of plot enables us to change the question or transform the data so that they more cleanly map onto each other. Now that we know why comparisons are important and what questions are connected to them, let's explore each kind of comparison with an example.

> **KNOWLEDGE CHECK: Identify questions that imply comparison.**

2. Which of the following questions usually involve two categorical variables?
   a. Is x more likely under conditions of y?
   b. Is the amount of x contingent on whether y exists?
   c. Is x related to y?
   d. Is a change in the amount of x related to a change in the amount of y?
3. Which of the following questions involve a dependent variable that is continuous and an independent variable that is categorical?
   a. Is x more likely under conditions of y?
   b. Is the amount of x contingent on whether y exists?
   c. Is x related to y?
   d. Is a change in the amount of x related to a change in the amount of y?
4. Which of the following questions involve two continuous variables?
   a. Is x more likely under conditions of y?
   b. Is the amount of x contingent on whether y exists?
   c. Is x related to y?
   d. Is a change in the amount of x related to a change in the amount of y?

Comparing Two Categorical Variables

As noted in Chapter 1, there are two primary kinds of variables central to analysis: continuous and categorical. When comparing categorical variables (party identification, geography, etc.), mosaic plots and cross-tabs are extremely useful.

Example: Attitudes Toward Police

Campaigns often want to know how different voters feel about public policy. It might be good to know, for example, whether Independents are more likely to support gay marriage relative to Republicans. It might also be helpful to know if income levels are associated with political ideology (liberal, conservative, libertarian). In this example, we want to know whether a relationship exists between party identification and attitudes toward the police.

Rising tensions between the police and Black communities have led to violence and growing distrust. Understanding how citizens view the police determines the level of cooperation or resistance officers face as they go about their jobs. Does party identification influence people's opinions on whether police discriminate against Blacks?

To get a simple and intuitive view of the relationship, we use a mosaic plot. A mosaic plot sizes cells according to how many individual cases reside in jointly determined categories. In this example, it indicates the number of Democrats—relative to Republicans and Independents—that feel police favor whites. According to the mosaic plot in Figure 10-1, a clear pattern emerges. Of those who think there is discrimination, Democrats are a majority. Of those who think police treat whites and Blacks equally, Democrats are a minority. The mosaic plot illustrates the correlation between party identification and attitudes toward the police. Democrats are more likely to think police are biased.

To draw the mosaic plot in Figure 10-1, a couple of steps are needed. First, we "clean up" the *pid3* variable in the NES data. If we're not interested in the categories, we can conveniently remove them by specifying them as missing data (NAs). I do this with an *ifelse()* command, which says if the variable *pid3* equals "other" or "not sure," then assign it an "NA"; otherwise make it equal to *pid3*. Once I've done that, I convert the variable to a factor with the *as.factor()* command and then relabel and reorder the categories (Code Chunk 10-1).

Code Chunk 10-1

```
nes$pid3.new <- ifelse(nes$pid3 == "Other", NA,
               ifelse(nes$pid3 == "Not sure", NA,
                      nes$pid3))

nes$pid3.new <- as.factor(nes$pid3.new)

levels(nes$pid3.new)=c("Democrat", "Rep.", "Indep.")

nes$pid3.new = factor(nes$pid3.new, levels(nes$pid3.new)[c(1,3,2)])
```

Once I have the variable in the desired form, I'm ready to make a mosaic plot. Notice that I use the same ggplot setup but the *aes()* command is located in the *geom_mosaic()* command and not in the *ggplot()* command. Also note that in the *ggplot()* command I specify the data set and I stipulate there should be no missing values (the *na.omit()* command). Finally, in the aesthetics command, I identify *pid3.new* as the x variable and *dpolice.new* as the y (Code Chunk 10-2).

Code Chunk 10-2

```
ggplot(data = na.omit(nes)) +
  geom_mosaic(aes(x = product(pid3.new, dpolice.new),
                  fill=pid3.new,na.rm=TRUE)) +
  theme_minimal() +
  theme(plot.title = element_text(size = 8, face = "bold"),
        axis.title = element_text(size = 8, face = "bold")) +
```

```
xlab("") +
ylab("") +
coord_flip() +
ggtitle("Figure 10-1: How Do Police Treat Whites and Blacks?") +
scale_fill_brewer(palette="Blues") +
theme(legend.position="none")
```

Since the share of Democrats, Independents, and Republicans changes significantly across the categories, we can say the two variables—attitudes toward the police and party identification—are related. While the mosaic plot gives us a nice view of the data and the ability to quickly recognize a pattern, it's hard to discern specific percentages. For more precision, we can consult a cross-tab, which indicates the percentage of Democrats, Independents, or Republicans that responded "treats whites much better," "treats whites moderately better," and so forth (Code Chunk 10-3).

Code Chunk 10-3

```
CrossTable(nes$dpolice.new, nes$pid3.new,
           prop.chisq=FALSE)
```

From the cross-tab, we see that 51.4% of Democrats think police treat whites much better, while only 16.6% of Democrats think the police treat both whites and Blacks equally. For Republicans, the numbers are reversed: while 7.1% of Republicans think police treat whites better, 57.6% feel police treat whites and Blacks equally. Each column in a cross-tab represents 100% of a particular population. We also learn from the cross-tab that of the 1,073 respondents in this sample, 440 were Democrats, 364 were Independents, and 269 were Republicans. Party identification is clearly related to how one perceives police behavior.

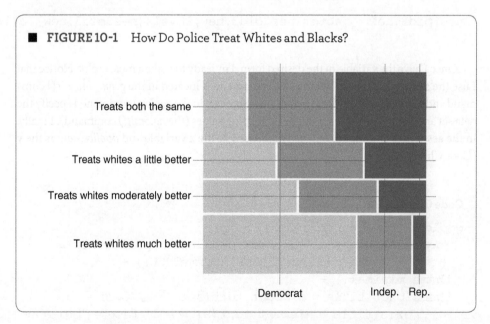

■ **FIGURE 10-1** How Do Police Treat Whites and Blacks?

TABLE 10-1 A Less Intuitive Way to View Categorical Comparisons

```
Cell Contents
------------------------------------
                    N
            N / Row Total
            N / Col Total
         N / Table Total
========================================================================
                          nes$pid3.new
nes$dpolice.new          Democrat   Independent   Republican    Total
------------------------------------------------------------------------
Treats whites much better    226          79            19        324
                           0.698       0.244         0.059      0.302
                           0.514       0.217         0.071
                           0.211       0.074         0.018
------------------------------------------------------------------------
Treats whites moderately      75          63            38        176
better                     0.426       0.358         0.216      0.164
                           0.170       0.173         0.141
                           0.070       0.059         0.035
------------------------------------------------------------------------
Treats whites a little        66          76            57        199
better                     0.332       0.382         0.286      0.185
                           0.150       0.209         0.212
                           0.062       0.071         0.053
------------------------------------------------------------------------
Treats both the same          73         146           155        374
                           0.195       0.390         0.414      0.349
                           0.166       0.401         0.576
                           0.168       0.136         0.144
------------------------------------------------------------------------
Total                        440         364           269       1073
                           0.410       0.339         0.251
========================================================================
```

Example: Religion and Politics

In the 1990s, the phrase "It's the economy stupid!" was used to describe how Americans voted. That simple phrase no longer explains electoral outcomes in the United States. Is partisanship associated with religion? Specifically, are Americans who self-identify as Republican, Democrat, or Independent more likely to attend church? The answer should influence how candidates conduct their campaigns.

To answer the question, we need to first eliminate the "skipped" and "don't know" categories from the variable that records church attendance: *nes$pew_churatd*. I offer another way to drop those two levels than used in the last example. First, I create a new data set that eliminates all of the NAs with the command *na.omit(nes)*. I use the *subset()* command to eliminate the cases that had "skipped" or "don't know" as values. Finally, I use the *droplevels()* command, which will eliminate those now empty categories (Code Chunk 10-4).

With the data in their desired form, I'm ready to create a mosaic plot (Code Chunk 10-5).

Code Chunk 10-4

```
nanes <- na.omit(nes)
nanes <- subset(nanes, pew_churatd!="Skipped"
                & pew_churatd!="Don't know" )
nanes$pew_churatd <- droplevels(nanes$pew_churatd)
```

Code Chunk 10-5

```
ggplot(nanes) +
  geom_mosaic(aes(x = product(pid3.new, pew_churatd),
                  fill=pid3.new,na.rm=TRUE)) +
  theme_minimal() +
  theme(plot.title = element_text(size = 8, face = "bold"),
        axis.title = element_text(size = 8, face = "bold")) +
  xlab("") +
  ylab("") +
  ggtitle("Figure 10-2: Who Goes to Church?") +
  scale_fill_brewer(palette="Blues") +
  theme(legend.position="none") +
  coord_flip()
```

The thickness of the horizontal bars in Figure 10-2 indicates that the largest category of church attendance is "never." Fortunately, the categories are aligned on a continuum from those who never attend (at the top) to those who frequently attend (the bottom). Generally, the pattern is as we would expect: the Republican share of respondents who frequently attend church seems to be larger than the Democrat or Independent share. As we move from the top to bottom, respondents identifying as Democrats decreases.

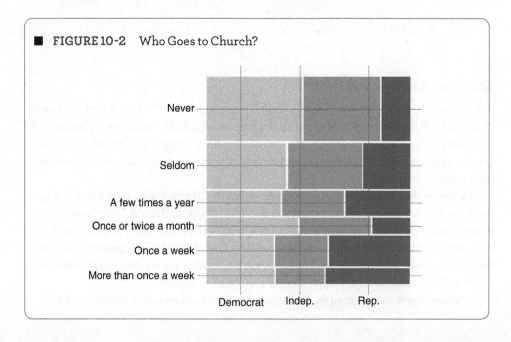

FIGURE 10-2 Who Goes to Church?

> **KNOWLEDGE CHECK: Construct and interpret figures when both variables are categorical.**

5. Compare party identification with the *nes$pid3* variable to the financial well-being variable *nes$finwell*. Is there a relationship between the two variables?
6. Are women more likely to feel their financial well-being improved over the last 20 years?
7. Are women more likely to identify as Republicans, Independents, or Democrats?

Art and Practice of Data Visualization
AXIS LABELS

Note that in Figure 10-2 I don't place the dependent variable on the y axis. While a good rule to follow involves placing the dependent variable along the y axis, the length of the labels is an important factor to consider with mosaic plots. Tilting the labels at 45 degrees is one option. However, if labels get too long on one side relative to the other, it might be helpful to place the longer labels on the y axis. This is easily achieved by the *coord_flip()* command in the ggplot expression. Note that the same concern with aesthetics is true when drawing bar plots: when the labels get long, consider flipping the axes for a cleaner, less cluttered look (Knaflic, 2015). When possible, try to keep words and labels at right angles.

Comparing Continuous and Categorical Variables

Sometimes we want to compare continuous and categorical variables. Boxplots and jitter plots are effective ways to understand the relationship between continuous and categorical variables. There are advantages and disadvantages to using each. Decide whether identifying specific cases is important to make your choice. For example, if we want to examine how life expectancy varies over the regions of the world, we use a jitter plot to identify specific countries. Consider using boxplots when identifying each observation is less important.

Example: Obama Feeling Thermometer

For this illustration, consider the relationship between attitudes toward President Obama and party identification. A common technique used in surveys is the **feeling thermometer**. Feeling thermometers are generated by asking respondents to give a number (usually between 0 and 100) indicating attitudes toward a policy, person, or thing (political party, unions, etc.). The higher the number, the "warmer" the respondent feels. Figure 10-3 includes a set of boxplots that reveal a very distinct pattern.

The code to generate the boxplots is straightforward (Code Chunk 10-6). Note that in order to avoid generating a boxplot for the NAs, I subset the data and eliminated missing values (NAs) by indicating *!is.na(pid7)*.

The boxplots indicate there is a strong association between the Obama feeling thermometer and partisan identification. We see that among respondents self-identifying as Democrats,

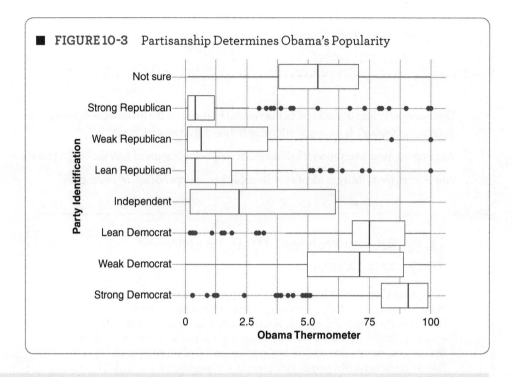

Code Chunk 10-6

```
ggplot(subset(nes, !is.na(pid7)), aes(pid7, ftobama)) +
  geom_boxplot(col="#0000bf") + theme_minimal() +
   theme(plot.title = element_text(size = 8, face = "bold"),
      axis.title = element_text(size = 8, face = "bold")) +
  ggtitle("Figure 10-3: Partisanship Determines Obama's Popularity") +
  ylab("Obama Thermometer") +
  xlab("Party Identification") +
  coord_flip()
```

the medians are higher. The boxplots not only give us the central tendency of the data, they indicate the data's dispersion. While there are some categories with relatively small variance, there are levels where it is quite large. As one might expect, the boxes on both extremes of the political spectrum are relatively small, indicating attitudes toward Obama do not vary much among these subsets of the population. Variation does, however, exist among Independents. Boxplots provide more information than simply comparing the medians between groups.

Jitter plots (first presented in Chapter 5) provide a nice alternative when examining continuous and categorical variables. Recall that jitter plots *jiggle* the observations, making it easier to see each observation. Consider the relationship between regions of the world and democracy. If we want a quick and informative illustration of democracy by region, we can construct jitter plots of democracy scores from the *Polity2* measure. Lower scores are associated with more authoritarian countries. The jitter plot makes clear that the most authoritarian regions are the Middle East/North Africa along with sub-Saharan Africa (Figure 10-4).

When we introduced jitter plots previously, we jittered the points by including an option in the *geom_point()* command. Here we use the *geom_jitter()* command to generate the plot (Code Chunk 10-7).

Code Chunk 10-7

```
ggplot(world, aes(region, polity2, col=region)) +
  geom_jitter() +
  theme_minimal() +
  theme(plot.title = element_text(size = 8, face = "bold"),
        axis.title = element_text(size = 8, face = "bold")) +
  theme(axis.text.x = element_text(size=8, vjust=.7),
        legend.position="none") +
  ggtitle("Figure 10-4: Most Authoritarian Regimes Reside in Africa") +
  ylab("Democracy (Polity2)") +
  xlab("") +
  scale_color_manual(values=c("#0000ff", "#008b00", "#bf0000",
                              "#ff7300", "#00e6e6", "#263333",
                              "#73e600")) +
  coord_flip()
```

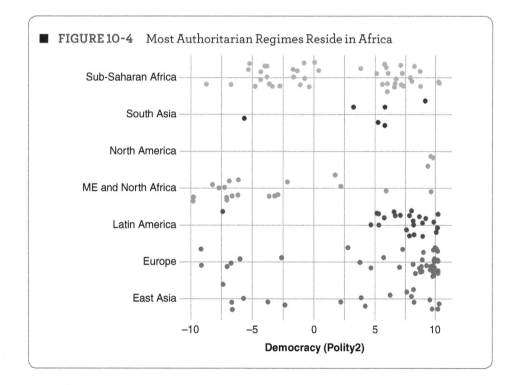

■ FIGURE 10-4 Most Authoritarian Regimes Reside in Africa

> **KNOWLEDGE CHECK: Construct and interpret figures with continuous and categorical variables.**

8. Draw a picture that illustrates the relationship between region and female state legislatures. Which region in the United States, on average, has the highest and lowest percentages in terms of women in the state legislature?

9. Draw a picture that illustrates the relationship between region and ethnolinguistic fractionalization. What region in the world has, on average, the highest and lowest rates of ethnolinguistic heterogeneity?

10. Create a jitter plot that shows the relationship between colonial heritage and voting turnout. Which colonial heritage seems, on average, to be least beneficial for voting turnout?

Comparing Two Continuous Variables

Scatter plots are ubiquitous in data analysis since many of the variables we use represent either counts or amounts. Scatter plots are ideal for determining whether a relationship exists between two continuous variables because they not only indicate whether two variables are related, but they also identify and locate where specific cases lie on the x and y axes.

Example: Gender and Education

The education of a country's female population figures prominently in studies on income and health among the world's developing countries. What is the relationship between the ratio of girls to boys in educational outcomes and infant mortality? Since we're interested in explaining health outcomes—in this case, infant mortality (the number of deaths per 1,000 live births)—we put infant mortality on the y axis and the ratio of girls to boys educational attainment on the x axis (Figure 10-5).

Note that I've identified several countries to highlight in the scatterplot (Code Chunk 10-8). To label a select few cases, I use the *ifelse()* command within the aesthetics of the *geom_text_repel()* command.

Code Chunk 10-8
```
ggplot(world, aes(gtbeduc, inf)) +
  geom_point(col="#bf0000") +
  geom_text_repel(size=3, vjust=2, col="black",
        aes(label= ifelse(iso3c=="PAK" |
                          iso3c=="YEM" |
                          iso3c=="AFG" |
                          iso3c=="SDN" |
                          iso3c=="SWZ" |
                          iso3c=="ETH",
                          as.character(iso3c), ''))) +
  ggtitle("Figure 10-5: Infant Mortality and Women's Education") +
  ylab("Infant Mortality: Deaths per 1,000 Live Births") +
  xlab("Ratio of Girls to Boys Educational Attainment") +
  theme_minimal() +
  theme(plot.title = element_text(size = 8, face = "bold"),
        axis.title = element_text(size = 8, face = "bold"))
```

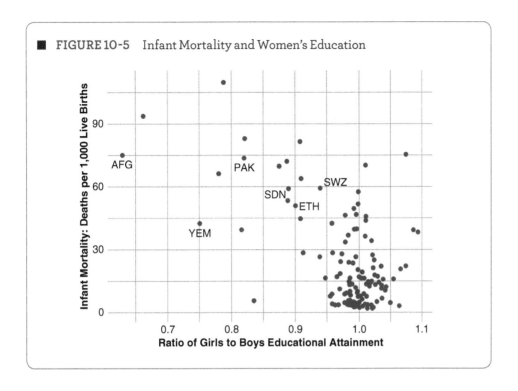

FIGURE 10-5 Infant Mortality and Women's Education

The resulting scatter plot indicates that most countries have ratios close to 1. And, in those cases, the infant mortality rates are lower. There are a few countries that have education ratios below .9, which indicates girls are provided fewer educational opportunities than boys. In those cases, the infant mortality rates are relatively high, generally 30 deaths per 1,000 live births and higher.

The scatter plot suggests that a relationship exists between the two variables. If the ratio of girls' education to boys' education is roughly 1, infant mortality rates are relatively low. While infant mortality does decrease as the educational ratio increases, one would be hard pressed to say that the relationship is strictly linear.

In addition to understanding whether a relationship exists between the two variables, labeling the points provides some information that suggests considering other causes. Most of the cases with a low girl-to-boy ratio tend to be primarily Muslim countries (Pakistan, Afghanistan, Yemen). Most are very poor (Sudan, Ethiopia, Swaziland). Consequently, in addition to girls' education, we may want to examine religion and income in order to explain the variation in infant mortality.

Example: Gender and Policymaking

Does gender influence policy making? Does having more women in state legislatures influence the legislation produced? Are women legislators more likely to favor policies on health, education, and welfare? Let's take a look at the percentage of seats held by women in state legislatures and the student/teacher ratios that exist in each state. Though it would be a more direct test of the question to consider actual policy, we'll use the student/teacher ratio as a proxy for education policy (Figure 10-6).

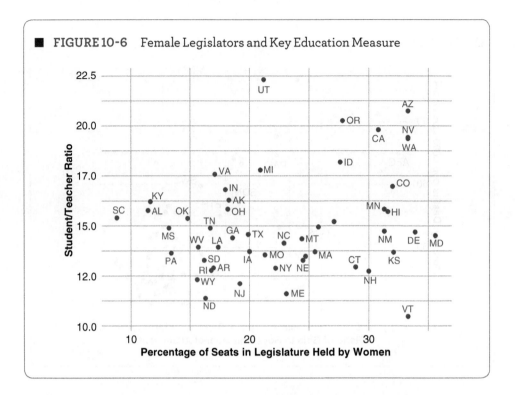

FIGURE 10-6 Female Legislators and Key Education Measure

The code to generate the plot is straightforward (Code Chunk 10-9). The only wrinkle here is that I made the gray labels somewhat lighter by setting the *alpha* setting in the *geom_text_repel()* command to .3.

Code Chunk 10-9
```
ggplot(states, aes(femleg, ptratio)) +
  geom_point(col="#bf0000") +
  geom_text_repel(size=3, vjust=2, col="#263333", alpha = .3,
          aes(label= st)) +
  ggtitle("Figure 10-6: Female Legislators and
                    Key Education Measure") +
  ylab("Student/Teacher Ratio") +
  xlab("Percentage of Seats in Legislature Held by Women") +
  theme_minimal() +
  theme(plot.title = element_text(size = 8, face = "bold"),
        axis.title = element_text(size = 8, face = "bold"))
```

Generally speaking, there does not seem to be a clear relationship (certainly not a linear one) between the two variables. It appears that most states hover between a ratio of 12.5 to 15 students per teacher regardless of what percentage of the legislature's seats are held by women. In other words, knowing the level of one variable does not help us predict the level of the other. If anything, the plot reveals some important outliers: Utah, Arizona, Oregon, California,

Nevada, Washington (at the high end of the student/teacher ratio), and Vermont (at the low end of the ratio).

Interestingly, the states with the highest ratios (meaning the larger class sizes) are all found west of Colorado. Population or population density might be an important factor to take into consideration when explaining student/teacher ratios.

Note that labeling all of the points in this scatter plot is extremely helpful to understand the relationship between legislatures and the student/teacher ratio. With a relatively few number of observations that are also relatively spread out, it's acceptable to label every point.

> **Art and Practice of Data Visualization**
> **PICK REPRESENTATIVE POINTS**
>
> When data points are numerous and clustered, it is better to pick out a few important, representative points to avoid superimposing labels. The *geom_text_repel()* command provides a nice feature that avoids superimposing labels; however, when we are dealing with numerous or clustered observations, scatter plots with every data point labeled begin to resemble porcupines or the Iron Throne from the *Game of Thrones*.

Armed with the tools used for comparisons and the questions they answer, let's see how the questions and comparisons interact to form good analysis. The following example combines what we've learned so far to illustrate how descriptions lead to questions, which then lead to comparisons. Once those comparisons are made, additional questions always result. Presented below is an example of a process that emphasizes the back-and-forth between theory and evidence: exploratory data analysis.

> **KNOWLEDGE CHECK: Construct and interpret figures with continuous variables.**

11. Generate a scatter plot of women's suffrage and infant mortality.
    a. Is there a relationship between the two variables?
    b. Does it appear to be linear?
    c. Are there any outlying values?
    d. If your answer to c is yes, which countries are they (identify three)?
12. Generate a scatter plot of *income* and *stuspend* in the United States.
    a. Is there a relationship between the two variables?
    b. Does it appear to be linear?
    c. Are there any outlying values?
    d. If your answer to c is yes, which state would you identify?

13. Generate a scatter plot of minimum wages and the ratio of women's to men's income in the United States.

    a. Is there a relationship between the two variables?

    b. Does it appear to be linear?

    c. Are there any outlying values?

    d. If your answer to c is yes, which state would you identify?

Exploratory Data Analysis: Investigating Abortion Rates in the United States

Let's proceed by analyzing abortions in the United States to demonstrate how exploration through visualization leads to interesting discoveries. Abortion rates are a public health concern and formulating good public policy demands we know how they relate to culture, the economy, and geography. Given the politicized nature of the issue, we might also wonder whether there is a strong relationship between partisanship and abortion rates. Our policy recommendations will depend on how much abortion rates are driven by socioeconomic factors (some of which we can change) and ideological factors (many of which we cannot). To begin, we examine a scatter plot of the abortion rate and party identification (Figure 10-7).

Again, like previous examples, in Code Chunk 10-10 we call out some specific cases to help build our intuition about which states have the highest, lowest, and most interesting values (values that deviate from the general pattern of the plot).

Code Chunk 10-10

```
ggplot(states, aes(democrat, abort)) + geom_point(col="#bf0000") +
  theme_minimal() +
  theme(plot.title = element_text(size = 8, face = "bold"),
        axis.title = element_text(size = 8, face = "bold")) +
  ggtitle("Figure 10-7: Plot of Abortion Rate and Party ID") +
  ylab("Abortions per 1,000 Women") +
  xlab("Percent That Self-Identifies as Democrat") +
  geom_text_repel(size=3, col="#04183d", aes(label= ifelse(st=="NY" |
                  st=="CA" |
                  st=="NJ" |
                  st=="WY" |
                  st=="UT" |
                  st=="DE" |
                  st=="ID", as.character(st), '')))
```

Several things jump out. First, there is a positive association between abortion rates and party identification. As we move across the x axis (increase the percentage identifying as Democrats), the abortion rate increases. Second, some of the most liberal states (New York,

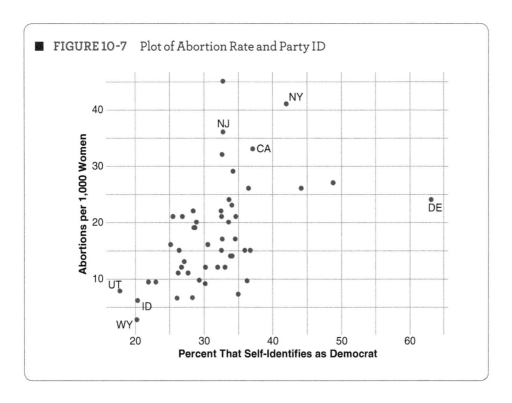

FIGURE 10-7 Plot of Abortion Rate and Party ID

California, and New Jersey) have the highest abortion rates, while Wyoming, Utah, and Idaho have the lowest. Third, the most populous states have the highest abortion rates. This observation alone evokes a number of questions. Do high rates of abortion only occur in states with large urban areas? Are abortions driven by the availability of doctors willing to perform the procedure? Is there a big city culture that differs substantially from rural areas? To get answers, let's be a little more systematic by adding another dimension to the plot: weighting the size of the points based on population density (Figure 10-8). Sizing the points by a different variable is easy, we simply include an *aes()* command in the *geom_point()* layer, specifying that we want to *size* the points by the variable *density*.

Code Chunk 10-11

```
ggplot(states, aes(democrat, abort)) +
  geom_point(aes(size=density), col="#0000bf") +
  theme_minimal() +
  theme(plot.title = element_text(size = 8, face = "bold"),
        axis.title = element_text(size = 8, face = "bold")) +
  theme(legend.position = c(.8, .2)) +
  ggtitle("Figure 10-8: Population Density Matters") +
  ylab("Abortions per 1,000 Women") +
  xlab("Percent of State That Are Democrat") +
```

```
geom_text(size=3, vjust=-1.5, col="#04183d",
          aes(label= ifelse(st=="NY" |
               st=="CA" |
               st=="NJ" |
               st=="WY" |
               st=="UT" |
               st=="DE" |
               st=="ID", as.character(st), ''))) +
  guides(size=guide_legend("Population Density"))
```

The bubble plot (Figure 10-8) indicates that density, party identification, and abortions are all related. This challenges the simple notion that abortion rates are higher in some states because of a more liberal culture. Perhaps abortion rates are higher in liberal states because of supply—the number of clinics. Clearly, we've got some more work to do.

Recap

It's time to step back and take inventory. First we looked at a simple plot of abortion rates and party identification. We noticed there was a positive correlation. A closer inspection of the plot revealed that states with the highest abortion rates were also the most densely populated, prompting us to add a third dimension to the plot—population density. The different view reveals abortion rates are related to demographic patterns. From that, we inferred abortion rates may not be influenced by political ideology alone but by proximity to abortion clinics. Let's keep going.

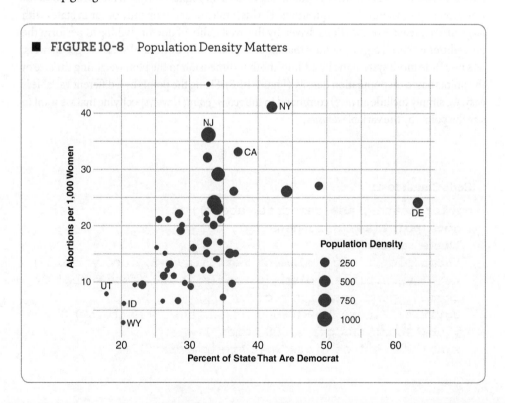

FIGURE 10-8 Population Density Matters

Given our new demographic hypothesis, divide the data accordingly. First, examine the same plot of abortion rates and party identification, only this time divide the data into three groups by population density (low, medium, and high). Second, fit lines to each of these groups to understand how party identification relates to abortion rates in each of these contexts (Figure 10-8). This is a commonly used and helpful approach with all kinds of data. Does the empirical relationship noted between abortion rates and party identification change depending on the state's population density? Before moving on, what do you expect to see if you think demographics or geography matter more than political ideology?

Two general steps are needed to draw the appropriate visualization. First, we need to take the *states$density* variable and cut it into three parts since we want low, medium, and high categories. In Code Chunk 10-12 we use the *cut()* command and specify that we want it cut in three parts.

Code Chunk 10-12

```
states$density3 <- cut(states$density, breaks=c(0,50,150,1200))

levels(states$density3)=c("Low", "Med", "High")

states$density3=as.ordered(states$density3)
```

Now that we have a new variable called *states$density3*, we're ready to draw the plot with three separate lines (Code Chunk 10-13).

Code Chunk 10-13

```
ggplot(states, aes(democrat, abort, col = density3)) +
  geom_point() + geom_smooth(method="lm", se=FALSE) +
  ggtitle("Figure 10-9: Partisanship Matters in Rural States") +
  ylab("Abortions per 1,000 Women") +
  xlab("Percent of State That Are Democrat") +
  geom_text_repel(size=3, aes(label= ifelse(st=="ND" |
                        st=="SD" |
                        st=="UT" |
                        st=="DE" |
                        st=="NV" |
                        st=="WY", as.character(st), ""))) +
  theme_minimal() +
  theme(plot.title = element_text(size = 8, face = "bold"),
        axis.title = element_text(size = 8, face = "bold")) +
  theme(legend.position = c(.7, .3)) +
  scale_color_manual(breaks = c("Low", "Med", "High"),
                     values=c("#bf0000", "#0000bf", "#008b00")) +
  guides(col=guide_legend("Population Density"))
```

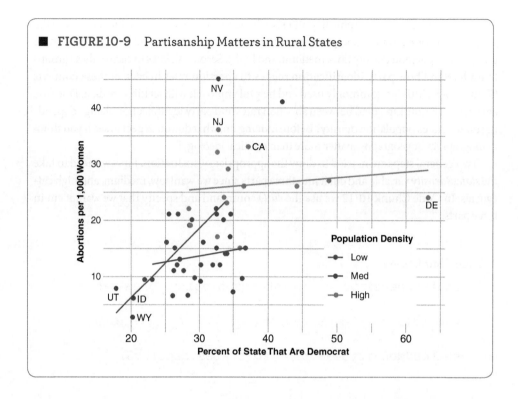

FIGURE 10-9 Partisanship Matters in Rural States

In Figure 10-9, densely populated states are colored green, states in the medium category are colored blue, and sparsely populated states are colored red. The slope of the green line may be unduly influenced by an important outlier: Delaware. One could imagine that the slope of the green line would be very different if Delaware were not in this sample. Let's file that away for now and move on.

Where population density is low, the relationship between party identification and abortion rates is relatively strong (of course, this result might be driven by Utah, Wyoming, North Dakota, and South Dakota). What does this tell us with respect to our demographic hypothesis? If population density was the important factor and not political ideology, we would expect that at each level of density there would be no relationship between party identification and abortion rates—the lines would be flat. Since not all of the lines are flat, political ideology matters but mainly in less populated areas. Why?

Perhaps politics matters among rural states because that category includes both sparsely populated states that are traditionally conservative (Utah, North Dakota, etc.) and states with liberal metropolitan areas (Colorado and Oregon). This begs the question, what is the relationship between abortion rates and population density?[2] Notice we are asking a different question having learned something new about the data.

To answer the question, I plot abortion and the log of population density (Figure 10-10). Rather than imposing linearity on the data, I use a loess smooth, which is easily accomplished by taking the *geom_smooth()* command and using the *method="loess"* feature (Code Chunk 10-14).

---

[2] I logged the density variable because of its skewed distribution.

Code Chunk 10-14

```
ggplot(states, aes(log(density), abort)) +
  geom_point(col="#bf0000") +
  theme_minimal() +
  theme(plot.title = element_text(size = 8, face = "bold"),
        axis.title = element_text(size = 8, face = "bold")) +
  geom_smooth(method="loess", se=FALSE, col="#0000ff") +
  geom_text_repel(size=3, col="#04183d", aes(label= ifelse(st=="NY" |
                            st=="CA" |
                            st=="NJ" |
                            st=="WY" |
                            st=="UT" |
                            st=="ID" |
                            st=="CO" |
                            st=="OR" |
                            st=="WI" |
                            st=="MO" |
                            st=="IN" |
                            st=="NV", as.character(st), ''))) +
  ggtitle("Figure 10-10: Abortion Rate and Party ID") +
  ylab("Abortions per 1,000 Women") +
  xlab("Percent of State That Are Democrat (logged)")
```

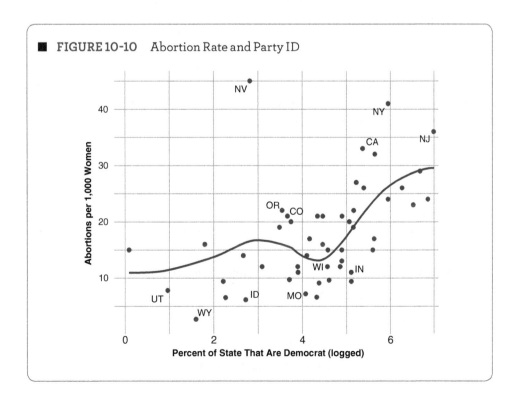

■ **FIGURE 10-10** Abortion Rate and Party ID

In sparsely populated states—the lower two-thirds of the sample—there is no relationship (the smooth is relatively flat). The smooth then rises quickly as we range over the more densely populated states. The different view raises new questions. Is there a threshold or minimum level of supply? Perhaps the low number of clinics dictates outcomes: where access is limited, abortions are relatively few. Nothing matters—political ideology, religion, culture—until there are enough doctors who can perform the procedure. The lowest abortion rates are all in very rural states (Utah, North Dakota, South Dakota, Idaho, and Wyoming). Abortion rates remain relatively low even in more populated states: Iowa, Missouri, Maine, Wisconsin, and Indiana. As the loess smooth indicates, however, the abortion rate jumps at a certain point as density increases.

Outliers can also help us uncover important information. Note that Nevada is an extreme outlier. Although it is not a very densely populated state, there are two possible explanations for its high abortion rate. First, it's a very urban state: a majority of the population lives in Las Vegas. Consequently, population density may not be a good way to measure clinic availability. Stated in more technical terms, population density may not be a *valid* measure of supply. Although Nevada has an average level of population density, it may more closely resemble states with very high urban populations like New York or New Jersey. Since Nevada is in the West and is much larger geographically than New York or New Jersey, its population density is much lower. Most Nevadans, however, experience the state as a large city. Second, the relatively close proximity of Los Angeles may influence either the availability of doctors or the number of women seeking the procedure.

Figure 10-10 raises two important concerns. First, population density doesn't fully capture what we're trying to measure: proximity to abortion clinics. Collecting data on the number of clinics in a state or data on the percent of the population living in urban areas would more directly test our hypothesis. Second, proximity to major population centers seems important. We need to account for neighbors or neighborhood effects.

By constructing different views of the data, we learned that while political ideology might be important, population density matters too. Not only did we find that population density is related to abortion rates, physical constraints like supply could be decisive. We also found that proximity to a large metropolitan area—even if in a different state—could have an important impact. Although more work remains, these last few pages have shown how combining data and theory through comparisons can advance our understanding.

> **KNOWLEDGE CHECK: Connect exploring data with generating hypotheses to make discoveries.**

14. What are the different ways the data were explored further in the example?

    a. The data were transformed.

    b. The data were grouped.

    c. Specific (interesting) cases were identified.

    d. Different mathematical functions were fit to the data.

15. Which of the following statements accurately describe exploratory data analysis?
    a. Knowing your subject is helpful.
    b. You must have all your hypotheses formulated in advance.
    c. Be willing to try things you're not sure will pan out.
    d. It is an iterative process, there's a back-and-forth between data and theory.

Good Analysis Generates Additional Questions

In the EDA example, note that through the process we identified a number of different questions. Why are there relatively more abortions in cities versus the rural areas? Why does politics seem to matter in rural areas but not in cities? Why is Delaware an outlier? Why is there a threshold with respect to population density? What does Nevada reveal about measuring rural/urban populations? And finally, what additional data should we collect?

Good analysis generates better questions. Better questions lead to deeper understanding. As I hope to have shown, the process never really ends: there is always the next question. Throughout the rest of the book, I hope you find that generating new and better questions is just as important as answering the ones we begin with.

> **KNOWLEDGE CHECK: Compose the next question after your analysis.**

16. What visualization led to the following "next" question?
    a. Why are there relatively more abortions in cities versus the rural areas?
    b. Why does politics seem to matter in rural areas but not in cities?
    c. Why is there a threshold with respect to population density?
    d. What does Nevada reveal about measuring rural/urban populations?

SUMMARY

Making interesting comparisons is a way to familiarize yourself with the data and to conduct scientific inquiry. Making comparisons allows us to not only understand the relationship between two variables but also to place what we observe in its proper context. Answering the question "compared to what?" helps us recognize the importance of an observation. Without placing our analysis in its proper context, we're more likely to exaggerate the importance of our finding or make the opposite mistake and fail to realize when an important discovery has been made.

Important discoveries come from the simple techniques learned in this chapter. The more we make comparisons, draw conclusions, and formulate new questions, the better the data analysis. Constructing interesting comparisons provides new insight and generates new questions.

Now that we've learned the basics of comparisons, it's time to branch out and introduce additional variables. Few things in the social sciences are caused by one thing. Consequently, we need to take a number of different factors into consideration when explaining most everything. The next chapter formally introduces and explains how to examine the relationship between two variables while holding a third variable constant. In the earlier example of EDA, we've already done it. For example, when looking at political ideology and abortions, we held population density constant. This is a controlled comparison, a fundamental concept that helps us untangle the complexity of the social phenomena we study.

COMMON PROBLEMS

- *Taking time.* Visualizing data has two purposes: (1) to deepen our understanding of a problem and (2) to familiarize ourselves with the data. While some visualizations are not designed for deep analysis (a histogram), others (a scatter plot) can be mined for all sorts of insights. All too often we move from data visualization to generating statistical models and their estimation. Careful analysis and scrutiny of a well-conceived visualization reveals important information otherwise obscured by a regression table or a simple statistical summary. An impatient and hurried approach manifests itself in an empty analysis lacking important detail, full of inaccuracy, and bereft of meaning.

- *Gaining experience.* You'll see a lot of things differently after you've seen a lot of things. Your ability to recognize patterns in the data and to understand their importance will increase with experience. If you can't wait for that, practice!

- *Using plots to both formulate more questions and answer them.* Often we infer more from a scatter plot than is recommended by the Surgeon General. Some visualizations may raise more questions than they answer. The temptation is to infer too many things from a scatter plot when the better approach is to turn those inferences into new questions.

- *Recognizing the difference between categorical and continuous variables.* When we compare, we have to know the difference between categorical and continuous variables. Sometimes it helps to continually make mistakes—try creating a scatter plot with two categorical variables—to understand the important difference between categorical and continuous variables.

REVIEW QUESTIONS

1. What is exploratory data analysis (EDA)?
2. What's a good way to compare two categorical variables?
3. How should one compare a continuous variable and a categorical one?
4. How can we add a third dimension to a scatter plot?
5. What does it mean when a measure is "valid"?
6. Why use a jitter plot?
7. How can you hold constant a variable in a scatter plot?
8. Why might abortion rates be so high in Nevada even though its overall population is fairly low?
9. Why is EDA useful? What can it accomplish?
10. How is the population density variable measured (what are the units)?

PRACTICE ON ANALYSIS AND VISUALIZATION

1. Compare the gender variable (*nes$gender*) to the immigration numbers variable (*nes$immig_numb*). Is there a relationship between the two variables?

2. Compare attitudes toward climate change (*nes$warmcause*) to party identification (*nes$pid3*). Is there a relationship between the two variables?

3. Compare attitudes on whether politicians should compromise (*nes$compromise*) to party identification (*nes$pid3.new*). Is there a relationship between the two variables?

4. Draw a picture that illustrates the relationship between regime type and the percentage of a country's population living in urban areas. What regime type, on average, has the biggest and smallest urban populations?

5. Draw a picture that illustrates the relationship between region and the age of the population (using the *world$young* variable). What region, on average, has the youngest and oldest populations?

6. Draw a picture that illustrates the relationship between region and political knowledge (use the *states$knowgov* variable). What region, on average, has the most and least political knowledge?

7. Generate a scatter plot of foreign direct investment (FDI) and the index of human capital per person (*world$pwthc*).
   a. Is there a relationship between the two variables?
   b. Does it appear to be linear?
   c. Are there any outlying values?
   d. If your answer to c is yes, which cases would you identify?

8. Generate a scatter plot of gun rights and homicide rates in the United States.
   a. Is there a relationship between the two variables?
   b. Does it appear to be linear?
   c. Are there any outlying values?
   d. If your answer to c is yes, which cases would you identify?

9. Generate a scatter plot of income and turnout in the United States.
   a. Is there a relationship between the two variables?
   b. Does it appear to be linear?
   c. Are there any outlying values?
   d. If your answer to c is yes, which cases would you identify?

10. Which of the following are useful ways to explore the data?
    a. Labeling the observations in a plot
    b. Coloring different groups in a plot
    c. Change the question being asked
    d. Generate new questions

ANNOTATED R FUNCTIONS

The following functions appear in this chapter. They are listed in order of their first appearance (with the code chunk number in parentheses) and annotated here to give a very brief description of their use. Some are not stand-alone functions and only work in combination with other commands. As a reminder, the code in every chapter will work properly if executed in the order it appears. Proper execution also depends on typing the author-defined *libraries()* command, which loads the required R packages.

ifelse(): logical function that allows you to construct an if-then statement. Useful for selecting specific cases to label and also for creating categorical variables from continuous variables. (10-1)

levels(): used to list categories (levels) in a variable. (10-1)

factor(): used to manipulate levels (categories) in a variable. (10-1)

ggplot(): defines the basic structure of a plot (usually the x and y variables). (10-2)

aes(): the aes (called "aesthetics") function is used in ggplot to define the basic structure of the plot, which often includes the variables you want to use and any shapes or colors. (10-2)

theme_minimal(): specifies a minimalist style for ggplot. (10-2)

theme(): specifies font, size, and so forth in a ggplot. (10-2)

xlab(): labels the x axis in ggplot. (10-2)

ylab(): labels the y axis in ggplot. (10-2)

coord_flip(): flips the x and y axes on a plot. (10-2)

ggtitle(): provides the title for a ggplot. (10-2)

scale_fill_brewer(): colors a ggplot with a specified palette in cases where you need multiple colors. (10-2)

CrossTable(): generates a cross-tab. (10-3)

subset(): chooses cases in the data specified in the argument. (10-4)

droplevels(): removes unused categories in a categorical variable. (10-4)

geom_boxplot(): generates a boxplot in ggplot framework. (10-6)

geom_jitter(): generates a jitter plot in the ggplot framework. (10-7)

scale_color_manual(): allows you to specify exactly what colors to use in a ggplot. (10-7)

geom_point(): draws points for a scatter plot in ggplot. (10-8)

geom_text_repel(): labels points in a scatterplot in the ggplot framework. (10-8)

guides(): provides detailed control of a plot's legend(s). (10-11)

cut(): takes a continuous variable as an argument and "cuts" it into the specified number of parts. (10-12)

ANSWERS

KNOWLEDGE CHECK

1. a, b, c, d
2. a, c
3. b, c
4. c, d
5. mosaic plot, yes

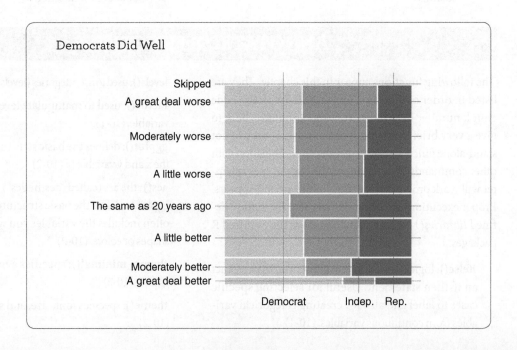

```
nanes$finwell <- droplevels(nanes$finwell)
ggplot(nanes) +
  geom_mosaic(aes(x = product(pid3.new, finwell),
                                fill=pid3.new,na.rm=TRUE)) +
  theme_minimal() +
  theme(plot.title = element_text(size = 8, face = "bold"),
        axis.title = element_text(size = 8, face = "bold")) +
  xlab("") +
  ylab("") +
  ggtitle("Democrats Did Well") +
  scale_fill_brewer(palette="Blues") +
  theme(legend.position="none") +
  coord_flip()
```

6. mosaic plot, no

```
nanes$finwell <- droplevels(nanes$finwell)
ggplot(nanes) +
  geom_mosaic(aes(x = product(gender, finwell),
                                fill=gender,na.rm=TRUE)) +
  theme_minimal() +
  theme(plot.title = element_text(size = 8, face = "bold"),
        axis.title = element_text(size = 8, face = "bold")) +
  xlab("") +
  ylab("") +
  ggtitle("Gender and Financial Well-Being") +
  scale_fill_brewer(palette="Blues") +
  theme(legend.position="none") +
  coord_flip()
```

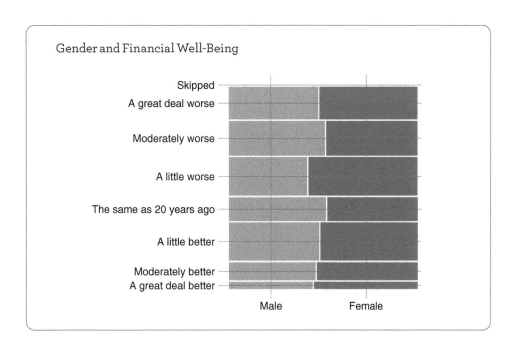

7. mosaic plot, yes

```
nanes$finwell <- droplevels(nanes$finwell)
ggplot(nanes) +
  geom_mosaic(aes(x = product(gender, pid3.new),
                              fill=gender,na.rm=TRUE)) +
  theme_minimal() +
  theme(plot.title = element_text(size = 8, face = "bold"),
        axis.title = element_text(size = 8, face = "bold")) +
  xlab("") +
  ylab("") +
  ggtitle("Independents Are Males") +
  scale_fill_brewer(palette="Blues") +
  theme(legend.position="none") +
  coord_flip()
```

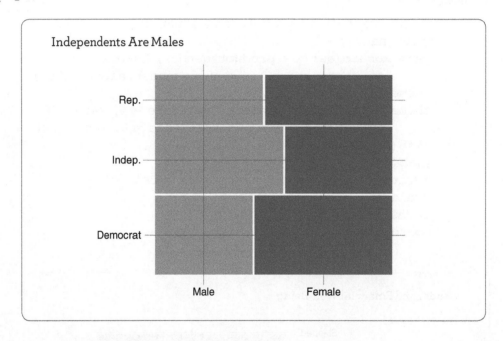

8. boxplot, highest = West, lowest = South

```
ggplot(states, aes(region, femleg, col=region)) +
  geom_boxplot() +
  theme_minimal() +
   theme(plot.title = element_text(size = 8, face = "bold"),
        axis.title = element_text(size = 8, face = "bold")) +
  theme(axis.text.x = element_text(size=8, vjust=.7),
        legend.position="none") +
  ggtitle("Women in Legislatures by Region") +
  ylab("Percentage of Legislators That Are Women") +
  xlab("")  +
  coord_flip()
```

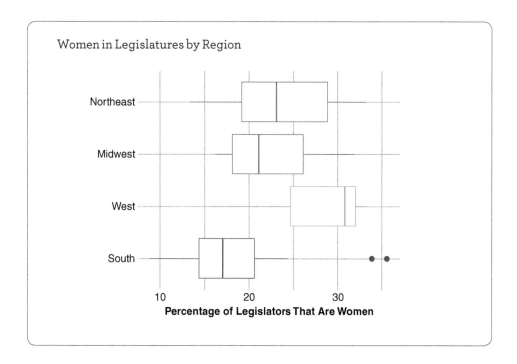

9. boxplot, highest = Africa, lowest = East Asia

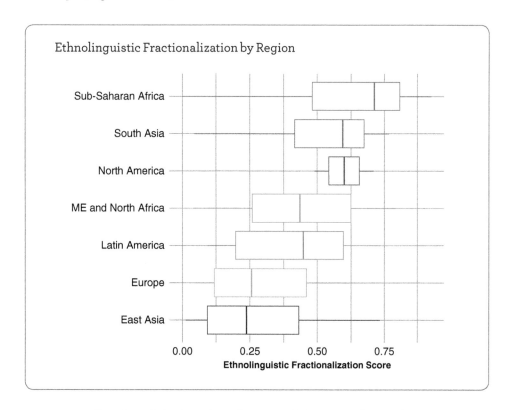

```
ggplot(world, aes(region, ethfrac, col=region)) +
  geom_boxplot() +
```

```
    theme_minimal() +
    theme(plot.title = element_text(size = 8, face = "bold"),
        axis.title = element_text(size = 8, face = "bold")) +
    theme(axis.text.x = element_text(size=8, vjust=.7),
        legend.position="none") +
    ggtitle("Ethnolinguistic Fractionalization by Region") +
    ylab("Ethnolinguistic Fractionalization Score") +
    xlab("")  +
    coord_flip()
```

10. jitter plot, French

```
    ggplot(world, aes(colony, turnout, col=colony)) +
    geom_jitter() +
    theme_minimal() +
     theme(plot.title = element_text(size = 8, face = "bold"),
          axis.title = element_text(size = 8, face = "bold")) +
    theme(axis.text.x = element_text(size=8, vjust=.7),
          legend.position="none") +
    ggtitle("Turnout and Colonial Past") +
    ylab("Percentage of Population Voting in Last Legislative Election") +
    xlab("")  +
    coord_flip()
```

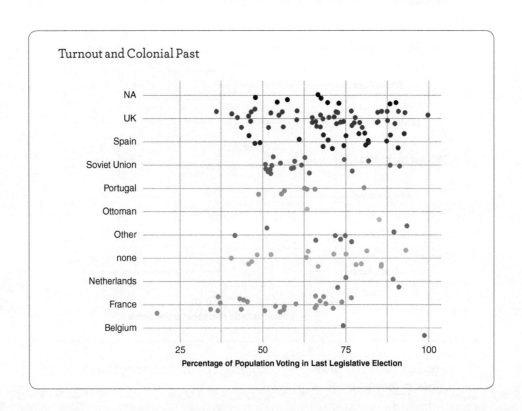

11. yes; no; yes; CAF, AGO, NAM

```
ggplot(world, aes(womyear, inf)) +
  geom_point(col="#bf0000") +
  geom_text(size=3, vjust=2, col="#263333", alpha = .3,
            aes(label= iso3c)) +
  ggtitle("Women's Suffrage and Infant Mortality") +
  ylab("Infant Mortality") +
  xlab("Year Women Were Granted Suffrage") +
  theme_minimal() +
  theme(plot.title = element_text(size = 8, face = "bold"),
        axis.title = element_text(size = 8, face = "bold"))
```

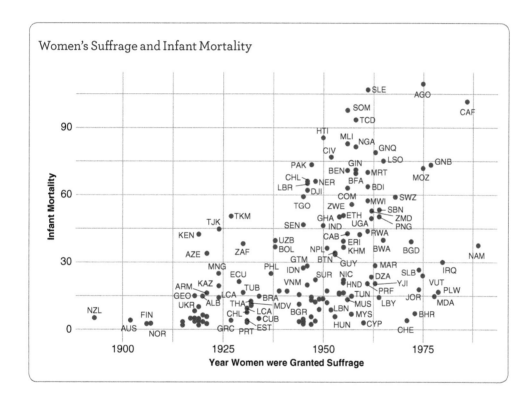

12. yes, yes, yes, NY

```
ggplot(states, aes(inc, stuspend)) +
  geom_point(col="#bf0000") +
  geom_text_repel(size=3, vjust=2, col="#263333", alpha = .3,
            aes(label= st)) +
  ggtitle("Income and Spending on Students") +
  ylab("Dollars per Student") +
  xlab("Income per Household") +
  theme_minimal() +
  theme(plot.title = element_text(size = 8, face = "bold"),
        axis.title = element_text(size = 8, face = "bold"))
```

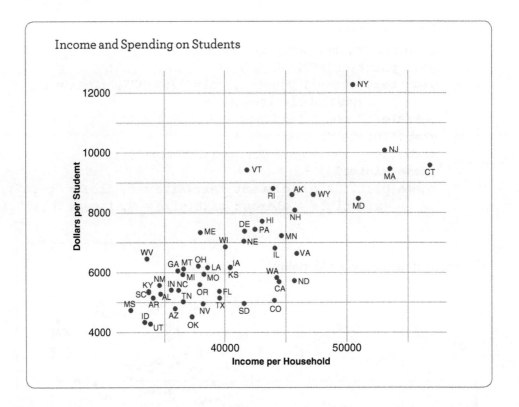

13. no; no; yes; CA, WA, WY, AR, GA

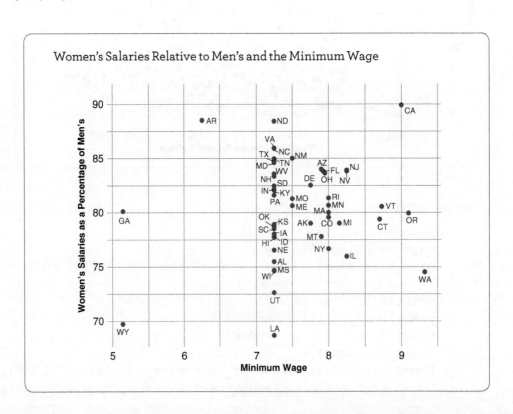

```
ggplot(states, aes(minwage, percwom)) +
  geom_point(col="#bf0000") +
  geom_text_repel(size=3, vjust=2, col="#263333", alpha = .3,
            aes(label= st)) +
  ggtitle("Women's Salaries Relative to Men's and the Minimum Wage") +
  ylab("Women's Salaries as a Percentage of Men's") +
  xlab("Minimum Wage") +
  theme_minimal() +
  theme(plot.title = element_text(size = 8, face = "bold"),
        axis.title = element_text(size = 8, face = "bold"))
```

14. b, c, d

15. a, c, d

16. Figure 10-6 (a), Figure 10-8 (b), Figure 10-9 (c), Figure 10-9 (d)

PRACTICE ON ANALYSIS AND VISUALIZATION

1. mosaic plot, no

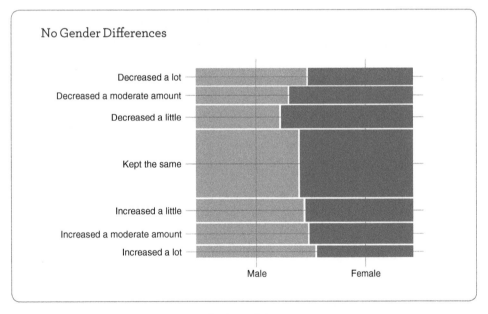

```
ggplot(data = na.omit(nes)) +
  geom_mosaic(aes(x = product(gender, immig_numb),
                  fill=gender,na.rm=TRUE)) +
  theme_minimal() +
  theme(plot.title = element_text(size = 8, face = "bold"),
        axis.title = element_text(size = 8, face = "bold")) +
  xlab("") +
  ylab("") +
```

```
        coord_flip() +
        ggtitle("No Gender Differences") +
        scale_fill_brewer(palette="Blues") +
        theme(legend.position="none")
```

2. mosaic plot, yes

```
        ggplot(data = na.omit(nes)) +
        geom_mosaic(aes(x = product(warmcause, pid3.new),
                                    fill=warmcause,na.rm=TRUE)) +
        theme_minimal() +
        theme(plot.title = element_text(size = 8, face = "bold"),
              axis.title = element_text(size = 8, face = "bold")) +
        xlab("") +
        ylab("") +
        ggtitle("Democrats Blame Humans for Climate Change") +
        scale_fill_brewer(palette="Blues") +
        theme(legend.position="none")
```

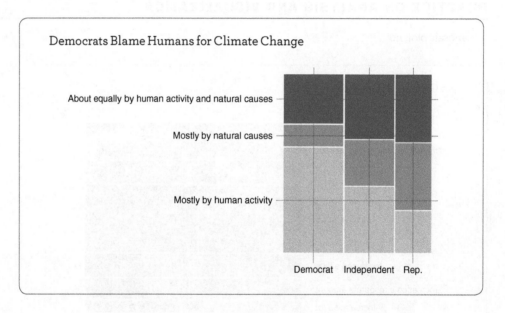

3. mosaic plot, yes

```
        ggplot(data = na.omit(nes)) +
        geom_mosaic(aes(x = product(compromise, pid3.new),
                                    fill=compromise,na.rm=TRUE)) +
        theme_minimal() +
        theme(plot.title = element_text(size = 8, face = "bold"),
              axis.title = element_text(size = 8, face = "bold")) +
        xlab("") +
        ylab("") +
        ggtitle("Democrats Favor Compromise") +
        scale_fill_brewer(palette="Blues") +
        theme(legend.position="none")
```

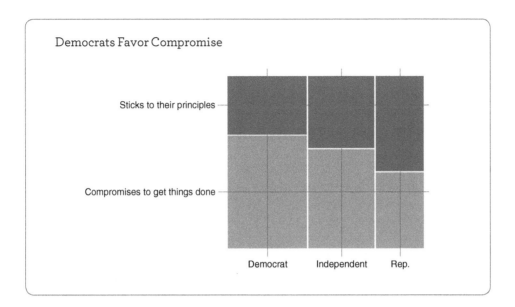

4. boxplot, biggest = royal dictatorships, smallest = civilian dictatorships

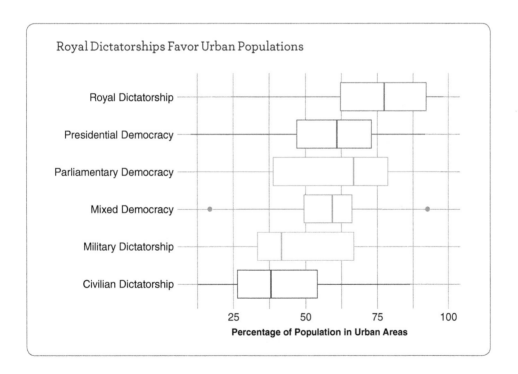

```
ggplot(world, aes(regime, urban, col=regime)) +
  geom_boxplot() +
  theme_minimal() +
  theme(plot.title = element_text(size = 8, face = "bold"),
        axis.title = element_text(size = 8, face = "bold")) +
  theme(axis.text.x = element_text(size=8, vjust=.7),
        legend.position="none") +
```

```
        ggtitle("Royal Dictatorships Favor Urban Populations") +
        ylab("Percentage of Population in Urban Areas") +
        xlab("") +
        coord_flip()
```

5. boxplot, youngest = Africa, oldest = Europe

```
        ggplot(world, aes(region, young, col=region)) +
        geom_boxplot() +
        theme_minimal() +
        theme(plot.title = element_text(size = 8, face = "bold"),
              axis.title = element_text(size = 8, face = "bold")) +
        theme(axis.text.x = element_text(size=8, vjust=.7),
              legend.position="none") +
        ggtitle("Youngest and Oldest Regions of the World") +
        ylab("Percentage of Population 0 to 21 Years Old") +
        xlab("") +
        coord_flip()
```

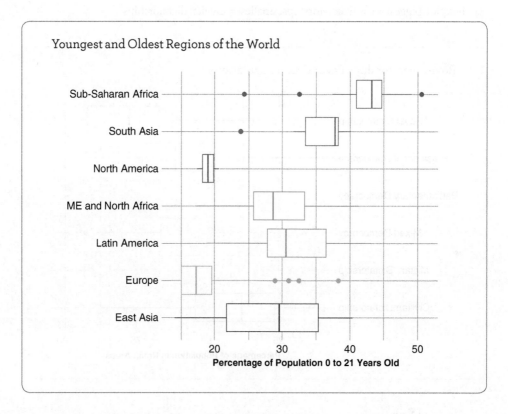

6. boxplot, most = NE, least = West

```
        ggplot(states, aes(region, knowgov, col=region)) +
        geom_boxplot() +
```

```
theme_minimal() +
theme(plot.title = element_text(size = 8, face = "bold"),
    axis.title = element_text(size = 8, face = "bold")) +
theme(axis.text.x = element_text(size=8, vjust=.7),
    legend.position="none") +
ggtitle("Northeast Most Knowledgeable") +
ylab("Percentage of Respondents Who Knew Governor's Name") +
xlab("")  +
coord_flip()
```

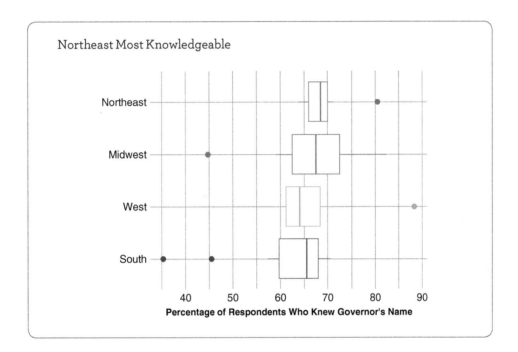

7. no (a), no (b), yes (c), Malta, Luxembourg, Cypress (d)

```
ggplot(world, aes(pwthc, fdi)) +
  geom_point(col="#bf0000") +
  geom_text_repel(size=3, vjust=2, col="#263333", alpha = .3,
        aes(label= iso3c)) +
  ggtitle("Question 7") +
  ylab("FDI") +
  xlab("Human Capital Index") +
  theme_minimal() +
  theme(plot.title = element_text(size = 8, face = "bold"),
      axis.title = element_text(size = 8, face = "bold"))
```

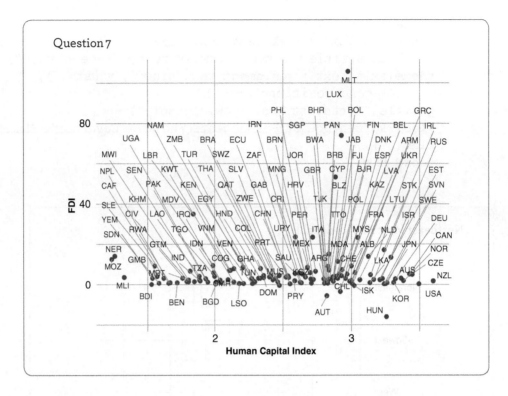

8. no (a), no (b), yes (c), LA, CA (d)

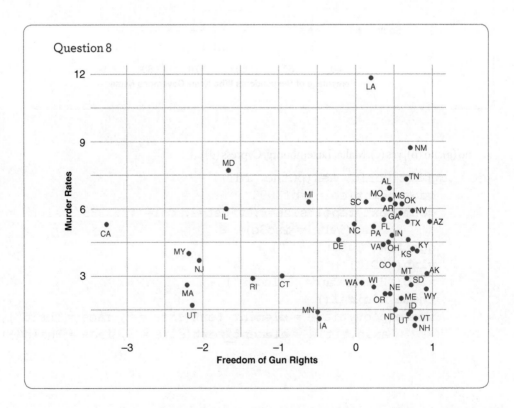

```
ggplot(states, aes(gunfree, murderrate)) +
  geom_point(col="#bf0000") +
  geom_text_repel(size=3, vjust=2, col="#263333", alpha = .3,
          aes(label= st)) +
  ggtitle("Question 8") +
  ylab("Murder Rates") +
  xlab("Freedom of Gun Rights") +
  theme_minimal() +
  theme(plot.title = element_text(size = 8, face = "bold"),
        axis.title = element_text(size = 8, face = "bold"))
```

9. yes (a), yes (b), yes (c), NY, HI (d)

```
ggplot(states, aes(inc, turnout)) +
  geom_point(col="#bf0000") +
  geom_text_repel(size=3, vjust=2, col="#263333", alpha = .3,
          aes(label= st)) +
  ggtitle("Question 9") +
  ylab("Turnout") +
  xlab("Median Income") +
  theme_minimal() +
  theme(plot.title = element_text(size = 8, face = "bold"),
        axis.title = element_text(size = 8, face = "bold"))
```

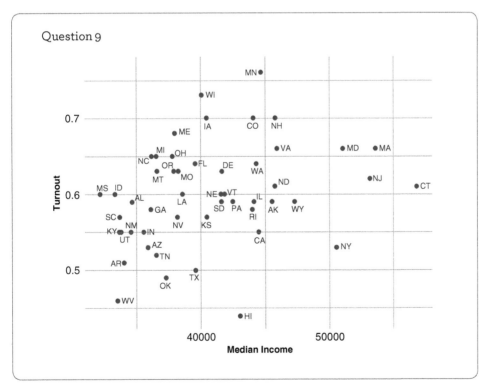

10. a, b, c, d

Access digital resources, including datasets, at http://edge.sagepub.com/brownstats1e.

11 Controlled Comparisons

CHAPTER OUTLINE

Learning Objectives
Overview
What Is a Controlled Comparison?
Comparing Two Categorical Variables, Controlling for a Third
Comparing Two Continuous Variables, Controlling for a Third
Arguments and Controlled Comparisons
Summary
Common Problems
Review Questions
Practice on Analysis and Visualization
Annotated R Functions
Answers

LEARNING OBJECTIVES

- Explain the difference between a comparison and a controlled comparison.
- Create a controlled comparison between categorical variables.
- Create a controlled comparison between continuous variables.
- Discuss how arguments and controlled comparisons are related.

Overview

In Chapter 10, we concentrated on understanding the relationship between two variables. We now add a third. Problems in the social sciences almost always involve more than one explanation. Consequently, our analysis usually involves more than two variables. Toward that end, it is useful to visualize how holding one variable constant—the third variable—influences the relationship between the other two. Drawing parallels between this exercise and making causal arguments is useful. Controlled comparisons simply recognize there is more than one cause; they allow us to *account* for more than one explanation. For example, party identification can explain perceived discrimination against Blacks, but that can't be the full story. Clearly, one's race should be considered.

This chapter demonstrates how controlled comparisons work. The goal is to build an intuition for making comparisons between two variables while holding a third variable constant. This will be extremely important in the next few chapters as we learn to build models and estimate them using multiple regression analysis.

You may not have noticed, but we've already used controlled comparisons. In the abortion rates example from Chapter 10, we examined the relationship between political ideology and abortion rates in the United States. while holding the population density constant (Figure 10-8). In this chapter we examine controlled comparisons in two different contexts—continuous variables and categorical variables. In the categorical context, we'll use two different examples—attitudes toward the police and attitudes toward immigration. With respect to the police, we will examine the relationship between party identification and opinions on police discrimination, controlling for gender and race. We will also examine the relationship between party identification and support for immigration, controlling for employment, education, and religion.

Through the exercise, you'll learn how to add a third dimension to your analysis in the context of continuous and categorical variables. The examples provide a way to systematize how to think about or structure problems with more than one explanation and how to represent that in scatter plots and mosaic plots. The chapter ends with two examples from contexts where the variables are continuous.

What Is a Controlled Comparison?

Controlled comparisons allow us to make comparisons between two variables at different levels of a third. Whether we're examining a mosaic plot (two categorical variables) or a scatter plot (two continuous variables), we introduce a third variable. Adding a third variable into consideration can do several things. What seems to be a strong relationship between two variables can disappear when we take a third variable into consideration. The opposite can occur as well: when introduced, a third variable may reveal that a strong relationship exists between two variables when at first there appeared to be none. Finally, additional variables may have no impact whatsoever on the relationship between the two variables of interest.

In the context of mosaic plots, we examine whether the relationship between two categorical variables depends on the introduction of a third categorical variable. For example, we may want to see if the relationship between gender and party identification—the x and y variables—changes depending on whether the individual lives in a city—the variable we control for or hold constant. In this example we would have two mosaic plots that each show the relationship between gender and party identification—one for respondents in the city and one for respondents in the country.

In the case of continuous variables, we plot the x and y variables and color the observations according to a third variable. Consider the example from Chapter 10 where we examined the percentage of the state that identifies as Democrats and abortion rates (our x and y variables). Recall that in Figure 10-8 we added a third variable to the mix—a categorical variable that indicated whether population density was low, medium, or high. By coloring the observations depending on density, we held density constant and observed the relationship between partisanship and abortion rates.

In both the categorical and continuous cases, we say that we're interested in the relationship between our x and y variables, *controlling for* a third. Controlling for a third variable allows us to begin the process of understanding complex phenomena that have two or more causes, the norm in the social sciences.

> **KNOWLEDGE CHECK: Explain the difference between a comparison and a controlled comparison.**

1. Which of the following statements describes a controlled comparison?
    a. It examines how one variable changes in relation to another.
    b. It examines the relationship between two variables, controlling for a third.

c. It examines if the relationship between two variables is dependent on a third.
   d. It examines the relationship between two variables, holding a third variable constant.
2. Indicate which variable is being held constant in the following descriptions.
   a. We want to know if the relationship between income and education varies by region.
   b. We examine the relationship between church attendance and income in states Trump won in 2016.
   c. Does population density influence crime differently in liberal versus conservative states?
   d. Is education more closely related to income in more populous states?
3. Indicate what adding a third variable can do to a relationship between two variables.
   a. It can reveal that there really was no relationship.
   b. It can reveal that there is a relationship when none appeared to exist at first.
   c. It does nothing.
   d. It can reveal that a relationship exists only under certain circumstances.

Comparing Two Categorical Variables, Controlling for a Third

In the following two examples (attitudes toward the police and attitudes toward immigration), we observe the relationship between two categorical variables, controlling for a third. In each example, look for whether consideration of a third variable influences the relationship between our two main variables. The exercise is analogous to determining whether an argument changes when additional factors are taken into consideration.

Example: Attitudes Toward the Police

In our social, political, and economic world, outcomes are rarely the result of just one cause. In Chapter 10, we found that when trying to explain abortion in the United States, both partisanship and population density were important. By making controlled comparisons, we're simply increasing the number of possible causal explanations from one to two. Let's first consider the relationship between opinions about the police and party identification.

In U.S. politics, the chasm between Republicans and Democrats has grown wider over the last few years. Evidence of their differences could not be more apparent than in their attitudes toward the police. Stop-and-frisk laws first used on the streets of New York City now pit the police against the Black community in many U.S. cities. With each police shooting, the gap between Democrats and Republicans widens. In the 2016 NES pilot survey, over 1,000 Americans were asked if police favored whites or Blacks. Some respondents feel police discriminate against Blacks, some feel everyone is treated equally, while others argue the police are encouraged to overcorrect and discriminate against whites. What explains the difference in opinion? Does party identification help explain the differences?

To answer those questions directly, we need to transform the data so that they more closely align with our interests. To analyze and visualize the data without any unnecessary distractions, remove any unwanted categories from the party identification data. In Code Chunk 11-1, I use the *ifelse()* command to replace the categories "other" and "not sure" in the *pid3* variable with "NAs." I then create a new factor variable called *pid3.new*, label the categories, and reorder them.

Code Chunk 11-1

```
nes$pid3.new <- ifelse(nes$pid3 == "Other", NA,
                       ifelse(nes$pid3 == "Not sure", NA,
                              nes$pid3))

nes$pid3.new <- as.factor(nes$pid3.new)

levels(nes$pid3.new)=c("Dem.", "Rep.", "Ind.")

nes$pid3.new = factor(nes$pid3.new,
                      levels(nes$pid3.new)[c(1,3,2)])
```

Now that the variable *pid3.new* is clean and reordered, we're ready to draw the mosaic plot. Notice that in the *ggplot()* command I use the *na.omit* option, which will remove the NAs from my graph. Since I'm not interested in those categories, they can be safely removed, producing a cleaner and clearer visualization of the data. The rest of the commands in Code Chunk 11-2 should by now be familiar.

Code Chunk 11-2

```
ggplot(data = na.omit(nes[,c("pid3.new","dpolice.new")])) +
  geom_mosaic(aes(x = product(pid3.new, dpolice.new),
                  fill=pid3.new,
                  na.rm=TRUE)) +
  guides(fill=guide_legend(title=NULL)) +
  ggtitle("Figure 11-1: Do Police Treat Whites and
                       Blacks the Same?") +
  theme_minimal() +
  theme(plot.title = element_text(size = 8, face = "bold"),
      axis.title = element_text(size = 8, face = "bold"),
      axis.text.x=element_text(size = 8),
      axis.text.y=element_text(size = 8)) +
  scale_fill_brewer(palette="Blues") +
  xlab("") +
  ylab("") +
  theme(axis.text.x = element_text(size=8, vjust=.5),
       legend.position = "none") +
  coord_flip()
```

■ **FIGURE 11-1** Do Police Treat Whites and Blacks the Same?

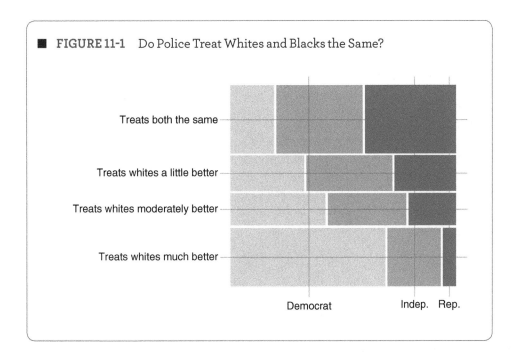

Consider the mosaic plot illustrating the relationship between party identification and opinions toward the police. In the plot, it is clear that the two are related: Democrats are more likely to think police discriminate against Blacks (Figure 11-1).

We might wonder, however, whether gender changes the relationship between attitudes toward police and party identification. Gender figures prominently in many contexts: how women view authority (e.g., the police) is certainly one of them. Once we account for gender, are we likely to see the same relationship between party identification and opinions of the police? To investigate, we produce two different mosaic plots—one for men and one for women respondents in the survey. What is the relationship between attitudes toward police discrimination and party identification, controlling for gender?

To answer the question, I created two separate data sets out of the NES survey, one for males and one for females (Code Chunk 11-3). This is a different way to create two objects in R. Once the two data sets are created, I use the *na.omit()* option in the *ggplot()* command to avoid having categories for NAs in the two mosaic plots drawn (p1 and p2). Finally, I use the by now familiar *grid.arrange()* command to arrange them on the page.

Code Chunk 11-3

```
nesmale <- subset(nes, nes$gender=="Male")
nesfemale <- subset(nes, nes$gender!="Male")

p1 <- ggplot(data =na.omit(nesmale[,
                c("pid3.new","dpolice.new")])) +
```

```
            geom_mosaic(aes(x = product(pid3.new, dpolice.new),
                            fill=pid3.new,na.rm=TRUE)) +
        guides(fill=guide_legend(title=NULL)) +
        ggtitle("Men") +
        xlab(label=NULL) +
        theme_minimal() +
        theme(plot.title = element_text(size = 8, face = "bold"),
              axis.title = element_text(size = 8, face = "bold")) +
        scale_fill_brewer(palette="Blues") +
        theme(axis.text.x = element_text(size=8, vjust=.5),
              axis.text.y = element_text(size=8, vjust=.5),
              legend.position="none") +
        xlab("") +
        ylab("") +
        coord_flip()

p2 <- ggplot(data =na.omit(nesfemale[,
                    c("pid3.new","dpolice.new")])) +
        geom_mosaic(aes(x = product(pid3.new, dpolice.new),
                    fill=pid3.new,na.rm=TRUE)) +
        guides(fill=guide_legend(title=NULL)) +
        ggtitle("Women") +
        xlab(label=NULL) +
        theme_minimal() +
        theme(plot.title = element_text(size = 8, face = "bold"),
              axis.title = element_text(size = 8, face = "bold")) +
        scale_fill_brewer(palette="Blues") +
        theme(axis.text.x = element_text(size=8, vjust=.5),
              axis.text.y = element_text(size=8, vjust=.5),
              legend.position="none") +
        xlab("") +
        ylab("") +
        coord_flip()

grid.arrange(p1, p2, nrow=2, top=textGrob("Figure 11-2: Men and Women
Hold Similar Attitudes", gp=gpar(fontsize=8)))
```

The two mosaic plots (Figure 11-2) show there's relatively little difference in attitudes between men and women. Save for a sizable difference in the category "treats whites moderately better," accounting for gender does not seem to influence the correlation between party identification and views toward the police.

Controlling for race might produce different results. The relationship between party identification and views toward the police might change dramatically if we account for race. Does race overwhelm the effects of being a Democrat or Republican? Following the same procedure

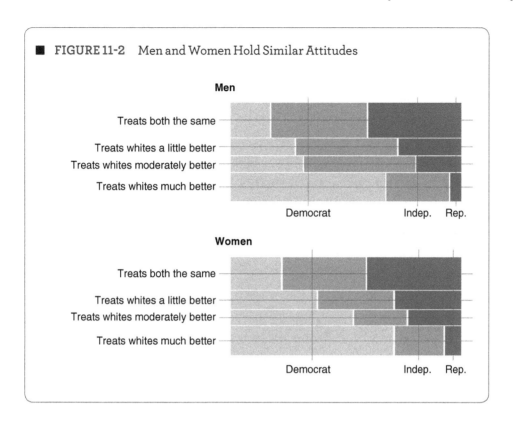

■ **FIGURE 11-2** Men and Women Hold Similar Attitudes

as in the previous code chunk, I produce two more mosaic plots of attitudes toward the police and party identification for two separate groups—white respondents and nonwhite respondents (Code Chunk 11-4). In Figure 11-3, we observe an important difference.

Code Chunk 11-4

```
neswhite <- subset(nes, nes$race=="White")
nesnonwhite <- subset(nes, nes$gender!="White")

p1 <- ggplot(data=na.omit(neswhite[,
                c("pid3.new","dpolice.new")])) +
        geom_mosaic(aes(x = product(pid3.new, dpolice.new),
                fill=pid3.new,na.rm=TRUE)) +
        guides(fill=guide_legend(title=NULL)) +
        ggtitle("Whites") +
        xlab(label=NULL) +
        theme_minimal() +
        theme(plot.title = element_text(size = 8, face = "bold"),
                axis.title = element_text(size = 8, face = "bold")) +
        scale_fill_brewer(palette="Blues") +
        theme(axis.text.x = element_text(size=8, vjust=.5),
                axis.text.y = element_text(size=8, vjust=.5),
                legend.position = "none") +
```

```
                xlab("") +
                ylab("") +
                coord_flip()

p2 <- ggplot(data=na.omit(nesnonwhite[,
            c("pid3.new","dpolice.new")])) +
            geom_mosaic(aes(x = product(pid3.new, dpolice.new),
                    fill=pid3.new,na.rm=TRUE)) +
            guides(fill=guide_legend(title=NULL)) +
            ggtitle("Black, Latino, Asian") +
            xlab(label=NULL) +
            theme_minimal() +
            theme(plot.title = element_text(size = 8, face = "bold"),
                axis.title = element_text(size = 8, face = "bold")) +

            scale_fill_brewer(palette="Blues") +
            theme(axis.text.x = element_text(size=8, vjust=.5),
                axis.text.y = element_text(size=8, vjust=.5),
                legend.position = "none") +
            xlab("") +
            ylab("") +
            coord_flip()

grid.arrange(p1, p2, nrow=2, top=textGrob("Figure 11-3: Do Police Favor
Whites?", gp=gpar(fontsize=8)))
```

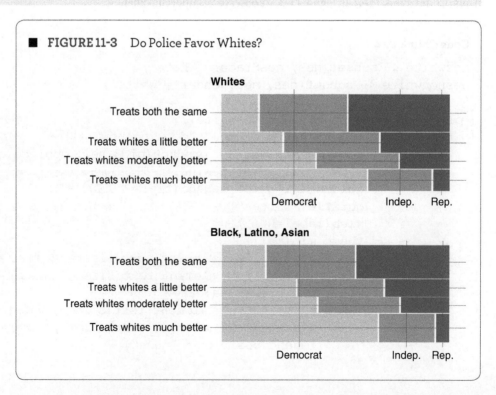

FIGURE 11-3 Do Police Favor Whites?

On one hand, note the thickness of the rows. Among Blacks, Latinos, and Asian Americans, a greater percentage of respondents feel police favor whites—the row for "treats whites much better" is thicker for Black, Hispanic, and Asian respondents. Conversely, among whites, the row for "treats both the same" is slightly thicker. On the other hand, race does not seem to change the relationship between party identification and attitudes toward police. Democrats, regardless of race, see police bias. Since the thickness of the rows changes but the patterns within each row stay the same, we would say that race matters but it does not seem to affect the relationship between partisanship and views toward the police.

Example: Attitudes Toward Immigration

The 2016 election laid bare the importance of immigration in U.S. politics, revealing yet another important partisan divide. Candidate Trump's emphasis on building a wall along the southern border represented an important symbol that divided the two campaigns. What is the relationship between party identification and views toward immigration? In the 2016 NES survey, respondents were asked whether they supported increasing or decreasing immigration. In Figure 11-4, views toward immigration are compared with party identification. The pattern is clear: Democrats are more likely than Independents and Republicans to favor more immigration. Moving down from the top to bottom row—placing fewer limits on immigration—produces a gradual increase in the relative share of Democrats and a corresponding decrease in Republicans.

Before drawing the two mosaic plots in Figure 11-4, I shorten the labels of each category so that the labels themselves don't take up too much space (Code Chunk 11-5).

Code Chunk 11-5

```
levels(nes$immig_numb)=c("Big increase", "Moderate increase",
                    "Small increase", "Same", "Small decrease",
                    "Moderate decrease", "Large decrease")
ggplot(data = subset(nes, nes$pid3.new!="NA")) +
    geom_mosaic(aes(x = product(pid3.new, immig_numb),
                            fill=pid3.new,
                            na.rm=TRUE)) +
    guides(fill=guide_legend(title=NULL)) +
    ggtitle("Figure 11-4: Democrats Favor More Immigration") +
    xlab("") +
    ylab("") +
    theme_minimal() +
    theme(plot.title = element_text(size = 8, face = "bold"),
            axis.title = element_text(size = 8, face = "bold")) +
    scale_fill_brewer(palette="Blues") +
    theme(axis.text.x = element_text(size=8, vjust=.5),
            axis.text.y = element_text(size=8, vjust=.5),
            legend.position = "none") +
    coord_flip()
```

As we discovered in the police example, the relationship between two variables can change depending on a third. We might wonder whether education has an impact on the relationship between attitudes toward immigration and party identification. Educated, highly skilled individuals may feel their jobs are less threatened by uneducated, low-skilled immigrants. In addition to employment concerns, education can change one's world view: education broadens horizons. As one learns more about the world, empathy with those beyond our borders could increase. The two mosaic plots in Figure 11-5 show the relationship between views toward immigration and party identification, controlling for education. The two plots separate respondents with a college degree from those without one.

To draw the two plots, I first create a variable using the *ifelse()* command that indicates whether a respondent graduated from college or not. The *ifelse()* command in this case says to assign a 1 if the respondent holds a degree from a 4-year institution and to assign a 1 if they attended postgraduate school. Everything else is assigned a 0 (Code Chunk 11-6).

Code Chunk 11-6

```
nes$college <- ifelse(nes$educ=="4-year", 1,
                      ifelse(nes$educ=="Post-graduate", 1, 0))
```

Once the distinction between college graduates and the rest has been defined and two separate data sets have been created, I then define the two mosaic plots (p1 and p2) and plot them (Code Chunk 11-7).

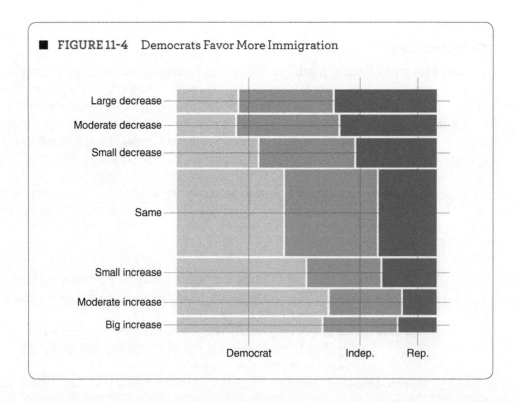

FIGURE 11-4 Democrats Favor More Immigration

Code Chunk 11-7

```
p1 <- ggplot(data = subset(nes, nes$pid3.new!="NA" & nes$college==1)) +
            geom_mosaic(aes(x = product(pid3.new, immig_numb),
                            fill=pid3.new,
                            na.rm=TRUE)) +
            guides(fill=guide_legend(title=NULL)) +
            ggtitle("College Educated") +
            xlab("") +
            ylab("") +
            theme_minimal() +
            theme(plot.title = element_text(size = 8, face="bold"),
                  axis.title = element_text(size = 8, face="bold")) +
            scale_fill_brewer(palette="Blues") +
            theme(axis.text.x = element_text(size=8, vjust=.5),
                  axis.text.y = element_text(size=8, vjust=.5),
                  legend.position = "none") +
            coord_flip()

p2 <- ggplot(data = subset(nes, nes$pid3.new!="NA" & nes$college!=1)) +
            geom_mosaic(aes(x = product(pid3.new, immig_numb),
                            fill=pid3.new,na.rm=TRUE)) +
            guides(fill=guide_legend(title=NULL)) +
            ggtitle("No College") +
            xlab("") +
            ylab("") +
            theme_minimal() +
            theme(plot.title = element_text(size = 8, face="bold"),
                  axis.title = element_text(size = 8, face="bold")) +
            scale_fill_brewer(palette="Blues") +
            theme(axis.text.x = element_text(size=7, vjust=.5),
                  axis.text.y = element_text(size=7, vjust=.5),
                  legend.position="none") +
            coord_flip()

grid.arrange(p1, p2, nrow=2,top=textGrob("Figure 11-5: Should
            Immigration Be Increased?", gp=gpar(fontsize=8)))
```

The level of education alters only very slightly the relationship between party identification and views toward immigration. The rows are a little wider for college-educated respondents that advocate for growth in immigration. We also notice that among the non-college educated, the share of Republicans is much larger in the category advocating for the largest increase. It appears that the biggest difference between the college and non-college educated lies at that extreme. Independents also make up a larger share of those arguing for large decreases among the non-college educated. Any other differences between the college and non-college

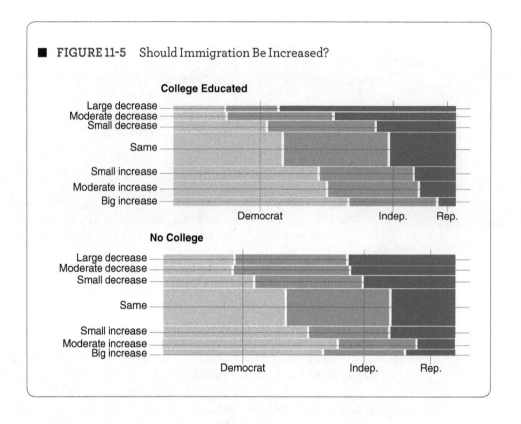

FIGURE 11-5 Should Immigration Be Increased?

educated seem minimal, and the general pattern between partisanship and views toward immigration holds.

So far, we've considered two possible mechanisms: skill level and world view. Let's examine skills in greater depth. Views toward immigration often coincide with fears of low-wage, low-skilled workers coming across the border, lowering wages, and increasing unemployment. If education influences views toward immigration because of skill level, then one's employment status might also be important. In fact, employment status might be more important than party identification or education.

To draw the two plots in Figure 11-6, simply make the distinction in the *subset()* command in each *ggplot()* command (Code Chunk 11-8). Specifically, in *p1* I subset the data by including the cases where *nes$employ*=="Full-time". Everything else is the same as the previous examples. For *p2*, I write *nes$employ*!="Full-time".

Code Chunk 11-8
```
p1 <- ggplot(data = subset(nes, nes$pid3.new!="NA" &
                           nes$employ=="Full-time")) +
       geom_mosaic(aes(x = product(pid3.new, immig_numb),
                       fill=pid3.new,na.rm=TRUE)) +
       guides(fill=guide_legend(title=NULL)) +
       ggtitle("Employed") +
```

```r
              xlab(label=NULL) +
              theme_minimal() +
              theme(plot.title = element_text(size = 8, face="bold"),
                  axis.title = element_text(size = 8, face="bold")) +
              scale_fill_brewer(palette="Blues") +
              theme(axis.text.x = element_text(size=8, vjust=.5),
                  axis.text.y = element_text(size=8, vjust=.5),
                  legend.position = "none") +
              xlab("") +
              ylab("") +
              coord_flip()

p2 <- ggplot(data = subset(nes, nes$pid3.new!="NA" &
                              nes$employ!="Full-time")) +
              geom_mosaic(aes(x = product(pid3.new, immig_numb),
                              fill=pid3.new,na.rm=TRUE)) +
              guides(fill=guide_legend(title=NULL)) +
              ggtitle("Unemployed") +
              xlab(label=NULL) +
              theme_minimal() +
              theme(plot.title = element_text(size = 8, face="bold"),
                  axis.title = element_text(size = 8, face="bold")) +
              scale_fill_brewer(palette="Blues") +
              theme(axis.text.x = element_text(size=8, vjust=.5),
                  axis.text.y = element_text(size=8, vjust=.5),
                  legend.position = "none") +
              xlab("") +
              ylab("") +
              coord_flip()

grid.arrange(p1, p2, nrow=2, top=textGrob("Figure 11-6: Employment
       Doesn't Change the Relationship", gp=gpar(fontsize=8)))
```

To examine the question, consider the two mosaic plots of party identification and views toward immigration for two different groups—one for respondents who are full-time employees and one for those who are not (Figure 11-6). In more exacting language, we're examining the relationship between views toward immigration and party identification, controlling for employment status. Interestingly, Independents seem to be the most influenced by changes in job status. If employed full-time, Independents are a relatively larger proportion of those favoring increased immigration compared to their unemployed counterparts—compare the bottom three rows of the mosaic plot for the employed and the unemployed. Controlling for employment status does seem to make a slight difference for Independents. Perhaps less partisan, Independents are relatively more sensitive to practical matters—like being employed or not. This raises a number of different questions about Independents and their attitudes. There is much more here that can be explored.

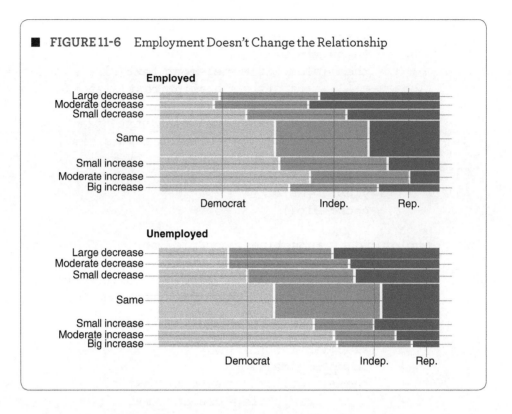

FIGURE 11-6 Employment Doesn't Change the Relationship

Having examined how economic factors—education and employment—affect the relationship between party identification and views toward immigration, we might be curious about culture. Characteristics we often associate with religion—charity, concern for the poor—could be more important. The NES survey asked respondents how often they attend church. In this example, we want to examine the relationship between views toward immigration and party identification, controlling for religion.

 Art and Practice of Data Visualization
RELIABILITY AND VALIDITY

The church attendance measure provides a nice opportunity to review validity and reliability (first discussed in Chapter 3). We say a measure is valid if it accurately represents the concept we're trying to measure. Does the number of times someone attends church represent their religiosity? Few would argue church attendance fully captures a respondent's religious devotion. It may not be the best (or valid) measure of religiosity. The reliability of a measure concerns its ability to register in a consistent and accurate fashion that which we are interested in measuring. In the church attendance example, the measure would be considered reliable if we kept polling the same respondents and kept getting the same answers. It would also be considered reliable if the respondents were accurately recounting the number of times they attended church. One could imagine inaccuracies in both dimensions of reliability.

We divide the sample between respondents who attend church more than once a week and those who attend once a week or less. If religion trumps politics, we would expect to see no relationship between party identification and views toward immigration among the avid churchgoers. The importance of politics, however, should reappear among those not attending church as frequently.

Since the dividing line separates one category in the church attendance variable from the rest of the categories, drawing the two plots requires little preparation other than using the *subset()* command where I stipulate that one plot includes those that answered "more than once a week" and the other includes everyone else (Code Chunk 11-9).

Code Chunk 11-9

```
p1 <- ggplot(data = subset(nes, nes$pid3.new!="NA" &
                    nes$pew_churatd=="More than once a week")) +
        geom_mosaic(aes(x = product(pid3.new, immig_numb),
                    fill=pid3.new,na.rm=TRUE)) +
        guides(fill=guide_legend(title=NULL)) +
        ggtitle("More than once a week") +
        xlab("") +
        ylab("") +
        theme_minimal() +
        theme(plot.title = element_text(size = 8, face="bold"),
            axis.title = element_text(size = 8, face="bold")) +
        scale_fill_brewer(palette="Blues") +
        theme(axis.text.x = element_text(size=8, vjust=.5),
            axis.text.y = element_text(size=8, vjust=.5),
            legend.position = "none") +
        coord_flip()

 p2 <- ggplot(data = subset(nes, nes$pid3.new!="NA" &
                    nes$pew_churatd!="More than once a week")) +
        geom_mosaic(aes(x = product(pid3.new, immig_numb),
                    fill=pid3.new,na.rm=TRUE)) +
        guides(fill=guide_legend(title=NULL)) +
        ggtitle("Once a week or less") +
        xlab("") +
        ylab("") +
        theme_minimal() +
        theme(plot.title = element_text(size = 8, face="bold"),
            axis.title = element_text(size = 8, face="bold")) +
        scale_fill_brewer(palette="Blues") +
        theme(axis.text.x = element_text(size=8, vjust=.5),
            axis.text.y = element_text(size=8, vjust=.5),
            legend.position = "none") +
        coord_flip()

grid.arrange(p1, p2, nrow=2, top=textGrob("Figure 11-7: Church
Attendance Changes the Relationship Between Party and Immigration",
gp=gpar(fontsize=8)))
```

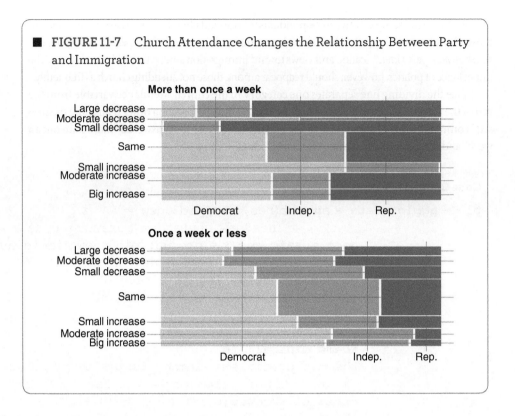

FIGURE 11-7 Church Attendance Changes the Relationship Between Party and Immigration

The partisan pattern we've observed in previous plots changes among those who attend church more than once a week. The biggest difference lies at both extremes. Frequent church-going Republicans are more likely to support large decreases or large increases compared to their more secular Republican friends. Although the pattern implies that church attendance changes the relationship between partisanship and immigration, more work is needed to determine whether we have discovered something real. Whether that finding stands up to further examination, note how our exploration generated a new question: why does religion influence Republican attitudes toward immigration?

> **KNOWLEDGE CHECK:** Create a controlled comparison between categorical variables.

4. Graph the relationship between partisanship (*nes$pid3.new*) and *nes$compromise*, controlling for church attendance (*nes$pew_churatd*).
    a. Is there a relationship between partisanship and compromise?
    b. Does church attendance influence the relationship?
    c. Does attending church influence Democrats, Republicans, or both?
    d. Is church attendance the dependent variable or an independent variable?

5. Graph the relationship between marital status—whether an individual is married (*nes$marstat*) or not—and having a college degree (*nes$educ*), controlling for whether the respondent is white.

    a. Is there a relationship be marital status and having a college degree?
    b. Does race influence the relationship?
    c. Does race influence the college educated, those without a degree, both, or neither?
    d. Is college education the dependent variable or an independent variable?

Comparing Two Continuous Variables, Controlling for a Third

In these next examples, since the two variables are continuous, we start with a scatter plot. First, we'll consider the relationship between infant mortality rates and ethnolinguistic fractionalization, controlling for GDP per capita. Second, we examine the relationship between homicide rates and GDP per capita, controlling for democracy. Finally, we will combine everything we've learned so far: how describing the data leads to making comparisons and controlled comparisons.

Example: Infant Mortality

Infant mortality rates—the number of deaths per 1,000 live births—are often used as a rough indicator to gauge the overall health of a population. For example, in countries where the rate is relatively low, a medical professional is present at the birth, the birth occurs in a hospital, and the mother has access to important educational and health-related services throughout the pregnancy. All of these contribute to the likelihood of a successful pregnancy and birth. The number of deaths per 1,000 live births, consequently, can provide a rough indication as to whether the medical and educational infrastructure is available to a wide segment of the population.

In addition to a country's medical infrastructure, birthing practices have deep cultural roots. How women and men relate to the pregnancy and birth of their child is determined, in part, by their cultural heritage. Consequently, how many women and men are educated about pregnancy and birth, how many births are attended by a health professional, and the population's access to health care in general might be influenced by the ethnic heterogeneity of the country. A scatter plot of infant mortality and ethnolinguistic heterogeneity shows that there is a strong relationship between the two variables (Figure 11-8).

Note that I use the *ifelse()* command within the *geom_text_repel()* command to highlight a small group of cases. Also note that I use the subtitle option to produce a useful subtitle that provides the country codes along with the associated countries (Code Chunk 11-10).

Code Chunk 11-10

```
ggplot(world, aes(ethfrac, inf)) +
  geom_point(col="#bf0000") +
  geom_smooth(method="loess", se=FALSE, col="#0000ff") +
  theme_minimal() +
  theme(plot.title = element_text(size = 8, face = "bold"),
        plot.subtitle = element_text(size = 7),
        axis.title = element_text(size = 8, face="bold")) +
```

```
xlab("Ethnolinguistic Fractionalization") +
ylab("Female Infant Mortality Rate") +
ggtitle(label = "Figure 11-8: Infant Mortality Increases as
                  Ethnic Heterogeneity Increases",
        subtitle = "Countries labeled are Haiti (HTI), Japan (JPN),
                    United States (USA), Kenya (KEN),
                    Equatorial Guinea (GNQ), Luxembourg (LUX),
                    and Somalia (SOM)") +
geom_text_repel(size=3, col="#04183d", aes(label=
                ifelse(iso3c=="HTI" |
                       iso3c=="JPN" |
                       iso3c=="KEN" |
                       iso3c=="LSO" |
                       iso3c=="GNQ" |
                       iso3c=="LUX" |
                       iso3c=="SOM", as.character(iso3c),"")))
```

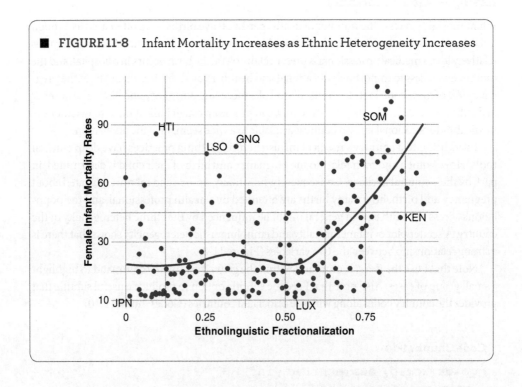

FIGURE 11-8 Infant Mortality Increases as Ethnic Heterogeneity Increases

While there is no relationship at low levels of ethnolinguistic heterogeneity (ELF), once the ELF score reaches .50, infant mortality rates increase dramatically. The simple argument that ethnic heterogeneity increases infant mortality rates has some merit, or at least some evidence behind it.

But, of course, there are a number of different factors that determine infant mortality. Some of those factors could also be related to ethnolinguistic fractionalization. Missing from the argument so far is income (i.e., GDP per capita). Unless we account for it or control for it, our argument is incomplete.

To illustrate how the relationship between two variables changes once we account for a third, I introduce GDP per capita into the argument. For illustrative purposes, I create a categorical variable that registers whether income is high or low (whether it is above or below the median of $9,863). I first create a variable in the *world* data set called *gp* and stipulate that when values are above 9,836 in the *gdppc* variable, assign a 1. If the value of *world$gdppc* is below 9,836, then assign it 0. I label the two levels "high" and "low."

Once the categorical variable is created, draw the scatter plot that controls for GDP per capita. Two separate lines are drawn for "high" and "low" cases by using the *color* option in the aesthetics command (*col=gd.f*).

Code Chunk 11-11
```
world$gp <- ifelse(world$gdppc > 9863, 1, 0)
world$gp.f <- as.factor(world$gp)
levels(world$gp.f)=c("Low", "High")
ggplot(subset(world, gp!="NA"), aes(ethfrac, infemale, col=gp.f)) +
  geom_point() +
  geom_smooth(method="lm", se=FALSE) +
  theme_minimal() +
  theme(plot.title = element_text(size = 8, face = "bold"),
        axis.title = element_text(size = 8, face = "bold")) +
  theme(legend.position = c(.6, .8)) +
  scale_color_manual(breaks = c("High", "Low"),
                     values=c("#bf0000", "#0000bf")) +
  guides(col=guide_legend("GDP per capita")) +
  xlab("Ethnolinguistic Fractionalization") +
  ylab("Female Infant Mortality Rate") +
  ggtitle("Figure 11-9: It's the Economy")
```

As Figure 11-9 illustrates, the relationship between ethnolinguistic fractionalization and infant mortality could be the result of income: heterogeneous countries with high infant mortality rates are also relatively poor. In fact, among the wealthier half of nations, there appears to be no relationship between ethnic heterogeneity and infant mortality whatsoever (the blue line is relatively flat). As we build a model or as we formulate our hypotheses and argument, we need to account for GDP per capita.

Example: Homicide Rates

Both sociologists and political scientists examine homicide rates throughout the world. Income levels go a long way toward explaining murder rates—the number of murders per 100,000 people. Poverty, lack of education, and unemployment are all correlated with low incomes and represent just some of the causes that explain homicide rates. There are, of course, other explanations. Democracy is a usual suspect. The rule of law, a professional

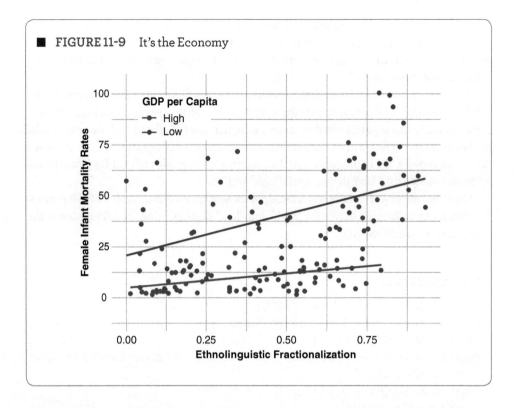

FIGURE 11-9 It's the Economy

police force, and an effective judicial system are all institutions that can influence homicide. They are also institutions associated with democratic governance. One complication, a serious one, is that income is related to democracy. Since democracies tend to be richer than authoritarian regimes, untangling the effects of democracy and income levels represents an added consideration.

With that in mind, we begin with a simple plot of the homicide rate (logged) and GDP per capita (logged), which reveals an interesting hump-shaped pattern (Figure 11-10). As we move from left to right along the x axis, homicide rates slowly increase until they reach a threshold—the hump. After reaching the apex—a GDP per capita logged value of 9, which is approximately $8,000 per capita—homicide rates decline. What explains this pattern?

Code Chunk 11-12
```
ggplot(world, aes(log(gdppc), log(homicide))) +
  geom_point(col="#bf0000") +
  geom_smooth(se=FALSE, col="#0000bf") +
  ggtitle(label = "Figure 11-10: Homicides Rise and Fall as Income Increases",
          subtitle = "Countries labeled are Honduras (HND), El Salvador (SLV), United States (USA), \nTonga (TON), Czech Republic (CZE), and Sweden (SWE)") +
  ylab("Murder Rate") +
  xlab("GDP per Capita (logged)") +
```

```
    geom_text_repel(size=3, aes(label= ifelse(iso3c=="TON" |
                                    iso3c=="USA" |
                                    iso3c=="UT" |
                                    iso3c=="HND" |
                                    iso3c=="SLV" |
                                    iso3c=="CZE" |
                                    iso3c=="SWE",
                                    as.character(iso3c), ''))) +
    theme_minimal() +
    theme(plot.title = element_text(size = 8, face = "bold"),
          axis.title = element_text(size = 8, face = "bold"),
          plot.subtitle = element_text(size = 7))
```

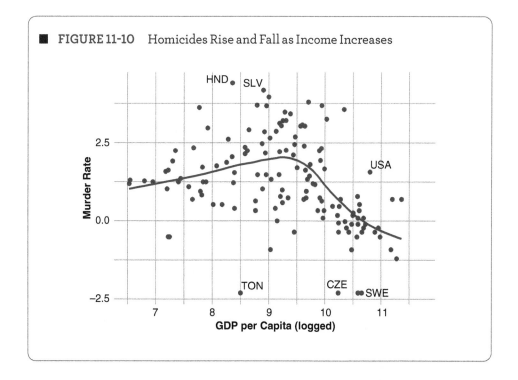

■ **FIGURE 11-10**   Homicides Rise and Fall as Income Increases

Previous scholarship on the question points to the importance of political institutions (Karstedt, 2006, 2013; LaFree & Tseloni, 2006). Specifically, democracy may have an impact on homicide since democratic regimes observe the rule of law, follow due process, and employ more professional police and judges. Consequently, any comprehensive theory on homicide rates should account for democracy. Toward that end, I constructed a **dichotomous variable** of democracy. The measure designates scores above 7 on the *Polity2* scale as democracies; countries with scores 7 and below are considered dictatorships (Code Chunk 11-13).

> **Art and Practice of Data Visualization**
> **DICHOTOMOUS VARIABLES**
>
> There are a number of terms used to describe variables that distinguish between two categories. The terms include dichotomous, binary, and dummy. They all refer to variables that include only ones and zeros or the names of two categories (which R ultimately interprets as ones and zeros).

Code Chunk 11-13

```
world$democ <- ifelse(world$polity2 > 7, 1, 0)
world$democ <- as.factor(world$democ)
levels(world$democ)=c("Dictatorship", "Democracy")
```

Using that dichotomous variable, I color the cases accordingly. In Code Chunk 11-14 I accomplish that by including the *col=* command in the aesthetics command, which tells R to color the plot by the categories in the new democracy variable I created. When using the color command in the main *ggplot()* aesthetics command, two separate lines are automatically generated for each color (i.e., regime type) when the *geom_smooth()* command is invoked. By coloring the graph according to my new democracy variable, I'm holding democracy constant as I examine the relationship between homicide rates and GDP per capita.

Code Chunk 11-14

```
ggplot(subset(world, democ!="NA"), aes(log(gdppc), log(homicide),
col=democ)) +
  geom_point() +
  ggtitle("Figure 11-11: Democracy Matters With Murder and Income") +
       ylab("Murder Rate (logged)") +
       xlab("GDP per Capita (logged)") +
  geom_smooth(method="lm", se=FALSE) +
  theme_minimal() +
  theme(plot.title = element_text(size = 8, face = "bold"),
       axis.title = element_text(size = 8, face = "bold")) +
  theme(legend.position = c(.3, .2)) +
  guides(col=guide_legend("Regime Type")) +
  scale_color_manual(breaks = c("Democracy", "Dictatorship"),
                    values=c("#bf0000", "#0000bf"))
```

The pattern we observe in Figure 11-11 is striking. We observe that when controlling for democracy, there is a marked difference in the relationship between homicides and GDP per capita. Among authoritarian regimes, GDP per capita seems to have relatively little influence on homicide. For the democratic countries of the world, GDP per capita has a sharply negative relationship with homicide rates. Why do homicide rates seem to be more sensitive to income in democracies?

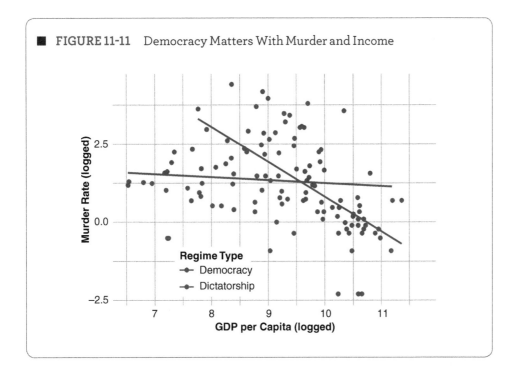

■ **FIGURE 11-11** Democracy Matters With Murder and Income

Democratic regimes may be more sensitive to demands made by the population. With respect to homicides, democracies might be more sensitive to the demand for public safety and security as incomes rise. There are a number of other related empirical implications that we could investigate. For example, are democracies sensitive to other demands relative to their more authoritarian counterparts? We now have a different question to pursue. One could argue that it is a more nuanced, if not slightly more informed, question. Controlling for democracy clearly has an impact on the relationship between income and homicide rates, an impact that generates questions we hadn't considered at the outset.

> **KNOWLEDGE CHECK: Create a controlled comparison between continuous variables.**

6. Draw a scatter plot that examines the relationship between income (*states$inc*) and the ratio of women's income to men's (*states$percwom*), controlling for whether Trump won the state.

   a. Is there a relationship between the wealth of a state (income) and the ratio between women's and men's incomes?

   b. Does the distinction between the states Trump won and lost influence the relationship?

   c. Is the ratio between women's and men's income better in states Trump won?

   d. Does it look like the states that supported Trump are generally wealthy or poor?

7. Draw a scatter plot that examines the relationship between GDP per capita and urban population, controlling for the democracy variable defined earlier in the chapter (Code Chunk 11-13).

    a. Is there a relationship between income and turnout?
    b. Does the distinction between dictatorship and democracy influence the relationship?
    c. Is turnout generally higher in democracies?
    d. What kind of states have the lowest levels of income and turnout?

Arguments and Controlled Comparisons

Arguments that focus on one cause to the exclusion of others are often wrong. While they might be useful in giving us purchase on the problem, they can be misleading. Note that in most every example in this chapter, when we included an additional variable, our understanding of the relationship between the two original variables changed. This is analogous to reformulating your argument once additional evidence is considered. While taking other things into consideration in an argument can leave the main hypotheses intact, the examples above indicate that they can radically alter how we think our original two variables relate.

With respect to attitudes toward the police, we found that taking gender into account didn't change the relationship between party identification and whether respondents felt police favored whites over Blacks. When we took the race of the respondent into account, we found that the perception of bias differed markedly between races. Accounting for race, however, did not seem to affect the relationship between party identification and attitudes toward the police. With respect to immigration, we found that employment status, education, and religion all had an observable influence on the relationship between party identification and views toward immigration. In the context of continuous variables, when taking democracy into account, our understanding of homicide rates and income changed dramatically. We also observed an appreciable change between infant mortality and ethnolinguistic fractionalization when accounting for GDP per capita.

As we'll see in the next few chapters, accounting for that third, fourth, or fifth variable can change our estimates of how two variables are related. Arguments and models are simply different ways to express the same idea. The arguments we make and the things we account for can be tied directly to the models we build and the estimates we generate. This will be a common theme throughout the rest of the book. Uninformed arguments can lead to mistaken conclusions. Uninformed models can lead to biased estimates. Making the connection between arguments and models allows you to use a set of tools in one context (making arguments) to help inform and improve your work in the other (building models).

> **KNOWLEDGE CHECK: Discuss how arguments and controlled comparisons are related.**

8. Which are examples of arguments that imply controlled comparisons?

    a. Famines throughout history have never happened in democracies.
    b. In tense situations, individuals rarely respond to threats with kindness.

c. Demands for health, education, and welfare are more likely to be met in parliamentary democracies.

d. The death penalty is more likely to be given out when the victim of the crime was white.

9. Which of the following describe arguments and models?

   a. When taking new information into account, both models and arguments can change.

   b. The better we know the data, the better the argument and model.

   c. Models are the same as arguments.

   d. Models are a simplification of an argument.

SUMMARY

Controlled comparisons allow us to account for more than one variable. The examples in this chapter started out with one cause: in both the police example and the immigration example we focused on party identification. Since we suspect gender, race, education, employment status, and religion might influence attitudes, like good detectives we broadened our investigation to include more suspects. If the simple scatter plot represents an argument with one cause, the controlled comparison accounts for a second. These examples show that our findings can change whenever we account for other variables.

Note that our findings didn't always change: gender did not appreciably alter the relationship between attitudes toward the police and party identification nor did race. Consequently, in those examples we looked elsewhere for other possible explanations.

Examining relationships while controlling for another variable is an important concept to understand. In effect, just like in a good argument, controlled comparisons allow us to account for more than one plausible explanation while we examine another.

COMMON PROBLEMS

- *Learning the lingo*. Making the connection between controlled comparisons, arguments, and models is new to many. The first step is understanding how to articulate the problem and how it is expressed either in data visualization, an oral argument, or a mathematical representation (an equation). We are interested in knowing the relationship between two variables, controlling for a third. In data visualization, it's handy to systematically place the first two variables on the x and y axes, and color the data points (if using a scatter plot) according to the third variable. In an argument we say we're examining the relationship between two variables, controlling for or accounting for a third. In a mathematical equation ($y = x$), we may simply be adding a term ($y = x + z$).

- *Considering controlled comparisons as a useful construct*. Using controlled comparisons to order variables in terms of what is the main hypothesis being tested (the relationship between x and y) and things we might want to control for (z) helps us keep things straight and allows us to communicate arguments more effectively. Having that kind of construct clearly in mind also helps us evaluate arguments made by others. When you hear someone making an argument or presentation, order what they are saying in terms of x and y, controlling for z.

- *Connecting hypotheses to graphs*. What will comparing x and y, controlling for z look like in a graph? When constructing effective and insightful views of the data, understanding what is being hypothesized and how it might be manifested in a graph is the goal. Before generating scatter plots or mosaic plots—or any plot, for that matter—actually draw out with pen and paper what you would expect to see if your hypothesis or argument is correct. What would you expect to see if it is wrong? Developing this intuition works both ways: it makes for better data analysis and it helps in our evaluation of arguments made by others in whatever form we encounter them.

REVIEW QUESTIONS

1. What are we doing when we add color to a scatter plot?

2. What is a controlled comparison?

3. In the example on immigration, what variables do we control for, hold constant, or account for?

4. If the slopes and intercepts of the lines in Figure 11-9 were similar, what would be your conclusion?

5. When is a measure valid?

6. When is a measure reliable?

7. What is a dichotomous, binary, or dummy variable?

8. Build a mosaic plot that illustrates the relationship between gender and feelings toward police (feeling thermometer for police), controlling for education (whether the respondent has a college degree). What do you observe?

9. What does the previous question imply about causation?

10. Does controlling for a third variable always change the relationship between the first two? Provide some evidence from examples in this chapter.

PRACTICE ON ANALYSIS AND VISUALIZATION

1. Why do we make "controlled" comparisons?
   a. Rarely is there only one cause.
   b. Controlling for one variable may alter the relationship between two others.
   c. The relationship between two variables may be determined by a third.
   d. A third variable may obscure the relationship between two variables.

2. Which of the following statements implies a controlled comparison is being made?
   a. Good players win games, coaching doesn't matter.
   b. Without good coaching, it doesn't matter how good the players are.
   c. You can make a lot of money if you're smart, it doesn't matter where you went to school.
   d. Nothing else matters, good players win games.

3. What are the independent and dependent variables in each component of question 2?
   a. Good players win games, coaching doesn't matter.
   b. Without good coaching, it doesn't matter how good the players are.
   c. You can make a lot of money if you're smart, it doesn't matter where you went to school.
   d. Nothing else matters, good players win games.

4. Draw mosaic plots to examine whether the relationship between contact with the police and perceived police bias is influenced by the race of the respondent (white or nonwhite). You'll need to create a variable that records whether the respondent has been stopped by the police or not. You'll also need to create a variable that records if a respondent answered "treats whites much better" or not (nes$stop_ever).

a. Does being white influence the relationship between having been stopped and views toward police bias?

b. What variable are you "controlling" for in this case?

c. What is the dependent variable?

d. Is there a relationship between being stopped and views toward police bias?

5. Draw mosaic plots to examine the relationship between having a college education and a respondent's belief that climate change is due mostly to human activity, controlling for church attendance (does the respondent attend church more than once a week?)

a. Does church attendance influence the relationship between being college educated and attributing climate change to mankind?

b. What variable are you "controlling" for in this case?

c. What is the dependent variable?

d. Is there a relationship between education and attitudes toward climate change?

6. Examine the relationship between spending on education and homicide rates, controlling for income (divide the states between rich and poor based on the variable *states$inc*).

a. Does being rich influence the relationship between spending on education and the homicide rate?

b. What variable are you "controlling" for in this case?

c. What is the dependent variable?

d. Is there a relationship between education spending and homicide?

7. Examine whether GDP per capita (logged) influences the log of foreign direct investment (FDI), controlling for regime type (use the earlier variable created for democracy and dictatorships).

a. Does being democratic influence the relationship between income and FDI?

b. What variable are you "controlling" for in this case?

c. What variable is the dependent variable?

d. Is there a relationship between GDP per capita and FDI?

8. Examine whether ethnolinguistic heterogeneity influences GDP per capita (logged), accounting for regime type (use the earlier variable created for democracy and dictatorships).

a. Does being democratic influence the relationship between ethnicity and income?

b. What variable are you "controlling" for in this case?

c. What variable is the dependent variable?

d. Is there a relationship between ethnolinguistic fractionalization and income?

9. Examine whether having more women in politics influences the relationship between GDP per capita (logged) and infant mortality rates for girls (*infemale*). Split the variable *womleg* into three parts (low, medium, and high) to answer the question.

a. Does having more women legislators influence the relationship between income and infant mortality?

b. What variable are you "controlling" for in this case?

c. What variable is the dependent variable?

d. Is there a relationship between infant mortality for girls and income?

10. Does region influence the relationship between education and voting Democratic in the United States?

a. Does region influence the relationship between education and voting?

b. What variable are you "controlling" for in this case?

c. What variable is the dependent variable?

d. Which region stands out?

ANNOTATED R FUNCTIONS

The following functions appear in this chapter. They are listed in order of their first appearance (with the code chunk number in parentheses) and annotated here to give a very brief description of their use. Some are not stand-alone functions and only work in combination with other commands. As a reminder, the code in every chapter will work properly if executed in the order it appears. Proper execution also depends on typing the author-defined *libraries()* command which loads the required R packages.

ifelse(): logical function that allows you to construct an if-then statement. Useful for selecting specific cases to label and also for creating categorical variables from continuous variables. (11-1)

factor(): used to manipulate levels (categories) in a variable. (11-1)

levels(): used to list categories (levels) in a variable. (11-1)

ggplot(): defines the basic structure of a plot (usually the x and y variables). (11-2)

na.omit(): removes all cases of a data set that have missing values. (11-2)

geom_mosaic(): draws a mosaic plot in ggplot. (11-2)

aes(): the aes (called "aesthetics") function is used in ggplot to define the basic structure of the plot, which often includes the variables you want to use and any shapes or colors. (11-2)

guides(): allows you to control features of the legends in a graph. (11-2)

ggtitle(): provides the title for a ggplot. (11-2)

theme_minimal(): specifies a minimalist style for ggplot. (11-2)

theme(): specifies font, size, and so forth in a ggplot. (11-2)

scale_fill_brewer(): colors a ggplot with a specified palette in cases where you need multiple colors. (11-2)

xlab(): labels the x axis in ggplot. (11-2)

ylab(): labels the y axis in ggplot. (11-2)

coord_flip(): flips the x and y axes on a plot. (11-2)

subset(): chooses cases in the data specified in the argument. (11-3)

grid.arrange(): allows you to arrange multiple figures on a page. (11-3)

geom_point(): draws points for a scatter plot in ggplot. (11-10)

geom_smooth(): draws lines or curves to a scatter plot. (11-10)

geom_text_repel(): labels points in a scatter plot in the ggplot framework. (11-10)

scale_color_manual(): allows you to specify exactly what colors to use in a ggplot. (11-11)

ANSWERS

KNOWLEDGE CHECK

1. b, c, d

```
p1 <- ggplot(data = subset(nes, nes$pid3.new!="NA" &
nes$pew_churatd=="More than once a week")) +
            geom_mosaic(aes(x = product(pid3.new, compromise),
                        fill=pid3.new,na.rm=TRUE)) +
            guides(fill=guide_legend(title=NULL)) +
            ggtitle("Attends Church Often") +
            xlab("") +
            ylab("") +
            theme_minimal() +
            theme(plot.title = element_text(size = 8, face="bold"),
                axis.title = element_text(size = 8, face="bold")) +
            scale_fill_brewer(palette="Blues") +
            theme(axis.text.x = element_text(size=8, vjust=.5),
                axis.text.y = element_text(size=8, vjust=.5),
                legend.position = "none") +
            coord_flip()

 p2 <- ggplot(data = subset(nes, nes$pid3.new!="NA" &
nes$pew_churatd!="More than once a week")) +
            geom_mosaic(aes(x = product(pid3.new, compromise),
                        fill=pid3.new,na.rm=TRUE)) +
            guides(fill=guide_legend(title=NULL)) +
            ggtitle("Does Not Attend Church Often") +
            xlab("") +
            ylab("") +
            theme_minimal() +
            theme(plot.title = element_text(size = 8, face="bold"),
                axis.title = element_text(size = 8, face="bold")) +
            scale_fill_brewer(palette="Blues") +
            theme(axis.text.x = element_text(size=7, vjust=.5),
                axis.text.y = element_text(size=7, vjust=.5),
                legend.position="none") +
            coord_flip()

grid.arrange(p1, p2, nrow=2,top=textGrob("Church Influences
                    Republicans", gp=gpar(fontsize=8)))
```

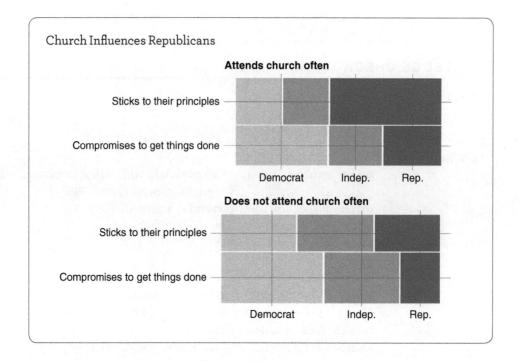

2. region, trumpwin, democrat (party ID), population

```
nes$hsgrad <- ifelse(nes$educ=="No HS", 0,
                ifelse(nes$educ=="High school graduate", 0, 1))

nes$hsgrad.f <- as.factor(nes$hsgrad)

levels(nes$hsgrad.f)=c("No College", "At least some college")

nes$White <- ifelse(nes$race=="White", 1, 0)

nes$White.f <- as.factor(nes$White)

levels(nes$White.f)=c("Non-White", "White")

nes$married <- ifelse(nes$marstat=="Married", 1, 0)

nes$married.f <- as.factor(nes$married)

levels(nes$married.f)=c("Not married", "Married")
```

```r
p1 <- ggplot(data = subset(nes, nes$White.f=="White")) +
        geom_mosaic(aes(x = product(hsgrad.f, married.f),
                        fill=hsgrad.f,na.rm=TRUE)) +
        guides(fill=guide_legend(title=NULL)) +
        ggtitle("White") +
        xlab("") +
        ylab("") +
        theme_minimal() +
        theme(plot.title = element_text(size = 8, face="bold"),
              axis.title = element_text(size = 8, face="bold")) +
        scale_fill_brewer(palette="Blues") +
        theme(axis.text.x = element_text(size=8, vjust=.5),
              axis.text.y = element_text(size=8, vjust=.5),
              legend.position = "none") +
        coord_flip()

p2 <- ggplot(data = subset(nes, nes$White.f!="White")) +
        geom_mosaic(aes(x = product(hsgrad.f, married.f),
                        fill=hsgrad.f,na.rm=TRUE)) +
        guides(fill=guide_legend(title=NULL)) +
        ggtitle("Non-white") +
        xlab("") +
        ylab("") +
        theme_minimal() +
        theme(plot.title = element_text(size = 8, face="bold"),
              axis.title = element_text(size = 8, face="bold")) +
        scale_fill_brewer(palette="Blues") +
        theme(axis.text.x = element_text(size=7, vjust=.5),
              axis.text.y = element_text(size=7, vjust=.5),
              legend.position="none") +
        coord_flip()

grid.arrange(p1, p2, nrow=2,top=textGrob("Race Doesn't Influence
                  Marriage and College", gp=gpar(fontsize=8)))
```

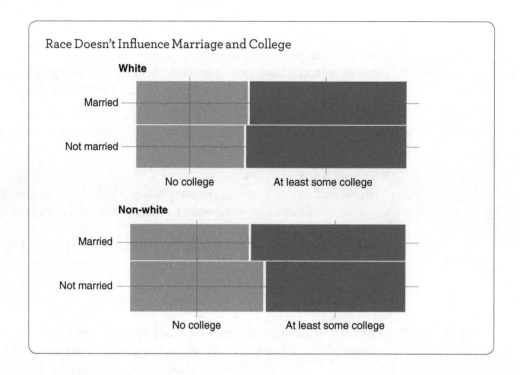

3. a, b, c, d

```
states$trumpwin.f <- as.factor(states$trumpwin)
levels(states$trumpwin.f)=c("Lost", "Won")

ggplot(states, aes(inc, percwom, col=trumpwin.f)) +
  geom_point() +
  geom_smooth(method="lm", se=FALSE) +
  theme_minimal() +
  theme(plot.title = element_text(size = 8, face = "bold"),
        axis.title = element_text(size = 8, face = "bold")) +
  theme(legend.position = c(.8, .2)) +
  scale_color_manual(breaks = c("Lost", "Won"),
                     values=c("#0000bf", "#bf0000")) +
  guides(col=guide_legend("Trump's Electoral Count")) +
  xlab("Median Household Income") +
  ylab("Ratio of Women's Income to Men's") +
  ggtitle("It's the Economy")
```

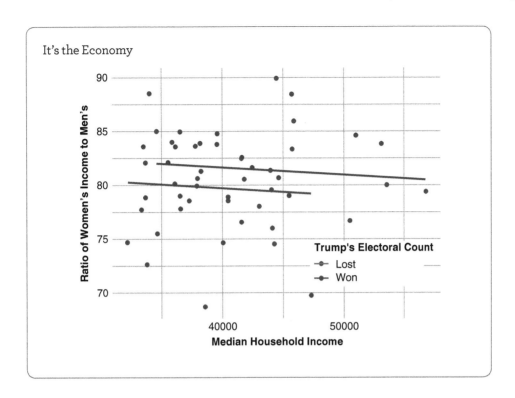

4. yes, yes, Republicans, Independent

```
ggplot(subset(world, democ!="NA"), aes(log(gdppc),turnout,col=democ)) +
  geom_point() +
  geom_smooth(method="lm", se=FALSE) +
  theme_minimal() +
  theme(plot.title = element_text(size = 8, face = "bold"),
        axis.title = element_text(size = 8, face = "bold")) +
  theme(legend.position = c(.8, .1)) +
  scale_color_manual(values=c("#bf0000", "#0000bf")) +
  guides(col=guide_legend("Regime Type")) +
  xlab("GDP per Capita (logged)") +
  ylab("Voter Turnout") +
  ggtitle("It's the Economy")
```

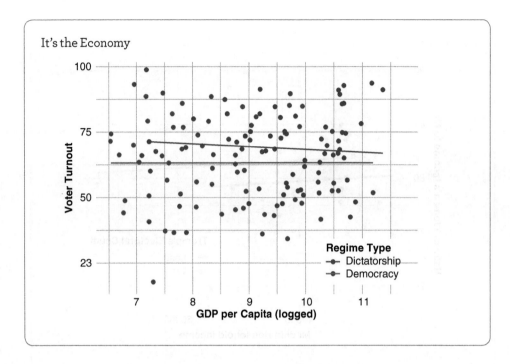

5. no (a), no (b), neither (c), independent (d)

6. no (a), no (b), no (c), poor (d)

7. no (a), no (b), yes (c), dictatorship (d)

8. b, c, d

9. a, b, c

PRACTICE ON ANALYSIS AND VISUALIZATION

1. a, b, c, d

2. a, b, c

3. games won (a), games won (b), income (c), games won (d)

4. no (a), race (b), attitudes (c), no (d)

```
nes$newpolice <-ifelse(nes$disc_police=="Treats whites much better", 1,
                ifelse(nes$disc_police=="Skipped", NA, 0))

nes$newpolice <- factor(nes$newpolice,
                    labels=c("Not Much Better", "Much Better"))
```

```r
nes$newstop <- ifelse(nes$stop_ever=="Has happened", 1,
                ifelse(nes$stop_ever=="Not asked", NA,
                    ifelse(nes$stop_ever=="Skipped", NA, 0)))

nes$newstop <-factor(nes$newstop, labels=c("Never Stopped", "Stopped"))

p1 <- ggplot(data = subset(nes, nes$newstop!="NA" &
            nes$White.f=="White")) +
        geom_mosaic(aes(x = product(newpolice, newstop),
                    fill=newpolice,na.rm=TRUE)) +
        guides(fill=guide_legend(title=NULL)) +
        ggtitle("White") +
        xlab(label=NULL) +
        theme_minimal() +
        theme(plot.title = element_text(size = 8, face="bold"),
            axis.title = element_text(size = 8, face="bold")) +
        scale_fill_brewer(palette="Blues") +
        theme(axis.text.x = element_text(size=8, vjust=.5),
            axis.text.y = element_text(size=8, vjust=.5),
            legend.position = "none") +
        xlab("") +
        ylab("") +
        coord_flip()

p2 <- ggplot(data = subset(nes, nes$newstop!="NA" & nes$newpolice!="NA"
                & nes$White!="White")) +
        geom_mosaic(aes(x = product(newpolice, newstop),
                    fill=newpolice, na.rm=TRUE)) +
        guides(fill=guide_legend(title=NULL)) +
        ggtitle("non-White") +
        xlab(label=NULL) +
        theme_minimal() +
        theme(plot.title = element_text(size = 8, face="bold"),
            axis.title = element_text(size = 8, face="bold")) +
        scale_fill_brewer(palette="Blues") +
        theme(axis.text.x = element_text(size=8, vjust=.5),
            axis.text.y = element_text(size=8, vjust=.5),
            legend.position = "none") +
        xlab("") +
        ylab("") +
        coord_flip()

grid.arrange(p1, p2, nrow=2, top=textGrob("", gp=gpar(fontsize=8)))
```

5. yes (a), church (b), attitudes (c), yes (d)

```
nes$newcause <- ifelse(nes$warmcause=="Mostly by human activity", 1, 0)

nes$newcause <- factor(nes$newcause,
                    labels=c("Natural Cause", "Man-made"))

nes$newchurch <- ifelse(nes$pew_churatd=="More than once a week", 1,
                    ifelse(nes$pew_churatd=="Once a week", 1,
                        ifelse(nes$pew_churatd=="Skipped", NA, 0)))

nes$newchurch <- factor(nes$newchurch,
                    labels=c("Not more than once a week",
                            "More than once a week"))

p1 <-ggplot(data=subset(nes, nes$newchurch=="More than once a week")) +
            geom_mosaic(aes(x = product(newcause, educ),
                        fill=newcause,na.rm=TRUE)) +
            guides(fill=guide_legend(title=NULL)) +
```

```r
            ggtitle("Frequent Church Attendance") +
            xlab(label=NULL) +
            theme_minimal() +
            theme(plot.title = element_text(size = 8, face="bold"),
                axis.title = element_text(size = 8, face="bold")) +
            scale_fill_brewer(palette="Blues") +
            theme(axis.text.x = element_text(size=8, vjust=.5),
                axis.text.y = element_text(size=8, vjust=.5),
                legend.position = "none") +
            xlab("") +
            ylab("") +
            coord_flip()

p2 <- ggplot(data = subset(nes, nes$newcause!="NA" &
                    nes$newchurch!="More than once a week")) +
            geom_mosaic(aes(x = product(newcause, educ),
                        fill=newcause, na.rm=TRUE)) +
            guides(fill=guide_legend(title=NULL)) +
            ggtitle("Non-Frequent") +
            xlab(label=NULL) +
            theme_minimal() +
            theme(plot.title = element_text(size = 8, face="bold"),
                axis.title = element_text(size = 8, face="bold")) +
            scale_fill_brewer(palette="Blues") +
            theme(axis.text.x = element_text(size=8, vjust=.5),
                axis.text.y = element_text(size=8, vjust=.5),
                legend.position = "none") +
            xlab("") +
            ylab("") +
            coord_flip()

grid.arrange(p1, p2, nrow=2, top=textGrob("", gp=gpar(fontsize=8)))
```

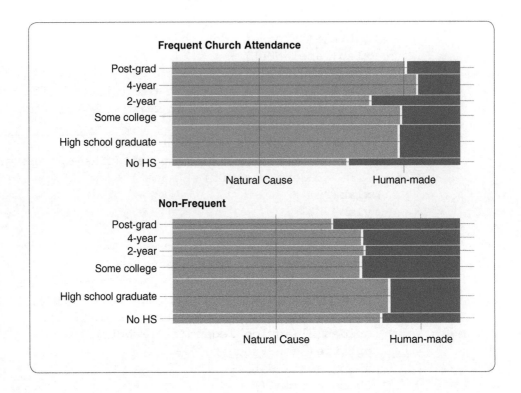

6. no (a), wealth (b), homicides (c), yes (d)

```
states$rich <- cut(states$inc, breaks=c(0,40000,60000))

states$rich <- factor(states$rich, labels = c("Poor", "Rich"))

ggplot(states, aes(stuspend, murderrate, col=rich)) +
  geom_point() +
  ggtitle("Murder, Income, and Education Spending") +
      ylab("Homicides") +
      xlab("Spending per Student") +
  geom_smooth(method="lm", se=FALSE) +
  theme_minimal() +
  theme(plot.title = element_text(size = 8, face = "bold"),
       axis.title = element_text(size = 8, face = "bold")) +
  theme(legend.position = c(.5, .8)) +
  guides(col=guide_legend("Wealth")) +
  scale_color_manual(breaks = c("Poor", "Rich"),
                    values=c("#bf0000", "#0000bf"))
```

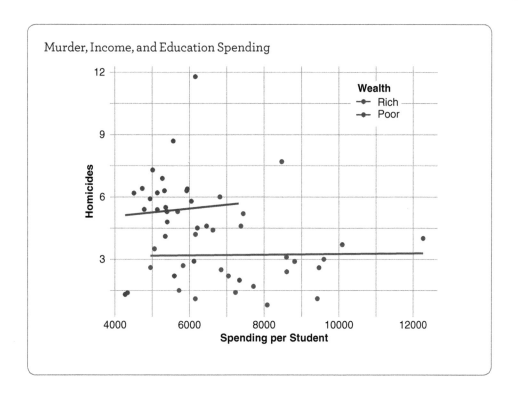

7. no (a), regime type (b), FDI (c), no (d)

```
ggplot(world, aes(log(gdppc), log(fdi), col=democ)) +
  geom_point() +
  ggtitle("Income, FDI, and Democracy") +
      ylab("FDI") +
      xlab("GDP per Capita (logged)") +
  geom_smooth(method="lm", se=FALSE) +
  theme_minimal() +
  theme(plot.title = element_text(size = 8, face = "bold"),
        axis.title = element_text(size = 8, face = "bold")) +
  theme(legend.position = c(.4, .9)) +
  guides(col=guide_legend("Regime Type")) +
  scale_color_manual(breaks = c("Dictatorship", "Democracy"),
                     values=c("#bf0000", "#0000bf"))
```

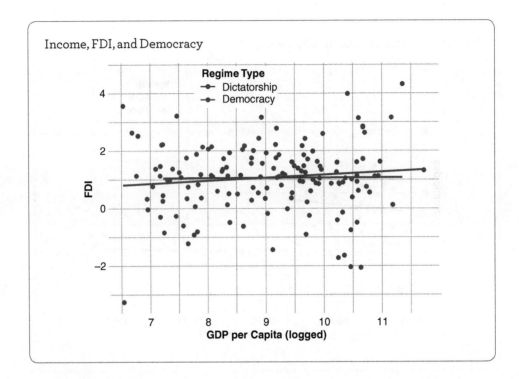

8. no (a), regime type (b), GDP per capita (c), yes (d)

```
ggplot(world, aes(ethfrac, log(gdppc), col=democ)) +
  geom_point() +
  ggtitle("Ethnicity, Income, and Democracy") +
      ylab("GDP per Capita (logged)") +
      xlab("Ethnolinguistic Fractionalization") +
  geom_smooth(method="lm", se=FALSE) +
  theme_minimal() +
  theme(plot.title = element_text(size = 8, face = "bold"),
        axis.title = element_text(size = 8, face = "bold")) +
  theme(legend.position = c(.9, .9)) +
  guides(col=guide_legend("Regime Type")) +
  scale_color_manual(breaks = c("Dictatorship", "Democracy"),
                     values=c("#bf0000", "#0000bf"))
```

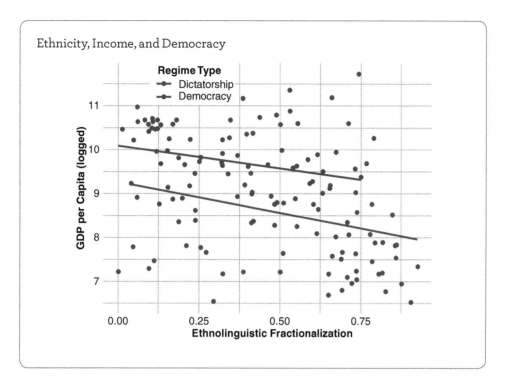

9. no (a), percentage of women in legislature (b), female infant mortality rate (c), yes (d)

```
world$women3 <- cut(world$womleg, breaks=c(0,10,18,57))

world$women3 <- factor(world$women3, labels = c("low", "medium", "high"))

ggplot(subset(world, women3!="NA"), aes(log(gdppc), infemale, col=women3)) +
  geom_point() +
  ggtitle("Women in the Legislature and Infant Health") +
      ylab("Infant Mortality for Girls") +
      xlab("GDP per Capita (logged)") +
  geom_smooth(method="lm", se=FALSE) +
  theme_minimal() +
  theme(plot.title = element_text(size = 8, face = "bold"),
      axis.title = element_text(size = 8, face = "bold")) +
  theme(legend.position = c(.9, .9))
```

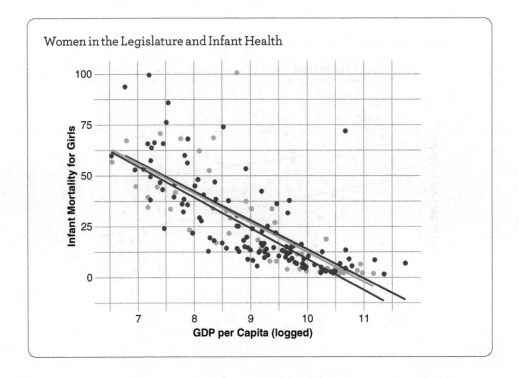

10. yes (a), region (b), voting (c), South (d)

```
ggplot(states, aes(hsdiploma, democrat, col=region)) +
  geom_point() +
  ggtitle("Education and Voting") +
      ylab("Percentage of the Electorate Identifying as Democrat") +
      xlab("Percentage of the Adult Population With High School Degrees") +
  geom_smooth(method="lm", se=FALSE) +
  theme_minimal() +
  theme(plot.title = element_text(size = 8, face = "bold"),
      axis.title = element_text(size = 8, face = "bold")) +
  theme(legend.position = c(.9, .9)) +
  guides(col=guide_legend("Region")) +
  scale_color_manual (values=c("#bf0000", "#0000bf", "#00bf00", "#0000bf"))
```

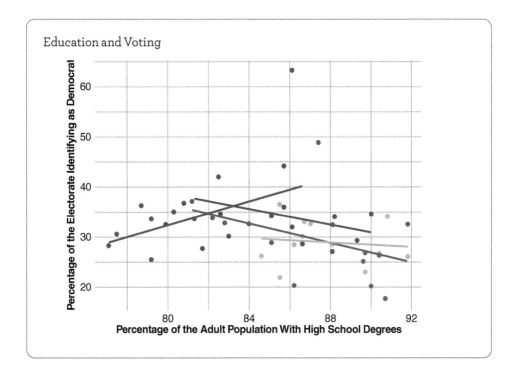

Access digital resources, including datasets, at
http://edge.sagepub.com/brownstats1e.

12 Linear Regression

CHAPTER OUTLINE

Learning Objectives
Overview
The Advantages of Linear Regression
The Slope and Intercept in Linear Regression
Goodness of Fit (R^2 Statistic)
Statistical Significance
Examples of Bivariate Regressions
Summary
Common Problems
Review Questions
Practice on Analysis and Visualization
Annotated R Functions
Answers

LEARNING OBJECTIVES

- Summarize the advantages of fitting a line to data.
- Identify the main features of linear regression.
- Describe model fit.
- Describe statistical significance.
- Interpret coefficients in a variety of contexts.

Overview

Having described and explored patterns in the data, we take the next step: regression analysis. We start with bivariate regression, calculating the slope and intercept of a line that describes the relationship between two variables. The concepts presented in this chapter form the foundation for understanding our next subject: multiple regression analysis. The difference between bivariate regression (two variables) and multiple regression (three variables or more) is analogous to the difference between comparisons and controlled comparisons. Controlled comparisons introduce additional variables for consideration. Multiple regression does the same.

Linear regression summarizes the relationship between two variables with a line. I start this chapter by identifying the advantages of fitting lines to data and explaining the mathematical and geometric properties of linear regression. Emphasis is placed on how to calculate a simple bivariate regression model along with visualizing its main features. Once the component parts of a line are explained, interpreting the *substantive significance* of the regression line's slope is the focus with special emphasis on its interpretation in different contexts.

Once we learn how to calculate the slope and intercept of a regression line, we learn to assess how well it *fits* the data. To evaluate model fit, we calculate the R^2 statistic. This statistic is a useful summary of how much variance in the dependent variable our regression model explains. Linear regression also provides ways to make probabilistic statements about our estimates, much like we did in Chapter 9 when we used the normal curve to determine how close our sample statistic was to the population parameter. Using the same machinery presented in Chapter 9, we can make similar statements about the *statistical* significance of our regression estimates. Toward that end, we learn to calculate the t-ratio for a regression coefficient.

The Advantages of Linear Regression

Let's start with defining **linear regression**. Put simply, linear regression summarizes the relationship between two variables with a line. Regression analysis differs from describing data in several ways. First, we make causal statements more explicit with regression analysis: we formally identify the dependent (y) and independent (x) variables. Second, we make predictions, generating predicted values of y given values of x. Third, we calculate the level of uncertainty surrounding those predictions. Finally, regression analysis reveals our ignorance—fitting lines demonstrates how the model diverges from reality, probably the most important and unappreciated aspect of the entire scientific enterprise. In sum, regression analysis does the following:

1. Distinguishes between cause and effect
2. Generates predictions
3. Indicates the level of uncertainty surrounding our predictions
4. Reveals our ignorance

Constructing linear models forces us to be explicit about cause and effect. While we imposed that discipline when constructing scatter plots (e.g., always placing the dependent variable on the y axis), summarizing the data with a line demands more commitment on the analyst's part. But while we are more explicit about causation, retaining a sense of humility is important: regression analysis doesn't prove anything. As with all data-analytic techniques, linear regression merely provides additional circumstantial evidence that helps solve the crime.

Making predictions clarifies our theories and hypotheses. Linear regression imposes a strict discipline: there is a corresponding predicted value (denoted $\hat{Y}$) for every value of x. Those predicted values form a line. The slope of the line is an explicit statement about how a change in the independent variable corresponds to a change in our dependent variable. The slope's magnitude states the relationship's **substantive significance**. Making predictions also allows us to compare our predictions to reality, a straightforward test of our hypothesis.

In addition to generating predictions, linear regression provides statistics that summarize how closely our predictions match reality: the **statistical significance** of our results. As you will see, this exercise resembles how we calculated the distance between the sample statistic and the population parameter in Chapter 9. Statistical significance provides an indication of how much confidence we can have in our results. How accurate are the predictions?

Finally, by being explicit about our predictions and understanding how closely they match reality, we gain a sense of our ignorance. Understanding how badly our predictions missed can motivate a search for more data or for better explanations. A firm grasp on what we don't know can be extremely enlightening.

> **KNOWLEDGE CHECK: Summarize the advantages of fitting a line to data.**

1. Which of the following are the advantages of fitting a line to data?
   a. It forces us to be explicit about cause and effect.
   b. It establishes the cause.
   c. It reveals the shortcomings of our hypothesis.
   d. It provides a sense of how confident we can be in our predictions.
2. Which statements refer to substantive significance?
   a. Retention rates for students rise 3 percentage points for every dollar spent on new faculty.
   b. Life expectancy is shortened by 2 years for every pack of cigarettes smoked in a day.
   c. The 3 percentage point rise in retention for every dollar spent could be as little as 1 percentage point or as much as 5.
   d. The decrease in life expectancy for smoking could be as little as 1 year or as much as 3.
3. Which statements refer to statistical significance?
   a. For every unit increase in x, there is a corresponding change in y.
   b. For every unit increase in x, we are confident that our estimate of the change in y is not due to random chance.
   c. The result in option (a) has important implications.
   d. The result in option (a) is fairly certain.

The Slope and Intercept in Linear Regression

We start with the equation of a line:

$$Y = a + B_1 X$$

where Y is the variable we're interested in explaining, a is the intercept, and B_1 is the slope of the line. The term X is what we hypothesize causes Y.

 Art and Practice of Data Visualization
THE LANGUAGE OF REGRESSING Y ON X

When we describe linear regression, it is always important to be precise. One way we keep things straight is to always state we're regressing our dependent variable on the independent variable. This is somewhat counterintuitive since we're stating that our independent variables have an effect on y, but it is important to remain consistent in our description of the linear model: y is always regressed on x.

The **slope** of the line, B_1, tells us the change in Y predicted with a one-unit change in X. The **intercept**, a, indicates where the line crosses the Y axis when X equals 0.

Interpretation of the Slope and Intercept

We know what the slope is (the change in y over the change in x) and we know what the intercept is (where the line crosses the y axis), but actually interpreting their significance is somewhat more involved. With respect to the slope, it's all about understanding the units of both the independent and dependent variables. With respect to the intercept, we need to decide whether knowing the value of y when x is zero is meaningful.

Example: Voter Turnout and Education

Education is considered a strong predictor of voter turnout: educated citizens are more likely to vote. Figure 12-1 depicts the relationship between turnout and education. Since the blue line in the plot (the regression line) slopes upward, there is a positive relationship between education and turnout. The line represents our prediction for turnout at each level of education. There are some significant outliers: Minnesota and Hawaii are far from the line. Minnesota has a much higher turnout than predicted and Hawaii has a much lower turnout than predicted.

Generating a regression line in R is extremely easy, we've been doing it every time a straight line is fit to a scatter plot. Specifically, we use the *geom_smooth()* command in ggplot, specifying that the method is *lm* for linear model. To make the scatter plot in Figure 12-1 more intuitive, I define a new variable *states$turnperc* by multiplying each value in the variable *states$turnout* by 100 to convert the decimals in voter turnout to whole numbers (Code Chunk 12-1). Once turnout is transformed, we're ready to construct the scatter plot.

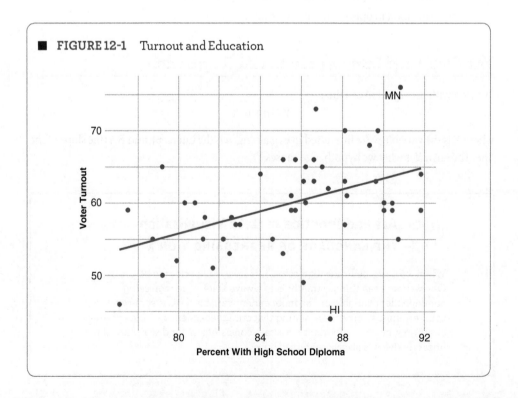

■ **FIGURE 12-1**   Turnout and Education

Code Chunk 12-1

```
states$turnperc <- states$turnout * 100

ggplot(states, aes(hsdiploma, turnperc)) +
    geom_text_repel(size = 3, aes(label = st,
            size = 1, hjust = 0, vjust = -1),
            col = ifelse(states$st == "MN", "#0000bf",
                ifelse(states$st == "HI", "#0000bf",
                    "white"))) +
    geom_smooth(method = "lm", se = FALSE, size = 0.25) +
    xlab("Percent With High School Diploma") + ylab("Voter Turnout") +
    ggtitle("Figure 12-1: Turnout and Education") +
    geom_point(col = "#bf0000", size = 0.8) +
    theme_minimal() +
    theme(plot.title = element_text(size = 8, face = "bold"),
        axis.title = element_text(size = 8, face = "bold"))
```

The code in Code Chunk 12-1 should by now be very familiar. We start with the *ggplot()* command and add the layers forming the plot. To that, we add some labels to the points along with some refinements to the themes and styles of the typeface.

Although Figure 12-1 is informative, we want to know the slope and intercept of the line. The regression output in Table 12-1 provides that information. To produce the output, define an object using the *lm()* function: the linear model function. The key part of the linear model function is the tilde (~), which separates the dependent variable from the rest of the variables. In this case, *turnperc* is the dependent variable and *hsdiploma* is the independent variable. As a matter of convenience, we can identify the data set with the *data=* option so that we can avoid typing the full variable names *states$turnperc* and *states$hsdiploma*. Once we've defined an object as a linear model and named it (e.g., *ed.lm*), we can use the command *stargazer()* to produce a regression table for that object.[1]

Code Chunk 12-2

```
ed.lm <- lm(turnperc ~ hsdiploma, data = states)

stargazer(ed.lm, title = "Table 12-1: Turnout and Education",
            type = "text", textheader = FALSE)
```

The **coefficient** in a regression is the slope of the line. For the *hsdiploma* variable, the slope is .76. With each *percentage point increase* in the population with a high school diploma, we predict a .76 *percentage point increase* in voter turnout. The intercept (or constant) is −5.4 (rounding from −5.39). In this example, when the percentage of people in the state with high

---

[1] The stargazer() command is from the package called 'stargazer.' It produces professional-quality tables for a variety of different kinds of objects. Although stargazer will produce ASCII text tables, they aren't as nice as those written in HTML or LATEX. You can specify "text," "latex," or "html" by using the type= option.

TABLE 12-1 Turnout and Education

```
=====================================================
                                  Dependent variable:
                                  -------------------
                                        turnperc
-----------------------------------------------------
hsdiploma                               0.764***
                                        (0.213)

Constant                                -5.390
                                        (18.258)

-----------------------------------------------------
Observations                              50
R2                                        0.211
Adjusted R2                               0.195
Residual Std. Error                 5.914 (df = 48)
F Statistic                     12.838*** (df = 1; 48)
=====================================================
Note:                       *p<0.1; **p<0.05; ***p<0.01
```

school diplomas is 0, we would expect −5.4% of the voting age population to vote. Of course, there are no states where the percentage of the population with high school diplomas is 0. It's also impossible to have a negative level of voter turnout. Consequently, the intercept is not informative in this example. In some cases the intercept represents a useful number, in this case it doesn't.

Figure 12-2 presents the anatomy of the regression line that summarizes the relationship between voter turnout and the percentage of the population with a high school diploma. Stated precisely, voter turnout is regressed on the percent of a state's population that completed high school.[2] The line helps identify several useful components of the relationship: (1) the **predicted value** ($\hat{Y}_i$) and (2) the **residual** ($Y_i - \hat{Y}_i$).[3] The residual indicates how well our model predicts the outcome Y_i. Large, positive residuals indicate our model underpredicts the outcome ($Y_i > \hat{Y}_i$). In this example, note how the value of Minnesota (MN) lies significantly above the line, indicating that we underpredicted its value. Large, negative residuals indicate our model overpredicts the outcome ($Y_i < \hat{Y}_i$). As we'll see in the next section, summaries of model fit start by summing and squaring all of the residuals ($\sum(Y_i - \hat{Y}_i)^2$).

To produce the visualization of a simple bivariate regression, I made liberal use of the *annotate()* command, which allows the placement of text and line segments in the ggplot (Code Chunk 12-3). The *annotate()* command has been used before, just not to the same extent. The only new addition to the code comes at the end with the *grid.brackets()* command, which provides a useful visual to highlight the distance between two points.

---

[2] Being precise with language in data analysis is important. Recall that when we describe the model, we start with the dependent variable. In the example here, voter turnout (the dependent variable) is *regressed on* education (the independent variable). When interpreting the coefficient for an independent variable, however, we start with the independent variable: "a one-unit change in the independent variable is associated with an *x* change in the dependent variable."

[3] These expressions involve operations that are performed on each observation *N*. The subscript *i* simply refers to a generic individual case. In the *states* data, our *N* is 50. The subscript *i* in that context refers to any of the 50 states.

Code Chunk 12-3

```
ggplot(states, aes(hsdiploma, turnperc)) +
    geom_point(col = "grey") + geom_text_repel(size = 3,
        aes(label = st, size = 1, vjust = 1),
        col = ifelse(states$st == "MN", "black", "white")) +
    geom_smooth(method = "lm", se = FALSE) +
    labs(x = "Percent With High School Diploma", y = "Voter Turnout",
        title = "Figure 12-2: Anatomy of a Line") +
    theme_minimal() +
    theme(plot.title = element_text(size = 8, face = "bold"),
        axis.title = element_text(size = 8, face = "bold")) +
    annotate("text", x = 85, y = 70, parse = T,
        label = "Residual~or~error~e==Y[i]~-~hat(Y[i])",
        size = 3) +
    annotate("text", x = 82, y = 67, parse = T,
        label = "'Equation for the
            line'~hat(Y[i])==a~+~beta[i]~x[i]", size = 3) +
    annotate("text", x = 89, y = 76, parse = T,
        label = "Y[i]") +
    annotate("text", x = 89, y = 64.5, parse = T,
        label = "~hat(Y[i])") +
    annotate("segment", x = 89, xend = 90.76, y = 64.3,
        yend = 64.3, size = 0.2) +
    annotate("segment", x = 87.3, xend = 90.76, y = 69.7,
        yend = 69.7, size = 0.2) +
    annotate("segment", x = 89.2, xend = 90.76, y = 76,
        yend = 76, size = 0.2) +
    annotate("segment", x = 82, xend = 82, y = 65.5,
        yend = 58, size = 0.2) +
    annotate("segment", x = 90.76, xend = 90.76, y = 76,
        yend = 64.5, col = "#bf0000", linetype = "dashed")
```

This regression line represents the best fit, or the line that minimizes the sum of the squared residuals $min\Sigma(Y_i - \hat{Y}_i)^2$.[4] First, the slope is calculated as follows:

$$\hat{\beta}_1 = \frac{\Sigma(X_i - \bar{X})(Y_i - \bar{Y})}{\Sigma(X_i - \bar{X})^2}$$

where X_i is the observation for the ith case, $\bar{X}$ is the mean of X, and $\bar{Y}$ is the mean of Y. Second, the intercept is calculated as follows:

$$\hat{\beta}_0 = \bar{Y} - \hat{\beta}_1 \bar{X}$$

where $\widehat{B}_1$ is the result of the previous equation.

---

[4] There are different criteria we can use to calculate the line of best fit. The most common framework or criterion is called ordinary least squares or OLS. The proceeding calculations form the building blocks of OLS. The bare bones of OLS are presented in this chapter so that we can build an intuition for regression analysis without getting bogged down in the mathematical, geometrical, and probabilistic infrastructure on which it is based.

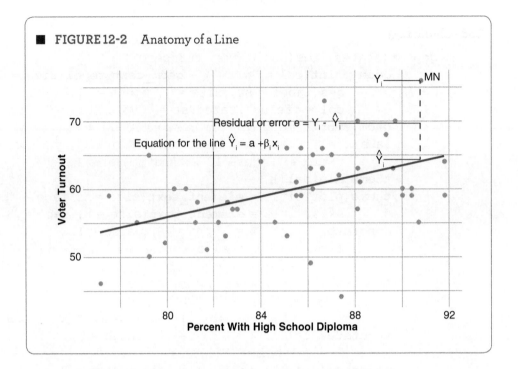

FIGURE 12-2 Anatomy of a Line

While it is a good exercise to calculate a bivariate regression line by hand, note that as the number of observations grows past 20 or 30, the exercise becomes extremely tedious and time-consuming. With the *states* data set we'd have to repeat all of our calculations 50 times in order to produce a slope and intercept. With the NES data, we'd have to repeat the operation more than 1,000 times!

> **KNOWLEDGE CHECK: Identify the main features of linear regression.**

4. Consider the following equation and specify the correct letter for each of the following.

$$\widehat{Y}_i = \alpha_i + \beta x_i$$

   a. The predicted value
   b. The intercept
   c. The slope
   d. The value of x for case i

5. When the slope of the line is negative, which of the following holds?

   a. Alpha is negative.
   b. The predicted value is negative.
   c. The value of x has to be negative.
   d. The value of beta has to be negative.

6. Which of the following statements describe a regression line?

    a. Alpha represents where the line crosses the y axis when x is zero.
    b. Beta represents the change in y given a unit change in x.
    c. The line represents the predicted values of y for different values of x.
    d. The line represents the predicted values for x given different values of y.

Goodness of Fit (R^2 Statistic)

In addition to providing a summary of the data, regression analysis indicates how well our model fits: the R^2 statistic. The **R^2 statistic** is simply the explained variance divided by the total variance of our dependent variable. To illustrate, Figure 12-3 shows the **total deviation** ($Y_i - \bar{Y}$), the **explained deviation** ($\hat{Y}_i - \bar{Y}$), and the **unexplained deviation** ($Y_i - \hat{Y}_i$) of one observation in the simple bivariate case. I use the term *deviation* in Figure 12-3 rather than *variance* since I'm referring to specific observations rather than their sum. The R^2 statistic sums the explained deviation for all of the observations and divides by the sum of the total deviation for all of the observations. The formula for the bivariate case is as follows:

$$R^2 = \frac{\sum(\hat{Y}_i - \bar{Y})^2}{\sum(Y_i - \bar{Y})^2} = \frac{\text{Explained Variation}}{\text{Total Variation}}$$

where $\hat{Y}$ is the predicted value for Y and $\bar{Y}$ is the mean of Y.

Depending on the statistics package, the R^2 statistic is expressed in percentage terms or as a number between 0 and 1. If the model correctly predicts all of the observations (all of the observations rest on the line), the R^2 will be equal to 1. If X does not help predict Y, the R^2 will be closer to 0.

Code Chunk 12-4

```
ggplot(states, aes(hsdiploma, turnperc)) +
    geom_text(size = 3, aes(label = st, size = 1, vjust = 0.8,
        hjust = 1.2), col = ifelse(states$st == "MN", "black",
            ifelse(states$st == "ME", "black", "white"))) +
    geom_smooth(method = "lm", se = FALSE) +
    labs(x = "Percent With High School Diploma", y = "Voter Turnout",
        title = "Figure 12-3: Goodness of Fit") +
    theme_minimal() +
    theme(plot.title = element_text(size = 8, face = "bold"),
        axis.title = element_text(size = 8, face = "bold")) +
    geom_hline(yintercept = 59, linetype = 3) +
    geom_point(col = "lightgrey") +
    annotate("segment", x = 88.4, xend = 90.76, y = 69, yend = 69,
        size = 0.2, col = "#bf0000") +
    annotate("text", x = 87.4, y = 69, parse = T,
        label = "'Unexplained deviation'",
        size = 3, col = "#bf0000") +
```

```
annotate("text", x = 88, y = 61, col = "#0000bf", parse = T,
        label = "'Explained deviation'", size = 3) +
annotate("text", x = 92.5, y = 67, parse = T,
        label = "'Total deviation'", size = 3) +
annotate("text", x = 94, y = 60, parse = T,
        label = "~bar(Y)==59") +
xlim(85, 95) +
annotate("segment", x = 90.76, xend = 90.76, y = 76,
        yend = 63, col = "#bf0000", size = 0.2,
        linetype = "dashed") +
annotate("segment", x = 90.76, xend = 90.76, y = 63,
        yend = 59, col = "#0000bf", size = 0.2,
        linetype = "dashed") +
annotate("segment", x = 91, xend = 91, y = 76, yend = 59,
        size = 0.5, linetype = "dashed") +
annotate("segment", x = 88.85, xend = 90.76, y = 61, yend = 61,
        size = 0.2, col = "#0000bf") +
annotate("segment", x = 91, xend = 91.8, y = 67,
        yend = 67, size = 0.2)
```

Figure 12-3 illustrates the mechanics of the R^2 statistic.[5] The further the observation is from the line, the smaller the explained deviation is relative to the total deviation. The explained deviation is small relative to the total deviation for the case of Minnesota (MN). If most cases were far

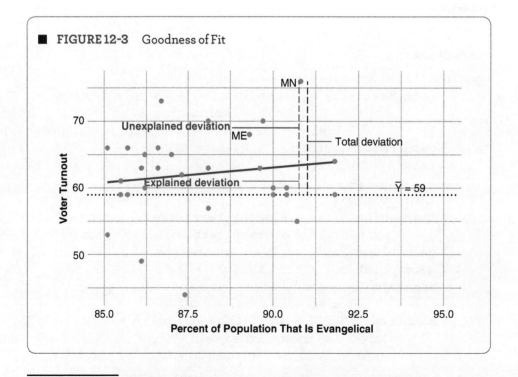

FIGURE 12-3 Goodness of Fit

---

[5] Figures 12-2 and 12-3 are inspired by similar figures from Tufte's essential text on data analysis (Tufte, 1974).

from the line, the R^2 would be relatively small—closer to zero. When the observation is closer to the line (Maine; ME), the explained deviation is relatively larger as a share of the total deviation. If most of the observations are close to the line, the R^2 would be relatively large—closer to 1.

Another view of the R^2 statistic is helpful. When our observations are close to the line, the fit is much better and the R^2 statistic is closer to 1. To visualize the difference, I create three variables in Code Chunk 12-5. Notice how one variable (*sd30*) has a standard deviation of 30 and the other (*sd5*) has a standard deviation of 5. This means that the scatter plot I draw using *sd30* will have much greater spread, and the observations will be further from the line. I then create a variable that simply goes from 1 to 100 in steps of 1. With those variables, I define *Y1* and *Y2* as linear combinations of x. Note that the spread in *Y1* will be much greater than the spread in *Y2*.

Code Chunk 12-5

```
sd30 <- rnorm(100, mean = 50, sd = 30)
sd5 <- rnorm(100, mean = 50, sd = 5)

xvar <- 1:100

Y1 <- 5 + 0.78 * xvar + sd30
Y2 <- 5 + 0.78 * xvar + sd5
```

Now we're ready to plot. I create two scatter plots (one for *Y1* and one for *Y2*). I then arrange them side by side using the *grid.arrange()* command (Code Chunk 12-6). The only new function used in Code Chunk 12-6 is one that limits the dimensions of the y axis: *ylim()*.

Code Chunk 12-6

```
p1 <- ggplot(, aes(xvar, Y1)) +
    geom_point(col = "red", size = 0.5) +
    geom_smooth(method = "lm", se = FALSE, size = 0.5) +
    theme_minimal() +
    theme(plot.title = element_text(size = 8, face = "bold"),
        axis.title = element_text(size = 8, face = "bold"),
        plot.subtitle = element_text(size = 8, face = "bold")) +
    labs(subtitle = expression(R^2 == .43), y = NULL, x = NULL) +
    ylim(0, 200)

p2 <- ggplot(, aes(xvar, Y2)) +
    geom_point(col = "red", size = 0.5) +
    geom_smooth(method = "lm", se = FALSE, size = 0.5) +
    theme_minimal() +
    theme(plot.title = element_text(size = 8, face = "bold"),
        axis.title = element_text(size = 8, face = "bold"),
        plot.subtitle = element_text(size = 8, face = "bold")) +
```

```
labs(subtitle = expression(R^2 = .95), y = NULL, x = NULL) +
ylim(0, 200)

grid.arrange(p1, p2, ncol = 2, top = textGrob("Figure 12-4:
            Comparison of R-Square Statistics",
            gp = gpar(fontsize = 8)))
```

In the left panel of Figure 12-4, the observations are relatively far from the line; the R^2 is .43. In the right panel, the observations are relatively close to the line; the R^2 is .95. Note that by definition (the mathematical expressions of $Y1$ and $Y2$ in Code Chunk 12-5) the lines have the same slope in each figure. The difference between the two lies solely in the size of the residuals. The closer the observations are to the line (the smaller the residuals), the higher the R^2 statistic.

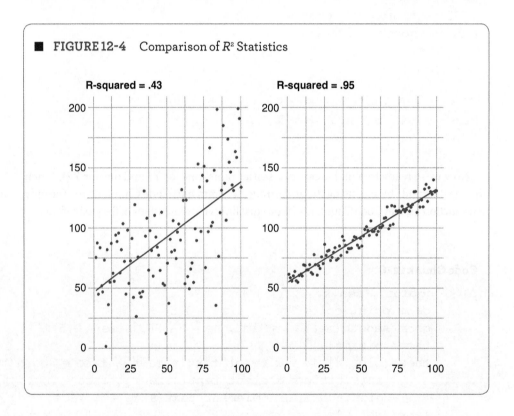

FIGURE 12-4   Comparison of R^2 Statistics

KNOWLEDGE CHECK: Describe model fit.

7. What happens to the R^2 statistic when the following occurs?

   a. The spread of the y variable increases.

   b. The spread of the x variable increases.

   c. The unexplained portion of the variance in y decreases.

   d. The explained portion of the variance in y increases.

8. The line's fit to the data is better when which of the following happen?

    a. The R^2 statistic is close to 1.

    b. Both the unexplained and explained portion of the variance increase.

    c. The observations are further from the line.

    d. The slope of the line is steeper.

Statistical Significance

Regression analysis also provides the means to establish the quality of our estimates. Can we be confident that our estimates are not the result of random chance? Our estimates of the slope and intercept are based on a sample of data we collected from the real world. Since there is an element of randomness that could influence our sample, we need to account for uncertainty. (Perhaps survey responses about a politician were obtained as a damaging news story broke, or opinions on personal safety were collected right after a major crime had been committed.) In our example of turnout and education, perhaps one of the candidates was a famous and much beloved high school administrator—not likely, but we are talking about random phenomena.

What is the probability our estimates were affected by those kinds of events? We want to avoid reporting a result that indicates a relationship exists when, in fact, it doesn't. The other mistake is generating a result that indicates no relationship exists when, in fact, it does. Depending on the context, both mistakes can be serious. In the first instance, we might undertake a very expensive public policy that is not necessary. In the second, we would fail to implement policy that could help us avoid a disastrous outcome. The quality of our estimates and the confidence we have in them can help us avoid either outcome.

We turn to the t-ratio to give us a rough indication of the confidence we can place in our estimates. While the t-ratio provides a very precise estimate of how much confidence to place in our estimates, the analyst would be well advised to treat the measure as a rough indicator. Low t-ratios would indicate we should not have very high expectations while we should save unbridled enthusiasm only for the highest t-ratios. And, even then, if they are too high, we should become suspicious.

Calculating the t-Ratio

In our example of voter turnout and education, a 1% increase in the population with a high school diploma was associated with a .76 percentage point increase in turnout. Given what we know about sample variance, the standard normal distribution, and confidence intervals, how likely is generating an estimate of .76 if the relationship is actually 0? If that probability is low, we can be more confident that there is a relationship between turnout and education. If the probability is high, we should remain cautious.

We need the following numbers to make this calculation: the coefficient $\hat{\beta}_1$ and its standard error. Since we have $\hat{\beta}_1$, we only need to calculate its standard error.

1. Calculate the residuals, the distance between the actual values Y_i and our prediction $\hat{Y}_i$. Sum the square of each N residual:

$$\Sigma\left(Y_i - \hat{Y}_i\right)$$

2. Then take the sum of the squared residuals and divide by N − 2 (the degrees of freedom)[6] to obtain the residual variation:

$$S^2_{Y|X} = \frac{\sum(Y_i - \hat{Y}_i)^2}{N-2}$$

3. With $S^2_{Y|X}$, calculate the standard error of the estimate:

$$S_{\hat{\beta}_1} = \frac{S_{Y|X}}{\sqrt{\sum(X_i - \bar{X})^2}}$$

With the standard error of the estimate in hand, we make the final calculation. The **t-ratio (for a regression coefficient)** provides an explicit test of how likely the real relationship is zero given our estimate. The t-ratio is obtained by dividing the estimate by its standard error:

$$\text{t-ratio} = \frac{\hat{\beta}_1}{S_{\hat{\beta}_1}}$$

Now that we know how to calculate the t-ratio, let's examine its properties. The larger the t-ratio associated with our estimate of the slope, the more confidence we have that the true slope of the line is not 0. Connecting this discussion back to random samples and confidence intervals (Chapter 9), the "true" slope in this context is synonymous with the population parameter and our estimate of it is synonymous with the sample statistic. Absolute values above 2 (they can be positive or negative) exceed conventional levels of confidence.[7]

Whether the t-ratio is greater than or less than the absolute value of 2 depends on what's in the numerator and denominator. The standard error of the estimate is in the denominator, meaning more variation in the data—large residuals—will result in a denominator that could be large relative to the numerator. Consequently, greater variation in the data will produce smaller t-ratios.

If the residuals $\sum(Y_i - \hat{Y}_i)$ are large, then the residuals' variance ($S^2_{Y|X}$) will be large. The larger the variance of the residuals $S_{\hat{\beta}_1}$, the larger the standard error of the estimate. The larger the standard error of the estimate $S_{\hat{\beta}_1}$, the smaller the resulting t-ratio. Run through each calculation to see how a change in one calculation affects the other.

Note what happens when we increase the number of observations (N). As N gets larger (as the number of our observations grows), the standard error decreases; in turn, this increases the t-ratio. Stated differently, as we get more information (increase N), the confidence in our estimates grows (t-ratio increases).

Art and Practice of Data Visualization
WITH MORE INFORMATION, CONFIDENCE INCREASES

It is worth emphasizing what the paragraph on the number of observations implies. It shows mathematically that as we gain more information (more data), our confidence in the results increases.

---

[6] The degrees of freedom are simply the number of observations less the number of variables in the model. Review the discussion of degrees of freedom in Chapter 9.
[7] The standard level (though somewhat arbitrary) is the 95% level of confidence. For a review of confidence levels, see the discussion in Chapter 9.

> **KNOWLEDGE CHECK: Describe statistical significance.**

9. How do the regression estimates (the slope and the intercept) parallel the concepts learned in Chapter 9?
    a. Regression estimates are like sample statistics.
    b. Regression estimates are like confidence intervals.
    c. Regression estimates indicate the level of significance.
    d. Regression estimates are influenced by N.

10. Which of the following statements are true.
    a. As N increases, the absolute value of the t-ratio decreases.
    b. As the standard error of the estimate increases, the t-ratio increases.
    c. As the magnitude of the coefficient increases, the t-ratio increases.
    d. As N increases, the absolute value of the t-ratio increases.

11. Which things in the regression context are synonymous with sample variance in the context of taking random samples?
    a. Random events that influence the regression estimate
    b. Mistakes made in collecting the data
    c. How changes in x influence changes in y
    d. Variance in the residuals

Examples of Bivariate Regressions

In the following examples, we learn how to interpret regression coefficients in a number of different contexts and with a number of different units of measurement. In terms of context, have the dependent or independent variables been transformed? In terms of measurement, how do we interpret variables measured in percentages, indices, or other units? We focus on the substantive significance of the relationship—the slope of the line. Unless we know the units of measure and whether the variables have been transformed, we can't understand the substantive significance of the relationship.

Variables are measured in different ways. Many variables (like *hsdiploma* and *democrat*) are measured in percentages. Other measures use indices, artificial constructs that convey an underlying concept like democracy (*Polity2* in the *world* data set) or political ideology (*pid* in the *NES* data set). Although indices and percentages are common, units come in many other forms: homicides per 100,000 population, people per square mile, or the number of infant deaths per 1,000 live births. To understand the substantive significance of regression results, we have to know the units of measurement.

The following series of bivariate regressions demonstrate how to correctly interpret the regression coefficient, the t-ratio, and the R^2 statistic.

Does the Level of Religiosity in a State Influence Abortion Rates?

Both Dependent and Independent Variables Are in Their Raw Form

In this example, both the dependent variable and the independent variable are in their raw form. Interpreting the coefficient is straightforward in this case: a 1 percentage point increase in the population identifying as evangelical *is associated with* a decrease of .3 abortions per 1,000 women (Table 12-2).

In Code Chunk 12-7 I define a regression model *abort.lm* and then place that object in a stargazer table. The corresponding scatter plot and regression line are provided in Figure 12-5.

TABLE 12-2 Evangelism Discourages Abortions

	Dependent variable:
	abort
evangel	−0.301***
	(0.089)
Constant	23.745***
	(2.105)
Observations	50
R2	0.191
Adjusted R2	0.174
Residual Std. Error	8.290 (df = 48)
F Statistic	11.324*** (df = 1; 48)
Note:	*p<0.1; **p<0.05; ***p<0.01

Code Chunk 12-7

```
abort.lm <- lm(abort ~ evangel, data = states)
stargazer(abort.lm, type = "text", main = "Table 12-2:
     Evangelism Discourages Abortions", header = FALSE)

ggplot(states, aes(evangel, abort)) +
    geom_point(col = "#bf0000") +
    geom_text_repel(size = 3, aes(label = st, size = 1,
        hjust = 0, vjust = -1), col = "grey", show.legend = FALSE) +
    ggtitle("Figure 12-5: Religion Reduces Abortion") +
    geom_smooth(method = "lm", se = F, fullrange = F,
                col = "#0000bf") +
    theme_minimal() +
    theme(plot.title = element_text(size = 8, face = "bold"),
        axis.title = element_text(size = 8, face = "bold")) +
    ylab("Abortions per 1,000 Women") +
    xlab("Percent of Population That Is Evangelical")
```

■ **FIGURE 12-5** Religion Reduces Abortion

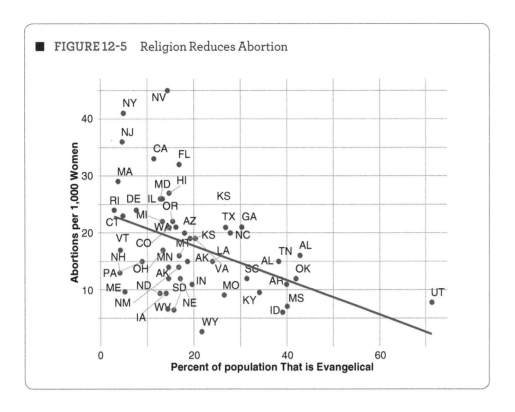

To calculate the t-ratio, we divide the estimated coefficient (−.301) by the standard error of the estimate (.089). The t-ratio for the estimate is −3.37. Since the t-ratio is greater than the absolute value of 2, the estimate is statistically significant.

Art and Practice of Data Visualization
STANDARD ERRORS OF THE ESTIMATE

Many software programs provide the standard errors of the estimate in parentheses just below or alongside the estimate. Others put the t-ratios themselves in parentheses. The convention I use here places the standard error of the estimates in parentheses. Many analysts, myself included, provide asterisks adjacent to the estimate to indicate the level of statistical significance. Usually one asterisk refers to .10, two refers to .05, and three refer to .01 (the 90%, 95%, and 99% levels of confidence). Some refer to the asterisks as stars, hence the name of the R 'stargazer' package that generates regression tables.

We can be confident that the slope of the line is not 0. Stated differently, even after taking into consideration possible random events, it's unlikely that we would get a coefficient of −.301 on evangelism if the true value were zero. The R^2 statistic is .19, indicating that we've managed to explain 19% of the total variance in abortion rates with one variable (evangelism).

Does Religiosity Affect Gun Laws?

Dependent Variable Is an Index

Here we regress an index that measures the freedom to own, sell, and carry guns (*gunfree* in the *states* data set) on the percentage of the state's population that is evangelical. Table 12-3 indicates that each percentage point increase in the evangelical population *is associated with* a .032 point increase in the gun freedom scale.

Code Chunk 12-8

```
gun.lm <- lm(gunfree ~ evangel, data = states)
stargazer(gun.lm, type = "text", main = "Table 12-3: Religiosity is
related to Gun Laws",
     header = FALSE)

ggplot(states, aes(evangel, gunfree)) +
     geom_point(col = "red") +
     geom_text_repel(size = 3, aes(label = st, size = 1,
          hjust = 0, vjust = -1), col = "grey",show.legend = FALSE) +
     ggtitle("Figure 12-6: Religion Bolsters Gun Freedom") +
     geom_smooth(method = "lm", se = F, fullrange = F) +
     theme_minimal() +
     theme(plot.title = element_text(size = 8, face = "bold"),
          axis.title = element_text(size = 8, face = "bold")) +
     ylab("Gun Freedom Index") +
     xlab("Percent of Population that is Evangelical")
```

TABLE 12-3 Religiosity Is Related to Gun Laws

	Dependent variable:
	gunfree
evangel	0.032***
	(0.010)
Constant	−0.606**
	(0.239)
Observations	50
R2	0.171
Adjusted R2	0.154
Residual Std. Error	0.942 (df = 48)
F Statistic	9.909*** (df = 1; 48)
Note:	*p<0.1; **p<0.05; ***p<0.01

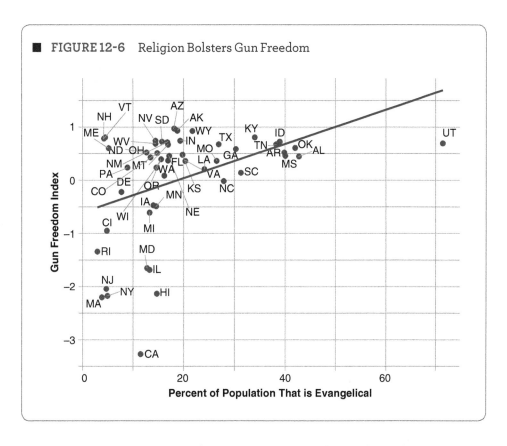

■ FIGURE 12-6 Religion Bolsters Gun Freedom

In this case, the estimated coefficient (.032) is statistically significant: the t-ratio is 3.15. Again, accounting for random events, it is unlikely our regression would generate a coefficient of .032 if there was no relationship between religion and gun freedom. The R^2 statistic is .17, indicating that our one-variable model explains 17% of the variation in gun freedom.

Are Ethnically Heterogeneous Countries Violent?

Dependent Variable Is Logged

Our dependent variable here is the homicide rate, which gives the number of homicides in each country per 100,000 people. Since in its raw form the variable is positively skewed, we take logs. Note that in the code, we take $log(homicide+1)$. By adding 1, we add a constant to the homicide variable so that there are no zero values.[8] If we were to take the log of 0, the value for that case would be undefined. Undefined cases are dropped from the regression and cause errors that prevent R from making the desired calculations.

The independent variable is ethnolinguistic fractionalization (commonly referred to as the ELF score), which indicates how likely two individuals chosen at random speak the same language or have the same religion. This scale ranges from 0 to 1, with 1 being extremely heterogeneous. Note that in Code Chunk 12-9 I transformed ELF to range between 0 and 100 to make interpretation easier. When the dependent variable is logged, an increase of one unit in the ELF score is associated with a $\left(e^{(beta)}-1\right)\times 100$ percentage change in homicides per 100,000 population. The estimates in Table 12-4 indicate our beta is .009. Therefore, a 1 point increase in the ELF score *is associated with* a .9% change in homicides per 100,000 population.

---

[8] If our raw data were skewed but ranged between 0 and 1, we could use .001 as our constant instead of 1.

Note that in Code Chunk 12-9 I transform the variable within the *lm()* command. This is a very helpful feature since we can forego the added step of creating a new variable. I also used a different method to label a select number of cases. Instead of using the vertical bar (|) operator in the *ifelse()* function, I use the operator %in% and specify a list of countries: United States (USA), Honduras (HND), Afghanistan (AFG), and Germany (DEU). If the labels variable (*world$iso3c*) meets that condition, the case will be labeled according to *world$country*, their country name.

Code Chunk 12-9

```
world$elf <- world$ethfrac * 100

murder.lm <- lm(log(homicide + 1) ~ elf, data = world)

stargazer(murder.lm, type = "text", title = "Table 12-4:
        Ethnic Heterogeneity and Violence Are Related",
        header = FALSE)

ggplot(world, aes(elf, log(homicide + 1))) +
    geom_point(col = "#bf0000") +
    geom_text_repel(size = 3, aes(label =
        ifelse(world$iso3c %in% c("USA", "HND", "AFG", "DEU"),
            as.character(world$country), ""),
            size = 1, hjust = 0, vjust = 1), col = "grey",
            show.legend = FALSE) +
    ggtitle("Figure 12-7: Heterogeneous Populations Spur Homicide") +
    geom_smooth(method ="lm", se = F, fullrange = F, col = "#0000bf") +
    theme_minimal() +
    theme(plot.title = element_text(size = 8, face = "bold"),
        axis.title = element_text(size = 8, face = "bold")) +
    ylab("Homicides per 100,000 People") +
    xlab("Ethnolinguistic Fractionalization")
```

The estimate has a t-ratio of 2.15, which is statistically significant: the likelihood of generating a coefficient of .009 when the relationship is actually 0 is fairly small. The R^2 statistic indicates we've explained 5% of the variation (.046 to be exact). So while there is a statistically significant relationship between the ELF score and homicide, we really haven't explained much of the variation in homicide rates throughout the world. There's clearly more work to do.

Art and Practice of Data Visualization
THE .LM SUFFIX FOR LINEAR MODELS

Note that I created names for each regression object with the ".lm" suffix in the last few code chunks. By placing ".lm" at the end of every object name defined by a linear model, I can keep my objects straight. As more and more objects are created in a session, having useful suffixes can be extremely helpful.

TABLE 12-4 Ethnic Heterogeneity and Violence Are Related

	Dependent variable:
	log(homicide + 1)
elf	0.009**
	(0.004)
Constant	1.223***
	(0.168)
Observations	122
R2	0.046
Adjusted R2	0.038
Residual Std. Error	0.974 (df = 120)
F Statistic	5.835** (df = 1; 120)
Note:	*p<0.1; **p<0.05; ***p<0.01

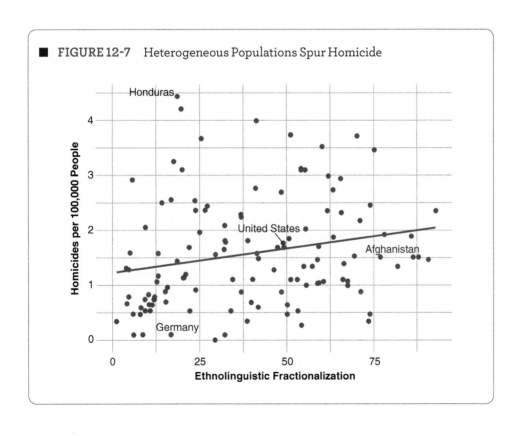

FIGURE 12-7 Heterogeneous Populations Spur Homicide

Does Violence Cause Political Instability?

Both Dependent and Independent Variables Are Logged

In this case, both the dependent variable and independent variable are positively skewed. The measure for political instability is the number of years a country's polity score has not changed by more than 3 points (the *durable* variable in the *world* data set). The homicide variable, as we observed in the last example, is also positively skewed. When the dependent variable and independent variables are logged, a 1% change in the independent variable is associated with an *X* percent change in the dependent variable. In our example, a 1% change in homicide rates is associated with a *beta* percent change in years of stability. Economists refer to this as an elasticity (the percent change associated with a percent change). In this example, a 1% increase in the homicide rate is associated with a .17% increase in the years since a significant change in the regime (Table 12-5).

Code Chunk 12-10

```
violence.lm <- lm(log(durable + 1) ~ log(homicide + 1), data = world)
stargazer(violence.lm, type = "text", title =
        "Table 12-5: Political Stability and Homicide",
            header = FALSE)

ggplot(world, aes(log(homicide + 1), log(durable + 1))) +
    geom_point(col = "#bf0000") +
    geom_text_repel(size = 3, aes(label =
        ifelse(world$iso3c %in% c("USA", "HND", "AFG", "SWE"),
        as.character(world$country), ""), size = 1, hjust = 1,
            vjust = 1), col = "grey", show.legend = FALSE) +
    ggtitle("Figure 12-8: Murder has Little Effect") +
    geom_smooth(method = "lm", se = F, fullrange = F, col= "#0000bf") +
    theme_minimal() +
    theme(plot.title = element_text(size = 8, face = "bold"),
        axis.title = element_text(size = 8, face = "bold")) +
    ylab("Years Since Significant Regime Change (logged)") +
    xlab("Homicide Rate (logged)")
```

The t-ratio is −1.5, indicating that at conventional levels of significance (below an absolute value of 2), we are not confident the estimate is different than zero. In other words, relative to the previous cases where the t-ratio was above the absolute value of 2, there's a higher probability that we could have a coefficient of .17 even when the real relationship is zero. The R^2 is also small—we're only explaining 2% of the total variance in the political stability.

Is Income (GDP per Capita) Related to Turnout?

Independent Variable Is Logged

In this example, we regress turnout on GDP per capita to understand the relationship between income levels and political participation. Since GDP per capita is positively skewed, we transform it by taking logs. Here we can make a choice whether we want to log it to the base 10 or to the base 2. If logged to the base 10, our interpretation is that a 10-fold increase in the independent variable is associated with a *beta* percentage point increase in turnout. When the independent variable is logged to the base 2, doubling the size of the independent variable is associated

TABLE 12-5 Political Stability and Homicide

	Dependent variable:
	log(durable + 1)
log(homicide + 1)	−0.173
	(0.112)
Constant	3.153***
	(0.211)
Observations	121
R2	0.020
Adjusted R2	0.012
Residual Std. Error	1.207 (df = 119)
F Statistic	2.405 (df = 1; 119)
Note:	*p<0.1; **p<0.05; ***p<0.01

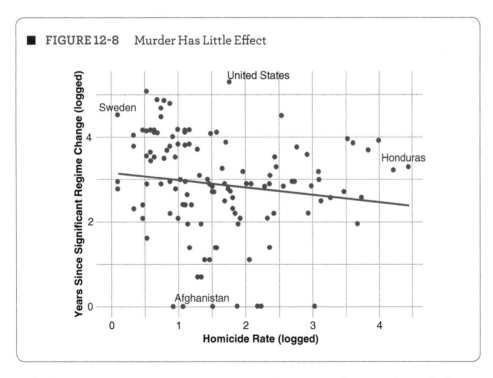

FIGURE 12-8 Murder Has Little Effect

with a *beta* percentage point increase in turnout. In our example, we use log to the base 2. Doubling GDP per capita is associated with a 1 percentage point (.99) increase in turnout, not much of an impact for such a significant change in income (Table 12-6).

As with the previous example, the t-ratio associated with our estimate is under the absolute value of 2—it is 1.34. The estimate is not statistically significant so the true slope

Code Chunk 12-11

```
wealth.lm <- lm(turnout ~ log2(gdppc), data = world)

stargazer(wealth.lm, type = "text", title =
          "Table 12-6: Income's Meek Influence on Turnout",
          header = FALSE)
ggplot(world, aes(log2(gdppc), turnout)) +
    geom_point(col = "#bf0000") +
    geom_text_repel(size = 3, aes(label =
        ifelse(world$iso3c %in% c("USA", "LUX", "RWA", "HTI"),
            as.character(world$country), ""), size = 1,
        hjust = 0, vjust = 1), col = "grey", show.legend = FALSE) +
    ggtitle("Figure 12-9: Income Can't Move Turnout") +
    geom_smooth(method = "lm", se = F, fullrange = F) +
    theme_minimal() +
    theme(plot.title = element_text(size = 8, face = "bold"),
          axis.title = element_text(size = 8, face = "bold")) +
    ylab("Voter Turnout") +
    xlab("GDP per Capita (logged)")
```

could be 0. As with the previous example, it's likely we could have an estimate of .99 even when the real relationship is 0. The R^2 is .01: GDP per capita explains only 1% of the variation in voter turnout.

TABLE 12-6 Income's Meek Influence on Turnout

	Dependent variable:
	turnout
log2(gdppc)	0.943
	(0.745)
Constant	53.911***
	(9.804)
Observations	150
R2	0.011
Adjusted R2	0.004
Residual Std. Error	15.912 (df = 148)
F Statistic	1.599 (df = 1; 148)
Note:	*p<0.1; **p<0.05; ***p<0.01

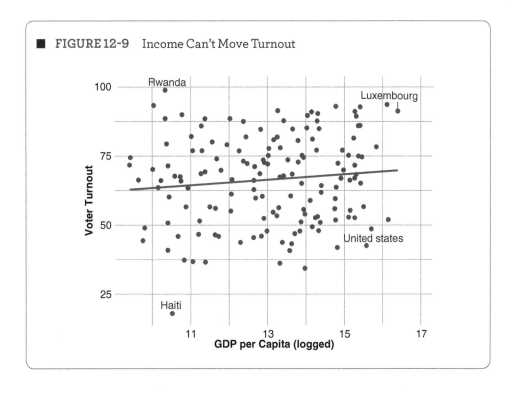

■ FIGURE 12-9 Income Can't Move Turnout

KNOWLEDGE CHECK: Interpret coefficients in a variety of contexts.

12. Identify the correct interpretation when the dependent variable is logged.
    a. A unit change in x is associated with a $\left(e^{(beta)}-1\right)\times 100$ percentage change in y.
    b. A percentage change in x is associated with a z percentage change in y.
    c. A 1 percentage point change in x is associated with an x point change in y.
    d. A 1 point change in x is associated with a unit change in y.

13. Identify the correct interpretation when both the dependent and independent variables are logged.
    a. A unit change in x is associated with a $\left(e^{(beta)}-1\right)\times 100$ percentage change in y.
    b. A percentage change in x is associated with a z percent change in y.
    c. A 1 percentage point change in x is associated with an x point change in y.
    d. A 1 point change in x is associated with a unit change in y.

14. Identify the correct interpretation when the independent variable is logged.
    a. A unit change in x is associated with a $\left(e^{(beta)}-1\right)\times 100$ percentage change in y.
    b. A percentage change in x is associated with a z percent change in y.

c. A 1 percentage point change in x is associated with an x point change in y.

d. None of the above

15. Identify the correct interpretation when the independent variable is an index.

    a. A unit change in x is associated with a $\left(e^{(beta)} - 1\right) \times 100$ percentage change in y.

    b. A percentage change in x is associated with a z percentage change in y.

    c. A 1 percentage point change in x is associated with an x point change in y.

    d. A 1 point change in x is associated with a z unit change in y.

Art and Practice of Data Visualization
BIVARIATE REGRESSIONS AND MULTIPLE REGRESSION

With the mechanics and conceptual scaffolding in hand for simple bivariate regression, moving to multiple regression is fairly straightforward. Analogous to the difference between comparisons and controlled comparisons, moving from bivariate regression to multiple regression allows us to introduce more possible causes into our explanation or theory. In terms of interpreting results from multiple regression analysis, the only difference is that after we state how a one-unit change in x is associated with a unit change in y, we state "controlling for the other variables included in the model."

SUMMARY

Regression analysis is one of the most powerful analytical tools at our disposal. Familiarity with the building blocks presented in this chapter will prove invaluable. A word of caution is, however, in order. Regression analysis is a powerful tool, but it must be used with care. As we proceed through the book, you'll learn more about its limitations and how it can be misused.

One simple way to avoid misuse is to describe regression results in exacting terms. For example, if we estimate a coefficient that is substantively significant—meaning a unit change in *x is associated with* a meaningful change in *Y*—we can say a linear relationship exists between our dependent variable *Y* and our independent variable *X*. We don't know if *x* causes *y*, but we have some circumstantial evidence that a change in *x* could cause a change in *y*. The following chapters explain why caution is important. They also demonstrate how regression analysis can be an important diagnostic and exploratory tool, never settling the matter but generating new and more informed questions.

COMMON PROBLEMS

- *Understanding the difference between substantive and statistical significance.* Although it's not difficult to understand what the two kinds of significance refer to, it's easy to get fixated on one and not the other. In particular, analysts get drawn into whether the t-ratio on an estimate is significant. They focus with laser-like determination on statistical significance only to find that after they've completed the study and written up the report, the substantive significance is actually fairly minimal.

- *Using the language of regression (regressing y on x).* This language is counterintuitive since we usually regard the independent variable x as having an impact on y. To therefore describe a regression as regressing y on x is not intuitive. However, like many things in statistics, being very exacting with language and following accepted practice is important. Therefore, we say that we regress y on x.

- *Paying attention to the units of measurement.* For reasons still unknown to the author, many beginners interpret coefficients as if all variables were measured in percentages. Even though an independent variable is measured in dollars per capita, the novice will still regard a unit change in that variable as a percentage change. I can't emphasize enough the importance of paying attention to units of measurement. Beginning analysts that struggle with this evince a level of impatience that shows up in many different contexts (not just statistical).

REVIEW QUESTIONS

1. What are the benefits of regression analysis?
2. What does the slope of the line convey in regression analysis?
3. What does the intercept mean in regression analysis?
4. What does the R^2 indicate?
5. What is the t-ratio?
6. How does the number of observations in a regression influence the t-ratio?
7. What happens to the standard error of the estimate when the residuals increase? How does that impact the t-ratio?
8. How do we interpret the coefficient when the dependent and independent variables are both logged?
9. Why do we often have to add a constant to a variable before taking logs?
10. What do we need to know in order to understand the substantive significance of the relationship between the dependent and independent variables?

PRACTICE ON ANALYSIS AND VISUALIZATION

1. Which of the following accurately describe comparison and linear regression?
   a. One involves fitting a line, the other doesn't.
   b. One involves specifying causation, the other doesn't.
   c. One involves making a prediction, the other doesn't.
   d. Only one involves analyzing the relationship between two variables.

2. Which of the following describe linear regression?
   a. The change in y is associated with a one-unit change in x.
   b. The mean of x and mean of y are integral parts of regression calculations.
   c. Goodness of fit is influenced by the size of the residuals.
   d. Goodness of fit is influenced by the variation in x.

3. Which of the following accurately describes the R^2 statistic?
   a. It indicates whether our regression results are different from zero.
   b. It represents the ratio of explained variance over total variance.
   c. Any R^2 statistic that is greater than .50 is significant.
   d. It is influenced by the spread of the dependent variable.

4. What does the goodness of fit tell us?

   a. It tells us how much confidence we can have in the results.
   b. It tells us how well our model explains the dependent variable.
   c. It gives the statistical significance of our estimates.
   d. It tells us how well our dependent variable explains the independent variable.

5. Which of the following statements accurately describe the t-ratio?

   a. It gives an indication of how much variation the associated variable explains.
   b. It provides a useful measure of statistical significance.
   c. A large t-ratio confirms that x causes y.
   d. A large t-ratio indicates that the coefficient we generated is not simply different than zero because of random chance.

6. Which of the following statements are accurate?

   a. Statistical significance means there will be substantive significance.
   b. Substantive significance always means there will be statistical significance.
   c. There is no relationship between substantive significance and statistical significance.
   d. Analysts often pay too much attention to statistical significance.

7. Regress income on education in the United States.

   a. Interpret the slope coefficient on the independent variable.
   b. Draw a scatter plot and regression line that summarizes the relationship.
   c. What is the R^2 statistic?
   d. What is the t-ratio for the independent variable?

8. What influence does political stability have on GDP per capita? To answer that question, use logged versions of each variable.

   a. Interpret the slope coefficient on the independent variable.
   b. Draw a scatter plot and regression line that summarizes the relationship.
   c. What is the R^2 statistic?
   d. What is the t-ratio for the independent variable?

9. How much does spending on education influence crime (homicides) in the United States?

   a. Interpret the slope coefficient on the independent variable.
   b. Draw a scatter plot and regression line that summarizes the relationship.
   c. What is the R^2 statistic?
   d. What is the t-ratio for the independent variable?

10. Does having a youthful population cause homicide rates to increase? To answer the question use the logged version of the homicide variable.

    a. Interpret the slope coefficient on the independent variable.
    b. Draw a scatter plot and regression line that summarizes the relationship.
    c. What is the R^2 statistic?
    d. What is the t-ratio for the independent variable?

ANNOTATED R FUNCTIONS

The following functions appear in this chapter. They are listed in order of their first appearance (with the code chunk number in parentheses) and annotated here to give a very brief description of their use. Some are not stand-alone functions and only work in combination with other commands. As a reminder, the code in every chapter will work properly if executed in the order it appears. Proper execution also depends on typing the author-defined *libraries()* command, which loads the required R packages.

ggplot(): defines the basic structure of a plot (usually the x and y variables). (12-1)

aes(): the aes (called "aesthetics") function is used in ggplot to define the basic structure of the plot, which often includes the variables you want to use and any shapes or colors. (12-1)

geom_text_repel(): labels points in a scatter plot in the ggplot framework. (12-1)

ifelse(): logical function that allows you to construct an if-then statement. Useful for selecting specific cases to label and also for creating categorical variables from continuous variables. (12-1)

geom_smooth(): draws lines or curves to a scatter plot. (12-1)

xlab(): labels the x axis in ggplot. (12-1)

ylab(): labels the y axis in ggplot. (12-1)

ggtitle(): provides the title for a ggplot. (12-1)

geom_point(): draws points on a ggplot grid. (12-1)

theme_minimal(): specifies a minimalist style for ggplot. (12-1)

theme(): specifies font, size, and so forth in a ggplot. (12-1)

stargazer(): generates a regression table with specified model. (12-2)

annotate(): places lines or text in the figure. (12-3)

grid.brackets(): places curly brackets in specified locations in the figure. (12-3)

grid.arrange(): allows you to arrange multiple figures on a page. (12-6)

xlim(): specifies the range of the x axis. (12-4)

rnorm(): randomly draws from a normal distribution. (12-5)

ANSWERS

KNOWLEDGE CHECK

1. a, c, d
2. a, b
3. b, d
4. $\hat{Y}$ (a), α_i (b), β (c), x_i (d)
5. d
6. a, b, c
7. increases (a), nothing (b), increases (c), increases (d)
8. a
9. a
10. c, d
11. a, b
12. a
13. b
14. d
15. d

PRACTICE ON ANALYSIS AND VISUALIZATION

1. b, c
2. a, b, c
3. b, d
4. b
5. b, d
6. d

7. A 1 percentage point increase in the size of the population with a high school degree is associated with a $554 increase in per capita income (a); scatter plot (b); .14 (c); 2.81 (d)

```
income.lm <- lm(inc ~ hsdiploma, data = states)

stargazer(income.lm, type = "text", title =
    "Education Increases Income", header = FALSE

ggplot(states, aes(hsdiploma, inc)) +
    geom_point(col = "#bf0000") +
    ggtitle("Income and Education in the United States") +
    geom_smooth(method = "lm", se = F, fullrange = F,
        col = "#0000bf") +
    theme_minimal() +
    theme(plot.title = element_text(size = 8, face = "bold"),
        axis.title = element_text(size = 8, face = "bold")) +
    ylab("Median per Capita Income") +
    xlab("Percentage of Population With High School Education")
```

```
Education Increases Income
=============================================================
                                         Dependent variable:
                                     ------------------------
                                                          inc
-------------------------------------------------------------
hsdiploma                                           553.929***
                                                    (196.954)

Constant                                           -6,626.114
                                                  (16,853.690)

-------------------------------------------------------------
Observations                                                50
R2                                                       0.141
Adjusted R2                                              0.124
Residual Std. Error                        5,459.255 (df = 48)
F Statistic                              7.910*** (df = 1; 48)
=============================================================
Note:                            *p<0.1; **p<0.05; ***p<0.01
```

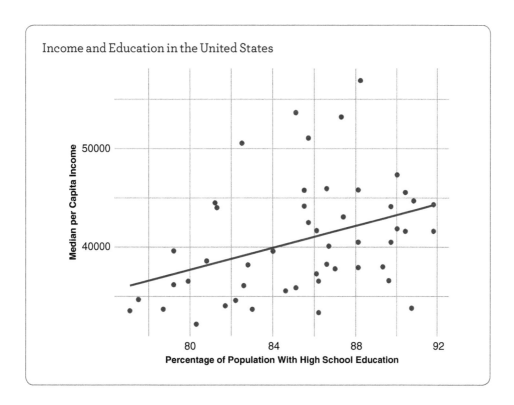

8. A 1% change in income is associated with a .52% change in the number of years of stability (a); scatter plot (b); .30 (c); 8.125 (d)

```
develop.lm <- lm(log(durable + 1) ~ log(gdppc), data = world)

stargazer(develop.lm, type = "text", main = "Political Stability and
Income",
    header = FALSE)

ggplot(world, aes(log(durable + 1), log(gdppc))) +
    geom_point(col = "#bf0000") +
    ggtitle("Political Stability and Income") +
    geom_smooth(method = "lm", se = F, fullrange = F,
        col = "#0000bf") +
    theme_minimal() +
    theme(plot.title = element_text(size = 8, face = "bold"),
        axis.title = element_text(size = 8, face = "bold")) +
    ylab("GDP per Capita (logged)") +
    xlab("Number of Years Since Political Change (logged)")
```

Political Stability and Income

	Dependent variable:
	log(durable + 1)
log(gdppc)	0.520***
	(0.064)
Constant	−1.929***
	(0.587)

Observations	154
R2	0.301
Adjusted R2	0.297
Residual Std. Error	0.990 (df = 152)
F Statistic	65.527*** (df = 1; 152)

Note: *p<0.1; **p<0.05; ***p<0.01

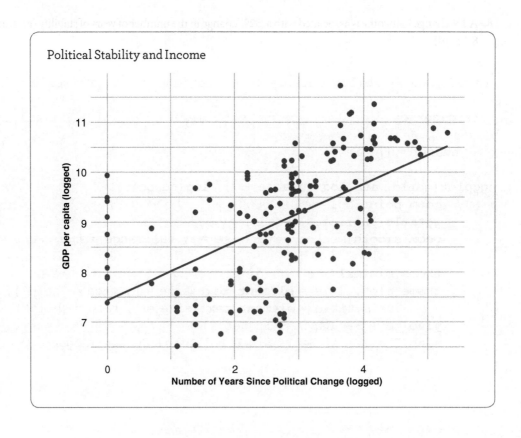

9. A $1 increase in spending per student is associated with a .0004 decrease in murders per 100,000 people (a); scatter plot (b); .08 (c); 2.04 (d)

```
murder.lm <- lm(murderrate ~ stuspend, data = states)

stargazer(murder.lm, type = "text", title =
          "Spend on Students to Save on Police", header = FALSE)

ggplot(states, aes(stuspend, murderrate)) +
    geom_point(col = "#bf0000") +
    ggtitle("Murder and Education in the United States") +
    geom_smooth(method = "lm", se = F, fullrange = F,
          col = "#0000bf") +
    theme_minimal() +
    theme(plot.title = element_text(size = 8, face = "bold"),
          axis.title = element_text(size = 8, face = "bold")) +
    ylab("Homicide Rate") +
    xlab("Dollars per Student")
```

```
Spend on Students to Save on Police
===============================================
                           Dependent variable:
                          ---------------------
                                 murderrate
-----------------------------------------------
stuspend                          -0.0004**
                                  (0.0002)

Constant                          6.724***
                                  (1.238)

-----------------------------------------------
Observations                         50
R2                                 0.080
Adjusted R2                        0.061
Residual Std. Error           2.200 (df = 48)
F Statistic                 4.174** (df = 1; 48)
===============================================
Note:                *p<0.1; **p<0.05; ***p<0.01

=======================================
Spend on Students to Save on Police
---------------------------------------
```

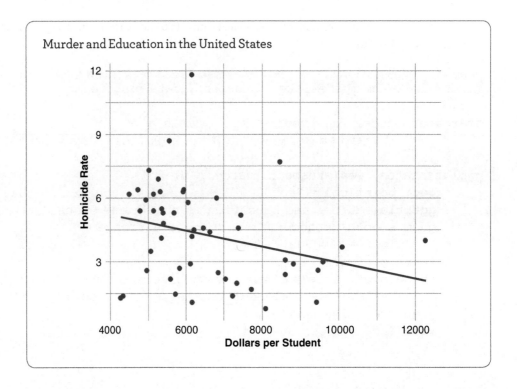

10. A 1 percentage point increase in the population aged 0–14 years is associated with a 6% increase in homicides (a); scatter plot (b); .18 (c); 4.9 (d)

```
youth.lm <- lm(log(homicide) ~ young, data = world)

stargazer(youth.lm, type = "text", title =
            "Youth and Homicide", header = FALSE)

ggplot(world, aes(young, log(homicide))) +
    geom_point(col = "#bf0000") +
    ggtitle("Youth Kills") +
    geom_smooth(method = "lm", se = F, fullrange = F,
                col = "#0000bf") +
    theme_minimal() +
    theme(plot.title = element_text(size = 8, face = "bold"),
          axis.title = element_text(size = 8, face = "bold")) +
    ylab("Homicide (murders per 100,000; logged)") +
    xlab("Percentage of Population Between 0 and 14 Years")
```

Youth and Homicide

	Dependent variable:
	log(homicide)
young	0.058***
	(0.012)
Constant	-0.499
	(0.349)
Observations	114
R2	0.177
Adjusted R2	0.170
Residual Std. Error	1.279 (df = 112)
F Statistic	24.125*** (df = 1; 112)
Note:	*p<0.1; **p<0.05; ***p<0.01

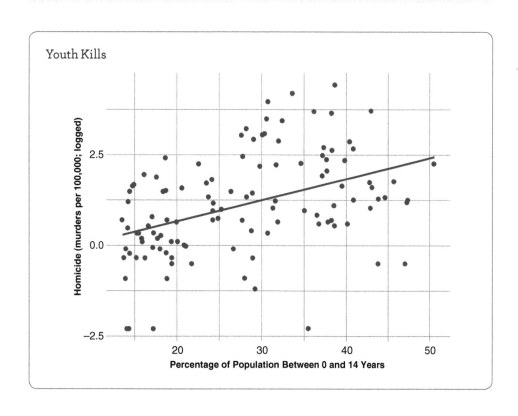

 Access digital resources, including datasets, at http://edge.sagepub.com/brownstats1e.

13 Multiple Regression

CHAPTER OUTLINE

Learning Objectives
Overview
What Is Multiple Regression?
Regression Models and Arguments
Regression Models, Theory, and Evidence
Interpreting Estimates in Multiple Regression
Example: Homicide Rate and Education
Summary
Common Problems
Review Questions
Practice on Analysis and Visualization
Annotated R Functions
Answers

LEARNING OBJECTIVES

- Define multiple regression.
- Identify the parallels between a model and an argument.
- Discuss the relationship between theory and evidence.
- Interpret coefficients, t-ratios, and measures of fit.
- Employ multiple regression analysis to examine a problem.

Overview

Multiple regression allows the incorporation of additional variables into our explanation. As was demonstrated in Chapter 11 (controlled comparisons), the relationship observed between two variables can change dramatically when we account for a third. In multiple regression, there is an independent variable of interest for which we want to obtain an accurate estimate. Toward that end, we formulate a convincing model, controlling for other variables we think matter. Arguments and opinions change when confronted with additional evidence. Arguments and opinions change when our regression model is confronted with additional variables.

This presentation of multiple regression analysis draws parallels between the arguments we make and their representation both mathematically and with code. A multiple regression model is simply the formalization of an argument that accounts for more than one variable. Translating arguments from English into mathematical expressions and code is the goal.

Regression models can take on many different sophisticated and exotic forms. In this chapter, simple linear regression models are the focus. We assume a linear relationship exists between the independent variables and the dependent variable. We also assume independence: the independent variables influence the dependent variable but not each other. For example, if we include education and income in a simple multiple regression model to explain crime, we assume education has no impact on crime through income. Since we know education does, in fact, have an important impact on income, assuming independence among the independent variables is fairly heroic. Fortunately, variants of the simple regression model provide some flexibility. The next chapter on dummies and interactions allows us to relax the assumption of independence.

Constructing models and estimating them is an iterative process in exploratory data analysis (EDA). There is, as a result, a back-and-forth between theory and evidence. Since *how* we look at things (the models we use) influences *what* we observe (the results we interpret), we need to approach the enterprise with honesty and humility. Examples in this chapter will show that the best way to understand a problem is to be honest about inconvenient results and be humble enough to admit when you're wrong.

Understanding the basics of multiple regression analysis includes learning its limitations, extracting information from the data, and communicating the results. Multiple regression analysis comes with an important warning: like operating a motor vehicle, things could go terribly wrong if you don't know what you're doing. When driving a car, we first check the oil light, glance at the gas gauge, and check all of the mirrors—important routines all performed before leaving the driveway. This chapter and the remainder of the book provides the checklist for regression analysis.

To some extent we've already been conducting multiple regression through graphical techniques, analyzing a relationship while holding other variables constant. With multiple regression we can expand the number of independent variables, accounting for a wider range of phenomena. This chapter offers an example of how to formulate regression models, interpret the estimates, and communicate the results.

What Is Multiple Regression?

In Chapter 12 we estimated the linear relationship between an independent variable and a dependent variable. Estimating a bivariate regression gave us the equation of a line: the slope and intercept. It also gave us a summary of how well the line fit the data and provided a sense of how confident we could be in the results.

Multiple regression analysis is a straightforward extension of bivariate regression. Similar to controlled comparisons, it allows us to account for additional variables when summarizing the relationship between two variables. For example, to explain a country's level of wealth, there are a number of causes to consider (e.g., institutions, culture, history, geography, etc.). Multiple regression analysis allows us to estimate the relationship between two variables while accounting for a number of others. The more we account for relevant variables, the more accurate are the estimates. For example, our estimate of democracy's relationship with GDP per capita will be more accurate if we account for culture, history, and geography—just to name a few.

With multiple regression, we can calculate the variation in the dependent variable our model explains (the same R^2 we calculated in the bivariate case). We can also calculate the statistical significance of our estimates, establishing what level of confidence we can have in the results.

Because with multiple regression we control for additional variables, we say "a unit change in x is associated with a change in y, *holding the other variables constant at their means.*"

Why Multiple Regression?

Multiple regression is one of the most valuable tools social scientists have at their disposal since it estimates the relationship between the dependent variable and a number of independent variables simultaneously. This is particularly useful since most social phenomena have many causes.

First consider a controversial example from public health. Regardless of their stance on abortion, both citizens and policymakers want effective policy. In a previous chapter, we observed there is a relationship between abortion rates, population density, religion, and political ideology. Because we only have a limited amount of resources, we need to understand exactly how population density, religion, and political ideology influence abortion rates. If population density's impact is strong, we'd know to concentrate our efforts on issues related to access, perhaps both on physicians who perform the procedure or on contraception in urban areas. If political ideology or religion register the biggest effects, our strategy changes. To more fully understand why some states have high abortion rates, we need to examine each possible explanation while accounting for the others.

Another example comes from the world of sports. Bill James, out of his garage in Lawrence, Kansas, started a revolution in baseball by collecting data on all aspects of the game and using that information to evaluate players: sabermetrics (Cukier & Mayer-Schönberger, 2013). Many teams have discovered that evaluating players based on regression analysis leads to a better win-loss record than listening to scouts who rely on their gut feelings. In the world of football, there are probably just as many theories of how to win the Super Bowl as there are fans. Should one emphasize defense? Should a coach draft the fastest or the smartest players? Should the emphasis be placed on discovering the next great quarterback? Surely, all of these things are important, but regression analysis can help.

Probably the most common—but no less crucial—example comes from business: how to increase profits? Important considerations might include streamlining production, buying more effective advertising, or producing a better product. Again, all of these surely play an important role in determining success. Unfortunately, companies operate under constraints: they do not have an endless amount of cash. Given what the CEO is willing to spend, what is the best strategy? Again, multiple regression can provide some answers.

> **KNOWLEDGE CHECK: Define multiple regression.**

1. What is multiple regression?
    a. It is analogous to controlled comparisons.
    b. It allows us to account for many different factors.
    c. It is a useful tool for policy analysis.
    d. It is a tool that can help identify priorities.
2. Which statements provide accurate descriptions of multiple regression?
    a. By accounting for additional factors, we can obtain accurate estimates.
    b. In multiple regression analysis we hold a number of variables constant.
    c. We can use goodness-of-fit measures with multiple regression.
    d. Understanding its limitations is key to its effective use.

Regression Models and Arguments

With a sense of what multiple regression is and why we use it, let's delve a little deeper into how regression models map onto arguments. Let's diagram an argument and link it to a multiple regression model.

In the example of abortion rates, we posited that population density, religion, education, and income all influence abortion rates in the United States. The simple, additive multiple regression model assumes that our independent variables all have an independent impact on abortion rates. The diagram in Figure 13-1A presents a picture of the **additive multiple regression model**. The mathematical representation of the model is given by the following equation:

$$Y_1 = a_1 + \beta_1 X_1 \text{PopulationDensity} + \beta_2 X_2 \text{Religion} + \beta_3 X_3 \text{Education} + \beta_4 X_4 \text{Income} + e$$

where Y_1 is the dependent variable, a_1 is the constant, the β represent the weights given to each variable, the Xs represent the independent variables, and e represents the error term. The error term accounts for random events that influence the outcome, events that can't possibly be identified systematically by any variables we include in the model. For example, perhaps there was a blizzard the day of the scheduled procedure, or second thoughts arose after running into a relative or friend before the procedure.

In this model, religion does not influence any of the other independent variables and vice versa. The model's functional form is strictly additive. We also assume the independent variables influence abortion rates and not vice versa (there is no reverse causality). While in some cases that is a proposition easily defended, in others it is not: one can imagine abortion affects the independent variables. For example, terminating a pregnancy can probably influence a woman's religious beliefs, her income, and whether she continues in school. Figure 13-1B represents a more complete and perhaps accurate model of reality. For now, our focus remains

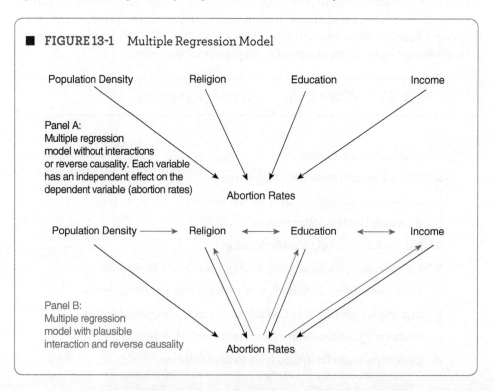

■ **FIGURE 13-1** Multiple Regression Model

Panel A:
Multiple regression model without interactions or reverse causality. Each variable has an independent effect on the dependent variable (abortion rates)

Panel B:
Multiple regression model with plausible interaction and reverse causality

on the simpler and admittedly more restrictive model (Figure 13-1A). Although in the next chapter I will introduce interactions to account for the relationships between our independent variables, presenting models designed to account for reverse causality is left to others.[1]

Diagramming arguments and translating them into regression models is a useful exercise for several reasons. First, translating back and forth between math and English helps clarify the argument. Through the exercise, you'll become much more aware of what you're actually arguing. Second, putting your argument into the form of a diagram or regression model is a useful way to organize your thoughts. Once on paper (or the computer screen), it's easy to play around with and change your argument while maintaining its overall logic and structure. Finally, once you've developed the skill, it's much easier to understand and evaluate arguments made by others. Diagramming arguments is a useful way to generate, improve, and communicate your ideas and to evaluate arguments made by others.

> **KNOWLEDGE CHECK: Identify the parallels between a model and an argument.**

3. Which of the following statements describe the benefits of modeling?
   a. Models can account for important factors in a relationship.
   b. Translating mathematical expressions into English helps clarify your theory.
   c. It's a useful way to organize your thoughts.
   d. It can help with evaluating arguments made by others.

4. Which statements describe the relationship between regression models and arguments?
   a. They both explicitly name cause and effect.
   b. Multiple regression always assumes independent variables are independent of each other.
   c. A regression model is simply the mathematical representation of an argument.
   d. Multiple regression analysis cannot account for reverse causality.

Regression Models, Theory, and Evidence

If we want an accurate estimate of the relationship between two variables, we need to construct an accurate model. Multiple regression models merely represent a well-defined and articulated argument that accounts for multiple causes. Consequently, we have to think—theorize—a bit to identify the most relevant independent variables. Whether our focus is on determining win-loss records in baseball, reducing abortion rates, or increasing profits, the best models produce the best estimates.

---

[1] Modeling one direction of causality implies the use of single-equation models, which is the emphasis of this book. A considerable number of problems in the social sciences involve reverse causality and selection effects that require estimating systems of equations. It is helpful, however, to build an intuition with single-equation models before diving into the deep end.

Let's begin with theory. If we're interested in reducing abortion rates, for example, what would constitute the best model? In this case, we don't need to explain every abortion in the United States; we need a short list of what most often determines the decision to terminate a pregnancy. A model, in other words, is an explicit attempt to simplify reality. We want to explain as much as we can with as little as possible.

We've already started interrogating the usual suspects. We think population density matters. Income, education, and religion are probably important as well. If we look at the *states* data set, we might see some other variables that should be included in our regression model. Let's start with the following theoretical model:

$$\text{Abortion} = f\left(\text{education}, \text{religion}, \text{density}, \text{income}\right)$$

We might want to add other variables, but this model provides a good start. Although we think these variables explain abortion rates, we should remain skeptical. This is where our honesty and humility (and our smarts) get tested. Honesty is involved because the estimates generated by the model can depend heavily on which variables we include. We might be tempted to present estimates from the model that furthers our agenda rather than from the model that best reflects reality.

To illustrate the point, I generated a regression table for four different models (Table 13-1). In Code Chunk 13-1 I define four different regression models. The only difference between the bivariate models introduced in the previous chapter and multiple regression models here is that there is more than one independent variable included. This is easily accomplished with the addition sign and the variables you want to include. To present the results, I then place the four models in the *stargazer()* command. Also note that to present a streamlined version of the table, I use the *omit.stat* option, telling stargazer not to include two statistics that we can ignore for now: the residual standard error and the F-statistic.

In the first model (Model 1), I estimated the model dictated by my theory. I then removed one variable at a time to illustrate how the coefficient on the *hsdiploma* variable changes when we alter the specification of the model. The observed change indicates the coefficient on education is not that stable with respect to the exclusion or inclusion of income. The magnitude of the coefficient on education (−.663) reduces by roughly one-third when we exclude income. Depending on what model we decide to adopt, our estimate on high school education can increase or decrease significantly.

Code Chunk 13-1

```
abort1 <-lm(abort~hsdiploma + evangel + density + inc, data=states)
abort2 <-lm(abort~hsdiploma + evangel + density, data=states)
abort3 <-lm(abort~hsdiploma + evangel, data=states)
abort4 <-lm(abort~hsdiploma, data=states)

stargazer(abort1, abort2, abort3, abort4, type = "text",
          title = "Table 13-1: Different Specifications
                    of the Abortion Rate Model",
          header=FALSE, omit.stat = c("ser", "f"))
```

TABLE 13-1 Different Specifications of the Abortion Rate Model

	Dependent variable: abort			
	(1)	(2)	(3)	(4)
hsdiploma	-0.663**	-0.430	-0.610**	-0.323
	(0.327)	(0.290)	(0.299)	(0.329)
evangel	-0.159	-0.222**	-0.346***	
	(0.105)	(0.096)	(0.089)	
density	0.008	0.014**		
	(0.006)	(0.005)		
inc	0.0005			
	(0.0003)			
Constant	57.621**	56.570**	76.799***	45.510
	(25.422)	(25.737)	(26.073)	(28.171)
Observations	50	50	50	50
R2	0.384	0.354	0.257	0.020
Adjusted R2	0.329	0.312	0.225	-0.001

Note: *p<0.1; **p<0.05; ***p<0.01

Art and Practice of Data Visualization
HONESTY

Suppose you work for the National Association of Education in Washington, D.C., regularly lobbying Congress for more education spending. Which of the models in Table 13-1 would you be more likely to feature in your pitch? Even if you present all of the models to let lawmakers understand the fragility of your results, you'll probably spend extra time promoting the model that makes education shine. Our normative priors shape, if not determine, our understanding of the world. Separating objective reality from our subjective viewpoint as human beings is difficult if not virtually impossible. We confront it directly as data analysts in the seemingly mundane choices or assumptions we make at almost every turn. It's a battle that can't be decisively won, but it should be continually fought.

Although an assortment of statistical tests can help (the R^2 statistic or the t-ratio), we shouldn't rely on them to choose the model for us. Often, there are important blind spots in statistical tests that favor one model over another for reasons not readily apparent to the analyst. At the end of the day, we have to rely on both theory and evidence.

In this example, there is evidence that education may not be that important when explaining abortion rates (Models 2 and 4). How do we decide which model provides the best estimate of education's impact? Short of anything we may learn when performing diagnostics on these regression models (diagnostics will be covered in Chapters 15 and 16), we have to rely on theory. Should income and population density be included or not? If there are good theoretical reasons to think population density and income influences abortion rates, they should be included.

While there is no magic formula, learning as much about the data as possible builds our intuition and can help inform our theory. In other words, the more you know about the subject, the better informed the theory. Good theory, in turn, influences how we examine the data. A conversation between the two not only provides better answers (estimates), it generates better questions (hypotheses).

> **KNOWLEDGE CHECK:** Discuss the relationship between theory and evidence

5. Which of the following statements best describe the relationship between theory and evidence?

    a. Which one receives primacy lies at the heart of many debates.

    b. Always start with theory.

    c. Always start with evidence.

    d. The honesty and humility of the analyst becomes important when deciding which one to emphasize.

6. What lessons are learned from Table 13-1?

    a. Sometimes the evidence obtained depends on the theory.

    b. Sometimes the evidence forces us to change our theory.

    c. Evidence doesn't always provide clear guidance.

    d. Evidence provides the best guide when developing theory.

7. Indicate whether an economist, a sociologist, a teacher's union lobbyist, or a policymaker would present the following from Table 13-1.

    a. Models 1 and 3

    b. Model 1

    c. Models 2 and 3

    d. Models 1, 2, 3, and 4

8. How should one decide on which model is best?
   a. Choose the model based on theory (your priors).
   b. Choose the model based on evidence (the results).
   c. Choose the model that explains the most variance in the dependent variable.
   d. If the choice matters, show both.

Interpreting Estimates in Multiple Regression

Fortunately, understanding the coefficients in multiple regression follows directly from the bivariate context. As in the bivariate case, we are still interested in the magnitude of the coefficients, their substantive significance. The coefficients' statistical significance also retains the same meaning from the bivariate context. The following sections rehearse interpreting results from multiple regression analysis by focusing on their substantive and statistical significance.

Substantive Significance

Let's interpret the coefficients in the regression in Table 13-1 to illustrate the difference in their substantive importance. Similar to the bivariate case, we have to know the units in order to interpret their substantive significance. The *abort* variable is the number of abortions per 1,000 women. The *hsdiploma* variable is the percentage of a state's population with a high school diploma, and *density* records the population per square mile. Interpreting coefficients in the bivariate and multiple regression contexts is the same except we add the phrase, "holding the rest of the variables in the model constant at their means." In what follows, you'll see that I drop the final three words "at their means" from the phrase. Most analysts simply say *holding the rest of the variables in the model constant*.

To illustrate, in the full model (Model 1 in Table 13-1), we find that a 1 percentage point increase in the state's population with a high school diploma is associated with a .66 decrease in the number of abortions per 1,000 women, holding the rest of the variables in the model constant. Note that in our simplest model (Model 4), a 1 percentage point change in the population with a high school degree is associated with a .32 decrease in the number of abortions per 1,000 women. In Model 1, we observe that for each additional person per square mile, we would predict a .008 increase in abortions per 1,000 women, holding the remaining variables in the model constant.

Recall that interpreting the coefficient gives the *substantive* significance for each independent variable, holding all of the remaining variables constant. Whether the variable is substantively important depends on how it is measured. Comparing the coefficient—the actual number—for different variables is like comparing apples to oranges. For the diploma variable, the coefficient is associated with a percentage point change in the population with a high school diploma. For the population density variable, the coefficient is associated with a change of one person per square mile. Those are very different things. Determining the substantive significance of a variable is a judgment call; a case has to be made whether the coefficient represents significant change.[2]

---

[2] To enable the comparison of coefficients, some analysts rely on standardized coefficients, placing all of the coefficients in a regression on the same scale. While this practice can be useful, it is rarely used. To maintain clarity in my understanding and communication of the results, I prefer to focus instead on the distinct units of analysis associated with each coefficient. Common practice indicates most analysts favor interpreting coefficients in their raw form.

To take it one step further, the policy implications for each variable and coefficient can be very different. Take the diploma variable as an example. A 1 percentage point increase in the population that has a high school diploma may be difficult to achieve. It may cost the state millions of dollars in revenue to raise the level by 1%. Similarly, it may be difficult for states like Wyoming, Montana, or Nebraska to increase population density by one person per square mile. To the degree those variables are difficult to manipulate through policy, they may or may not represent opportunities to effectively lower abortion rates.

Statistical Significance

Multiple regression analysis comes with an important added feature: the statistical machinery that indicates the accuracy of our estimates. With multiple regression, we can construct confidence intervals around our estimates to determine their accuracy. This parallels the exercise of drawing a random sample, calculating a sample statistic, and using the standard normal curve to establish how close our sample statistic is to the population parameter.

There is an important difference, however, between calculating sample statistics from a sample drawn randomly and multiple regression analysis. Remember that the nice properties of the standard normal curve and the inferences we make based on it derive from random samples in which each observation drawn is independent of the next: each observation in the sample has an equal chance of being drawn. Unfortunately, the same is seldom true of the observations used in multiple regression analysis. Rare is the multiple regression model based on data from a truly random sample. Instead, data are collected from objects or individuals that are not randomly selected. For example, national-level statistics may only be available from countries with the bureaucratic wherewithal to collect them. The regression, as a result, would be based solely on observations from highly functioning states, a biased sample.

In regression analysis we limit the sample in a number of different ways. Perhaps we limit our sample to developing countries. Or we remove outlying cases to make better comparisons. In each instance, the sample statistic we calculate (in this case the regression coefficient) is not based on a random sample. To the extent we stray from random selection, view all regression results with a healthy dose of skepticism. While inferences and probabilistic statements can be made, their accuracy is dependent on how far we stray from the ideal: the random sample.

With that disclaimer, we can still use the statistical machinery introduced in Chapters 8 and 9 to inform, but not necessarily confirm, our analysis. The asterisks—or stars—next to the coefficients in the regression tables do provide some indication of how confident we can be that the estimate is not the result of some random event. These asterisks are based on the t-ratios, which were introduced in the previous chapter. They indicate how likely we'd obtain that estimate from our sample if the real relationship were 0. The larger the absolute value of the t-ratios, the less likely the real slope (think population parameter) of the line is 0.

In this book I use the following convention: one asterisk indicates a 90% level of confidence, two asterisks indicate a 95% level of confidence, and three asterisks represent a 99% level of confidence. Although this is an arbitrary standard in many respects (more an artifact of custom than statistical principle), the 95% level of confidence is judged to be *statistically significant* in most scholarly journals. Results that are not significant at the 95% level of confidence are rarely taken seriously (i.e., they are rarely published). This is unfortunate since in some contexts we might settle for more or less. For example, much would be lost if a scientist found a therapeutic drug that reduced the chance of dying from COVID-19 by 40%, but the results were never published because the estimates were only significant at the 90% level.

In addition to the asterisks, the numbers in parentheses placed under the estimate in the stargazer-generated table represent the *standard errors* of the estimate. The estimate divided by its standard error produces the t-ratio. In regression output, the t-ratios themselves can be reported instead of or in addition to the standard errors.

Art and Practice of Data Visualization
LEVELS OF SIGNIFICANCE

Different software packages report different levels of significance. While one package may assign one asterisk to a coefficient at the .10 level, others might not assign any. R assigns a period (.) when the significance level is .10. Regardless of their differences, the levels are always indicated somewhere in the regression table, usually in the notes at the bottom.

Goodness of Fit: R^2

Just as the bivariate regression in Chapter 12 produced an R^2 statistic, multiple regression analysis produces the same statistic. In multiple regression, the R^2 statistic represents the percentage of total variance explained by the model. In Table 13-1, the R^2 for Model 1 is .384 (33%). The model explains 38.4% of the variance in the dependent variable. This means roughly 62% of the variance remains unexplained. Note how the R^2 varies over the different models. When explaining abortion rates using Model 4, only 2% of the variation is explained by education (*hsdiploma*). That number jumps quickly when we add religion (Model 3).

The R^2 statistic can be manipulated by adding additional variables, without any thought as to whether there is a good theoretical reason to include the variable in the model. Increasing the R^2 is not hard to do: mathematically, each variable added to the model will increase the R^2 statistic.[3] This mathematical property of the R^2 statistic often induces analysts to employ the kitchen sink method, throwing all of the available variables into the model—also referred to as "overfitting" the model. Since models with too many independent variables come with pathologies of their own, stay out of the kitchen! Construct models based on your expertise and knowledge of the problem rather than the R^2 statistic.

Happily, there is an antidote. Note the **adjusted R^2 statistic** in the regression table. When regressing a dependent variable on more than one independent variable, use the adjusted R^2 statistic since it "adjusts" for the number of variables included in the model. With the adjusted R^2, adding a variable to the model does not automatically increase the statistic.

---

[3] For a mathematical treatment of how adding additional variables to a regression model automatically increases the R^2 and whether they actually aid in the explanation of Y, see Greene (2000, p. 238).

> **KNOWLEDGE CHECK: Interpret coefficients, t-ratios, and measures of fit.**

9. What are the lessons learned when we realize comparing the magnitude of the coefficients is like comparing apples to oranges?
    a. Always look to the t-ratio to decide a variable's importance.
    b. Always compare the relative magnitude of the coefficients.
    c. Knowing the units of analysis is key.
    d. Comparing the substantive importance of variables in the model is a judgment call.

10. Interpreting coefficients in multiple regression requires which of the following:
    a. Adding the phrase "holding all other variables in the model constant"
    b. Knowing the units of analysis
    c. Knowing whether the variable has been transformed
    d. The t-ratio

11. t-Ratios in multiple regression indicate which of the following:
    a. How likely it is that we obtained that estimate even if the real slope is zero
    b. Whether we can reject the null hypothesis: there is no correlation between the independent variable and the dependent variable
    c. Whether there is a substantively significant relationship between x and y.
    d. Whether there is a statistically significant relationship between x and y.

12. Which of the following accurately describes goodness of fit in multiple regression?
    a. It answers the question of how much of the unexplained variance of our dependent variable the model explains.
    b. It answers the question of how much of the total variance of our dependent variable the model explains.
    c. The adjusted R^2 statistic provides a more accurate measure since it accounts for the number of observations.
    d. The adjusted R^2 statistic provides a more accurate measure since it accounts for the number of variables.

Example: Homicide Rate and Education

The following example includes the main components of multiple regression analysis we've learned so far: (1) describing the data to build models and (2) estimation and interpretation. To those we add a third, generating new questions and hypotheses. Since the majority of technical issues and analytics were covered in the previous chapter on linear regression, this example focuses on how describing data, model building, estimation, and generating the next question all fit together.

Theory

Although EDA argues for a conversation between theory and evidence, we have to start the conversation somewhere. Generally speaking, even in the most open-ended situations, we have some notion about how things work in the world, otherwise we wouldn't have collected the data in the first place. We begin with what we suspect are the main causes of homicide. I posit there are three main causal factors that help explain the variation in homicide rates among the 50 states. I hypothesize education has a strong, negative influence on the number of murders, but to obtain accurate estimates of that relationship I will account for income and population density. Income is clearly an important factor in accounting for crime in general and homicides in particular. Joblessness, lack of education, and distrust of the judicial system reside primarily among the poor. Poverty limits options, encouraging individuals to solve problems with violence. Population density is another important variable to account for in the model. People who live close to one another experience more human contact. Increased encounters with others are more likely to generate situations that lead to violence.

To understand the role education plays, I focus on the stock of human capital: the amount. I hypothesize that a more educated population—a higher percentage of a state's population with high school degrees—will witness less crime, controlling for income and population density. Those with high school educations are more likely to be employed. Stable incomes provide opportunity, reduce stress, and decrease substance abuse, all conditions thought to reduce crime. To measure education, we use the *hsdiploma* variable from the *states* data set (the percentage of the population with high school diplomas). My basic model is as follows:

$$\text{Murderrate} = f(\text{income}, \text{education}, \text{populationdensity})$$

Description of Data

To orient the reader, I generate a histogram (Figure 13-2) of the homicide variable to show its distribution. As can be observed from the histogram, homicides per 100,000 range from 0 to 12 among the 50 states. The mean seems to hover around four to five murders per 100,000. The variable seems normally distributed: the mean and median will be roughly the same. A quick look at a scatter plot between homicides and education (Figure 13-3) shows that Louisiana has the highest rate at about 12 murders per 100,000. Other states with relatively high rates are New Mexico and Maryland. New Hampshire has the lowest rates and shares low levels of violence with states like Vermont, Utah, and Minnesota. The scatter plot also reveals there is a strong negative relationship between education and homicide, confirming we're on the right track with our hypotheses.

Code Chunk 13-2

```
ggplot(states, aes(murderrate)) +
  geom_histogram(bins=6, colour = "#0000bf", fill = "white") +
  ggtitle("Figure 13-2: Histogram of Murder") +
  xlab("Homicides per 100,0000") +
  theme_minimal() +
  theme(plot.title = element_text(size = 8, face = "bold"),
        axis.title = element_text(size = 8, face = "bold"))

ggplot(states, aes(hsdiploma, murderrate)) +
  geom_point(col="#bf0000") +
```

```
geom_text_repel(size=3, aes(label = st, size = 1,
                hjust = 0, vjust=-1),col="grey", show.legend=FALSE) +
ggtitle("Figure 13-3: Education Prevents Murder") +
geom_smooth(method="lm", se=F, fullrange=F, col="#0000bf") +
theme_minimal() +
theme(plot.title = element_text(size = 8, face = "bold"),
      axis.title = element_text(size = 8, face = "bold")) +
ylab("Homicides per 100,0000") +
xlab("Percentage of Population With High School Diploma")
```

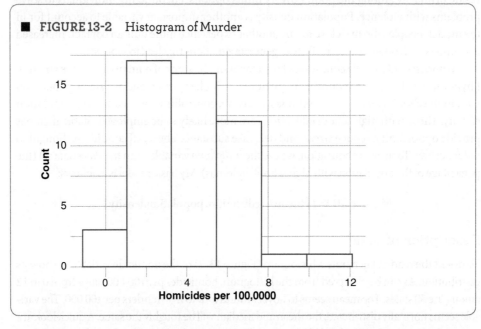

■ FIGURE 13-2   Histogram of Murder

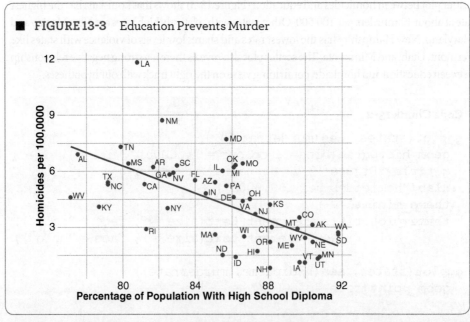

■ FIGURE 13-3   Education Prevents Murder

In addition to education (our main hypothesis), we suspect income and population density might also influence homicides. In Code Chunk 13-3 I generate scatter plots for homicides and income along with homicides and population density.

Code Chunk 13-3

```
ggplot(states, aes(medinc, murderrate)) + geom_point(col="#bf0000") +
  geom_text_repel(size=3, aes(label = st, size = 1,
                  hjust = 0, vjust=-1),col="grey", show.legend=FALSE) +
  ggtitle("Figure 13-4: Wealth and Violence Don't Mix") +
  geom_smooth(method="lm", se=F, fullrange=F, col="#0000bf") +
  theme_minimal() +
  theme(plot.title = element_text(size = 8, face = "bold"),
        axis.title = element_text(size = 8, face = "bold")) +
  ylab("Homicides per 100,0000") +
  xlab("Median Household Income")

ggplot(states, aes(density, murderrate)) + geom_point(col="#bf0000") +
  geom_text_repel(size=3, aes(label = st, size = 1, hjust = 0,
                  vjust=-1), col="grey", show.legend=FALSE) +
  ggtitle("Figure 13-5: Homicides Happen in Crowded Places") +
  geom_smooth(method="lm", se=F, fullrange=F, col="#0000bf") +
  theme_minimal() +
  theme(plot.title = element_text(size = 8, face = "bold"),
        axis.title = element_text(size = 8, face = "bold")) +
  ylab("Homicides per 100,0000") +
  xlab("Person per Square Mile")
```

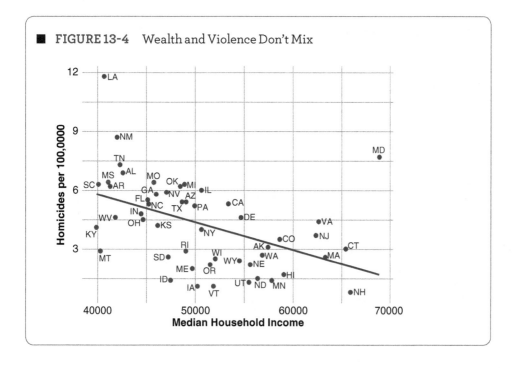

FIGURE 13-4 Wealth and Violence Don't Mix

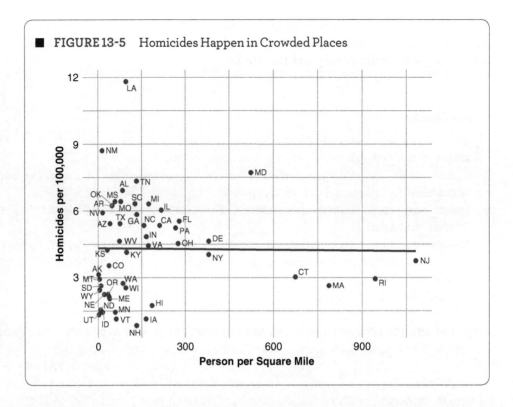

FIGURE 13-5 Homicides Happen in Crowded Places

The scatter plot in Figure 13-4 suggests that a relationship exists between income and homicides. As the median household income in a state increases, the number of homicides per 100,000 population declines in a linear fashion. The scatter plot in Figure 13-5 suggests something different. At first glance, the scatter plot of homicide rates and population density indicates there is no relationship between the two variables. The scatter plot suggests, however, that the population density variable needs to be transformed: there are a few relatively large values (Connecticut, Massachusetts, Rhode Island, and New Jersey) while the bulk of the distribution is well below 300 people per square mile.

Code Chunk 13-4

```
h1 <- ggplot(states, aes(density)) +
  geom_histogram(bins=6, colour = "#0000bf", fill = "white") +
  ggtitle("Raw Form") +
  xlab("People per Square Mile") +
  theme_minimal() +
  theme(plot.title = element_text(size = 8, face = "bold"),
        axis.title = element_text(size = 8, face = "bold"))

h2 <- ggplot(states, aes(log2(density))) +
  geom_histogram(bins=6, colour = "#0000bf", fill = "white") +
  ggtitle("Logged Form") +
  xlab("People per Square Mile (logged)") +
```

```
    theme_minimal() +
    theme(plot.title = element_text(size = 8, face = "bold"),
          axis.title = element_text(size = 8, face = "bold"))

grid.arrange(h1, h2, ncol=2,
             top=textGrob("Figure 13-6: Raw and Logged
                      Form of Density", gp=gpar(fontsize=8)))
```

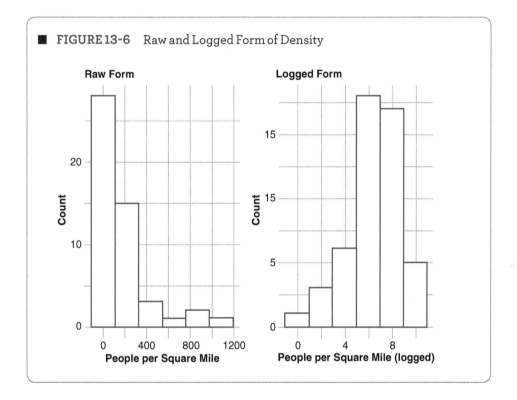

FIGURE 13-6   Raw and Logged Form of Density

A quick look at the histogram of population density confirms the need to transform the variable. When performing the regression analysis, we should check to see if including the logged form of the density variable alters our results in any way.

Estimation

Let's proceed to estimate the relationship between education and homicide, controlling for income and population density.

To estimate the regressions and present the estimates in tabular form, I first create two objects *murder.lm* and *murder1.lm*. I then insert those two model names into the *stargazer()* command. Note that the two regressions in Code Chunk 13-5 are only different in that one uses the log form of the density variable.

Code Chunk 13-5

```
murder.lm <- lm(murderrate ~ medinc + hsdiploma + density, data=states)

murder1.lm <- lm(murderrate ~ medinc + hsdiploma +
                log2(density), data=states)

stargazer(murder.lm, murder1.lm, title = "Table 13-2:
        Logging Density and Its Impact on Regression Results",
        type = "text", header=FALSE)
```

TABLE 13-2 Logging Density and Its Impact on Regression Results

	Dependent variable: murderrate (1)	murderrate (2)
medinc	−0.0001 (0.00005)	−0.0001 (0.00004)
hsdiploma	−0.300*** (0.083)	−0.271*** (0.088)
density	0.0002 (0.001)	
log2(density)		0.104 (0.139)
Constant	32.915*** (5.943)	30.374*** (6.751)
Observations	50	50
R2	0.423	0.429
Adjusted R2	0.385	0.392
Residual Std. Error (df = 46)	1.780	1.770
F Statistic (df = 3; 46)	11.220***	11.533***

Note: *p<0.1; **p<0.05; ***p<0.01

The regression results indicate that after having controlled for income and population density, the stock of education (*hsdiploma*) seems to yield the strongest estimates. For every percentage point increase in the population with a high school degree, there is a .30 decrease in homicides per 100,000 population, holding the other variables constant. For every person we add per square mile to a state's population, there is an associated .0002 increase in homicides per 100,000, holding the other variables constant.

Of the variables in the regression, only education (*hsdiploma*) is *statistically significant*. We can be confident that if we were to repeat the experiment (our regression with different samples) 100 times, the confidence interval around our estimate would not include zero in approximately 99 of those experiments. The estimates also indicate that while we have a fairly good fit, there is much more to explain: the adjusted R^2 statistic is .38, meaning that our model explains 38% of the variation in homicides throughout the 50 states. So while we've made a good start here, there is still over 60% of the variation in homicides left to be explained. Table 13-2 indicates that logging density has relatively little impact on the coefficient for education.

Empirical Implication

There is a strong relationship between education and violent crime (homicides). Using that result to generate an **empirical implication** is a helpful device to obtain additional evidence. An empirical implication is simply a hypothesized empirical relationship that follows logically from the existence of another. For example, *if* the level of education is an important factor in deterring homicides, *then* education could have a similar influence on another violent crime (incidents of rape per 100,000 population). (I chose that variable simply because it's the only other variable measuring crime in the data.) In other words, *if* we think education is an important deterrent to violent crime, *then* it should have a similar association with sexual violence (the number of rapes per 100,000 population). Empirical implications give us additional hypotheses we can test to determine whether we have actually discovered something. I use the same model designed to estimate education's association with murder to estimate the relationship between education and sexual violence.

First, we can simply look at a scatter plot of sexual violence (the number of rapes per 100,000 population) and education (Figure 13-7). From that plot we observe there is a surprising pattern: there is a slightly positive relationship between the two. Looking at the plot, we see that the states with the highest levels of sexual violence are Alaska, South Dakota, and New Mexico, all very sparsely populated states. Might population density be associated with the crime? I plotted the log of population density against sexual violence (Figure 13-8), finding that there's a significantly negative relationship between the two: the more densely populated the state, the lower the crime.

Code Chunk 13-6

```
ggplot(states, aes(hsdiploma, raperate)) +
  geom_point(col="#bf0000") +
  geom_text_repel(size=3, aes(label = st, size = 1,
          hjust = 0, vjust=-1), col="grey", show.legend=FALSE) +
  labs(title = paste("Figure 13-7: Education Increases Rape Rate")) +
  geom_smooth(method="lm", se=F, fullrange=F, col="#0000bf") +
  theme_minimal() +
  theme(plot.title = element_text(size = 8, face = "bold"),
        axis.title = element_text(size = 8, face = "bold")) +
  ylab("Reported Incidents of Rape per 100,0000 Population") +
  xlab("Percentage of Population With High School Diploma")

ggplot(states, aes(log2(density), raperate)) +
  geom_point(col="#bf0000") +
```

```
geom_text_repel(size=3, aes(label = st, size = 1,
        hjust = 0, vjust=-1), col="grey", show.legend=FALSE) +
labs(title =paste("Figure 13-8: Population Density and Rape Rate")) +
geom_smooth(method="lm", se=F, fullrange=F, col="#0000bf") +
theme_minimal() +
theme(plot.title = element_text(size = 8, face = "bold"),
      axis.title = element_text(size = 8, face = "bold")) +
ylab("Reported Rapes per 100,000 population") +
xlab("People per square mile")
```

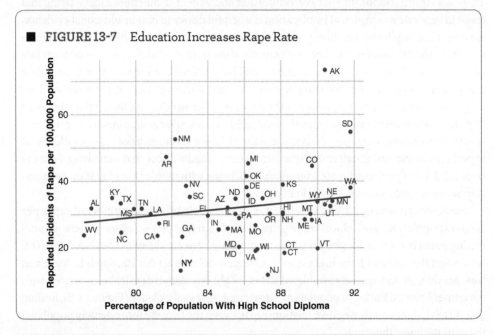

■ **FIGURE 13-7**   Education Increases Rape Rate

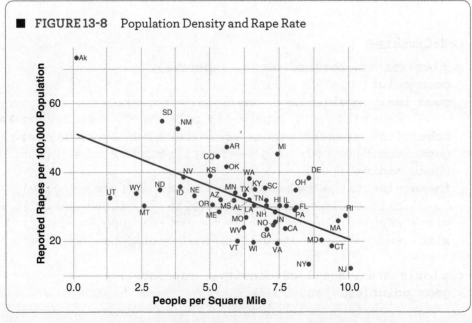

■ **FIGURE 13-8**   Population Density and Rape Rate

When I perform the regression that tests the relationship between education and sexual violence—controlling for population density and income—I find there is no relationship. There is, however, a strong, negative relationship between population density and sexual violence.

Code Chunk 13-7

```
rape.lm <- lm(raperate ~ medinc + hsdiploma +
                  log2(density), data=states)

stargazer(rape.lm, title = "Table 13-3: Rape Rate
          Regressed on Education", type = "text", header=FALSE)
```

The regression analysis tells us the following. First, each percentage point increase in the population with a high school diploma is associated with a .18 increase in the number of incidents of sexual violence per 100,000 population, holding the other variables constant. Note that the coefficient is not statistically significant so we have less confidence in the result.[4] The

TABLE 13-3 Rape Rate Regressed on Education

	Dependent variable:
	raperate
medinc	−0.0002
	(0.0002)
hsdiploma	0.181
	(0.420)
log2(density)	−2.866***
	(0.666)
Constant	43.811
	(32.272)
Observations	50
R2	0.404
Adjusted R2	0.366
Residual Std. Error	8.459 (df = 46)
F Statistic	10.410*** (df = 3; 46)
Note:	*p<0.1; **p<0.05; ***p<0.01

---

[4] If we repeated the experiment with a different sample 100 times, 0 would be included in our estimate's confidence interval more times than what is acceptable. In this case, roughly 33 times.

population density variable, however, is both *substantively* and *statistically* significant. Each time we double population density, there is a 2.9 decrease in the number of violent incidents per 100,000 population. The coefficient is statistically significant at the 99% level. We can be fairly confident in our estimate. Finally, this model's fit is similar to the homicide model's: our model explains 37% of the variance in reported incidents of sexual violence across the 50 states.

Discussion

An important part of regression analysis involves translating the estimates and findings into policy implications. The implications here are several. First, we shouldn't lump all violent crimes together because they are different and have very different causes. Second, while education seems to help for deterring homicide, it seems less effective in deterring other violent crimes such as rape. It is interesting that population density brings about a very unexpected change in one kind of violent crime and not the other. Interestingly, in more isolated areas, women may be more exposed to crime than had been expected. In urban areas, the results suggest, there may be more support networks or awareness about the problem, bringing its relative numbers down. We could test whether the relationship between density and sexual violence holds over different geographic regions to help identify other possible causes. Does the relationship hold up when we control for regional effects?

In this example I found that education has a very important deterrent effect on murder rates. Education's influence is not present, however, for all violent crimes. While regression results showed that education has a strong association with murder, controlling for income and population density, it does not have a strong association with sexual violence.

> **KNOWLEDGE CHECK: Employ multiple regression analysis to examine a problem.**

13. What components of data analysis were included in the example of homicide rates and education?
    a. Describing data
    b. Diagnostics
    c. Model estimation
    d. Generating the next question

14. What was the empirical implication derived from the regressions reported in Table 13-2?
    a. Income should affect sexual violence.
    b. Education levels in a state should influence sexual violence.
    c. Population density and sexual violence are related.
    d. There should be no relationship between education and sexual violence.

15. Which of the following illustrate the interactive nature between theory and evidence when performing regression analysis?

    a. The scatter plot of homicides and density suggested we needed to log population density.

    b. Our theory that income has an impact on homicide caused us to include it in the regression model.

    c. Results from Table 13-2 led to the regression reported in Table 13-3.

    d. The scatter plot of income and homicide (Figure 13-4) confirmed it should be included in the regression.

SUMMARY

Multiple regression allows us to evaluate in more exacting terms the relationship between more than two variables simultaneously. Consequently, we can hold a number of variables constant so that we can untangle the many causes of social phenomena. Although multiple regression analysis is extremely powerful, it can be misused: the specification of the model—which variables we include—as well as its functional form can have a huge impact on the actual results. Moreover, statistical tests can't be relied on to choose our model for us. A deep knowledge of both the substantive issue at hand and an understanding of the limitations associated with regression analysis can prevent us from making important and, at times, very costly mistakes. Viewed in a more positive light, regression analysis can help us discover important empirical patterns that once understood can help move progress forward.

In this chapter we introduced multiple regression in its simplest form. We assumed that the independent variables had independent effects on the dependent variable. In the next chapter we relax that assumption, allowing the independent variables to interact with each other. Mathematically, the additive model presented in this chapter will include terms where variables are multiplied (interacted).

COMMON PROBLEMS

- *Coefficients in a regression table are like apples and oranges.* Multiple regression results, especially coefficients of the estimates, cause the newly initiated to compare their relative magnitude without taking the units each variable represents into account. For example, a .008 coefficient on the variable for population density means something very different than a .008 coefficient on a variable representing the logged version of income per capita. In the former, a one-person increase per square mile is associated with a .008 change in y. In the latter, a 10-fold increase in income is associated with a .008 change in y. While the coefficient is the same, the relative importance of each can be very different. Each coefficient needs to be interpreted with its respective units of measure in mind.

- *Should theory or evidence should be the guide?* When embarking on your first model-building exercise, whether to include one variable or another can make for a very difficult choice since its inclusion can influence the overall results of the regression dramatically. When there is no clear answer, the best approach is transparency and humility. If the results vary dramatically from model to model and there's no obvious reason to pick one or the other, be transparent and share that finding with the audience. Since uncertainty underlies much of the scholarly enterprise, admitting as much (being humble) is the best approach.

- *Bivariate regression results can mislead.* Newly born regression analysts often like to show a series of bivariate regressions with the hope that by doing so, they're getting the unbiased estimate of each variable (sounds like a good place to start). Unfortunately, those bivariate estimates can be extremely biased since important variables that should be included aren't. Remember, the best estimates are obtained when you have accounted for the variables that should be included in a good model of the phenomenon.

- *What's a good R^2 statistic?* Unfortunately, there is no such guide. Experience will show that in certain contexts you can expect the statistic to hover around .90. The statistic is usually much smaller when regressions are performed on survey results (usually not much higher than .20). In sum, the R^2 varies tremendously depending on the dependent variable you're trying to explain. Fortunately, comparisons between your estimates and previous work along with comparing models within your own study make the R^2 statistic a helpful guide.

REVIEW QUESTIONS

1. What are the advantages to using multiple regression analysis?
2. How do honesty and integrity become a consideration in data analysis?
3. How does one go about constructing a model?
4. What aspects of the model concern functional form?
5. What is substantive significance?
6. What is statistical significance?
7. How does the adjusted R^2 differ from the R^2?
8. Why is articulating possible empirical implications a useful exercise?
9. What is the kitchen sink method of model selection?
10. Why not just rely on either theory or evidence exclusively?

PRACTICE ON ANALYSIS AND VISUALIZATION

1. Which of the following are advantages associated with multiple regression analysis relative to bivariate regression?
   a. Multiple regression analysis explains more variance.
   b. It allows us to account for additional variables.
   c. It provides a measure of model fit.
   d. Specific predictions can be generated for the dependent variable.

2. Which of the following expressions represents the regression of education on income, controlling for religion and partisan identity?
   a. Income = f(education, religion, party ID)
   b. Education = f(Party ID, religion, income)
   c. Education = f(Party ID, religion)
   d. Religion = f(Party ID, income, education)

3. Regress infant mortality on GDP per capita (logged) and voting turnout using the *world* data set.
   a. Interpret the coefficient on voting turnout.
   b. Interpret the coefficient on GDP per capita.
   c. What is the t-ratio for voting turnout?
   d. Is the coefficient on voting turnout statistically significant?

4. Using the *states* data set, regress *hsdiploma* on income, controlling for *infant* and *democrat*.
   a. Interpret the coefficient on *democrat*.
   b. Interpret the coefficient on *inc*.
   c. What is the t-ratio for *inc*?
   d. Is the coefficient on *inc* statistically significant?

5. Using the *world* data set, find out if the year women gained suffrage (*womyear*) influences the ratio of girls' to boys' education (*gtbeduc*), controlling for GDP per capita (logged).

   a. Interpret the coefficient on the independent variable of interest.
   b. What is the corresponding t-ratio?
   c. Is there a statistically significant relationship?
   d. How much variance in y is explained by the model?

6. Use a regression to evaluate whether the year women gained suffrage (*womyear*) influences the ratio of female to male infant mortality, controlling for GDP per capita (logged).

   a. Interpret the coefficient on the independent variable of interest.
   b. What is the corresponding t-ratio?
   c. Is there a statistically significant relationship?
   d. How much variance in y is explained by the model?

7. Regress per capita income on education, population, and spending on students in the *states* data set. Compare that regression to the same one that adds *democrat* to the regression and answer the following.

   a. Interpret education's relationship with income in the first regression.
   b. Interpret education's relationship with income in the second regression.
   c. Is education's estimate statistically significant?

8. Does the inclusion of partisanship appreciably change the estimate on education? Do the youth of a country's population influence its carbon emissions? Answer that question controlling for GDP per capita (logged) and population.

   a. Interpret the coefficient on the youth of the population.
   b. Is the coefficient statistically significant?
   c. Interpret GDP per capita's relationship with CO_2 emissions.
   d. Is GDP per capita statistically significant?

9. Repeat the regression in question 4 with New York, California, and New Jersey removed from the regression and answer the following.

   a. Interpret the coefficient on *democrat*.
   b. Interpret the coefficient on *inc*.
   c. What is the t-ratio for *inc*?
   d. Does your overall conclusion of income's effect on education change depend on whether those cases are included?

10. Repeat the regression you ran in question 5 without the Middle Eastern cases.

    a. Interpret the coefficient on the independent variable of interest.
    b. What is the corresponding t-ratio?
    c. Is there a statistically significant relationship?
    d. Are your answers contingent on the inclusion or exclusion of the Middle Eastern cases?

ANNOTATED R FUNCTIONS

The following functions appear in this chapter. They are listed in order of their first appearance (with the code chunk number in parentheses) and annotated here to give a very brief description of their use. Some are not stand-alone functions and only work in combination with other commands. As a reminder, the code in every chapter will work properly if executed in the order it appears. Proper execution also depends on typing the author-defined *libraries()* command, which loads the required R packages.

stargazer(): generates a regression table with specified model. (13-1)

ggplot(): defines the basic structure of a plot (usually the x and y variables). (13-2)

aes(): the aes (called "aesthetics") function is used in ggplot to define the basic structure of the plot, which often includes the variables you want to use and any shapes or colors. (13-2)

geom_histogram(): generates a histogram of a variable. (13-2)

ggtitle(): provides the title for a ggplot. (13-2)

xlab(): labels the x axis in ggplot. (13-2)

theme_minimal(): specifies a minimalist style for ggplot. (13-2)

theme(): specifies font, size, and so forth in a ggplot. (13-2)

geom_point(): draws points on a ggplot grid. (13-2)

geom_text_repel(): labels points in a scatter plot in the ggplot framework. (13-2)

geom_smooth(): draws lines or curves to a scatter plot. (13-2)

ylab(): labels the y axis in ggplot. (13-2)

grid.arrange(): allows you to arrange multiple figures on a page. (13-4)

ANSWERS

KNOWLEDGE CHECK

1. a, b, c, d
2. a, b, c, d
3. a, b, c, d
4. a, c
5. a, d
6. a, b
7. lobbyist (a), economist (b), sociologist (c), policymaker (d)
8. d
9. c, d
10. a, b, c
11. a, b, d
12. b, d
13. a, c, d
14. b
15. a, c, d

PRACTICE ON ANALYSIS AND VISUALIZATION

1. b
2. b
3. A 1 percentage point increase in voter turnout is associated with a .112 decrease in the number of infant deaths per 1,000 live births, holding GDP constant (a); doubling the amount of income is associated with a decrease in 11.8 deaths per 1,000 live births, holding voting turnout constant (b); −1.339 (c); no (d)

```
summary(lm(inf ~ turnout + log2(gdppc), data=world))

Call:
lm(formula = inf ~ turnout + log2(gdppc), data = world)

Residuals:
    Min      1Q  Median      3Q     Max
-26.887  -9.988  -2.745   4.712  80.095
```

```
Coefficients:
             Estimate Std. Error t value Pr(>|t|)
(Intercept) 189.84427   11.02530  17.219   <2e-16 ***
turnout      -0.10620    0.08423  -1.261    0.209
log2(gdppc) -11.76875    0.76794 -15.325   <2e-16 ***
---
Signif. codes:  0 '***' 0.001 '**' 0.01 '*' 0.05 '.' 0.1 ' ' 1

Residual standard error: 16.31 on 147 degrees of freedom
  (32 observations deleted due to missingness)
Multiple R-squared:  0.6231,    Adjusted R-squared:  0.618
F-statistic: 121.5 on 2 and 147 DF,  p-value: < 2.2e-16
```

4. A 1 percentage point increase in the share of a state's population that identifies as democrats is associated with a .99 percentage point decrease in the percentage of population with high school degrees, holding income and infant mortality constant (a); a $1 increase in per capita income is associated with a .0002 percentage point increase in the percentage of population with a high school degree, holding party identification and infant mortality constant (b); 1.5 (c); no (d)

```
summary(lm(hsdiploma ~ inc + infant + democrat, data=states))

Call:
lm(formula = hsdiploma ~ inc + infant + democrat, data = states)

Residuals:
    Min      1Q  Median      3Q     Max
-7.1201 -2.6018  0.4804  2.5909  5.6360

Coefficients:
              Estimate Std. Error t value Pr(>|t|)
(Intercept) 88.9641683  5.8437041  15.224   <2e-16 ***
inc          0.0001588  0.0001055   1.505   0.1391
infant      -0.9939696  0.4315306  -2.303   0.0258 *
democrat    -0.0949175  0.0743142  -1.277   0.2079
---
Signif. codes:  0 '***' 0.001 '**' 0.01 '*' 0.05 '.' 0.1 ' ' 1

Residual standard error: 3.385 on 46 degrees of freedom
Multiple R-squared:  0.3142,    Adjusted R-squared:  0.2694
F-statistic: 7.023 on 3 and 46 DF,  p-value: 0.0005504
```

5. For every year of delay in granting women's suffrage, there is a .00026 increase in the ratio of girls' education to boys' education, holding GDP per capita constant (a); .641 (b); no (c); 14.5% (d)

```
summary(lm(gtbeduc ~ womyear + log(gdppc), data=world))
```

```
Call:
lm(formula = gtbeduc ~ womyear + log(gdppc), data = world)

Residuals:
      Min        1Q    Median        3Q       Max
-0.279994 -0.021910  0.005049  0.034179  0.134674

Coefficients:
             Estimate Std. Error t value Pr(>|t|)
(Intercept) 0.2384604  0.8299714   0.287 0.774462
womyear     0.0002608  0.0004068   0.641 0.522941
log(gdppc)  0.0255863  0.0066863   3.827 0.000225 ***
---
Signif. codes:  0 '***' 0.001 '**' 0.01 '*' 0.05 '.' 0.1 ' ' 1

Residual standard error: 0.06162 on 101 degrees of freedom
  (78 observations deleted due to missingness)
Multiple R-squared:  0.1617,    Adjusted R-squared:  0.1451
F-statistic: 9.742 on 2 and 101 DF,  p-value: 0.0001353
```

6. For every year delayed in granting women's' suffrage, there is a .06 percentage point increase in the ratio of female/male infant deaths per 1,000 live births, holding income per capita constant (a); 3.18 (b); yes (c); 5% (d)

```
summary(lm((infemale/infmale)*100 ~ womyear +
                log(gdppc), data = world))

Call:
lm(formula = (infemale/infmale) * 100 ~ womyear + log(gdppc),
    data = world)

Residuals:
    Min      1Q  Median      3Q     Max
-14.593  -2.255  -0.034   2.311  20.786

Coefficients:
            Estimate Std. Error t value Pr(>|t|)
(Intercept) -42.20388   39.74096  -1.062  0.29008
womyear       0.06203    0.01965   3.157  0.00195 **
log(gdppc)    0.44781    0.31416   1.425  0.15627
---
Signif. codes:  0 '***' 0.001 '**' 0.01 '*' 0.05 '.' 0.1 ' ' 1

Residual standard error: 3.903 on 140 degrees of freedom
  (39 observations deleted due to missingness)
Multiple R-squared:  0.06671,   Adjusted R-squared:  0.05337
F-statistic: 5.003 on 2 and 140 DF,  p-value: 0.007967
```

7. A 1 percentage point increase in the state's share of the population with high school degrees is associated with a $443 increase in per capita income, holding population and student spending constant (a); a 1 percentage point increase in the state's share of the population with a high school degree is associated with a $446 increase in per capita income, holding population, student spending, and party identification constant (b); yes (c); no (d)

```
wout.lm <- lm(inc ~ hsdiploma + poptotal + stuspend, data = states)

with.lm <- lm(inc ~ hsdiploma + poptotal + stuspend +
              democrat, data = states)

stargazer(wout.lm, with.lm, header=FALSE, type="text", title =
          "Party Identification Is Relatively Inconsequential")
```

Party Identification Is Relatively Inconsequential

==

	Dependent variable:	
	inc	
	(1)	(2)
hsdiploma	442.701***	446.097***
	(129.234)	(136.083)
poptotal	0.0002***	0.0002***
	(0.0001)	(0.0001)
stuspend	2.428***	2.417***
	(0.282)	(0.311)
democrat		6.444
		(72.236)
Constant	−14,200.750	−14,619.970
	(10,941.430)	(12,018.260)
Observations	50	50
R2	0.701	0.701
Adjusted R2	0.682	0.675
Residual Std. Error	3,288.938 (df = 46)	3,324.987 (df = 45)
F Statistic	36.015*** (df = 3; 46)	26.431*** (df = 4; 45)

Note: *p<0.1; **p<0.05; ***p<0.01

8. For every 1 percentage point increase in the share of the population aged between 0 and 14 years, CO_2 emissions decline by 1.29%, holding GDP per capita constant (a); no (b); a 1% change in GDP per capita is associated a .08% decline in carbon emissions, holding the youth population constant (c); no (d)

```
summary(lm(log(co2) ~ young + log(gdppc), data=world ))

Call:
lm(formula = log(co2) ~ young + log(gdppc), data = world)

Residuals:
    Min      1Q  Median      3Q     Max
-1.9582 -0.4761 -0.1397  0.4456  2.1894

Coefficients:
             Estimate Std. Error t value Pr(>|t|)
(Intercept)  0.117069   0.956081   0.122    0.903
young       -0.012545   0.009248  -1.356    0.177
log(gdppc)  -0.078424   0.078757  -0.996    0.321

Residual standard error: 0.7238 on 141 degrees of freedom
  (38 observations deleted due to missingness)
Multiple R-squared:  0.01301,   Adjusted R-squared:  0.0009911
F-statistic: 0.9292 on 2 and 141 DF,  p-value: 0.3973
```

9. A 1 percentage point increase in the share of the state's population identifying as democrat is associated with a .05 percentage point decrease in the share of the population with high school degrees, holding income and infant mortality constant (a); a $1 increase in per capita income is associated with a .0002 percentage point increase in the share of the population holding a high school diploma (b); 1.83 (c); yes (d). *Note: Stargazer rounds its numbers such that a more accurate estimate of t-ratios and significance can be gained by using the *summary()* command to observe the estimates.

```
without.lm <- lm(hsdiploma ~ inc + infant + democrat,
          data = subset(states, st!="NY" & st!="CA" & st!="NJ"))

withall.lm <- lm(hsdiploma ~ inc + infant + democrat, data = states)

stargazer(without.lm, withall.lm, header=FALSE, type="text",
        title = "Absences of Cases Changes Results")
```

Absences of Cases Changes Results

```
===============================================================
                        Dependent variable:
                    -------------------------------------
                                hsdiploma
                        (1)                 (2)
---------------------------------------------------------------
inc                   0.0002*              0.0002
                     (0.0001)             (0.0001)

infant               -1.223***            -0.994**
                      (0.422)              (0.432)

democrat              -0.051               -0.095
                      (0.073)              (0.074)

Constant             88.094***            88.964***
                      (5.715)              (5.844)

---------------------------------------------------------------
Observations            47                   50
R2                    0.390                0.314
Adjusted R2           0.347                0.269
Residual Std. Error  3.234 (df = 43)     3.385      (df = 46)
F Statistic       9.156*** (df = 3; 43) 7.023*** (df = 3; 46)
===============================================================
Note:                           *p<0.1; **p<0.05; ***p<0.01
```

10. With each year delayed in granting women's suffrage, there is a .04 percentage point increase in the ratio of girls' to boys' education, holding GDP per capita constant (a); .91 (b); no (c); no (d)

```
woutME.lm <- lm(gtbeduc*100 ~ womyear + log(gdppc), data=subset(world,
region!="ME and North Africa"))

withME.lm <- lm(gtbeduc*100 ~ womyear + log(gdppc), data=world)

stargazer(woutME.lm, withME.lm, header=FALSE, type = "text", title =
"Including the Middle Eastern Cases Makes Little Difference")
```

Including the Middle Eastern Cases Makes Little Difference

```
===============================================================
                                    Dependent variable:
                              ---------------------------------
                                       gtbeduc * 100
                                   (1)              (2)
---------------------------------------------------------------
womyear                           0.039            0.026
                                 (0.043)          (0.041)

log(gdppc)                       2.818***         2.559***
                                 (0.686)          (0.669)

Constant                         -3.892           23.846
                                (86.951)         (82.997)

---------------------------------------------------------------
Observations                        97              104
R2                                0.191            0.162
Adjusted R2                       0.174            0.145
Residual Std. Error     6.058    (df = 94)    6.162  (df = 101)
F Statistic            11.087*** (df = 2; 94) 9.742*** (df = 2; 101)
===============================================================
Note:                                *p<0.1; **p<0.05; ***p<0.01
```

 Access digital resources, including datasets, at http://edge.sagepub.com/brownstats1e.

14 Dummies and Interactions

CHAPTER OUTLINE

Learning Objectives

Overview

What Is a Dummy Variable?

Additive Models and Interactive Models

Bivariate Dummy Variable Regression

Multiple Regression and Dummy Variables

Interactions in Multiple Regression

Summary

Common Problems

Review Questions

Practice on Analysis and Visualization

Annotated R Functions

Answers

LEARNING OBJECTIVES

- Describe a dummy variable.

- Discuss the difference between additive and interactive models.

- Explain what questions a dummy variable helps answer.

- Interpret dummy variables in multiple regression.

- Interpret results from multiple regression with an interactive term.

Overview

This chapter introduces two extensions of multiple regression: the use of dummy variables and interactions. Dummy variables are binary, registering 1 if a condition is true and 0 if false. An example might be a variable the indicates whether a respondent in a survey on police and race is Black or not. Dummy variables help answer questions that involve distinct characteristics. For example, are attitudes toward the police influenced by race? Interactions are two independent variables multiplied, allowing us to determine whether the impact of one is dependent on the other. Interactions ask a different kind of question: is the relationship between two variables affected by a third? For example, is the relationship between attitudes toward the police and race influenced by income?

Our goal throughout this book has been to understand the connection between the variables we use and the questions we ask. Dummy variables allow us to ask questions involving characteristics of people, places, or things that are distinct. Survey respondents either have kids or they don't. They either own a home or they don't. They either live in Nebraska or they live somewhere else. Does parenting, home ownership, or living in Nebraska increase political participation? Interactions model a different kind of question. Interactions allow us to determine whether parenting, home ownership, or being from Nebraska changes the relationship between things like income, age, party identification, and political participation.

In this chapter you'll learn how to interpret coefficients for dummy variables in the multiple regression context. The chapter then explains how to use dummy variables to form an interaction, testing whether the slope of a line changes appreciably between two categories. Beware, interpreting the coefficients in an interactive model requires care and a solid understanding of the more rudimentary models introduced in previous chapters.

What Is a Dummy Variable?

We often want to know if race, gender, party affiliation, education, or immigrant status influences attitudes or outcomes. We use **dummy variables** to indicate whether a voter is Republican or not, whether a test subject in a drug trial is a man or woman, or whether a beneficiary of the food stamp program is a citizen or immigrant. Do states experience more crime when marijuana use is legal? Are men or women more likely to vote, join unions, or participate in protests? Does your parents' political affiliation influence yours? These are all questions that require making an important distinction between people, places, or things. Also note that many of these questions involve "yes" or "no" answers.

Questions that generate answers other than "yes" or "no" also use dummy variables. How much do men typically make compared to women? How many homicides typically occur in states with or without the death penalty? How many cases of infection are associated with the decision to allow restaurants to remain open during a pandemic? All of these questions involve dummy variables.

Dummy variables record whether the condition is true or not by assigning a 1 or 0 to each case (best practice is assigning a 1 if the condition is true). In R, you'll see dummy variables either as a series of ones and zeros, or the variable will have labels assigned to each condition or category (i.e., graduate or undergraduate, Republican or Democrat, male or female, etc.). If the variable has more than two categories, it is not considered a dummy variable. Variables with more than two categories are categorical variables. Dummy variables used in the context of multiple regression allow us to estimate whether a specific state of being influences our dependent variable, controlling for the other variables included in the model.

> **KNOWLEDGE CHECK:** Describe a dummy variable.

1. Which questions involve using dummy variables?
    a. Should I stay or should I go?
    b. How many games will the Chiefs win this year?
    c. Will the Patriots make the playoffs?
    d. Will you vote in the next election?
2. Which of the following describe dummy variables?
    a. A variable that indicates if someone has brown hair
    b. A variable indicating the color of one's hair
    c. Does the respondent own a home or rent?
    d. Is the respondent a homeowner?

Additive Models and Interactive Models

Dummy variables are used in additive and interactive models. In an **additive model**, we want to know whether a kind of person, place, or thing is associated with our dependent variable. In an **interactive model**, we want to know if the relationship between our dependent variable and an independent variable *depends* on a kind of person, place, or thing. For example, an additive model would test whether, on average, President Obama received more support in the 2012 election from women or men (Figure 14-1A). An interactive model asks whether the relationship between support for Obama and another variable (education) depends on gender (Figure 14-1B).

FIGURE 14-1 Additive and Interactive Models

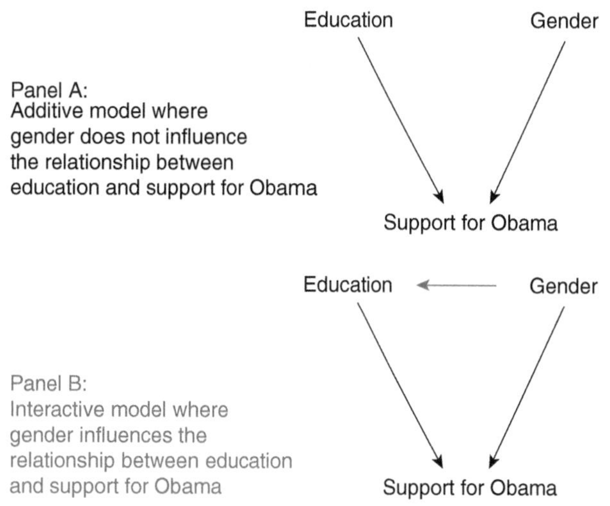

Panel A:
Additive model where gender does not influence the relationship between education and support for Obama

Panel B:
Interactive model where gender influences the relationship between education and support for Obama

Additive and interactive models are easily distinguishable. Additive models are a series of terms separated by the addition sign. Interactive models include at least two terms that are multiplied. Dummy variables can be multiplied by other dummy variables. Continuous variables can be multiplied by other continuous variables. In this chapter, we'll focus on dummy variables and their interaction with continuous variables to help build our intuition. We've already seen interactions between dummy variables and continuous variables when we asked whether the slope summarizing the relationship between two continuous variables changes over different levels of a categorical variable (controlled comparisons in Chapter 11). Models that include dummy variables allow us to measure the difference between the intercepts of two lines. Models that include interactive terms allow us to estimate the difference between their slopes.

> **KNOWLEDGE CHECK:** Discuss the difference between interactive and additive models.

3. Which of the following statements describe interactive models?
   a. They contain a multiplicative term.
   b. They contain a dummy variable.
   c. They allow independent variables to influence each other.
   d. They relax an important assumption associated with additive models.

Bivariate Dummy Variable Regression

A dummy variable—sometimes called a dichotomous variable—only records two values: 0 or 1. A variable recording whether a respondent self-identifies as male or female can be used to ask whether gender shapes attitudes about immigration, the poor, or the environment. A dummy variable that indicates whether a country is democratic or authoritarian can be used to determine whether democracy promotes economic growth. Are college-educated adults more likely to vote? Will former colonized nations experience relatively more political instability? Will government spending on health increase once a country has witnessed a case of Ebola? These are all questions that can be answered by employing dummy variables in a multiple regression model. The advantage of using a dummy variable with multiple regression is that we can control for a number of other variables.

Let's begin with and build on a simple model. Suppose we're interested in attitudes toward Muslims in the United States. Specifically, we use the feeling thermometer on views toward Muslims from the NES survey—a continuous variable that ranges from 0 to 100. Let's ask what the average score on the thermometer is for all respondents and then ask whether a difference exists between men and women.

$$Y = a_1 + e$$
$$Y = a_1 + b_1(Gender) + e$$

In the first equation, the intercept a_1 gives the mean of Y for all respondents, both men and women. In the second equation, we add the dummy variable. We assign the value of 1 for women and 0 for men. In this example, men are the **reference category** since they are assigned 0. Since the dummy variable registers 1 when the respondent is a woman and 0 when the respondent is a man, the b_1 drops out when the respondent is a man—multiplying b_1 by 0 equals 0, leaving a_1. Consequently, the term a_1 provides the mean value of Y when the respondent is a man. When the respondent is a woman, the gender variable registers a 1. Consequently, to calculate the mean value of Y for women, we take $a_1 + b_1$. Note the coefficient b_1 gives us *the difference* between men and women with respect to their attitudes toward Muslims. As a bonus, regression analysis indicates whether the difference between men and women (b_1) is statistically significant. Consider the following regression table for the corresponding regressions (Table 14-1).

The code in Code Chunk 14-1 should look familiar. I define two linear models, one with a constant of 1 and the other including the dummy variable *nes$gender*. I then use the *stargazer()* command to present the results from both regressions.

Code Chunk 14-1

```
genmus1 <- lm(ftmuslim ~ 1, data=nes)
genmus2 <- lm(ftmuslim ~ gender, data=nes)

stargazer(genmus1, genmus2, type = "text",
          title = "Table 14-1: Women View Muslims More Positively",
          header=FALSE)
```

First, note the gender variable is called *genderFemale*, indicating that the female category of gender is set to 1. This happens automatically in R: the category assigned a 1 is appended to the name of the variable. The difference between men and women (b_1) is 5.89. Women, on average,

TABLE 14-1 Women View Muslims More Positively

```
===============================================================
                         Dependent variable:
                         ---------------------------------------
                                   ftmuslim
                         (1)                       (2)
---------------------------------------------------------------
genderFemale                                       5.892***
                                                   (1.749)

Constant                 45.439***                 42.383***
                         (0.877)                   (1.259)

---------------------------------------------------------------
Observations             1,178                     1,178
R2                       0.000                     0.010
Adjusted R2              0.000                     0.009
Residual Std. Error      30.118 (df = 1177)        29.986 (df = 1176)
F Statistic                                        11.355*** (df = 1; 1176)
===============================================================
Note:                            *p<0.1; **p<0.05; ***p<0.01
```

record feelings toward Muslims 5.9 points higher than men. The difference is statistically significant: we are 99.9% confident that the difference between men and women is not zero. As a matter of completeness, the constant in Model 2 gives us the mean Muslim thermometer score for men—the reference category. So from the regression, we find that the average score for men is 42.38 and that the difference between men and women is 5.89, which implies the average for women is 42.38 + 5.89 = 48.27.

> **KNOWLEDGE CHECK: Explain what questions a dummy variable helps answer.**

4. What question does a model with no independent variables (just a constant) answer?
    a. What is the mean for the category indicated by zero?
    b. What is the mean of y for the entire sample?
    c. What is the mean for the reference category?
    d. What is the difference between the mean and the median?
5. Which of the following are true in the context of the single dummy variable model?
    a. The coefficient on the dummy variable gives the mean of the indicated category.
    b. The coefficient on the dummy variable gives the mean of the reference category.

c. The coefficient on the dummy variable gives the difference between the means of the two categories.

d. The coefficient on the constant gives the mean score for the reference category.

6. Which are the advantages associated with including dummy variables in the bivariate context?

   a. It calculates the difference between two categories.

   b. It calculates whether the difference is statistically significant.

   c. We can calculate options a and b while controlling for other variables.

   d. The coefficients associated with dummy variables give the difference in slopes between two lines.

Multiple Regression and Dummy Variables

Using dummy variables in regression analysis allows us to ask the same questions posed earlier while accounting for other variables. Using dummy variables in multiple regression analysis also allows us to test statistically whether the mean in one group is different from the mean in another, holding a number of other variables constant. Consequently, it provides a more accurate estimate of the differences between groups while giving us a sense of its statistical significance. We can use the t-ratio on the coefficient of a dummy variable to test whether the estimated difference is statistically significant. In the example here, we examine gender's impact on attitudes toward Muslims, controlling for other variables that might explain attitudes toward Muslims. Gender is clearly not the only variable that influences attitudes toward Muslims and multiple regression analysis allows us to account for that, ultimately generating a more accurate estimate of gender's effect.

When adding variables to the model, the interpretation stays the same, while adding the obligatory phrase *holding everything else constant*. Let's add a few more controls to the bivariate model: education and political ideology. The education variable ranges from 1 to 7, with 1 being no high school diploma and 7 being a postgraduate degree. The political ideology variable ranges from 1 to 7 as well, with 1 being the most liberal and 7 being the most conservative.

To illustrate, both variables are converted to continuous variables in Code Chunk 14-2. Once expressed as continuous variables, they are added to the linear model.

Code Chunk 14-2

```
nes$educ.n <- as.numeric(nes$educ)
nes$pid7.n <- as.numeric(nes$pid7)

muslim.lm <- lm(ftmuslim ~ gender + educ.n + pid7.n, data=nes)

stargazer(muslim.lm,
  title = "Table 14-2: Gender Influences Attitudes Toward Muslims",
  header = FALSE, type="text")
```

Although political ideology and education are both strongly associated with views toward Muslims, their inclusion in the model does not appreciably change the coefficient on the

TABLE 14-2 Gender Influences Attitudes Toward Muslims

```
=========================================================
                         Dependent variable:
                    -------------------------------
                              ftmuslim
---------------------------------------------------------
genderFemale                  5.039***
                              (1.607)

educ.n                        3.179***
                              (0.526)

pid7.n                       -4.978***
                              (0.358)

Constant                     51.058***
                              (2.526)

---------------------------------------------------------
Observations                   1,164
R2                             0.178
Adjusted R2                    0.176
Residual Std. Error    27.335 (df = 1160)
F Statistic         83.873*** (df = 3; 1160)
=========================================================
Note:                 *p<0.1; **p<0.05; ***p<0.01
```

gender dummy variable. Gender seems to have a statistically significant association with feelings toward Muslims, controlling for political ideology and education.

Before we move on to the next case, let's consider what would happen if we add another dummy variable to the regression. Instead of using political ideology as a continuous variable, let's use *nes$partyid3*, which classifies respondents as Democrat, Independent, or Republican. When a categorical variable with three levels is added to a model in R, two dummy variables are automatically created (in this case, one for Independents and one for Republicans). The resulting equation and regression output are discussed next.

Art and Practice of Data Visualization
DUMMY VARIABLES IN R

When adding categorical variables to a regression in R, the program creates n − 1 dummy variables where n equals the number of categories in the variable. The category not included in the regression is called the reference category and is captured by the constant.

In this example, the omitted category or the reference category is "Democrat" since it is the category not included as a dummy variable in the regression. Consequently, the regression output has the gender variable and two additional dummy variables: one for Republicans and one for Independents. In the third equation, when the respondent is a Democrat (the omitted category or reference category), b_2 and b_3 will be zero. As a result, the constant in the regression gives the average Muslim thermometer score for men who consider themselves Democrats—remember that men are the omitted category for the gender dummy. Being a Democrat and a male are the two reference categories.

$$Y = a_1 + e$$
$$Y = a_1 + b_1(Gender) + e$$
$$Y = a_1 + b_1(Gender) + b_2(Independent) + b_3(Republican) + e$$

Before we add the party identification variable to the regression, a little tidying is necessary (Code Chunk 14-3). First, we convert the "other" and "not sure" responses to "NA" since we want to restrict our sample to those that answered either Democrat, Republican, or Independent. We accomplish this with the *ifelse()* command. Translating the R code into English, if *nes$pid3* is equal to "other," then assign it an "NA." If *nes$pid3* equals "not sure," then assign an "NA"; otherwise it gets the value of *nes$pid3*. I then designate it as a factor, I label the levels, and I put them in the order Democrat-Independent-Republican.

Code Chunk 14-3

```
nes$pid3.new <- ifelse(nes$pid3 == "Other", NA,
              ifelse(nes$pid3 == "Not Sure", NA,
                     nes$pid3))

nes$pid3.new <- as.factor(nes$pid3.new)

levels(nes$pid3.new)=c("Democrat", "Republican", "Independent")

nes$pid3.new = factor(nes$pid3.new, levels(nes$pid3.new)[c(1,3,2)])
```

Now that we have the party identification variable specified exactly how we'd like it, we're ready to include it in the regression. To illustrate the difference it makes, I first define one regression without it (*muslim3.lm*) and then one regression with it (*muslim4.lm*). I then place the two models into the *stargazer()* command (Code Chunk 14-4).

Code Chunk 14-4

```
muslim3.lm <- lm(ftmuslim ~ gender, data=nes)
muslim4.lm <- lm(ftmuslim ~ gender + pid3.new, data=nes)

stargazer(muslim3.lm, muslim4.lm, type = "text",
          title = "Table 14-3: Views Toward Muslims", header = FALSE)
```

TABLE 14-3 Views Toward Muslims

	Dependent variable: ftmuslim	
	(1)	(2)
genderFemale	5.892***	4.345**
	(1.749)	(1.718)
pid3.newIndependent		-12.953***
		(1.991)
pid3.newRepublican		-25.449***
		(2.163)
Constant	42.383***	54.391***
	(1.259)	(1.668)
Observations	1,178	1,100
R2	0.010	0.122
Adjusted R2	0.009	0.119
Residual Std. Error	29.986 (df = 1176)	28.253 (df = 1096)
F Statistic	11.355*** (df = 1; 1176)	50.590*** (df = 3; 1096)

Note: *p<0.1; **p<0.05; ***p<0.01

In our example, Republicans record thermometer scores 25.45 points less than Democrats and Independents record thermometer scores 12.95 points less than Democrats, holding gender constant.

> **KNOWLEDGE CHECK: Interpret dummies in multiple regression.**

7. What is the advantage to using dummy variables in a multiple regression model?
   a. We gain the same advantage as when we move from the bivariate to the multiple regression contexts.
   b. With dummy variables we can test whether the difference in mean values from each category is substantively significant.
   c. With dummy variables we can test whether the difference mean values from each category are statistically significant.
   d. With multiple regression analysis we can hold a number of different variables constant while examining the distinction defined in a dummy variable.

8. Identify the reference category in the following regressions containing the following dummy variables.

   a. Men = 0, women = 1; Catholic = 0, other = 1
   b. Republican = 0, other = 1; college educated = 0, other = 1
   c. Republican = 1, other = 0; college educated = 1, other = 0
   d. College educated = 1, other = 0; U.S. citizen = 0, noncitizen = 1; employed = 1, unemployed = 0

9. Consider the regression results in Table 14-3 and answer the following questions.

   a. What is the mean value for females when controlling for other variables?
   b. What is the mean value for females without controlling for other variables?
   c. What is the mean value for men when controlling for other variables?
   d. Is the value for men significantly different (statistically) when controlling for other variables?

Interactions in Multiple Regression

Sometimes the relationship between two variables is affected by a third. When we included a dummy variable in multiple regression, we were able to tell whether being in one category versus another (e.g., male or female) resulted in a sizable or statistically significant shift in our predictions. Specifically, we were measuring whether there was a difference in intercepts. With interactions, we're interested in whether there is a difference in slopes. Including interaction terms in multiple regression allows us to estimate the difference between the slopes of two lines and to ascertain whether the difference is statistically significant, while holding all other variables in the model constant.

For example, the relationship between income and education could be influenced by gender: we know that gender discrimination exists, lowering the amount that women make relative to men. The question is whether the relationship between income and education is different for men and women. In other words, if we plotted income against educational achievement and fit two lines to the data (one for women and one for men), would the lines have the same slope? If men gained more with each additional year of education than women did, there is an interaction. If the lines had the same slope but the men's line had a larger intercept, we'd say there's no interaction—it's just that men, on average, always make more than women by the same amount at all levels of education. Consider the plot in Figure 14-2.

In Figure 14-2, I plotted income (*nes$faminc*) against education (*nes$educ*) and fit two separate lines to the data based on gender. As with the previous example, there is some tidying to do before we plot.

Remember that rather than registering an exact amount of income, the survey assigns respondents to one of 16 categories—16 being the highest level of income. In Code Chunk 14-5, to preserve the original variable I create a copy of *nes$faminc* and call it *nfam*. I then recode all values above 16 as "NA." (The categories above 16 recorded either nonresponses or were not available.) I also transform the educational variable from a factor to a numeric value.

Code Chunk 14-5

```
nes$nfam <- nes$faminc
  nes$nfam[nes$nfam > 16] <- NA

nes$educ.n <- as.numeric(nes$educ)
```

Having "cleaned up" the *nes$faminc* variable and converted the *nes$educ* variable into a numeric vector, I'm ready to plot (Code Chunk 14-6). Although you have seen it before, note a few extras I've included in this plot: (1) I manually determined the colors of the graph with *scale_color_manual()* and (2) I placed the legend of the graph within the graph itself using the option *legend.position*.

Code Chunk 14-6

```
ggplot(nes, aes(educ.n, nfam, col=gender)) +
  geom_smooth(method="lm", se=FALSE) +
  theme_minimal() +
  theme(plot.title = element_text(size = 8, face = "bold"),
      axis.title = element_text(size = 8, face = "bold"),
      legend.title = element_text(size = 10),
      legend.position = c(.8,.3)) +
  scale_color_manual(values = c("#0000bf", "#bf0000"), "Gender") +
  ggtitle("Figure 14-2: Interaction Between Education and Gender") +
  ylab("Income (ordinal scale)") +
  xlab("Education Level (ordinal scale)")
```

The plot illustrates there is an interaction: as education increases, the difference between men's and women's income increases. The slopes are different, indicating there is an interaction between education and gender when explaining income. Estimating interactions in the multiple regression framework allows us to calculate exactly what the difference in slopes is and to determine whether that difference is statistically significant. Finally, note that the y and x axes indicate we are comparing education and income, but the units are left somewhat vague. Since I'm using the variables from a survey that asked respondents to

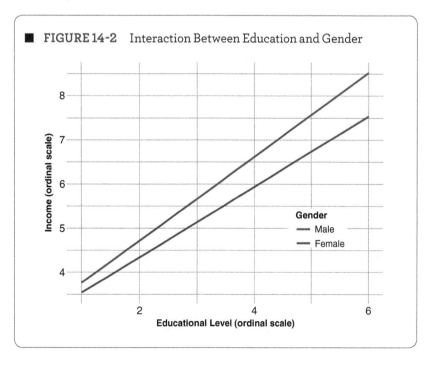

■ **FIGURE 14-2** Interaction Between Education and Gender

identify what their levels of education and income were rather than exact amounts (of years or dollars), we're only left with a vague sense that income is increasing when we move up the y axis and the level of education increases when we move from left to right on the x axis.

Example: Bernie Sanders, Education, and Income

To help build our intuition with dummies and interactions, it's important to note that we've done this before. In a previous example we plotted homicides versus GDP per capita (logged) and fit two separate lines—one for democracies and one for dictatorships (Figure 11-9). Let's consider another example.

Senator Bernie Sanders from Vermont fought hard for the democratic nomination in the 2016 election cycle, challenging Hillary Clinton's seemingly inevitable path to the White House. Senator Sanders's support appeared to come from the youngest voters, primarily college-educated men and women. An important emphasis in his campaign centered on the 2008 financial crisis and the ability of bank executives and hedge-fund managers to emerge unscathed from the pain and suffering caused when people's savings were cut in half overnight.

Understanding exactly how income and education interacted to bolster support for the Vermont senator would not only have helped inform campaign officials in the Clinton camp, but it could also help inform strategists contemplating campaigns in 2018 and beyond. How is the relationship between income and support for Sanders influenced by the level of a voter's education? To answer that question, consider the following model:

$$Y =_1 + \beta_1 \text{Income} + \beta_2 \text{CollegeDummy} + \beta_3 \text{CollegeDummy} \times \text{Income} + \epsilon$$

In this regression, Y is a feeling thermometer for Bernie Sanders, β_1 is the coefficient for a continuous income variable, β_2 is the coefficient for a dummy variable registering one if the respondent has a college degree, and β_3 is the coefficient for the interaction between education and income. The error terms is represented by ϵ.

Similar to the previous examples in this chapter, we need to do some cleaning (Code Chunk 14-7). Given our question, we need to re-express the *nes$educ* variable so that it registers 1 when the respondent has received a college degree and 0 otherwise. We also need to transform the variable *nes$faminc* into a continuous variable that ranges from 1 to 16, producing a continuous variable.

Code Chunk 14-7

```
nes$college <- nes$educ
nes$college <- ifelse(nes$college=="4-year", 1,
                      ifelse(nes$college=="Post-grad", 1, 0))

nes$inc <- nes$faminc
nes$inc <- ifelse(nes$inc > 16, NA, nes$inc)
nes$inc.n <- as.numeric(nes$inc)
```

Now that we've created a dummy variable that indicates whether a respondent is college educated or not and we have a continuous variable that measures the level of a respondent's income, we're ready to estimate a multiple regression model with an interaction. The model tests whether the relationship between income and support for Sanders is influenced by having a college education (Code Chunk 14-8).

Code Chunk 14-8

```
sanders <- lm(ftsanders ~ inc.n + college + inc.n:college, data=nes)

stargazer(sanders, title = "Table 14-4: Income, Education, and Bernie",
          header = FALSE, type = "text")
```

TABLE 14-4 Income, Education, and Bernie

```
=============================================
                     Dependent variable:
                 ----------------------------
                           ftsanders
---------------------------------------------
inc.n                        0.594
                            (0.417)

college                      3.244
                            (5.286)

inc.n:college               -0.719
                            (0.712)

Constant                   47.961***
                            (2.390)

---------------------------------------------
Observations                 1,033
R2                           0.002
Adjusted R2                 -0.001
Residual Std. Error   33.408 (df = 1029)
F Statistic          0.716 (df = 3; 1029)
=============================================
Note:              *p<0.1; **p<0.05; ***p<0.01
```

The estimates are surprising (Table 14-4). First, none of the coefficients in the model are substantively or statistically significant. There do not seem to be any strong associations between income, education, and support for Senator Sanders. The coefficient on the college dummy variable is 3.24, indicating that college-educated respondents held only a slightly higher view of Sanders. The coefficient on the income variable (*inc.n*) is positive but is not statistically significant. For each 1-point increase in the *inc.n* variable, there is just over a .5-point increase in the Sanders feeling thermometer. Second, the interaction term indicates that while the slope for college-age respondents is −.719 less than the slope for those without degrees, the difference is not statistically significant. The variance explained by the interactive model (the R^2) is also unimpressive (.002). According to this model, it does not seem income and education had much to do with Bernie's appeal.

While regression tables give the precise estimates, an effective way to fully understand the substantive and statistical significance of an interaction is through visualization. Fortunately, there is a package in ggplot called 'interactions' that produces high-quality graphical representations of interactions. In Code Chunk 14-9 I take the regression from Code Chunk 14-8 and generate a visualization using the 'interactions' package. Since the *interact_plot()* command uses ggplot as its back end, the same ggplot features I've used throughout the book can be used to add layers to the figure. In Figure 14-3 I show that the slopes and the difference in the intercepts correspond exactly to the regression Table 14-4.

Code Chunk 14-9
```
interact_plot(sanders, pred = inc.n, modx = college,
              legend.main = "College Graduate",
              modx.labels = c("no", "yes")) +
  ylim(20,80) +
  theme_minimal() +
  theme(plot.title = element_text(size = 8, face = "bold"),
        axis.title = element_text(size = 8, face = "bold"),
        legend.title = element_text(size = 10),
        legend.position = c(.3,.3)) +
  annotate("segment", x = 12, xend = 12, y = 65, yend = 55, size = .2) +
  annotate("text", x = 12, y = 67, parse = F,
           label = 'Slope = .59', size = 2) +
  annotate("segment", x = 14, xend = 14, y = 49, yend = 40, size =.2) +
  annotate("text", x = 14, y = 37, parse = F,
           label = 'Slope = .59 - .719', size = 2) +
  annotate("segment", x = 1, xend = 1, y = 48, yend = 58, size = .2) +
  annotate("text", x = 4, y = 60, parse = F,
           label = 'When x = 0, the difference between the
                    intercepts is 3.24', size = 2) +
  ggtitle("Figure 14-3: Interaction Between Education and Income") +
  ylab("Feeling Thermometer for Bernie Sanders") +
  xlab("Income (ordinal scale)")
```

The two lines depicted in Figure 14-3 show there is an interaction. We observe that at the income level of 4, the model predicts a very similar feeling thermometer score for both those with and without college educations (roughly 50). As we proceed along the x axis (as incomes rise), the difference approaches 8 points. In terms of feeling thermometer scores, that may or may not be much. Is the difference in slopes between these two lines statistically significant? According to the regression table, the estimate on the interaction is not

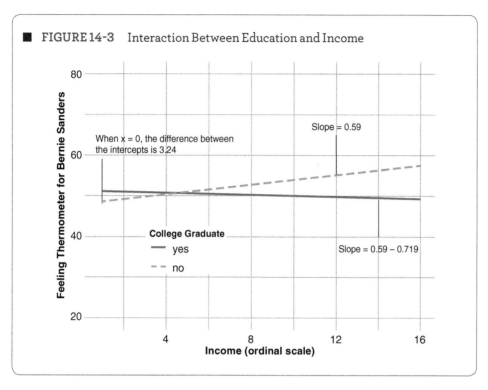

FIGURE 14-3 Interaction Between Education and Income

statistically significant. To visualize that result, I added 95% confidence bands to the estimates in Code Chunk 14-10.

Since the model only includes the variables that we interact, the lines depicted in Figure 14-3 correspond directly to the coefficients in Table 14-2. The slope of the line for those without a college education is .59. The slope for those with a college education is .59 minus .719. Remember that the coefficient on the interaction term gives the difference between the two slopes and not the actual slope of the line for the category indicated (in this case, those with a college education).

Code Chunk 14-10

```
interact_plot(sanders, pred = inc.n, modx = college,
              interval = TRUE, int.width = .95,
              legend.main = "College Graduate",
              modx.labels = c("no", "yes")) +
ylim(20,80) +
theme_minimal() +
theme(plot.title = element_text(size = 8, face = "bold"),
      axis.title = element_text(size = 8, face = "bold"),
      legend.title = element_text(size = 10),
      legend.position = c(.3,.3)) +
ggtitle("Figure 14-4: Confidence Bands Overlap") +
ylab("Feeling Thermometer for Bernie Sanders") +
xlab("Income (ordinal scale)")
```

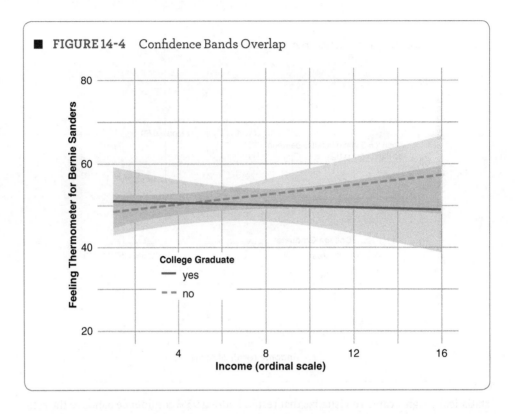

FIGURE 14-4 Confidence Bands Overlap

Notice how the two confidence bands overlap significantly. This indicates that our predictions for the college-educated voter are not really that different compared to the predictions for those without a college education. This corresponds with the lack of statistical significance associated with the interaction term in the regression table. While there is a difference in the slopes of the two lines, it is not substantively all that significant and we cannot reject the null hypothesis that the difference between the two slopes is zero.

Of course, there may be other variables that could be included in this model. Missing, perhaps, is age. Also missing is geography. It is entirely possible that if age and geography were added to the regression, the coefficients reported in Table 14-4 would change.[1]

Example: Foreign Aid, GDP per Capita, and Democracy

Scholars in economics and political science are interested in explaining foreign aid levels. Some argue a country's wealth determines the amount of aid it receives. Others argue that regime type is an important factor. Perhaps there's an interaction between regime type and income.

---

[1] The lines drawn in Figures 14-2 and 14-3 and the regression coefficients correspond because this is an analysis of covariance (ANCOVA) model that interacts each continuous variable with the same dummy. Other than income, education, and their interaction, there are no other variables in the model. Consequently, the estimates (coefficients) in the regression correspond directly to the slopes and intercepts we observe in the figure. When additional variables are added to a regression with interactions, presenting the interaction is not as straightforward. Fortunately, packages like 'interactions' in R exist, which make rendering the marginal effects of a model extremely easy.

Does regime type influence the relationship between a country's income and the foreign aid it receives? Let's first look at the equation.

$$Y = \alpha_1 + \beta_1 \text{GDP per capita}(\text{logged}) + \beta_2 \text{Democracy} + \beta_3 \text{GDP} \times \text{Democracy} + \beta_4 \text{Durable}(\text{logged}) + \beta_5 \text{ELF} + \epsilon$$

In this equation, α_1 represents the intercept for dictatorships, β_1 gives the slope of the line for dictatorships, β_2 registers the difference between the dictatorship intercept and the democracy intercept, and β_3 gives the difference between the slopes of the lines for dictatorships and democracies. Finally, ϵ represents the error term. Note how we're also able to control for the number of years since a significant regime change (*durable*) and the ethnolinguistic fractionalization score (ELF).

To execute the regression, we need to transform some variables (Code Chunk 14-11). First, I create a logged version of foreign aid. Since there are some countries that receive no aid, I add a constant of 1 to the variable before logging (remember, the log of 0 is undefined). I also create a logged version of GDP per capita. Finally, I create a dummy variable that classifies parliamentary and presidential democracies as "democratic." All other regime types are classified as authoritarian.

Code Chunk 14-11

```
world$lnaid <- log(world$aid +1)
world$lngdppc <- log(world$gdppc)

world$dem <- ifelse(world$regime=="Parliamentary Democracy", 1,
            ifelse(world$regime=="Presidential Democracy", 1, 0))

world$dem <- as.factor(world$dem)

levels(world$dem)=c("Dictatorship", "Democracy")
```

Once I have performed the necessary transformations, we're ready to estimate the regression. There are a number of different ways to specify an interaction in R. In Code Chunk 14-12, I multiply the logged version of GDP per capita and democracy by placing an asterisk between the two variables.

Code Chunk 14-12

```
aid.lm <- lm(lnaid ~ lngdppc * dem + log(durable + 1) +
                ethfrac, data=world)

    stargazer(aid.lm, title = "Table 14-5: Democracy Interacts
            With Income", header = FALSE, type = "text")
```

TABLE 14-5 Democracy Interacts With Income

```
===============================================
                    Dependent variable:
                    ------------------------
                    lnaid
-----------------------------------------------
lngdppc             -0.780***
                    (0.075)

demDemocracy        2.765***
                    (1.055)

log(durable + 1)    -0.088
                    (0.059)

ethfrac             -0.239
                    (0.257)

lngdppc:demDemocracy -0.330***
                    (0.123)

Constant            8.361***
                    (0.675)

-----------------------------------------------
Observations        99
R2                  0.736
Adjusted R2         0.722
Residual Std. Error 0.587 (df = 93)
F Statistic         51.823*** (df = 5; 93)
===============================================
Note:               *p<0.1; **p<0.05; ***p<0.01
```

The intercept for authoritarian regimes (the reference category) is 8.36. Since aid and GDP per capita are both logged, a 1% change in GDP per capita is associated with a .78% decrease in aid. Since the coefficient on the dummy term is positive, democracies receive more aid than do their authoritarian counterparts when all other variables are 0. Calculating the difference in intercepts is somewhat more involved since the dependent variable is logged. To convert that coefficient to proper units, we exponentiate 2.765, which generates 15.88.[2] When all other

---

[2] To exponentiate a number in R, simply type 'exp()' at the console prompt and put the number in the parentheses. In this example, *exp*(2.765) = 15.88.

independent variables are at zero, the difference in aid between democracies and dictatorships is 15.88 percentage points. While significant as a percentage of GDP per capita, remember that there are no cases where the other variables are at 0.

We find that the slope for the GDP per capita line is less for democracies than for dictatorships. Specifically, the slope on the GDP per capita line is −.330 points less for democracies than it is for dictatorships. We also learn that the coefficient is statistically significant, indicating that the difference between the slopes is not likely zero. This indicates that as a country's GDP per capita increases, foreign aid for democratic regimes decreases at a faster rate than it does for authoritarian regimes.

To visualize the relationship, consider Figure 14-5. See if you can connect the figure's features with the regression output. Since not all of the variables in the model are interacted, the coefficients on the intercepts won't match exactly with the lines drawn in the figure, but they should be fairly close. Aid flows, our results suggest, are more sensitive to GDP per capita in democracies than in authoritarian regimes: very poor democratic countries receive relatively more aid, but their richer counterparts receive comparatively less than do authoritarian regimes.

The only difference in the code for this example is my use of colors. Note that I specify exactly which colors to use in the *interact_plot()* command (Code Chunk 14-13). Everything else is the same as the previous example.

Code Chunk 14-13

```
interact_plot(aid.lm, pred = lngdppc, modx = dem, data = world,
              legend.main = "Regime Type",
              modx.labels = c("Dictatorship", "Democracy"),
              plot.points = TRUE,
              colors = c("#bf0000", "#0000bf")) +
  xlab("GDP per Capita (logged)") +
  ylab("Aid as a Percentage of GDP (logged)") +
  ggtitle("Figure 14-5: Interaction Between Democracy and
                      GDP per Capita") +
  theme_minimal() +
  theme(plot.title = element_text(size = 8, face = "bold"),
        axis.title = element_text(size = 8, face = "bold"),
        legend.title = element_text(size = 10),
        legend.position = c(.8,.8))
```

The steeper line—more rapidly declining—for democratic states indicates rich authoritarian states receive more aid than do their democratic counterparts. Perhaps richer authoritarian regimes are more likely to possess important natural resources or occupy strategic positions around the globe. Or wealthy authoritarian regimes will more likely purchase military equipment. All are possibilities worth exploring.

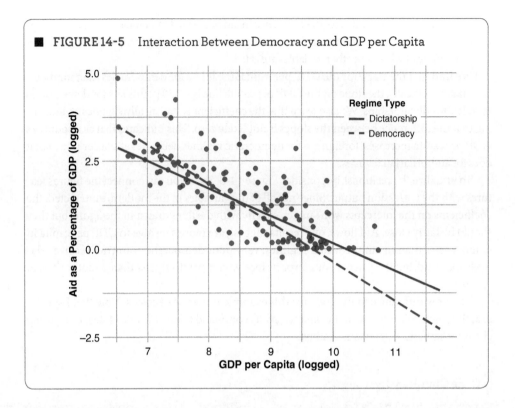

FIGURE 14-5 Interaction Between Democracy and GDP per Capita

KNOWLEDGE CHECK: Interpret results from a multiple regression with an interactive term.

10. Which are accurate statements regarding interactions in multiple regression analysis?

    a. The coefficient on the interaction term indicates the slope of the nonreference category.

    b. The coefficient on the interaction term indicates the slope of the reference category.

    c. The coefficient on the interaction term indicates the difference in slopes of the two lines.

    d. The coefficient on the interaction term indicates the difference in intercepts of the two lines.

11. Which are accurate statements regarding interactions in multiple regression analysis?

    a. The coefficient on the dummy term indicates the slope of the nonreference category.

    b. The coefficient on the dummy term indicates the slope of the reference category.

    c. The coefficient on the dummy term indicates the difference in slopes of the two lines.

    d. The coefficient on the dummy term indicates the difference in intercepts of the two lines.

12. Which are accurate statements regarding interactions in multiple regression analysis?

    a. The coefficient on the continuous variable (without the interaction) indicates the slope of the nonreference category.

    b. The coefficient on the continuous variable (without the interaction) indicates the slope of the line for the reference category.

    c. The coefficient on the continuous variable (without the interaction) represents the difference in slopes of the two lines.

    d. The coefficient on the constant indicates the intercept of the line for the reference category.

SUMMARY

In this chapter we considered dummy variables and interactions. Dummies and interactions help answer some important questions we might ask as we explore the data. Dummy variables help us understand whether simple distinctions between people, places, or things matter. Interactions of the kind we've studied here are particularly helpful when we want to see if a relationship depends on those simple distinctions. In addition to calculating the substantive differences generated by the distinction, regression analysis provides estimates of their statistical significance. By placing dummies and their interactions in a convincing model with other controls, we can obtain relatively accurate estimates of their effects.

Dummies and interactions are not intuitive: their mastery requires practice. The trick is to associate the different kinds of interactions with the questions they ask, the regressions needed to run them, and the results. The more you gain an intuition, you'll find that there are a multitude of interesting questions out there you've not yet thought of or explored.

COMMON PROBLEMS

- *Interpret the coefficient on dummy terms or interactions as differences between two intercepts and two slopes.* Hands down, the most common problem is making the mistake of interpreting the coefficient on dummy terms or interactions as actual intercepts or slopes rather than the differences between two intercepts and two slopes. The coefficients on dummy terms and interactions always represent the difference between intercepts or the difference between slopes. The coefficient for the constant term represents the intercept for the reference category (the category assigned 0 in a dummy variable). The coefficient on the continuous variable when it is by itself (not interacted) represents the slope of the line for the reference category.

- *Always graph the interaction.* While there may be a large substantive and statistically significant difference between intercepts and slopes, where the two lines intersect holds important substantive information. Where the two lines intersect represents a point where the predictions for both categories under examination are exactly the same. Knowing if most of the cases exist to the left or right of that point of intersection provides a better substantive sense of what is really going on.

- *How does one interpret the constant when there is either one or multiple dummy variables in the regression?* When there is one dummy variable in the regression, the constant represents the y intercept for the category of the dummy variable assigned a zero. For each additional dummy variable, simply add the reference category as an additional descriptor. For example, consider a regression model explaining how much a respondent can bench press with the following dummies: no PhD = 1, age is under 60 = 1, starred in high school sports = 1. In that regression, the constant would represent professors, over the age of 60, who did not star in high school sports.

REVIEW QUESTIONS

1. What is a dummy variable?
2. What's the difference between an additive model and one with an interaction?
3. What question does the use of a dummy variable answer?
4. What question does the use of an interaction answer?
5. What is a reference category?
6. If a categorical variable has six categories and is placed in a regression, how many dummy variables will it create?
7. What is the advantage of using a dummy variable in a multiple regression over a simple means test?
8. What does the coefficient on the interactive tell us in a multiple regression?
9. If there is a dummy variable in the regression, what does the intercept represent?
10. What does the coefficient associated with a dummy variable indicate?

PRACTICE ON ANALYSIS AND VISUALIZATION

1. Which of the following are dummy variables?
   a. *world$regime*
   b. *nes$gender*
   c. *states$trumpwin*
   d. *nes$pid3*

2. Indicate whether the description represents an interactive or additive model.
   a. $y = a + b$
   b. $y = a \times b$
   c. $y = a + b + c + a \times b$
   d. $y = a + b + c$

3. What is the correct interpretation of a dummy variable in a multiple regression model?
   a. The average value of the indicated category, holding all other variables constant
   b. The average value of the indicated category
   c. The difference between the reference category and the indicated category when the other variables are zero
   d. The average difference between the reference category and the indicated category

4. Which of the following models is represented by the question "Does gender matter when considering the relationship between education and income?"
   a. *Income = gender + education*
   b. *Income = gender × education*
   c. *Income = gender + education + gender × education*
   d. *Income = gender + education × gender*

5. Which of the following models is represented by the question "Does gender influence the relationship between education and income?"
   a. *Income = gender + education*
   b. *Income = gender × education*
   c. *Income = gender + education + gender × education*
   d. *Income = gender + education × gender*

6. Which of the following models is represented by the question "Does gender and education influence income?"
   a. *Income = gender + education*
   b. *Income = gender × education*
   c. *Income = gender + education + gender × education*
   d. *Income = gender + education × gender*

7. Determine whether democracy (using the dummy variable defined in Code Chunk 14-11) influences the relationship between income and ethnolinguistic heterogeneity (income being the dependent variable) and answer the following:

   a. What is the intercept for dictatorships?
   b. What is the intercept for democracies?
   c. What is the slope for dictatorships?
   d. What is the slope for democracies?

8. Determine whether having some marijuana legislation on the books (*weed*) changes the relationship between income (*inc*) and the homicide rate in U.S. states and answer the following:

   a. What is the intercept for states without legalized marijuana?
   b. What is the intercept for states with legalized marijuana?
   c. What is the slope for states without legalized marijuana?
   d. What is the slope for states with legalized marijuana?

9. Draw a picture of the interaction described in question 7.

10. Draw a picture of the interaction described in question 8.

ANNOTATED R FUNCTIONS

The following functions appear in this chapter. They are listed in order of their first appearance (with the code chunk number in parentheses) and annotated here to give a very brief description of their use. Some are not stand-alone functions and only work in combination with other commands. As a reminder, the code in every chapter will work properly if executed in the order it appears. Proper execution also depends on typing the author-defined *libraries()* command, which loads the required R packages.

stargazer(): generates a regression table with specified model. (14-1)

ifelse(): logical function that allows you to construct an if-then statement. Useful for selecting specific cases to label also for creating categorical variables from continuous variables. (14-3)

levels(): specifies the labels of levels for a categorical variable. (14-3)

ggplot(): defines the basic structure of a plot (usually the x and y variables). (14-6)

aes(): the aes (called "aesthetics") function is used in ggplot to define the basic structure of the plot, which often includes the variables you want to use and any shapes or colors. (14-6)

geom_smooth(): draws lines or curves to a scatter plot. (14-6)

theme_minimal(): specifies a minimalist style for ggplot. (14-6)

theme(): specifies font, size, and so forth in a ggplot. (14-6)

scale_color_manual(): allows you to specify which colors you want to use in a ggplot. (14-6)

ggtitle(): provides the title for a ggplot. (14-6)

ylab(): labels the y axis in ggplot. (14-6)

xlab(): labels the x axis in ggplot. (14-6)

interact_plot: an effective tool to visualize interactions from a regression model. (14-9)

ylim(): specifies the range of the x axis. (14-9)

annotate(): places lines or text in the figure. (14-9)

ANSWERS

KNOWLEDGE CHECK

1. a, c, d
2. a, c, d
3. a, c, d
4. b
5. c, d
6. a, b
7. a, b, c, d
8. Catholic men (a), Republican with college education (b), not Republican or college educated (c), a noncollege-educated citizen without a job (d)
9. 58.736 (a), 48.2 (b), 54.391 (c), yes (d)
10. c
11. d
12. b, d

PRACTICE ON ANALYSIS AND VISUALIZATION

1. b, c
2. additive (a), interactive (b), interactive (c), additive (d)
3. c, d
4. a
5. c
6. a
7. 9.818 (a), 9.819 (b), −2.01 (c), −1.44 (d)
8. 10.31 (a), .6 (b), −.0001579 (c), −.000132 (d)

9.
```
mod1.lm <- lm(lngdppc ~ ethfrac + dem + ethfrac:dem, data=world)

interact_plot(mod1.lm, pred = ethfrac, modx = dem, data = world,
          legend.main = "Regime Type",
          modx.labels = c("Dictatorship", "Democracy"),
          plot.points = TRUE,
          colors = c("#bf0000", "#0000bf")) +
xlab("Ethnolinguistic Fractionalization") +
ylab("GDP per Capita (logged)") +
ggtitle("Interaction Between Democracy and ELF") +
theme_minimal() +
theme(plot.title = element_text(size = 8, face = "bold"),
      axis.title = element_text(size = 8, face = "bold"),
      legend.title = element_text(size = 10),
      legend.position = c(.9,.9))
```

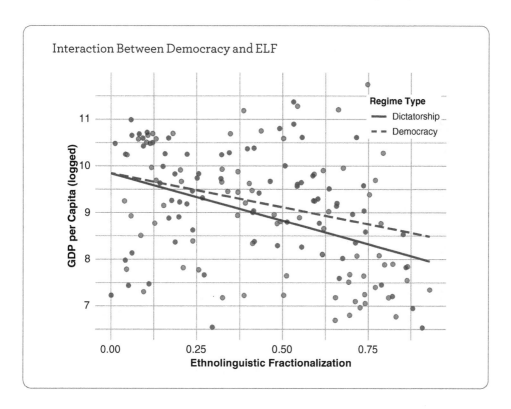

10.
```
mod2.lm <- lm(murderrate ~ weed + inc + weed:inc, data=states)

interact_plot(mod2.lm, pred = inc, modx = weed, data =states,
            legend.main = "Marijuana Laws",
            modx.labels = c("Not Legal", "Legal"),
            plot.points = TRUE,
            colors = c("#bf0000", "#0000bf")) +
  xlab("Income per capita") +
  ylab("Homicide Rate (murders per 100,000)") +
  ggtitle("Interaction Between Marijuana Laws and Income") +
  theme_minimal() +
  theme(plot.title = element_text(size = 8, face = "bold"),
        axis.title = element_text(size = 8, face = "bold"),
        legend.title = element_text(size = 10),
        legend.position = c(.9,.9))
```

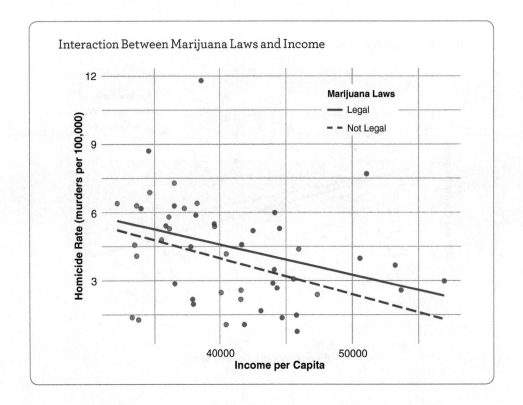

 Access digital resources, including datasets, at http://edge.sagepub.com/brownstats1e.

15 Diagnostics I: Is Ordinary Least Squares Appropriate?

CHAPTER OUTLINE

Learning Objectives
Overview
Diagnostics in Regression Analysis
Properties of Statistics and Estimators
The Gauss-Markov Assumptions
The Residual Plot
Summary
Common Problems
Review Questions
Practice on Analysis and Visualization
Annotated R Functions
Answers

LEARNING OBJECTIVES

- Describe why diagnostics are so important.
- Identify properties of estimators.
- Discuss the Gauss-Markov assumptions.
- Create residual plots to evaluate Gauss-Markov assumptions.

Overview

There are any number of potential problems that lurk behind the clean veneer of a regression table. Diagnostics help identify those problems. Diagnostics also lead to important discoveries. In this chapter, you'll learn what diagnostic tools are at your disposal, how to use them, and what pathologies they identify. Some diagnostic tests evaluate how closely we've met the assumptions associated with ordinary least squares regression (OLS). Others help identify important observations in cases. Ultimately, diagnostics help us learn more about the data. Generating a regression table and interpreting its coefficients is not the end of our journey. The best work takes regression diagnostics seriously. Diagnostic tests force us to question our results, help us to construct better models, and in most cases help us ask better questions.

Diagnostics quickly reveal if the assumptions that underpin linear regression have been violated. To understand those assumptions, we need to be acquainted with the properties of estimators. After a brief presentation of those properties, I list the assumptions that undergird OLS: the Gauss-Markov assumptions.

In this first installment of diagnostics, we concentrate on examining whether our use of OLS is appropriate. Specifically, I will formally introduce the residual plot, a useful view of our model's predictions. In this chapter, we'll see how residual plots can be used to evaluate our use of OLS. Are we using OLS in a context such that it is the best possible option? With a clear understanding of properties of estimators and the Gauss-Markov assumptions, we can use residual plots to identify potential problems.

Diagnostics in Regression Analysis

Regression diagnostics are analogous to the blood work performed in an annual medical exam. While the patient may have the outward appearance of a healthy individual, things could be going wrong inside. Regression tables may look fine and give no apparent outward appearance of a problem. Regression tables won't necessarily indicate if there's a typo in the data, identify a strange group of cases that are driving the results, or indicate that the assumptions we've made have been violated. Diagnostics protect against falling prey to those maladies. Diagnostics provide some assurance that using OLS is appropriate and that the estimates generated from our regression analysis are sound, stable, and don't mislead.

While describing data, theorizing, building a model, and applying it to data are all important components of data analysis, performing diagnostics is equally essential. The more time spent on the procedures contained in this chapter and the next, the more likely your analysis will stand the test of time. While diagnostics are employed to confirm results, they can help make ground-breaking discoveries as well. Before we continue, let me explain the motivation behind diagnostics.

Diagnostics are the best way to test whether the assumptions associated with linear regression are valid. Every estimator we use, whether it's the regression coefficient, mean, median, standard deviation, and so forth, can be evaluated based on its properties. OLS regression coefficients hold many nice properties, but they depend on whether certain assumptions are met. An important reason why we perform diagnostics is to check whether those assumptions apply.

Diagnostics help identify what is driving the results. While we may find, for example, that there is a strong statistical relationship between GDP per capita and democracy, a regression table won't tell us which cases are responsible for that result. Is it the large differences between the industrialized West and sub-Saharan Africa? Do the oil states obscure the relationship? Is it a truly linear relationship or is it driven by large regional differences? Diagnostics allow us to see and better understand how the results are generated. Answers to any of the questions posed above provide important information that can influence our explanations and conclusions.

The techniques learned in this chapter and the next are designed to identify mistakes. Data are produced in all sorts of ways by all sorts of people. In the process of data generation, human error abounds. Whether through simple typos or programming errors, mistakes in the data can be catastrophic, generating regression estimates that are highly misleading. Good diagnostic work is perhaps the most effective way to identify problems in the data that would have otherwise gone undetected.

Finally, the best data analysis recognizes that the question an analyst starts out with won't be the last. In fact, confronting data with an open mind often leads to better questions than those posed at the outset. Taking advantage of diagnostic techniques is an extremely useful endeavor in this regard. Subtle patterns in the residuals can generate a wide range of new questions and discoveries.

This chapter concentrates on using diagnostics to determine whether our use of OLS is appropriate. OLS estimates are ubiquitous because they have very nice properties. If certain assumptions are met, OLS is the best linear unbiased estimator (BLUE). This chapter describes the properties that make it a good estimator and the assumptions required for those properties to hold. First I discuss the properties of estimates in general and then discuss the Gauss-Markov assumptions that undergird OLS regression.

> **KNOWLEDGE CHECK: Describe why diagnostics are so important.**

1. Which of the following statements accurately describes why diagnostics are important?
   a. They help identify the next question.
   b. They identify mistakes in the data.
   c. They indicate whether the assumptions that underlie the model are met.
   d. They help make important discoveries.

Properties of Statistics and Estimators

The sample mean is a statistic we calculate from a random sample to understand something about the population. There are, of course, other statistics that describe a population. Regardless of which statistic used—whether it's the mean, the standard deviation, or the slope of a line—we hope the statistic or estimator offers a good approximation. In statistics we have some very specific criteria in mind when evaluating estimators: bias, **efficiency**, and **consistency**. Ideally, we want the estimate to be unbiased, efficient, and consistent. The estimator/statistic with the greatest number of these properties is called *sufficient*. Let's describe and illustrate each of these criteria.

- *Unbiased.* The sample statistic is unbiased if the mean of the sampling distribution equals the population parameter.
- *Efficient.* If two different statistics are unbiased, choose the one that has a smaller variance in its sampling distribution.
- *Consistent.* A statistic is consistent if as the sample size n increases, the sample statistic gets closer to the population parameter.
- *Sufficient.* The statistic that has the most desirable properties is the sufficient statistic, while the other statistics are insufficient.

To illustrate these properties, consider the density plots in Figure 15-1, each one representing a different property. Assume that the population we want to draw from has a normal distribution with mean of 0 and a standard deviation of 1. To illustrate an unbiased, biased, and efficient estimator, I manufacture what those curves would look like. The unbiased estimator's sampling distribution will have a mean of 0 and a standard deviation of 1. The biased estimator's sampling distribution will not have the correct mean. The efficient estimator's sampling distribution will have a smaller standard deviation.

To generate Figure 15-1, I first create a data frame with three variables I label *unbiased, biased,* and *efficient.* As the code in Code Chunk 15-1 indicates, the *unbiased* variable is composed of 1 million observations drawn from a normal distribution with a mean of 0 and a standard deviation of 1. The *biased* variable is also composed of 1 million observations; however, it has a mean of 1 and a standard deviation of 1. Finally, the *efficient* variable is generated from 1 million observations with a mean of 0 and a standard deviation of .5. Once I generate the three variables, I use the *melt()* command to stack them on top of each other in order to generate the figure.

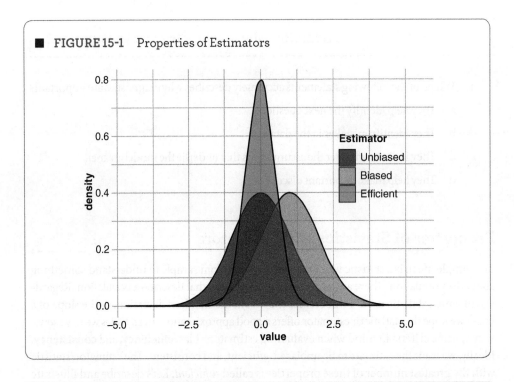

FIGURE 15-1 Properties of Estimators

Code Chunk 15-1

```
x <- data.frame(Unbiased=rnorm(1000000,0,1),
                Biased=rnorm(1000000,1,1),
                Efficient=rnorm(1000000,0,.5))

data<- melt(x)
```

Having created the data, which is now a data frame *x* consisting of two variables, I am ready to plot three different density plots on top of each other. Note that in the *geom_density()* command I use the option *alpha=0.5*, which adds transparency to the colors of each density plot (Code Chunk 15-2). I also use the command *scale_fill_discrete()* to assign colors, create the legend title, and assign labels for each density curve.

Code Chunk 15-2

```
ggplot(data,aes(x=value, fill=variable)) +
  geom_density(alpha=0.50) +
  ggtitle("Figure 15-1: Properties of Estimators") +
  theme_minimal() +
  theme(plot.title = element_text(size = 8, face = "bold"),
        axis.title = element_text(size = 8, face = "bold"),
        legend.position = c(.8,.6)) +
  scale_fill_discrete(name = "Estimator",
          labels = c("Unbiased", "Biased", "Efficient"))
```

The curve in red depicts the sampling distribution of an unbiased estimator of the mean, another depicts a biased estimator (green), and finally, the blue plot illustrates efficiency. Recall that the data generating the plots were constructed from random draws with n = 1,000,000 observations. The red density plot has a mean of 0 and a standard deviation of 1, representing what the sampling distribution of an unbiased estimator would look like since that's how we defined the population parameters. We use that to represent the true population parameters. A biased estimator is depicted by the green density plot, showing that the mean of the sampling distribution diverges from zero, the population parameter. The blue density plot represents efficiency since it is not biased and has less variation than the other sampling distributions. The blue density plot, therefore, represents the sampling distribution of an efficient estimator.

It's worth familiarizing yourself with these properties since you'll run into them time and again. Understanding what is meant when an estimator is efficient or biased will clarify much of what you encounter in more advanced statistics texts.

Now that we've seen representations of bias and efficiency, let's consider consistency. Consistency simply means the estimate of the population parameter should improve as n increases. Consistency implies that as our samples increase in size, the accuracy of the sample statistic we generate from it (in this case, the mean) improves.

To illustrate consistency, I generate three density plots composed of observations drawn from a uniform distribution. The number of observations in each draw is 5 for one density plot, 25 in a second, and 100 in a third.

To generate those observations, I use a for-loop designed to draw a sample from a uniform distribution, the *runif()* command (Code Chunk 15-3). I instruct the loop to make a million draws. The first for-loop directs R to take 1 million draws of five observations each, the next is 1 million draws of 25 observations, and the third is 1 million draws of 100 observations. Once I've generated those three variables and created three separate data frames, I use the *melt()* command to stack them on top of each other.

Code Chunk 15-3

```
set.seed(3376)
min=-40; max=40; n=5
xbar1=rep(0,1000000)
for (i in 1:1000000) {xbar1[i]=mean(runif(n, min, max))}

x1 <- data.frame(v1=xbar1)

set.seed(3376)
min=-40; max=40; n=25
xbar2=rep(0,1000000)
for (i in 1:1000000) {xbar2[i]=mean(runif(n, min, max))}

x2 <- data.frame(v1=xbar2)

set.seed(3376)
min=-40; max=40; n=100
xbar3=rep(0,1000000)
```

```
for (i in 1:1000000) {xbar3[i]= mean(runif(n, min, max))}

x3 <- data.frame(v1=xbar3)

x <- list(v1=x1,v2=x2,v3=x3)
data <- melt(x)
```

Now we're ready to plot. To help make sense of what's going on with the *melt()* command, view the data that now reside in the data set called *data*. You'll see that the data set is made of three variables: *data$variable*, *data$value*, and *data$L1*. The ggplot command in Code Chunk 15-4 simply draws three separate density plots, one for each of the categories in variable *data$L1*: v1, v2, and v3.

Code Chunk 15-4
```
ggplot(data,aes(x=value, fill=L1)) +
 geom_density(alpha=.50) +
 ggtitle("Figure 15-2: Increased Draws, Increased Accuracy") +
 theme_minimal() +
 theme(plot.title = element_text(size = 8, face = "bold"),
       axis.title = element_text(size = 8, face = "bold"),
       legend.position = c(.8,.6)) +
      scale_fill_discrete(name = "Number of Observations",
                          labels = c("5", "25", "100"))
```

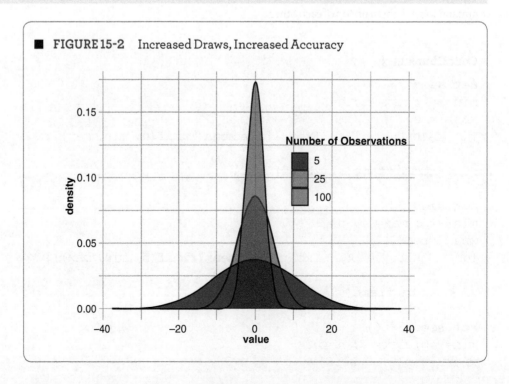

■ **FIGURE 15-2**   Increased Draws, Increased Accuracy

The density plots in Figure 15-2 indicate that as the number of observations increases, the accuracy of our estimate increases. Note how the tails in each distribution get smaller and smaller as the number of observations increases. This is the property of consistency.

Finally, it's important to reiterate that the estimator (the mean, in our case) that has the most properties (bias, efficiency, and consistency) is deemed *sufficient* and all other estimators are considered *insufficient*.

To summarize, we learned the main properties of estimators: bias, efficiency, and consistency. Estimators that are unbiased, efficient, and consistent are deemed sufficient. OLS is so ubiquitous in regression analysis because it meets these criteria. OLS estimates meet these criteria, however, only if certain assumptions hold. In the next section I discuss each of those assumptions and provide the tools to determine if they are realistic.

> **KNOWLEDGE CHECK: Identify properties of estimators.**

2. Assume that the population mean is 30. Indicate whether the specified curve pictured in the figure below represents the sampling distribution of an unbiased, biased, or efficient estimator.

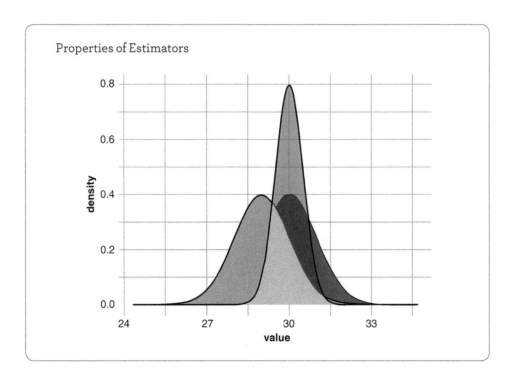

    a. Red

    b. Green

    c. Blue

The Gauss-Markov Assumptions

The same properties—bias, efficiency, and consistency—that apply to the mean of a population apply to estimates in a regression. When generating regression estimates, we want an estimate that's unbiased, efficient, and consistent.

As noted earlier, OLS regression is ubiquitous in empirical analysis. Its use is pervasive because it provides the best linear unbiased estimator (BLUE). That exalted status depends, however, on a set of assumptions. The assumptions themselves are derived from what is called the **Gauss-Markov theorem**. The Gauss-Markov theorem says that if three important assumptions are met, OLS regression provides the best linear unbiased estimate. If the following assumptions are met, "OLS is best."

Before articulating the three main assumptions, we need to make a distinction between errors and residuals. Often they are used interchangeably and, in truth, are very similar. Nevertheless, keeping the distinction between them is extremely important. **Errors** are the difference between the observed values of y and the true model, the true line. You can think of the true line as synonymous with the population parameter. The line we estimate in OLS is synonymous with the sample statistic. Knowing the true line requires we know the true model. Since we really never know the true model, we just have our best guess. Consequently, we really never know the errors. However, we do know the residuals. The **residuals** are the distance between the actual value of y and our OLS model's prediction. The Gauss-Markov assumptions are based on statements about the errors and not the residuals. In each case, I'll make clear the distinction between errors and residuals.

It is also helpful to visualize the assumptions being made in OLS. Let's start by creating two variables whose relationship we can specify as a linear function. Since we are creating the data ourselves, we actually know what the true line is. The differences between the line and the observations in this case are the errors.

In Code Chunk 15-5, I generate two random variables, then I define one of them (y) as a linear combination of the other (x) along with an error term (e). I then plot the relationship between x and y.

Code Chunk 15-5

```
happydf <- data.frame(id = 1:1000, x = rnorm(1000, 0, 1), e = rnorm(1000, 0, 3))
happydf$y <- 4.5*happydf$x + happydf$e

ggplot(happydf, aes(x, y)) +
  geom_smooth(method="lm", se=FALSE) +
  geom_point(col = "#bf0000", alpha = .4) +
  theme_minimal() +
  theme(plot.title = element_text(size = 8, face = "bold"),
        axis.title = element_text(size = 8, face = "bold"),
        legend.title = element_text(size = 10)) +
  ggtitle("Figure 15-3: The True Model") +
  ylab("y") +
  xlab("x")
```

■ **FIGURE 15-3** The True Model

Figure 15-3 represents the ideal case for OLS regression. I constructed a variable x and e and I then made y a linear combination of the two. Since we know exactly how the data were generated, the points (in red) represent the observed values and we can think of the line as the true line.

The following are the main assumptions associated with the Gauss-Markov theorem. First let's express the OLS regression model mathematically. Note how we use the Greek letters β and ϵ to represent the slope and the errors of the model. This is analogous to using Greek letters to represent the true population parameters we discussed in Chapters 8 and 9.

$$y_i = x_i \beta_i + \epsilon_i$$

In the mathematical model, y_i is a linear combination of the x variable and , the errors. The following assumptions refer to this theoretical model.

Gauss-Markov Assumption 1: the errors in the theoretical model sum to zero. This assumption is straightforward: the size of the errors above the true line should be counterbalanced by those lying beneath the line. The true model shouldn't systematically overpredict or underpredict. If the errors don't sum to zero, this means our estimate of the intercept will be biased. Remember, the error term is there to capture random events that we simply can't account for in the model. If they truly are random, their mean should be zero. If the mean is not zero, there is something in that error term that needs to be included in the model.

Unfortunately, we can't use diagnostics to check this assumption. Since, by definition, the estimation of OLS minimizes the sum of the residuals, when we estimate a regression line the residuals will always sum to zero. Here we have to lean on theory. Is our model convincing?

Have we constructed a model that accounts for the important variables, leaving only random chance for the error term? Building a model is like building an argument. If an argument is incorrect, there will be serious questions about its conclusions. If a model is incorrect, there will be serious questions about its estimates.

Gauss-Markov Assumption 2: the errors are independently distributed. This means the size of one error is not influenced by the size of the error for the case next to it. When the error for one case influences the size of the error next to it, the errors are correlated and not randomly generated. If the size of one error can help us predict the size of the one next to it, it implies the errors are not generated randomly and there is something in the error term that should be accounted for. In more concrete terms, we shouldn't systematically overpredict or underpredict.

Fortunately, we can examine the residuals from our regression model to determine whether the errors implied in our theoretical model meet this condition. If the errors are not independent of each other, our estimates might systematically underpredict or overpredict. This occurs in seasonal data where data points close to each other might be systematically higher than average (think temperatures in July) or systematically lower (think temperatures in January). The problem could also be the result of fitting a straight line to data that would be better summarized by a curve.

To illustrate the problem, I generate random data based on a formula to create a curved line (Code Chunk 15-6). I then draw a scatter plot to illustrate what fitting a straight line to curved data looks like in the context of a regression. I use the *exp()* command to generate the nonlinear data since it is a nonlinear function (it exponentiates the value for each x). Remember, the *geom_smooth()* command with the *method = "lm"* option fits an OLS regression line to the scatter plot.

Code Chunk 15-6

```
curveddf <- data.frame(id = 1:1000, x = rnorm(1000, 0, 1),
                       e = rnorm(1000, 0, 1))

curveddf$y <- exp(curveddf$x) + curveddf$e

curvedmodel.lm <- lm(y ~ x, data = curveddf)

ggplot(curveddf, aes(x, y)) +
  geom_smooth(method="lm", se=FALSE, col="#0000bf") +
  geom_point(col = "#bf0000", alpha = .2) +
  theme_minimal() +
  theme(plot.title = element_text(size = 8, face = "bold"),
        axis.title = element_text(size = 8, face = "bold"),
        legend.title = element_text(size = 10)) +
  ggtitle("Figure 15-4: Fitting a Straight Line to Curved Data") +
  ylab("y") +
  xlab("x")
```

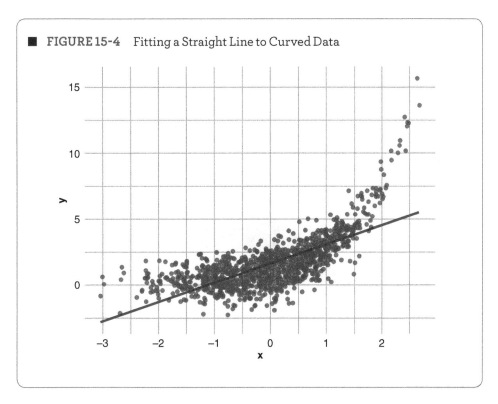

FIGURE 15-4 Fitting a Straight Line to Curved Data

Note the curvature in the plot: at the lower values of x the line is below the observations, in the middle values of x the line is above the observations, and at the upper reaches of x the line is below. Consequently, our predictions (the points along the line) systematically underpredict and overpredict values of y over different values of x. As Figure 15-5 makes clear, the straight line we fit to these data is not a good summary of the relationship between y and x; it is not the appropriate model. A better model would fit a curved line to these data.

Gauss-Markov Assumption 3: the errors have constant variance. The true model should provide predictions that are equally accurate across the range of our predictions. For example, if we're trying to predict income levels for countries (GDP per capita), our predictions should be just as good for poor countries, middle-income countries, and rich countries. The degree to which we miss our target should not change over the range of our predictions.

Fortunately, we can examine the residuals to see if this condition holds. To illustrate non-constant variance in the errors, I changed the data to increase their variance as x increases. In Code Chunk 15-7 I alter the definition of y by multiplying the error term by (*1 + df$x/2*), which generates a larger error as x increases. I then plot it and fit a regression line to illustrate what the regression look like fitted to the data.

Figure 15-5 illustrates that as x increases, the variance of y increases. While the line does a pretty good job of predicting y at levels below 0 in terms of x, our predictions worsen considerably once we move between values of 0 and 2.

Notice the fan-like shape of the plot. This is another tell-tale sign that all is not right with using OLS in this situation. It's clear we've violated the assumption that there is constant variance in our errors. Nonconstant variance in the residuals is another indication that the errors are not randomly distributed and that there may be an important variable we have not accounted for in the model.

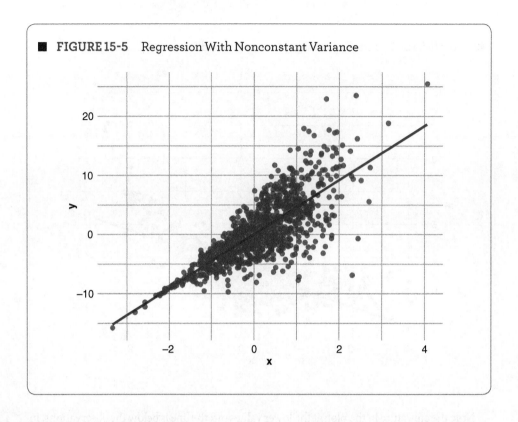

FIGURE 15-5 Regression With Nonconstant Variance

Code Chunk 15-7
```
vardf <- data.frame(id = 1:1000, x = rnorm(1000, 0, 1),
                    e = rnorm(1000, 0, 3))
vardf$y <- 4.5*vardf$x + (1 + vardf$x / 2)*vardf$e

varmodel.lm <- lm(y ~ x, data = vardf)

ggplot(vardf, aes(x, y)) +
  geom_smooth(method="lm", se=FALSE, col="#0000bf") +
  geom_point(col = "#bf0000", alpha = .2) +
  theme_minimal() +
  theme(plot.title = element_text(size = 8, face = "bold"),
        axis.title = element_text(size = 8, face = "bold"),
        legend.title = element_text(size = 10)) +
  ggtitle("Figure 15-5: Regression With Nonconstant Variance") +
  ylab("y") +
  xlab("x")
```

Fortunately, there are a number solutions to the problems that arise when the Gauss-Markov assumptions are violated. One approach is to rethink which variables you want to include in your model. Violating the Gauss-Markov assumption often results from an important (omitted) variable left out of the model whose impact is being absorbed in the error term, generating interesting patterns in the residuals. The other approach is to consider transforming either your dependent or independent variables. Transforming your data or introducing additional variables to your model can help in many instances. Finally, in some cases adding interactive terms to the model provides a solution. There are, however, situations where these solutions are not adequate. OLS may not be the appropriate estimator in many situations. It may be necessary to use entirely different methods of estimation. Violating the Gauss-Markov assumptions and applying the appropriate estimator will, unfortunately, have to wait for your next class in data analysis. Fortunately, changing the model specification, transforming data (skills you now already have), or adding interactions will carry you a long way when the Gauss-Markov assumptions are not met.

Let's briefly review what we've learned in this section. If the three conditions of the Gauss-Markov theorem hold, then our use of the OLS model is appropriate. Put simply, if any of these three assumptions is violated, then our use of OLS is problematic. The three assumptions concern the error term in our theoretical model: (1) the errors must sum to zero, (2) the errors must be independent of each other, and (3) the errors must have constant variance. We're left to use theory to validate the assumption that the errors sum to zero. Assumptions 2 and 3, however, can be validated by looking at our predicted values (the regression line) and examining how well they predict the actual values. While we've merely glanced at the problems when considering how well the regression line fits the data, we're ready to take the next step. This is where diagnostics comes in: we can assess whether the second and third conditions hold by analyzing the residuals from our regression estimates.

The illustrative examples in this section were based on data we generated ourselves. This was helpful since by generating the data ourselves, we knew what the true, underlying model was: we knew exactly how the data were generated. In the next section we combine artificially generated data and real data, showing how we use diagnostics not only to check on the Gauss-Markov assumptions but also to identify any outlying or influential cases that can provide more information and clues about our problem.

> **KNOWLEDGE CHECK:** Discuss the Gauss-Markov assumptions.

3. Which statement best describes why the Gauss-Markov assumptions are so important?
   a. If they are met, we know that OLS is the best linear unbiased estimator.
   b. They guarantee that the OLS estimates will be accurate.
   c. If they are not met, we should have less confidence in the inferential statistics generated by the model.
   d. All of them help describe what patterns we should see in the model's residuals.

4. Which of the following are true statements about errors and residuals?
   a. We really never know what the actual residual is.
   b. We really never know what the actual error is.

c. Errors are to regression analysis as parameters are to sampling.

d. Errors and residuals are the same thing.

5. How do we know that the errors sum to zero?

    a. If the residual plot looks like a formless cloud

    b. If the residual plot resembles a fan-shaped pattern

    c. If the errors are purely the result of random chance

    d. If our theoretical model is correct

The Residual Plot

Two concerns orient the examination of residuals as a diagnostic tool. We look for patterns in the residuals and we look for influential cases (cases that influence our results). We look for patterns to confirm whether we've violated the Gauss-Markov assumptions. The size of one residual e_i should not have any impact on the residual next to it. If there is a pattern, our analysis of the residuals suggests that either the errors are not independent of each other or that they have nonconstant variance. If either is the case, we have violated one of the assumptions we make when using OLS regression.

To help determine if either assumption is violated, we examine the residual plot. The **residual plot** is simply a scatter plot of the residuals from a regression on the y axis and the predicted values on the x axis. It also usually features a horizontal dotted line indicating where the residual is zero, the point at which the predicted value perfectly predicts the actual value. If errors are independent and they have constant variance, we should observe a formless cloud of points when examining a residual plot. Patterns in the residual plot indicate the OLS estimator may not be appropriate.

Figure 15-6 provides a view of a simple bivariate regression and the resulting residual plot. The side-by-side presentation builds an intuition for what residual plots represent. The

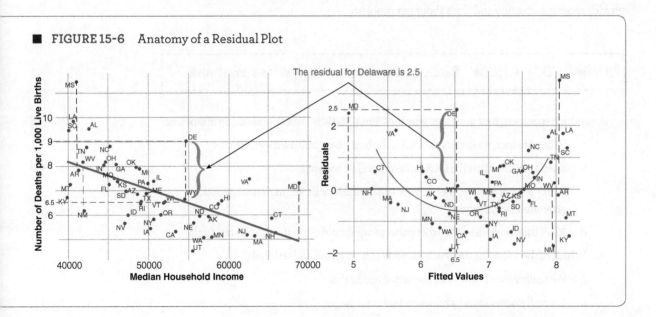

■ **FIGURE 15-6** Anatomy of a Residual Plot

regression has the infant mortality rate for the 50 states as the dependent variable and median household income as the independent variable. In the regression (left panel) we see that Mississippi (MS), New Mexico (NM), Delaware (DE), and Maryland (MD) are all states that lie a relatively large distance from their predicted values (the blue line). Consequently, their residuals are relatively large in the right panel.

To construct the plot that links the regression line to the residual plot, I generate two scatter plots. To produce the plot of the bivariate regression, I need only produce a scatter plot that includes the command *geom_smooth(method="lm")*, which places an OLS regression line over the scatter plot (Code Chunk 15-8). I also use the *annotate()* function to highlight the residual for Delaware.

Code Chunk 15-8

```
p1 <- ggplot(states, aes(medinc, infant)) +
  geom_point(col="#bf0000") +
  theme_minimal() +
  geom_smooth(method="lm", se=FALSE, col="#0000bf") +
  geom_text_repel(size=2, data=states, aes(label=
          ifelse(states$st %in% c("DE", "MS", "MD", "NM"),
                    as.character(states$st), "")),
             hjust=-.5, vjust=0, show.legend=FALSE) +
  ylab("Number of Deaths per 1,000 Live Births") +
  xlab("Median Household Income") +
  theme_minimal() +
  theme(plot.title = element_text(size = 8, face = "bold"),
        axis.title = element_text(size = 8, face = "bold")) +
  annotate("segment", x = 54570, xend = 54570, y = 9,
           yend = 6.5, size = .2, col="#bf0000", linetype = "dashed") +
  annotate("text", x = 60000, y = 8, parse = F,
           label = 'Residual for Delaware = 2.5', size = 2)
```

To produce the residual plot, I estimate the bivariate regression model with the *lm()* command and then create two variables that contain the predicted values and the residuals from the model (Code Chunk 15-9). I then plot the residuals against the predicted values. Once I generate those two figures, I put them into Adobe Illustrator to make some additional refinements.

Code Chunk 15-9

```
model <- lm(infant ~ medinc, data=states)

states$infant.p <- predict(model)
states$infant.r <- resid(model)
```

```
p2 <- ggplot(states, aes(infant.p, infant.r)) +
  geom_point(col="#bf0000") +
  geom_smooth(method="lm", se=FALSE, col="#0000bf") +
  geom_smooth(col="grey", se=FALSE, linetype = "dashed") +
  geom_text_repel(size=2, data=states, aes(label=
          ifelse(states$st %in% c("DE", "MS", "MD", "NM"),
                  as.character(states$st), "")),
              hjust=-.5, vjust=0, show.legend=FALSE) +
  ylab("Residuals") +
  xlab("Fitted Values") +
  theme_minimal() +
  theme(plot.title = element_text(size = 8, face = "bold"),
        axis.title = element_text(size = 8, face = "bold")) +
  annotate("segment", x = 6.52, xend = 6.52, y = 0,
        yend = 2.5, size = .2, col="#bf0000", linetype = "dashed") +
  annotate("text", x = 5.9, y = 1.2, parse = F,
        label = 'Residual for Delaware = 2.5', size = 2)
```

The left panel in Figure 15-6 presents a graphical representation of a bivariate regression and the right panel features a residual plot of that regression. Note how the axes for both plots are different. The y axis in the right panel records the residuals and the x axis records the fitted values. To better understand how the two plots correspond, consider Delaware (DE). The regression plot indicates that the predicted value for Delaware is 6.5 but its actual value is 9 (the difference between the two is 2.5). In the residual plot, we see that the residual is 2.5 and that its fitted value is approximately 6.5. If the two plots show essentially the same information, why use the residual plot?

To answer that question, I drew by hand a red curve to the residual plot in Figure 15-6. As the curve illustrates, it's somewhat easier for the eye to detect a pattern as we move across the x axis because of the new horizontal orientation. In the residual plot, rather than showing the predicted values, we're showing the residuals: the blue line represents where the residual is zero. Cases close to the blue line in the residual plot represent cases we came close to predicting with greater accuracy.

Let's continue with this aspect of examining residuals: identifying patterns that might alert us to violations with the Gauss-Markov assumptions. First consider the ideal condition: the

Code Chunk 15-10

```
happymodel.lm <- lm(y ~ x, data=happydf)

p1 <- ggplot(happydf, aes(x, y)) +
  geom_smooth(method="lm", se=FALSE) +
```

```
  geom_point(col = "#bf0000", alpha = .2) +
  theme_minimal() +
  theme(plot.title = element_text(size = 8, face = "bold"),
        axis.title = element_text(size = 8, face = "bold"),
        legend.title = element_text(size = 10)) +
  ggtitle("Regression Line") +
  ylab("y") +
  xlab("x")
```

case where y is a linear combination of x and there is constant variation in the errors. Alongside that happy situation, I drew the resulting residual plot.

To produce the plot, I simply copied the code from Code Chunk 15-5 and named the ggplot *p1*.

Code Chunk 15-11

```
happydf$res <- resid(happymodel.lm)
happydf$pred <- predict(happymodel.lm)

p2 <- ggplot(happydf, aes(pred, res)) +
  geom_smooth(method="lm", se=FALSE, col="black",
              linetype = "dashed") +
  geom_point(col = "#bf0000", alpha = .2) +
  theme_minimal() +
  theme(plot.title = element_text(size = 8, face = "bold"),
        axis.title = element_text(size = 8, face = "bold"),
        legend.title = element_text(size = 10)) +
  ggtitle("Residual Plot") +
  ylab("y") +
  xlab("x")

grid.arrange(p1, p2, ncol=2,
  top=textGrob("Figure 15-7: Gauss-Markov Is Met",
               gp=gpar(fontsize=8)))
```

I then took the regression implied by the scatter plot and extracted the residuals *resid(model.lm)* and the predicted values *predict(model.lm)* to obtain the components of the residual plot, which is simply a scatter plot of a model's residuals on its predicted values. I then used the *grid.arrange()* command to place the two scatter plots side by side for comparison.

Note how the residuals take on the shape of a formless cloud (residual plot). This is the ideal situation for OLS regression.

Let's move on to the problematic regressions and their associated residual plots. Again, here we're using the scatter plot to identify troublesome patterns that indicate whether the

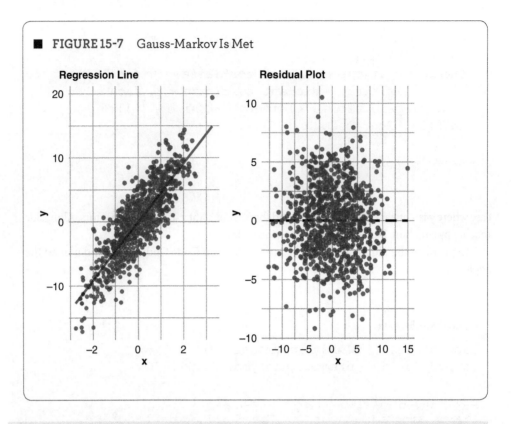

■ FIGURE 15-7   Gauss-Markov Is Met

Code Chunk 15-12
```
p1 <- ggplot(curveddf, aes(x, y)) +
  geom_smooth(method="lm", se=FALSE, col="#0000bf") +
  geom_point(col = "#bf0000", alpha = .1) +
  theme_minimal() +
  theme(plot.title = element_text(size = 8, face = "bold"),
        axis.title = element_text(size = 8, face = "bold"),
        legend.title = element_text(size = 10)) +
  ggtitle("OLS Regression") +
  ylab("y") +
  xlab("x")
```

Gauss-Markov assumptions are being met: the errors are independent of each other and they have constant variance.

Consider the scatter plot I constructed when we try to fit a straight line to data that resembles a curve. This case highlights what a residual plot will look like when we systematically overpredict and underpredict. Following the example of the perfect case (Gauss-Markov assumptions appear to be met), I reproduce the code from Code Chunk 15-7: the scatter plot of a line fit to curved data.

With the new model defined as *model.lm*, I then collect the residuals and predicted values from the model and plot them side by side with the plot that describes the regression (Code Chunk 15-13).

> **Art and Practice of Data Visualization**
> **COLOR TRANSPARENCY WITH *ALPHA***
>
> When you are dealing with a large number of observations, use the *alpha* option to alter the transparency of the colors. By doing so, it becomes clearer where a majority of the cases actually lie. In Code Chunk 15-9 note that I specify *alpha = .1* so that we can gain a better understanding of the underlying pattern.

Code Chunk 15-13

```
curveddf$res <- resid(curvedmodel.lm)
curveddf$pred <- predict(curvedmodel.lm)

p2 <- ggplot(curveddf, aes(pred, res)) +
  geom_smooth(method="lm", se=FALSE, col="black", linetype = "dashed") +
  geom_point(col = "#bf0000", alpha = .1) +
  theme_minimal() +
  theme(plot.title = element_text(size = 8, face = "bold"),
        axis.title = element_text(size = 8, face = "bold"),
        legend.title = element_text(size = 10)) +
  ggtitle("Residual Plot") +
  ylab("y") +
  xlab("x")


grid.arrange(p1, p2, ncol=2,
   top=textGrob("Figure 15-8: Systematic Overprediction and Underprediction",
                 gp=gpar(fontsize=8)))
```

Remember that the dotted line at zero in the residual plot indicates where prediction was perfect: there is no difference (zero) between the predicted value and actual value. We can see more easily in the residual plot that as we move along the x axis from −2.5 to 0, most of our predictions underpredict the true values (most of the points lie above the dotted line). Between 0 and 2.5, most of the observations lie below our predictions. Finally, as we continue moving from left to right between 2.5 and 6, we go back to underpredicting. Since the underprediction and overprediction happens systematically in some regions of the plot, it appears that our residuals may not be independent of each other: if we underpredict one case, it's likely that the case next to it will meet the same fate. From this, we should be worried that the errors are not independent of each other.

Finally, let's consider what the pattern in the residual plot looks like when we violate the assumption that the errors have constant variance over the range of the predicted values.

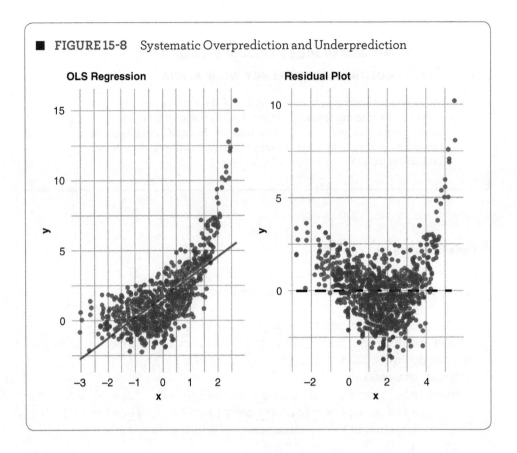

FIGURE 15-8 Systematic Overprediction and Underprediction

To produce this view, I go back to Code Chunk 15-8 and reproduce the scatter plot of y on x. Only in this case I've built in greater variance in y as we range over the value of x. I place the code in Code Chunk 15-14. Note again the mathematical expression I use to define *df$y*: I multiply *e* by an expression that increases with x. This produces a growing error term as x increases.

Code Chunk 15-14
```
df <- data.frame(id = 1:1000, x = rnorm(1000, 0, 1),
                 e = rnorm(1000, 0, 3))
df$y <- 4.5*df$x + (1 + df$x / 2)*df$e

model.lm <- lm(y ~ x, data = df)

p1 <- ggplot(df, aes(x, y)) +
  geom_smooth(method="lm", se=FALSE, col="#0000bf") +
  geom_point(col = "#bf0000", alpha = .2) +
  theme_minimal() +
  theme(plot.title = element_text(size = 8, face = "bold"),
        axis.title = element_text(size = 8, face = "bold"),
```

```
        legend.title = element_text(size = 10)) +
  ggtitle("Regression With Nonconstant Variance") +
  ylab("y") +
  xlab("x")
```

Having defined the relationship between y and x with increasing variance in the error term, let's plot the representation of the regression and the resulting residual plot side by side (Code Chunk 15-15).

Code Chunk 15-15

```
df$res <- resid(model.lm)
df$pred <- predict(model.lm)

p2 <- ggplot(df, aes(pred, res)) +
  geom_smooth(method="lm", se=FALSE, col="black", linetype = "dashed") +
  geom_point(col = "#bf0000", alpha = .2) +
  theme_minimal() +
  theme(plot.title = element_text(size = 8, face = "bold"),
        axis.title = element_text(size = 8, face = "bold"),
        legend.title = element_text(size = 10)) +
  ggtitle("Residual Plot Shows Nonconstant Variance") +
  ylab("y") +
  xlab("x")

grid.arrange(p1, p2, ncol=2,
    top=textGrob("Figure 15-9: Nonconstant Variance in the Residuals",
                  gp=gpar(fontsize=8)))
```

When nonconstant variance is the problem, the residual plot will resemble a fan shape. Warning: the fan shape will rarely look as symmetrical and well defined as in this made-up example. Nonetheless, the illustration in Figure 15-9 shows the archetypal situation that will occur when there is nonconstant variance in the errors.

As I hope to have demonstrated in the last few examples, by rendering the regression plot to a horizontal orientation, we can better evaluate whether the residuals systematically change as we range over the x axis. The residual plot also allows us to directly examine whether there is constant variance in the residuals over the predicted (fitted) values.

When patterns appear, there are potential problems. First, if the errors are not randomly distributed over all values of our predictions, it suggests that there could be a variable we should have included in the model. Second, we may have gotten the functional form of the model wrong. We may need to transform one or more of the variables to generate a more linear form. Finally, introducing an interactive term may provide a better specification of the model. If adding the possible omitted variable, transforming the data, or adding an interaction does not fix the problem, there are many different models available that are designed to address the

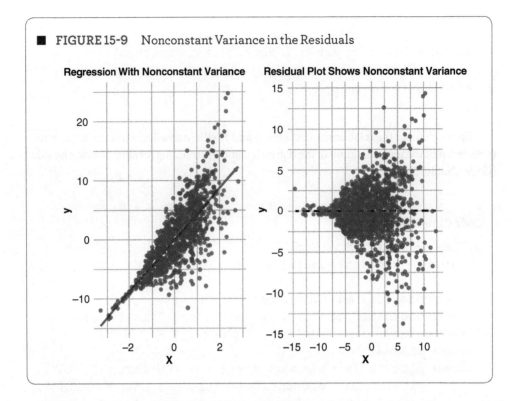

FIGURE 15-9 Nonconstant Variance in the Residuals

issue. We will explore one such model—logistic regression—in Chapter 17. Fortunately, you have many tools at your disposal and can look forward to the next class in data analysis that will cover additional methods of estimation designed to produce the best estimates in a variety of different contexts.

> **KNOWLEDGE CHECK: Connect residual plots to Gauss-Markov assumptions.**

6. How do we know we have violated the assumption that we have constant variance in our errors?
   a. The plot of the errors forms a fan-shaped pattern.
   b. The plot of the residuals forms a fan-shaped pattern.
   c. There is systematic underprediction and overprediction in the residual plot.
   d. The residual plot resembles a formless cloud.

7. How do we know if we violated the assumption that the errors are independent of each other?
   a. The plot of the errors forms a fan-shaped pattern.
   b. The plot of the residuals forms a fan-shaped pattern.

c. There is systematic underprediction and overprediction in the residual plot.

d. The residual plot resembles a formless cloud.

8. What are some of the remedies when one of the Gauss-Markov assumptions is violated?

a. Include a variable that may have been omitted.

b. Transform your data.

c. Include interactions in the model.

d. Take the next class in data analysis.

SUMMARY

Bias, efficiency, and consistency are criteria we use to evaluate estimators. The Gauss-Markov theorem states that if three important conditions are met, OLS regression provides the best linear unbiased estimates of a line. Diagnostics are essential to data analysis since they help evaluate whether some of the Gauss-Markov assumptions hold. As we will see in the next chapter, diagnostics also help identify important cases by examining residual plots, Cook's distance, leverage plots, and added variable plots. Not only do they identify potential problems with our regression results, they can generate additional clues that help deepen our understanding of the question at hand.

Being on the lookout for stranger things in our regression takes vigilance. Diagnostics are usually given short shrift—many statistics books don't include a section on it. Diagnostics prevent the analyst from making predictions that are terribly wrong. Spend *at least* as much time with diagnostics as with the rest of your analysis combined. Not only can it keep you from making silly mistakes, important discoveries are made along the way and produce a more persuasive set of results, results that are likely to endure.

COMMON PROBLEMS

- *Residuals versus errors*. Keeping the distinction between errors and residuals straight is important for the same reason we needed to distinguish between sample statistics and parameters (Chapters 8 and 9). Much of the difficulty may stem from the terms' usage: the two terms are often used interchangeably. If a clear distinction is made, however, the first Gauss-Markov assumption that states the error terms sum to zero becomes less confusing. If errors and residuals are confused, mixed, and intermingled, confusion results. For example, stating that all of the residuals sum to zero is self-evident since OLS regression is based on making calculations that minimize the sum of squared residuals, guaranteeing the residuals will sum to zero. The residuals from our OLS estimation will always sum to zero. The errors from our theoretical model may, however, not sum to zero if we've omitted an important variable.

- *Bias, efficiency, and consistency in an estimator*. Although one can perform sophisticated analysis without having an intuitive and deep understanding of these concepts, almost all intermediate and advanced texts reference these characteristics extensively. Keeping them straight and knowing intuitively what they mean will help make sense of what you will encounter. Mastering the basics here will take a lot of the mystery out of more advanced statistical texts. To review, estimators are unbiased if the sampling distribution's mean approaches the population parameter. Estimators are efficient if the variance of the sampling distribution is relatively small. Finally, estimators are consistent if they become more accurate as you increase the n in the sample.

REVIEW QUESTIONS

1. Why are diagnostics so important?
2. What are the Gauss-Markov conditions underpinning OLS?
3. What is a residual plot?
4. What can a residual tell us?
5. How would you describe a residual plot that implies there is nonconstant variance in the errors?
6. How would you describe a residual plot that implies the errors are not independent?
7. How would you describe a residual plot that implies the assumptions of the Gauss-Markov theorem are met?
8. How would you describe a residual plot that implies the errors sum to zero?
9. What is the difference between residuals and errors?
10. What can be done if we've violated one of the Gauss-Markov assumptions?

PRACTICE ON ANALYSIS AND VISUALIZATION

1. When comparing the sampling distributions of two estimators, if the tails of the distribution are wider for one estimator compared to the other, we say one estimator is more of what?

   a. Biased
   b. Efficient
   c. Consistent
   d. BLUE

2. When examining the sampling distributions of an estimator and the sampling distribution's variance decreases as n increases, we say that estimator is

   e. Unbiased
   f. Consistent
   g. Efficient
   h. Biased

3. If the population parameter for the mean is 50, indicate which property describes each estimator that produced the following density plot (sampling distribution).

   a. Red
   b. Blue
   c. Green

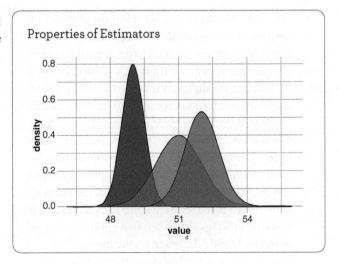

4. If the population parameter for the mean is 100, indicate which property describes each estimator that produced the following density plot (sampling distribution).

   a. Red
   b. Blue
   c. Green

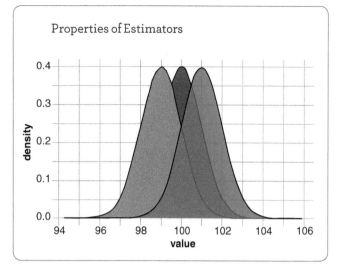

5. If the population parameter for the mean is 100, indicate which property describes each estimator that produced the following density plot (sampling distribution).

   a. Red
   b. Blue
   c. Green

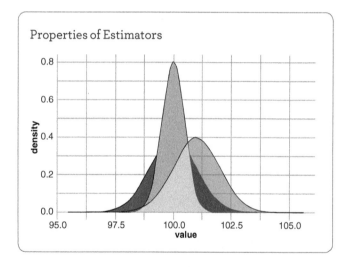

6. Which of the following are assumptions associated with the Gauss-Markov theorem?

   a. There is constant variance in the errors.
   b. The errors sum to zero.
   c. The residuals sum to zero.
   d. The residuals are independent of each other.

7. Which of the following describes a residual plot indicating the errors are not independent of each other?

   a. A fan-like shape in the residuals
   b. A formless cloud
   c. No perceptible pattern
   d. Systematic underprediction and overprediction in some regions of the plot

8. Which of the following describes a residual plot indicating nonconstant errors?

   a. A fan-like shape in the residuals
   b. A formless cloud
   c. No perceptible pattern
   d. Systematic underprediction and overprediction in some regions of the plot.

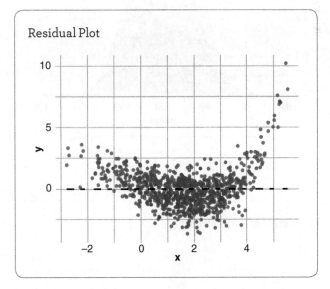

9. What assumption seems to be violated when looking at the following residual plot?

   a. Nonconstant variance in the errors
   b. Nonconstant variance in the residuals
   c. Independence of the errors
   d. Independence of the residuals

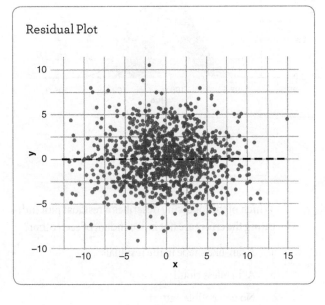

10. What assumption seems to be violated when looking at the following residual plot?

    a. Nonconstant variance in the errors
    b. Nonconstant variance in the residuals
    c. Independence of the errors
    d. None

ANNOTATED R FUNCTIONS

The following functions appear in this chapter. They are listed in order of their first appearance (with the code chunk number in parentheses) and annotated here to give a very brief description of their use. Some are not stand-alone functions and only work in combination with other commands. As a reminder, the code in every chapter will work properly if executed in the order it appears. Proper execution also depends on typing the author-defined *libraries()* command, which loads the required R packages.

data.frame(): takes an object or list of objects and creates a data frame which then can be used in a number of different commands that requires the data frame structure. (15-1)

rnorm(): randomly draws from a normal distribution. (15-1)

melt(): takes columns in a data frame and stacks them on top of each other. This is used often to plot certain objects in the same figure. (15-1)

ggplot(): defines the basic structure of a plot (usually the x and y variables). (15-2)

aes(): the aes (called "aesthetics") function is used in ggplot to define the basic structure of the plot, which often includes the variables you want to use and any shapes or colors. (15-2)

ggtitle(): provides the title for a ggplot. (15-2)

theme_minimal(): specifies a minimalist style for ggplot. (15-2)

theme(): specifies font, size, and so forth in a ggplot. (15-2)

scale_fill_discrete: allows you to color discrete objects in a plot. (15-2)

set.seed(): sets the random number generator. Often used so that results generated randomly can be reproduced. (15-3)

rep(): short command that tells R to replicate values in a vector. Helpful for creating empty variables that can be filled in with data generated with for-loops. (15-3)

for (i in x): for-loop command that specifies how many times a series of commands are to be repeated. (15-3)

runif(): randomly draws from a uniform distribution. (15-3)

list(): combines objects into a list. (15-3)

geom_smooth(): draws lines or curves to a scatter plot. (15-5)

geom_point(): draws points on a ggplot grid. (15-5)

ylab(): labels the y axis in ggplot. (15-5)

xlab(): labels the x axis in ggplot. (15-5)

lm(): linear model or regression command. (15-6)

geom_text_repel(): labels points in a scatter plot in the ggplot framework. (15-8)

ifelse(): logical function that allows you to construct an if-then statement. Useful for selecting specific cases to label and also for creating categorical variables from continuous variables. (15-8)

annotate(): places lines or text in the figure. (15-8)

predict(): when applied to an object described by a linear model, it will generate the predicted values. (15-9)

resid(): when applied to an object described by a linear model, it will generate the residuals. (15-9)

grid.arrange(): allows you to arrange multiple figures on a page. (15-11)

ANSWERS

KNOWLEDGE CHECK

1. a, b, c, d
2. unbiased (a), biased (b), efficient (c)
3. a, c
4. b, c
5. d
6. b
7. c
8. a, b, c, d

PRACTICE ON ANALYSIS AND VISUALIZATION

1. b
2. b
3. biased (a), biased (b), unbiased (c)
4. unbiased (a), biased (b), biased (c)
5. unbiased (a), biased (b), efficient (c)
6. a, b
7. d
8. a
9. c
10. d

 Access digital resources, including datasets, at http://edge.sagepub.com/brownstats1e.

16 Diagnostics II: Residuals, Leverages, and Measures of Influence

CHAPTER OUTLINE

Learning Objectives
Overview
Outliers
Leverages
Measures of Influence
Added Variable Plots
Summary
Common Problems
Review Questions
Practice on Analysis and Visualization
Annotated R Functions
Answers

LEARNING OBJECTIVES

- Describe what we can learn from residuals.
- Explain the difference between residuals and leverages.
- Employ measures of influence to make discoveries.
- Explore results using added variable plots.

Overview

In the previous chapter, I introduced diagnostics, focusing on the residual plot. The residual plot helps identify patterns that indicate whether we've met the Gauss-Markov theorem's assumptions. If the assumptions are met, OLS is the best linear unbiased estimator. In this chapter, our attention turns to identifying outlying cases. Outliers are observations that stand out from the rest of the data. We are interested in identifying outliers since they could influence our regression results. Outliers could also be the result of a mistake made in collecting or entering the data. Finally, outliers could provide important information that helps solve our puzzle.

There are a number of different tools we have at our disposal to identify outliers. Each analyst has their preferred method. There is no single agreed-upon set of views or measures appropriate for every situation. Analysts can agree, however, on one principle: more is better. In other words, don't rely on just one diagnostic tool. Multiple views provide invaluable information.

In this installment of diagnostics, you'll learn how to identify outliers, calculate leverages, and understand how they differ from measures of influence. You'll not only learn how to produce useful visualizations of outliers, leverages, and measures of influence, but you'll also gain an intuition for how they all can affect regression results. You'll also learn about added variable plots, useful visualizations of what underlies the coefficients reported in a regression table.

As we discussed in Chapter 15, diagnostics are used to check whether the estimator we're using (OLS) is appropriate. In this chapter, diagnostics are used to determine whether our results are generated by a few distinctive cases. Diagnostics also help *explain* our estimates. The results reported in a regression table may support our hypothesis. Diagnostics may reveal, however, that the coefficients, t-ratios, and R^2 statistics are there for completely different reasons than had been suspected.

Outliers

Let's begin this section with some important definitions. When speaking generally about data, analysts use the term *outlier* to mean any observation or case that stands out from the rest along any number of dimensions. When making distinctions between residuals, leverages, and measures of influence, I will refer specifically to observations that stand out from the predicted value. **Outliers** are large residuals that indicate our prediction for the *i*th observation was particularly bad.

Understanding the difference between outliers, leverages, and measures of influence is important since each provides slightly different information. Outliers help us understand the accuracy of our predictions, giving a frank evaluation of how well our model predicts the outcomes of interest. Outliers are unique cases that stand out from other observations in relation to the y axis. Leverages identify observations that are distinctive along the x axis. Finally, measures of influence indicate how the presence of each observation in the regression affects our regression estimates. Because each measures something different, each can provide important clues that can help solve our mystery.

Understanding the distinction between all three is also important because while they are all related, a large value in one doesn't necessarily imply it will be large in another. For example, we could have an observation that is an extreme outlier, or has a high leverage value, but does not have much influence on our results. To keep things straight, think of an outlier as being extreme in the y variable, leverages indicate how far an observation is from most values in the independent variables, and influence measures indicate how much an observation "influences" our results.

Whether we're interested in outliers, leverages, or measures of influence, we start with residuals. Recall that errors are the difference between the true model (or true line) and the observed values of y. Residuals are similar in that they are the difference between the predicted value generated by our regression model $\hat{Y}_i$ and Y_i. More formally, we use the following equation:

$$e_i = y_i - \hat{y}_i$$

Outliers, leverages, and measures of influence are based on the residuals generated by the estimation of our regression model. Let's start with identifying large residuals that we're calling outliers.

Let's consider an example from a model that predicts the percentage of a state legislature's seats that are held by women. The variables in the regression model are as follows:

$$femleg = f(density, evangel, democrat, hsdiploma)$$

The functional form of the model is rendered into code in Code Chunk 16-1. As with the previous example, I first define the linear regression model using the *lm()* command and then

I create the required residuals and predicted values. Note that in the first two rows I create a new data frame and direct R to use the variable *states$st* as the *rowname* variable. This will come in handy when we want to use commands associated with different packages since they return the row name as the identifying variable. It's more informative to see the observation for New York labeled as the state code (NY) rather than the observation number (32).

Code Chunk 16-1

```
nstate <- states
row.names(nstate) <- nstate$st

model.lm <- lm(femleg ~ density + evangel + democrat + hsdiploma, data = nstate)

states$res <- resid(model.lm)
states$pred <- predict(model.lm)


ggplot(states, aes(pred, res)) +
  geom_point(col="#bf0000") +
  geom_text_repel(size=3.5, col = "black", aes(label=
           ifelse(states$st %in% c("AZ", "PA", "ID", "SD"),
                      as.character(states$st), "")),
                 hjust=-.5, vjust=0, show.legend=FALSE) +
  geom_hline(yintercept = 0, linetype = 2) +
  theme_minimal() +
   theme(plot.title = element_text(size = 8, face = "bold"),
        axis.title = element_text(size = 8, face = "bold")) +
  xlab("Predicted or "Fitted" Values") +
  ylab("Residuals") +
  ggtitle("Figure 16-1: A Formless Cloud")
```

Residual plots throw into stark relief the relationship between our residuals and fitted values so that we can decide whether our analysis meets the Gauss-Markov conditions. Recall that observations above the dotted line represent cases we've underpredicted. Observations below the dotted line represent cases our model overpredicted. The x axis records the level of our predictions, allowing us to quickly see whether the accuracy of our predictions—the distance of the observations from the dotted line—is related to our predicted values. This provides a quick and easy way to see if our residuals are independently distributed and/or have constant variance, two important Gauss-Markov conditions.

There are a couple of things to observe in Figure 16-1. First, notice that the residuals don't seem to vary with the fitted values: we're not systematically overpredicting or underpredicting. Moreover, the size of the residuals don't seem to vary systematically as we range over the predicted values. There are a couple of cases worth mentioning: Idaho (ID), Arizona (AZ), South Dakota (SD), and Pennsylvania (PA). Note that three of the four reside in the middle of the data (along x) and are relatively close to other cases. Consequently, their exclusion or inclusion

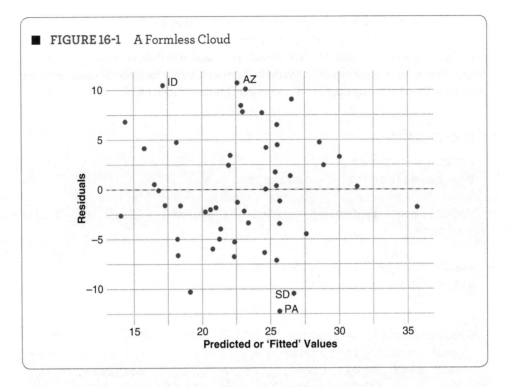

FIGURE 16-1 A Formless Cloud

from the regression is not likely to change the results by much. Therefore, the size of the outlier does not entirely determine that observation's impact or influence on the result. The Idaho case is, however, located at the lower end of the fitted values and is somewhat off by itself in terms of both y and x. Given the size of the Idaho outlier, it might make sense to see if our results are *stable with respect to the inclusion or exclusion of Idaho*. For now, let's simply make a note of how poorly our model predicts that case.

The observations made about Idaho deserve more discussion. We know our prediction for Idaho was poor, making it an outlier. We also see that it is somewhat removed from the center of the data in terms of x, giving it some leverage. Consequently, it would be nice to know what influence it has on our results. Leverages begin to move us in that direction.

> **KNOWLEDGE CHECK: Describe what we can learn from residuals.**

1. Which of the following accurately describe what we can learn from residuals?
    a. Have we met the Gauss-Markov assumptions?
    b. Which cases could have the most influence on our regression results?
    c. Which cases represent our best and worst predictions?
    d. Were there mistakes in data collection or data entry?
2. Which of the following best describes residuals?
    a. The difference between the true line and the observed value
    b. The difference between the true line and the predicted value

c. The difference between the regression line and the observed value

d. The difference between the regression line and the predicted value

3. Which of the following accurately describe the differences between outliers, leverages, and measures of influence?

   a. Outliers are cases that influence the regression results.

   b. Outliers are cases that could influence the regression results.

   c. Cases with high leverages help identify cases that are removed from the rest of the data along the x axis.

   d. Cases with high leverages help identify cases far removed from the regression line.

Leverages

Points with high leverage are said to be relatively far from the center of the regressor space, taking account of the correlational pattern among the regressors (Fox & Weisberg, 2011). In other words, **leverage** values indicate how far a case is relative to the mean of the other independent variables considered together. It's also helpful to think of values with high leverages as being outliers on the x axis. Since leverages are calculated using standardized versions of the independent variables (we assume their mean is 0 and they have a standard deviation of 1), the leverage values we calculate are bounded between 0 and 1. Most expositions of how to calculate leverage values involve matrix algebra, which I will avoid here. The simplest way to present the calculation of leverage is as follows:

$$h_i = 1 - \frac{y_i - \hat{y}_i}{(y_i - \hat{y}_i)_{(i)}}$$

where the leverage value h_i for the ith observation is equal to 1 minus the following quotient: the calculated residual for the ith case when it is included in the model $y_i - \hat{y}_i$, divided by the calculated residual when the observation is removed $y_i - \hat{y}_{(i)}$. To be clear, in the numerator we have the residual calculated when all observations are included. The denominator is the residual calculated between y_i and $\hat{y}_i$ when the ith case is removed from the model, which is denoted by the parentheses with the (i) subscript. Consequently, leverages provide an indication of how much the predicted value for the ith observation changes when the ith observation is removed from the calculation.

R makes it easy to produce graphics that allow us to spot observations with high leverages. Using the method to produce residual plots, we can pull leverage values from the regression model much in the same way we collected residuals and predicted values. In Code Chunk 16-2 you'll see I define a new variable called *states$hat* using the command *hatvalues()*, which collects the leverage for each observation in the model. Once I've extracted those values from the model, I'm ready to plot.

Code Chunk 16-2

```
states$hat <- hatvalues(model.lm)

ggplot(states, aes(hat, res)) + geom_point(col="#bf0000") +
  geom_text_repel(size=3.5, col = "black", data=states, aes(label=
           ifelse(states$st %in% c("UT", "DE", "NJ", "RI"),
```

```
                            as.character(states$st), "")),
            hjust=-.5, vjust=0, show.legend=FALSE) +
geom_hline(yintercept = 0, linetype = 2) +
theme_minimal() +
theme(plot.title = element_text(size = 8, face = "bold"),
      axis.title = element_text(size = 8, face = "bold")) +
xlab("Leverages or 'Hat' Values") +
ylab("Residuals") +
ggtitle("Figure 16-2: Residuals Versus Leverages")
```

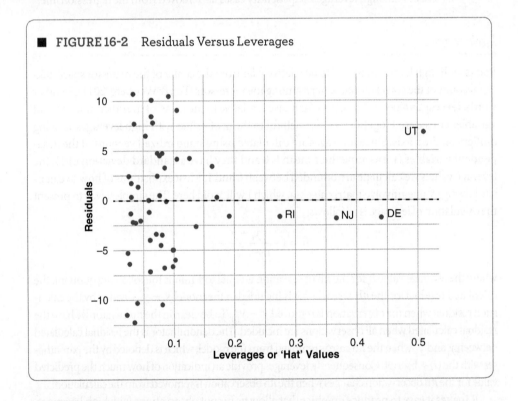

■ FIGURE 16-2 Residuals Versus Leverages

As Figure 16-2 indicates, there are a number of potentially influential cases: Utah (UT), Delaware (DE), Rhode Island (RI), and New Jersey (NJ). We see that of all the outliers, Utah seems to perhaps be the most important: it has the highest leverage value and a fairly large residual.

Perhaps it should be no surprise that these four states are distant from the other observations. First, Utah is a strange case as we've noted earlier: it's a Western state yet has a very high level of evangelism. The three remaining cases (Rhode Island, New Jersey, and Delaware) are all on the East Coast, have populations that identify as Democrat, and are densely populated. The plot of leverages suggests those cases could affect our estimates. In particular, the conclusions we draw from the regression estimates may be dependent on those states.

Let's continue to build our intuition. Leverages provide an indication of how far the observation is from the other points in terms of the x axis. Consider the bivariate regression between the percentage of the population with high school diplomas (*states$hsdiploma*) and median household income (*states$medinc*). In that regression, note that Maryland (MD) has the highest x value. It seems to be the most distant from the "middle" of the data along the x axis. As a result, it has a relatively high leverage. It's leverage is .14 compared to the leverage value of .02 for Pennsylvania (PA), which is closer to the mean of x.

To produce Figure 16-3, I first defined the bivariate regression as an object (*leverage.lm*; Code Chunk 16-3). I then extracted the leverages from the model with the *hatvalues()* function. Finally, I constructed the ggplot, using the *ifelse()* command to label the points with their state code and the associated leverage value.

Code Chunk 16-3

```
leverage.lm <- lm(hsdiploma ~ medinc, data = states)
states$lev <- hatvalues(leverage.lm)

ggplot(states, aes(medinc, hsdiploma)) +
  geom_point(col="#bf0000") +
  geom_smooth(method="lm", col="#0000bf", size = .5, se=FALSE ) +
  geom_text_repel(size=3.5, col = "black", data=states, aes(label=
          ifelse(states$st %in% c("MD", "NY"),
                      as.character(states$st), "")),
              hjust=1.5, vjust=1.5, show.legend=FALSE) +
  geom_text(size=3.5, col = "#0000bf", data=states, aes(label=
          ifelse(states$st %in% c("MD", "NY"),
                      round(states$lev, digits = 2), "")),
              hjust=1, vjust=-1, show.legend=FALSE) +
  xlab("Median Household Income") +
  ylab("Percentage of Population With High School Diploma") +
  ggtitle("Figure 16-3: Maryland Has Largest Value in x") +
  theme_minimal() +
  theme(plot.title = element_text(size = 8, face = "bold"),
        axis.title = element_text(size = 8, face = "bold"))
```

Now let's see what happens to the leverage for Maryland when we increase its level of income in x (median household income) by $20,000. I'll also draw a scatter plot next to it showing what happens when we increase Maryland by both x and y. I then plot the leverage value next to Maryland (in blue) showing that increasing the value of y has no impact on leverage.

To produce the illustration, I create a new variable I'll call *states$nmedinc* and a new variable *states$nhsdiploma* that changes the value for Maryland (Code Chunk 16-4). Once those are created, I can create two different objects defined by regressions where Maryland changes just in x (*xlev.lm*) and in both x and y (*xylev.lm*). Once the regression objects are created, I define two new variables that collect the leverages from each (*xlev* and *xylev*).

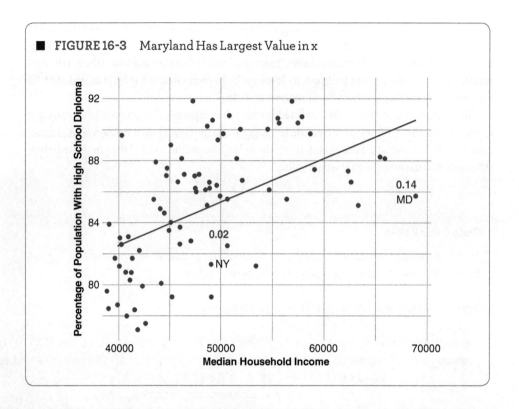

FIGURE 16-3 Maryland Has Largest Value in x

Code Chunk 16-4

```
states$nmedinc <- ifelse(states$st=="MD", 20000 +
                         states$medinc, states$medinc)
states$nhsdiploma <- ifelse(states$st=="MD", 7 + states$hsdiploma,
                         states$hsdiploma)
xlev.lm <- lm(hsdiploma ~ nmedinc, data = states)
xylev.lm <- lm(nhsdiploma ~ nmedinc, data = states)

states$xlev <- hatvalues(xlev.lm)
states$xylev <- hatvalues(xylev.lm)
```

Once those objects and variables are created, I'm ready to plot. Although there's a lot of code, it should all be familiar by now. I define two scatter plots (p1 and p2) and then plot them side by side using the *grid.arrange()* command (Code Chunk 16-5).

Code Chunk 16-5

```
p1 <- ggplot(states, aes(nmedinc, hsdiploma)) +
  geom_point(col="#bf0000") +
  geom_smooth(method="lm", col="#0000bf", size = .5, se=FALSE ) +
  geom_text(size=3.5, col = "black", data=states, aes(label=
        ifelse(states$st == "MD",
```

```r
                         as.character(states$st), "")),
             hjust=1.5, vjust=1.5, show.legend=FALSE) +
  geom_text_repel(size=3.5, col = "#0000bf", data=states, aes(label=
             ifelse(states$st == "MD",
                         round(states$xlev, digits = 2), "")),
             hjust=1, vjust=-1, show.legend=FALSE) +
  xlab("Median Household Income") +
  ylab("Percentage of Population With High School Diploma") +
  ggtitle("Panel A: Maryland's Income Is Changed") +
    theme_minimal() +
  theme(plot.title = element_text(size = 8, face = "bold"),
        axis.title = element_text(size = 8, face = "bold"))

p2 <- ggplot(states, aes(nmedinc, nhsdiploma)) +
  geom_point(col="#bf0000") +
  geom_smooth(method="lm", col="#0000bf", size = .5, se=FALSE ) +
  geom_text(size=3.5, col = "black", data=states, aes(label=
             ifelse(states$st == "MD",
                         as.character(states$st), "")),
             hjust=1.5, vjust=1.5, show.legend=FALSE) +
  geom_text_repel(size=3.5, col = "#0000bf", data=states, aes(label=
             ifelse(states$st == "MD",
                         round(states$xylev, digits = 2), "")),
             hjust=1, vjust=-1, show.legend=FALSE) +
  xlab("Median Household Income") +
  ylab("Percentage of Population With High School Diploma") +
  ggtitle("Panel B: Education and Income Are Changed") +
    theme_minimal() +
  theme(plot.title = element_text(size = 8, face = "bold"),
        axis.title = element_text(size = 8, face = "bold"))

grid.arrange(p1, p2, ncol=2, top=textGrob("Figure 16-4: Changing Value's
Impact on Leverage",
                         gp=gpar(fontsize=8)))
```

As you can observe in Figure 16-4A, increasing Maryland's income (the x variable) increases Maryland's leverage from .14 to .39. In Figure 16-4B, we increased both Maryland's income (by $20,000) and education (by 7 percentage points). Note that the leverage value calculated from each regression is the same (.39) regardless of Maryland's position on the y axis.

Figure 16-4 demonstrates that leverages do not indicate influence directly. It's true that in both regressions Maryland *could* be influential. In fact, Maryland has greater influence on the regression line in Figure 16-4A than in Figure 16-4B where it is closer to the line. Since leverages register the distance from the mean x value along the x axis, the leverage value in both

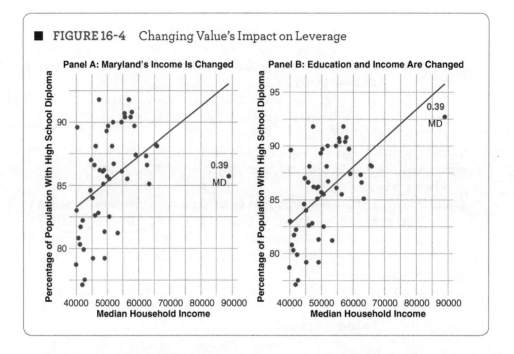

FIGURE 16-4 Changing Value's Impact on Leverage

panels for Maryland is exactly the same (.39). We therefore say that leverages identify points with potential influence, but we don't know for sure. That's why in the next section we turn to measures of influence.

To review, we've identified cases that are outliers (distant from the predicted value) or have leverage (distant from the majority of cases on the x axis). Some cases could be relatively large outliers and have high leverage values. Observations with large residuals, however, will not necessarily have much influence on our regression results. Similarly, observations with high leverage values will not necessarily influence our estimates. However, cases that are both outliers and have high leverages are likely to influence our results.

Influence measures are the topic of the next section. There are two measures of **influence** commonly used that I'll introduce next: Cook's distance and dfbetas. Cook's distance, otherwise known as Cook's D, is based on the residual and the leverage. It indicates the ith observation's influence on all of the regression model estimates. Dfbetas are a measure of influence that indicate how much each individual model estimate (coefficient) changes when we exclude the ith observation from the regression. Let's first consider Cook's distance.

> **KNOWLEDGE CHECK: Explain the difference between residuals and leverages.**

4. Which of the following statements accurately describe residuals and leverages?

   a. Leverages are a better indicator of influence.

   b. Residuals are a better indicator of influence.

   c. Leverages indicate if the observation represents a distinct value along the x axis.

   d. Residuals indicate if the observation is distinct along the x and y axes.

5. Which of the following statements is true?

   a. Leverages indicate how much the *i*th observation affects a model's predicted value for the *i*th observation.

   b. Leverages indicate how far an observation is from the centroid of the independent variables along the x axis.

   c. An observation might have a high leverage value and little influence on the slope of a line.

   d. Observations with high leverage values always have considerable influence on the slope of a line.

Measures of Influence

Cook's Distance

Cook's distance is a commonly used measure of influence. Cook's distance indicates the relative influence the *i*th observation has on all of the regression model's estimates. There are two components to that calculation. One component of Cook's distance registers the distance of an observation from the values of the independent variables (the leverage). The bigger the leverage, the bigger Cook's distance. The other component of Cook's D is based on the size of the observation's residual. The larger the residual, the larger Cook's distance. Cook's D is measured as follows:

$$D_i = \frac{(y_i - \hat{y}_i)^2}{p \times MSE} \left(\frac{h_i}{(1-h_i)} \right)$$

where y_i is the observed value of y, $\hat{y}_i$ is the predicted value, p is the number of parameters in the model including the intercept, MSE is the regression's mean squared error, and h_i is the leverage for the *i*th observation. To gain a conceptual familiarity with Cook's D, let's take Figure 16-2 and size the observations according to Cook's D.

To produce the plot, we follow the same procedure used to collect values for residuals and for leverages. Once the Cook's distances have been extracted from the model, we're ready to plot. Note the use of the size option in the *geom_point()* command, which allows us to size the observations according to Cook's D (Code Chunk 16-6). Note also that I use the *ifelse()* command to label the top-ranked observations in terms of Cook's D.

Art and Practice of Data Visualization
ONE COMMAND, MULTIPLE PLOTS

As an alternative to the plot in Code Chunk 16-6, I modified some existing code from the 'olsrr' package to identify the cases with the three largest Cook's distances in a residual plot, a simple bar plot, and a scatter plot that plots residuals against leverages. The command is *db_diag()*. The command asks for the model name and title. Remember to declare which variable you want to use as labels as the row name, otherwise the points will be identified by the number of the row (not that helpful). Also note that in some cases resizing the window displaying Cook's distances is necessary to view the plots.

Code Chunk 16-6

```
states$cook <- cooks.distance(model.lm)

ggplot(states, aes(hat, res)) +
  geom_point(aes(size = cook), col="#bf0000") +
  geom_text_repel(size=3.5, col = "black", data=states, aes(label=
        ifelse(rank(-cook) < 5, st, NA)),
              hjust= 1.5, vjust= 2, show.legend=FALSE) +
  geom_hline(yintercept = 0, linetype = 2) +
  theme_minimal() +
  theme(plot.title = element_text(size = 8, face = "bold"),
        axis.title = element_text(size = 8, face = "bold")) +
  xlab("Leverages or 'Hat' Values") +
  ylab("Residuals") +
  ggtitle("Figure 16-5: Previous Plot Sized by Cook's D")
```

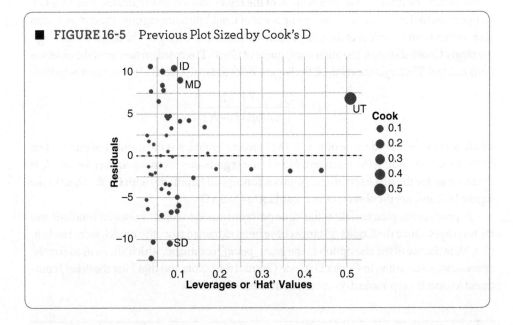

FIGURE 16-5 Previous Plot Sized by Cook's D

There are several things to observe. First, the top-ranked observations in terms of Cook's distance are somewhat different than the top leverage values labeled in Figure 16-5. Rather than Utah, New Jersey, Rhode Island, and Delaware, the top-ranked observations in terms of Cook's D are Utah, Idaho, Maryland, and South Dakota. As the plot makes clear, there is an observation that stands out from the rest in this regression: Utah. Utah's high value is no doubt the result of its religiosity. While most of the states with large evangelical populations are in the South, Utah's location in the West makes it stand out. We might also want to consider whether classifying Utah's largely Mormon population as evangelical is an accurate representation of its population.

Dfbetas

By combining leverage values with residuals, Cook's distance provides a useful indication of an observation's influence on our estimates. While Cook's distance measures the influence of an observation on the predicted values in combination with the size of the residual, we still don't know how much each observation actually affects each individual regression coefficient. Fortunately, there is a measure of influence based on how much each coefficient in the model changes when we remove the ith observation: the dfbeta.

Dfbetas are calculated by recording the difference in the coefficients with and without each observation. That difference is then standardized by dividing by the mean squared error of the regression when the case is deleted and multiplying that by the kth diagonal element of the covariance matrix. So far, I have avoided matrix algebra in this book and will not introduce it here. Put simply, dfbetas record the difference in a regression coefficient produced when excluding the ith observation. That difference is transformed (standardized) so that we can have a common reference for the impact each observation has on all of the coefficients in our model (including the intercept). The formula for dfbetas is as follows:

$$dfbeta_{ik} = \frac{\beta_k - \beta_{k(i)}}{\sqrt{MSE_{(i)} c_{kk}}}$$

where β_k is the coefficient for variable k with all of the observations and $\beta_{k(i)}$ is the coefficient for variable k when the ith observation is removed. The numerator is divided by the square root of the regression's mean squared error when the ith case is removed, multiplied by the diagonal element of the unscaled covariance matrix. Explaining the diagonal element of the unscaled covariance matrix will have to wait for your next statistics class or a class on matrix algebra. For now, it's important to understand that we're measuring how much each observation changes each coefficient in the regression. That value is standardized so that we can understand its relative impact across all of the model's coefficients.

While the number of residuals, leverages, and Cook's distances equals the number of observations in a regression, the number of dfbetas is larger since there is a dfbeta for each observation multiplied by the number of parameters in the model. In other words, a dfbeta is calculated for each observation and coefficient so that we know the influence of each observation on each coefficient.

Standard practice to examine dfbetas is to plot them for each parameter. Also standard practice is to provide a threshold that indicates whether the dfbeta has significant influence on the parameter. In general, when it comes to residuals, leverages, or Cook's distances, I prefer to examine actual visualizations of the data to identify observations that are different from the others rather than rely on an arbitrary threshold. Nevertheless, the threshold commonly used for dfbetas is 2 divided by the square root of the number of observations in the regression:

$$\frac{2}{\sqrt{n}}$$

To examine the dfbetas from a regression, there are several options. The 'car' package can quickly draw dfbeta plots with the command *dfbetaPlots()*. An important feature of the *defbetaPlots()* command is you can specify a variable that will supply labels to the outliers, which is crucial for identification. I've found that while a number of diagnostic tools will plot various measures of influence, not all of them make it easy to label points with a variable that is useful. Outlying cases labeled with the row number (i.e., observation number), for example, are less than helpful.

To place the plots in the familiar ggplot framework, I created a function that plots the dfbetas and allows the user to label the points that surpass the threshold and an option to provide the title of the figure (Code Chunk 16-7). The function is called *db_dfb()* and it was defined in the data environment for this book. I use it to produce the dfbeta plots for the object *model.lm*.

An important step before using the functions associated with the *db_dfb()* command is to establish the variable you want to use to label the points before you plot them. In Code Chunk 16-7, you'll see that I use the *row.names()* function to identify *st* as the labeling variable. This command tells R to use the state codes as labels rather than the row numbers.

Code Chunk 16-7

```
nstate <- states
row.names(nstate) <- nstate$st

db_dfb(model.lm, "Figure 16-6: Dfbetas for Each Coefficient")
```

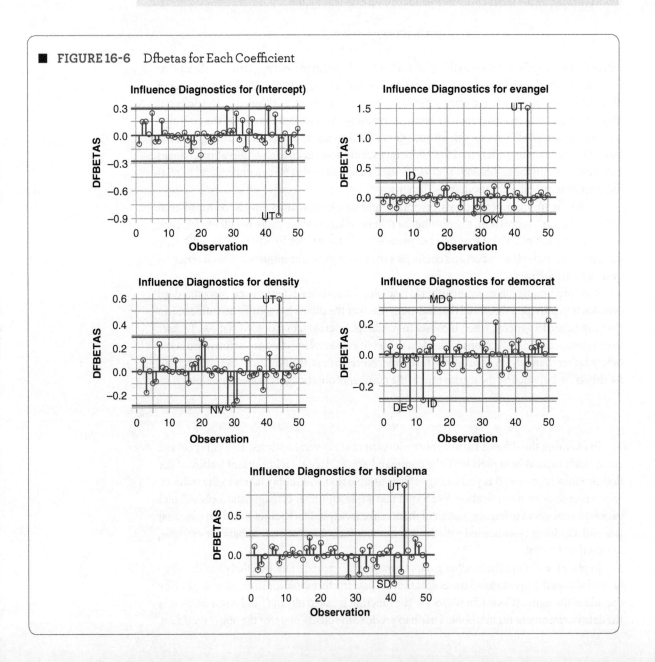

■ **FIGURE 16-6** Dfbetas for Each Coefficient

In Figure 16-6 there is a dfbeta plot for each of the parameters (including the intercept) in the model we used to explain the percentage of legislative seats held by women. To gain a feel for how to use the dfbeta plots, consider the plot for the variable *evangel*. Utah is considerably above the threshold (the red line). Its oversized impact probably results from having such a high number of citizens classified as evangelicals. Utah shows up again as having considerable influence for the density variable. It also has a dfbeta above the threshold level for the *hsdiploma* variable. According to the dfbeta plots, we know that when Utah is removed from the regression, the coefficients on the *evangel*, *density*, and *hsdiploma* variables will change considerably. The coefficient on the intercept should change noticeably as well.

In Code Chunk 16-8 I reproduce the regression with and without Utah in the data to see its impact on the coefficients of the regression. To make an easy comparison, I simply use the *update()* command, which allows me to specify that I'd like to update the model by removing Utah. I then put the two models into the *stargazer()* command to display the regression results from the regression model (Code Chunk 16-8).

Code Chunk 16-8

```
without.lm <- update(model.lm, data = subset(nstate, st!="UT"))

stargazer(model.lm, without.lm, header = FALSE, type = "text",
          title = "Table 16-1: Utah's Important Impact on
                                 the Regression")
```

As the regression table shows, removing Utah influences the coefficients (and their statistical significance) noticeably for three variables: *evangel*, *density*, and *hsdiploma* (Table 16-1). Also note that as indicated, Utah has relatively little influence on the coefficient for *democrat*. Now that we've identified an influential case, what do we do?

The proper course of action is to investigate further. We've already mentioned the odd nature of the evangelical variable (*evangel*), which must be registering the Mormon faith as evangelical. Perhaps a better measure of religion is in order. When Utah is removed, density and religion seem to become more important variables in the model (the magnitude of their coefficients grows as does their statistical significance). While tempting to discard the case and only report the results without Utah (the R^2 is higher in the model without Utah), a better course of action would be to alert the reader that Utah makes a difference, show how it influences the results, and identify a number of possible courses of investigation implied by this finding.

Art and Practice of Data Visualization

THREE DIMENSIONS

Why not just focus on measures of influence? That is, after all, the measure that will indicate which observations matter most with respect to our estimates. While concentrating our efforts solely on influence measures can expedite our analysis in some ways, we give up significant information by not considering all three dimensions. We may learn that an observation has significant influence on our results, but knowing whether that influence stems from being an outlier or having a high leverage value can provide important information.

TABLE 16-1 Utah's Important Impact on the Regression

	Dependent variable: femleg	
	(1)	(2)
density	−0.006	−0.009*
	(0.004)	(0.004)
evangel	−0.152*	−0.268**
	(0.079)	(0.104)
democrat	0.377***	0.385***
	(0.132)	(0.129)
hsdiploma	0.680***	0.479*
	(0.237)	(0.262)
Constant	−43.152*	−23.770
	(22.615)	(25.030)
Observations	50	49
R2	0.376	0.412
Adjusted R2	0.320	0.359
Residual Std. Error	5.933 (df = 45)	5.819 (df = 44)
F Statistic	6.777*** (df = 4; 45)	7.723*** (df = 4; 44)

Note: *p<0.1; **p<0.05; ***p<0.01

> **KNOWLEDGE CHECK: Employ measures of influence to make discoveries.**

6. Indicate which of the following indicates the distance of an outlier in terms of y, in terms of x, and its influence on regression coefficients.

   a. Dfbetas
   b. Leverages
   c. Cook's distance
   d. Residuals

7. Which of the following are true?
   a. Leverages measure the influence of an observation on the slope of the line.
   b. Dfbetas measure the distance an observation is from the center of the data in x.
   c. An observation with a relatively high Cook's distance will always be more influential than one with a relatively low Cook's distance.
   d. An observation with a relatively high residual will always be more influential than one with a relatively low residual.

Added Variable Plots

Added variable plots, sometimes called partial regression plots, plot the relationship between an independent variable and the dependent variable from a regression, holding all other variables in the regression constant at their means. Added variable plots are useful since they reveal whether the relationship between an independent variable and a dependent variable is linear and provide an indication of how the regression coefficient might be unduly influenced by a single case or group of cases.

Added variable plots are constructed by regressing the dependent variable on all of the independent variables except for the independent variable of interest. We then save the residuals from that regression. Next, we regress our independent variable of interest against the other independent variables and save those residuals. We can then plot the first set of residuals against the second. Fitting a regression line to that plot produces a line with a slope that matches the regression coefficient on the independent variable of interest.

Consider the added variable plots in Figure 16-7 that graph the partial effects of each independent variable in the regression model we used earlier:

$$femleg = f(hsdiploma, democrat, density, evangel)$$

To produce the added variable plots for the regression, I modified code from the 'olsrr' package to produce meaningful labels for the plots.

Code Chunk 16-9

```
db_avp(model.lm, "Figure 16-7: Added Variable Plots for 'model.lm'")
```

In Figure 16-7, the two largest values along the x (leverages) and y (residuals) axes are labeled, giving a very quick but comprehensive view of each independent variable and its relationship with the percentage of a legislature's seats held by women. Also note the x and y axes in these plots; they no longer represent the units of each variable. Instead, the y axis represents the residuals from the regression of y on all of the x values (except for the independent variable of interest) and the x axis represents the residuals from the regression of the independent variable of interest on the other independent variables. We look for two things in added variable plots: (1) Is the relationship linear? (2) Are there any cases with undue influence on the line? We observe several things from the added variable plots in Figure 16-7.

■ **FIGURE 16-7** Added Variable Plots for 'model.lm'

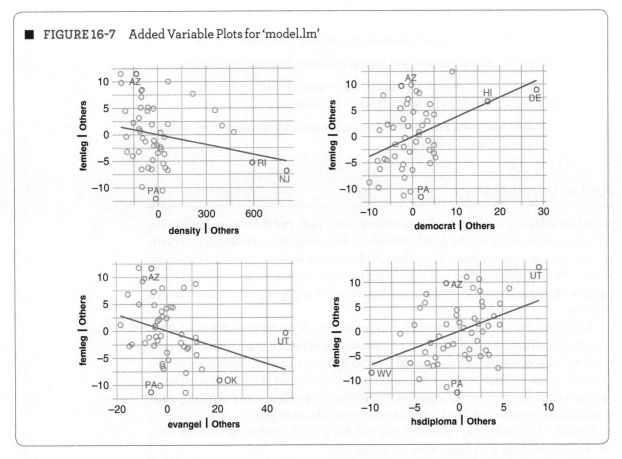

First, we note that the relationship between *hsdiploma* and *femleg* is linear (relative to the other plots). Second, Utah's influence on several of the coefficients is readily obvious since it is both distant from the other x values and somewhat removed from the line. Third, the added variable plot for density indicates that the density variable is skewed. While we should have caught that earlier, it's clear that most of the cases are located at the low end of the distribution with a few cases with high values: Rhode Island and New Jersey seem to have large leverage values. Finally, the added variable plot of religion (captured by *evangel*) indicates that Utah may be exerting undue influence on the line. Since the blue line in each of these graphs corresponds directly with the coefficients in the multiple regression, we can see how various cases and patterns might influence our estimates.

Added variable plots are underused in diagnostics. They are, however, extremely useful since they provide a very quick and intuitive look at the data. They also provide a much greater amount of information in a relatively small space.

> **KNOWLEDGE CHECK:** Identify potential cases that influence regression results.

8. Which statements best describe added variable plots?
   a. Added variable plots are partial regression plots.
   b. Added variable plots are based on measures of influence.

c. Added variable plots show the bivariate relationship between an independent variable and the dependent variable in a regression.

d. Added variable plots show the relationship between an independent variable and the dependent variable, accounting for the other independent variables.

9. Using the world data set, regress infant mortality (*inf*) on GDP per capita (*gdppc*) and population (*pwtpop*). Answer the following.

    a. What country has the biggest residual?

    b. What countries seem to have the most influence on a population's relationship with infant mortality?

    c. What country has the biggest influence on GDP per capita's association with infant mortality?

    d. Do the Gauss-Markov conditions seem to be met? (yes/no)

10. Using the world data set, regress turnout (*turnout*) on ethnolinguistic fractionalization (*ethfrac*) and human capital (*pwthc*). Answer the following.

    a. What country has the biggest residual?

    b. What country has the most influence on ethnolinguistic fractionalization's relationship with turnout?

    c. What two countries have the biggest influence on human capital's association with turnout?

    d. Do the Gauss-Markov conditions seem to be met? (yes/no)

SUMMARY

In this installment of diagnostics, we learned about the important differences between residuals (outliers), leverages, and measures of influence. Residuals indicate the distance between an observation and the predicted value. They help identify outliers in relation to the y axis. Leverages indicate how much the predicted value for the *i*th observation changes when the *i*th observation is removed from the regression. They indicate how distinct an observation is relative to others along the x axis. Cook's distance, a measure of influence, combines residuals and leverages, providing a helpful measure of how much the model estimates change when a specific observation is excluded from the regression. Dfbetas are another measure of influence discussed, indicating how much each coefficient in a regression changes when the *i*th observation is removed. Finally, we introduced added variable plots, an efficient method we can use to check whether the relationship between an independent variable and a dependent variable is linear and to identify cases that might be exerting undue influence on each coefficient in the model.

While each diagnostic provides important information, don't limit yourself to one measure or one view. Not all influential observations will reveal themselves in a residual plot. While Cook's distances can provide indications of overall influence, insight can be gained by understanding the influence each observation has on each coefficient: dfbetas. To repeat the advice given in the conclusion of Chapter 15, afford diagnostics as much time as describing data, model building, and estimation. Think of diagnostics as continued exploration. Not only do diagnostics identify potential problems with our regression results, but they can also generate additional clues that help solve the puzzle or answer our question. They also lead to additional questions.

COMMON PROBLEMS

- *Understanding the distinction between outliers, leverages, and measures of influence.* The problems most have with distinguishing between outliers, leverages, and measures of influence could arise because many use the terms interchangeably. Outliers (as I distinguish them from leverages) indicate the distance in y between the observation and the predicted value (the regression line). Leverages give the distance in x between the observation and the centroid of all the independent variables. Finally influence measures indicate each observation's impact on the regression estimates. Cook's D indicates an observation's impact all of the estimates combined. Dfbetas register an observation's impact on each individual regression coefficient.

- *Going down the rabbit hole.* Describing data, generating hypotheses, estimating models, performing diagnostics, and generating new questions can be a never-ending process. There will always be new discoveries, additional insights, and new things to explore. An important skill to develop is to know when to stop. Referring back to the original question can be helpful to ascertain whether it has been sufficiently answered or whether the analyst has strayed too far. Another useful anchor is to determine whether the analysis has produced any concrete, actionable recommendations for solving the problem.

REVIEW QUESTIONS

1. What can we learn from studying residuals?
2. Why examine measures of influence?
3. If we have measures of influence, why examine residuals or leverages?
4. What are some influence measures and how are they different than residuals?
5. Why can a large residual have no influence?
6. What is Cook's D?
7. What is a leverage or hat-value?
8. How can you construct an added variable plot?
9. What should you look for in an added variable plot?
10. Why examine more than one kind of diagnostic plot?

PRACTICE ON ANALYSIS AND VISUALIZATION

1. Indicate which of the following statements are true.
   a. Large leverages indicate an observation has influence.
   b. Large residuals indicate an observation has influence.
   c. A large Cook's distance indicates an observation has influence.
   d. A Large leverage could have relatively little influence.

2. Examine the following bivariate regression plot and answer these questions.
   a. Which state has the highest leverage?
   b. Which state has the highest residual?
   c. Which state is likely to have the largest influence?
   d. Will Hawaii (HI) have more leverage or influence?

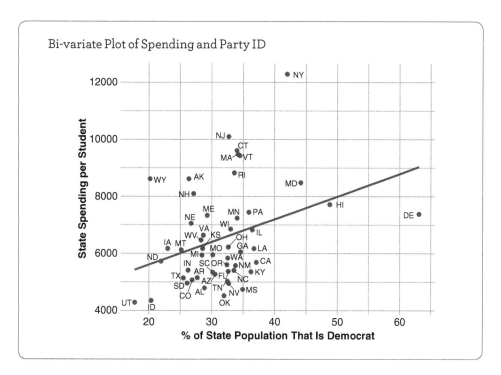

3. Examine the following added variable plots and answer the following questions.

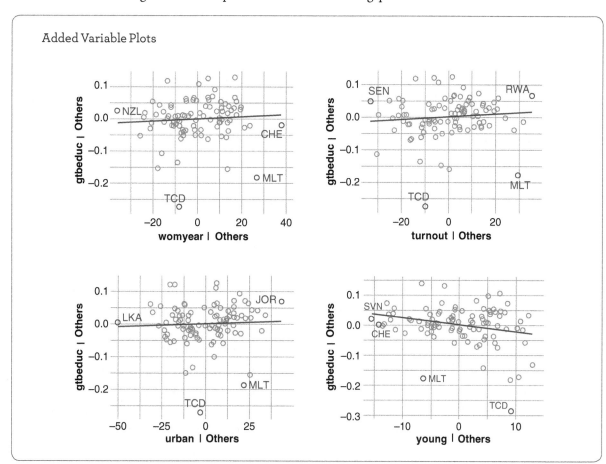

a. What variable indicates there is a negative relationship?

b. Which countries have the highest leverage values when examining the relationship between *gtbeduc* and *urban*?

c. Which countries have the largest residuals when examining the relationship between *gtbeduc* and *young*?

d. Which country has the most influence when examining the relationship between *gtbeduc* and *turnout*?

4. Using the *world* data set, regress *gtbeduc* on *womyear*, *turnout*, *urban*, and *young* and answer the following.

   a. What case has the biggest influence on the coefficient for *urban*?

   b. What case has the biggest influence on the coefficient for *young*?

   c. Which case seems to impact the most coefficients?

   d. On what coefficient does Jordan have the most influence?

5. Examine the Cook's distances generated from the regression in the previous question and answer the following (hint: use *db_diag()* to generate the proper diagnostic views).

   a. What country has the largest Cook's distance?

   b. Why is Malta's Cook's distance larger than Chad's?

   c. Does Cook's distance roughly reflect your answer in 4c?

   d. Why is Chad's Cook's distance more than Angola's?

6. Which of the following best describes leverages?

   a. They indicate the observations furthest from the predicted values.

   b. They indicate the observations furthest from the center of the independent variables.

   c. They indicate the influence the observation has on the regression coefficients.

   d. They represent a combination of options a and b.

7. Which of the following best describes Cook's distances?

   a. They indicate the observations furthest from the predicted values.

   b. They indicate the observations furthest from the center of the independent variables.

   c. They indicate the influence the observation has on the regression coefficients.

   d. They represent a combination of options a and b.

8. Which of the following best describes dfbetas?

   a. They indicate the observations furthest from the predicted values.

   b. They indicate the observations furthest from the center of the independent variables.

   c. They indicate the influence the observation has on the regression coefficients.

   d. They represent a combination of options a and b.

9. Regress *world$gtbeduc* on *world$womyear*, *world$urban*, and *world$womleg*. Answer the following.

   a. What two countries have the biggest Cook's distances?

   b. What two countries have the most influence on *world$urban*'s impact on *world$gtbeduc*?

   c. Do any of the countries have a significant impact on the relationship between *world$gtbeduc* and *world$womleg*?

   d. Is there any evidence of nonconstant variance in the errors?

10. Regress *states$democrat* on *states$medinc*, *states$evangel*, and *states$region*. Answer the following.

    a. What two states have the biggest Cook's distances?

    b. What three states have the most influence on income's relationship with partisanship?

    c. What state has the most influence on religion's impact on partisanship?

    d. According to the added variable plot, which coefficient seems most unstable?

ANNOTATED R FUNCTIONS

The following functions appear in this chapter. They are listed in order of their first appearance (with the code chunk number in parentheses) and annotated here to give a very brief description of their use. Some are not stand-alone functions and only work in combination with other commands. As a reminder, the code in every chapter will work properly if executed in the order it appears. Proper execution also depends on typing the author-defined *libraries()* command, which loads the required R packages.

row.names(): a useful command that designates a variable as the first row in a data frame. Helps with labeling observations when using certain functions. (16-1)

lm(): linear model or regression command. (16-1)

resid(): when applied to an object described by a linear model, it will generate the residuals. (16-1)

predict(): when applied to an object described by a linear model, it will generate the predicted values. (16-1)

ggplot(): defines the basic structure of a plot (usually the x and y variables). (16-1)

aes(): the aes (called "aesthetics") function is used in ggplot to define the basic structure of the plot, which often includes the variables you want to use and any shapes or colors. (16-1)

geom_point(): draws points on a ggplot grid. (16-1)

geom_text_repel(): labels points in a scatter plot in the ggplot framework. (16-1)

ifelse(): logical function that allows you to construct an if-then statement. Useful for selecting specific cases to label and also for creating categorical variables from continuous variables. (16-1)

geom_hline(): will produce a horizontal line in a figure. (16-1)

theme_minimal(): specifies a minimalist style for ggplot. (16-1)

theme(): specifies font, size, and so forth in a ggplot. (16-1)

xlab(): labels the x axis in ggplot. (16-1)

ylab(): labels the y axis in ggplot. (16-1)

ggtitle(): provides the title for a ggplot. (16-1)

hatvalues(): when applied to an object described by a linear model, it will generate the leverage values. (16-2)

geom_smooth(): draws lines or curves to a scatter plot. (16-3)

grid.arrange(): allows you to arrange multiple figures on a page. (16-5)

cooks.distance(): when applied to an object described by a linear model, it will generate the Cook's distance for each observation. (16-6)

db_dfb(): an author-defined function that produces ggplot versions of dfbetas. (16-7)

update(): a useful command that allows you to change an aspect of a regression model without having to retype the entire model formula. (16-8)

stargazer(): useful command to produce regression tables. (16-8)

db_avp(): author-defined function that generates ggplot version of added variable plots. (16-9)

ANSWERS

KNOWLEDGE CHECK

1. a, b, c, d
2. c
3. b, c
4. a, c

5. b, c

6. regression coefficients (a), distance in x (b), predicted values (c), distance in y (d)

7. c

8. a, d

9. Angola (AGO) (a), India (IND) and China (CHN) (b), Qatar (QAT) (c), no (d)

10. Rwanda (RWA) (a), Rwanda (RWA) (b), Rwanda (RWA) and United States (USA) (c), yes (d)

PRACTICE ON ANALYSIS AND VISUALIZATION

1. b

2. Delaware (a), New Jersey (b), Delaware (c), leverage (d)

3. young (a), Sri Lanka and Jordan (b), Malta and Chad (c), Chad (d)

4. Malta (a), Chad (b), Malta (c), urban (d)

5. Malta (a), leverage (b), yes (c), residual (d)

6. b

7. b

8. c

9. AFG and MMR (a), TCD and AGO (b), TCD and AGO (c), yes (d)

10. Utah (UT) and Delaware (DE) (a), DE, HI, and VA (b), Delaware (c), *evangel* (d)

 Access digital resources, including datasets, at http://edge.sagepub.com/brownstats1e.

17 Logistic Regression

CHAPTER OUTLINE

Learning Objectives
Overview
Questions and Problems That Require Logistic Regression
Logistic Regression Violates Gauss-Markov Assumptions
Working With Logged Odds
Working With Predicted Probabilities
Model Fit With Logistic Regression
Summary
Common Problems
Review Questions
Practice on Analysis and Visualization
Annotated R Functions
Answers

LEARNING OBJECTIVES

- Discuss problems best summarized by a nonlinear framework.
- Explain how logistic relationships violate the Gauss-Markov assumptions.
- Calculate a logistic regression.
- Explain how to convert logged odds to predicted probabilities.
- Evaluate the ability of the logistic model to predict outcomes.

Overview

Will you vote? Will there be a coup? Will war breakout? Will the company turn a profit? No matter the discipline, yes/no questions are central to what we study in the social sciences. Unfortunately, the Gauss-Markov framework doesn't apply when our dependent variable is binary: fitting a straight line to a binary outcome is less than optimal. Consequently, we must turn to **logistic regression**. Logistic regression is tantamount to fitting an S-shaped curve to the data rather than the straight line we employ when using ordinary least squares (OLS) regression. Although there are some similarities between linear and logistic regression, our interpretation of the coefficients and the diagnostics requires extra work.

This chapter begins by introducing the kinds of questions best answered with logistic regression, followed by its calculation and interpretation. Estimating logistic regression is straightforward; understanding the estimates is not. Since we're entering the nonlinear world, our interpretation of the coefficients is different. We can no longer say a one-unit increase in the independent variable is associated with a fixed amount of change in the dependent variable, at least not in an intuitive way. Consequently, we need to convert our results to present them in a way that is useful. Finally, you will learn the different tools used to assess how well logistic models describe reality. Fortunately, R provides the code to make the enterprise relatively easy.

Questions and Problems That Require Logistic Regression

Suppose we want to understand the relationship between income and voter turnout. First, we note that the dependent variable describes two distinct states: the individual either voted or they did not. This is the key characteristic of logistic regression. The dependent variable is a dichotomous variable. We might suspect that the relationship between income and the probability of voting is best described by a curve: an S-shaped curve.

People with very low incomes, the theory might go, simply don't have the time or energy to be actively engaged in politics. Long hours at work, no time off, and transportation challenges all make voting prohibitive. Higher incomes, however, afford individuals the time and energy to engage, attend meetings, contact politicians, and vote. At both extremes, a $10,000 increase in annual income probably won't matter. For those with little, an extra $10,000 may simply go toward meeting basic needs. For society's wealthiest, $10,000 will make little difference. However, an additional $10,000 for those in the middle might significantly raise the probability of voting.

An S-shaped curve (logistic curve) might also describe the relationship between education and voting. Increasing someone's education by four years—from kindergarten to fourth grade—will not matter. Increasing education by four years for someone who already has a master's degree is not likely to change voting behavior either. A four-year increase after high school—completing college—could influence voting significantly.

The kind of relationship described in the income and education examples exists in a number of other contexts. Decisions to go to college, join a union, buy a house, or join a political party often involve the same jump from one state of being to another brought on by an incremental change at a certain level of an independent variable. Whether countries engage in war involves the same kind of functional considerations. Usually a series of smaller developments presage the event that sparks the full conflagration. Protests and civil unrest represent another context where a series of events spur relatively little action. One incident, however, finally sparks nationwide and global protests. All of these examples describe a process that follows roughly the same functional form.

In all of the examples given so far, the relationship theorized is nonlinear: increments of the independent variable at the low, medium, and high end have different effects. The scenario described necessitates the use of logistic regression. Since these scenarios are explicitly nonlinear, using a linear estimator (OLS) is not appropriate.

> **KNOWLEDGE CHECK:** Discuss problems best summarized by a nonlinear framework.

1. Which of the following questions imply a logistic functional form?
    a. What is the relationship between temperature and a liquid's boiling point?
    b. What is the relationship between income and the amount spent on college?
    c. How are education and income related?
    d. How are graduating from college and income related?

2. What clues suggest using a logistic regression?
   a. The dependent variable is continuous.
   b. The dependent variable is dichotomous.
   c. The hypothesized relationship is linear.
   d. The hypothesized relationship is nonlinear.

Logistic Regression Violates Gauss-Markov Assumptions

To illustrate how the Gauss-Markov assumptions are violated in this context, consider the following example. Suppose we want to explain why some states have "stand your ground" rules and others don't. Stand your ground laws permit citizens to defend themselves with lethal force if they feel threatened. Currently, half of the U.S. states have these laws on the books in one form or another. What explains their adoption? Does a state's wealth encourage or discourage this kind of law? Consider a scatter plot of the relationship (Figure 17-1).

In what should be familiar code by now (Code Chunk 17-1), I used ggplot to construct a scatter plot of the stand your ground variable (*states$stand*) against the log of median household income variable (*states$medinc*).

Code Chunk 17-1
```
ggplot(states, aes(log(medinc), stand)) +
  geom_point(col="#bf0000", cex=.7) +
  theme_minimal() +
  theme(plot.title = element_text(size = 8, face = "bold"),
        axis.title = element_text(size = 8, face = "bold")) +
  geom_smooth(method="lm", se=FALSE, col = "#0000bf") +
  geom_text_repel(size=3, aes(label = st, size = 1,
                  hjust = 0, vjust=-1),
        col="grey") +
  ggtitle("Figure 17-1: More Income Lowers Defensiveness") +
  ylab("Stand Your Ground = 1") +
  xlab("Median Household Income")
```

The plot looks odd because we've plotted a dichotomous variable (stand your ground law) against a continuous variable (median income). The line does a poor job of fitting the data. At the lower end of income, the line consistently underpredicts the level of Y—our predicted values lie underneath the states in the upper-left quadrant of the plot (Oklahoma, Nevada, and Texas). As income increases, our line overpredicts all of the cases (California, Wyoming, and Wisconsin). To illustrate the point further, let's run a bivariate regression of these two variables and generate a residual plot from the model (Figure 17-2). Remember that in the previous chapter we established that any patterns observed in the residual plot indicate we are violating Gauss-Markov assumptions.

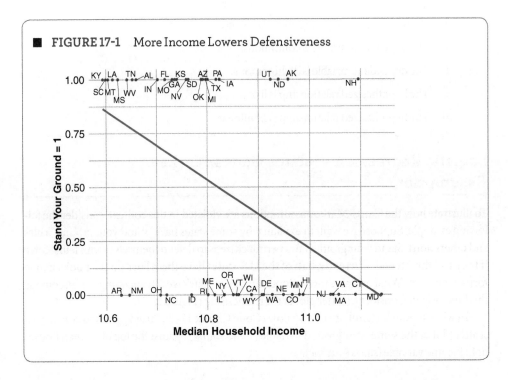

FIGURE 17-1 More Income Lowers Defensiveness

In Code Chunk 17-2, I define a regression object using the *lm()* command and then produce the associated regression table using the *stargazer()* command.

Code Chunk 17-2

```
badreg.lm <- lm(stand ~ medinc, data = states)

stargazer(badreg.lm, title = "Table 17-1: OLS Regression With
                    Binary Dependent Variable",
        header = FALSE, type = "text")
```

Once I have the results from the regression *badreg.lm*, I can collect the residuals and the predicted values from the regression (the first two lines in Code Chunk 17-3) and use them to create a residual plot.

Code Chunk 17-3

```
states$res <- resid(badreg.lm)
states$pred <- predict(badreg.lm)

ggplot(states, aes(pred, res)) + geom_point(col="red", cex=.7) +
  geom_hline(yintercept = 0, linetype = 2) +
  theme_minimal() +
```

```
    theme(plot.title = element_text(size = 8, face = "bold"),
        axis.title = element_text(size = 8, face = "bold")) +
    geom_text_repel(size=3,
        aes(label = st, size = 1, hjust = 0, vjust=-1),
        col="grey") +
    xlab("Predicted or 'Fitted' Values") +
    ylab("Residuals") +
    ggtitle("Figure 17-2: Residual Plot")
```

TABLE 17-1 OLS Regression With Binary Dependent Variable

	Dependent variable:
	stand
medinc	−0.00003***
	(0.00001)
Constant	2.037***
	(0.442)
Observations	50
R2	0.205
Adjusted R2	0.188
Residual Std. Error	0.455 (df = 48)
F Statistic	12.351*** (df = 1; 48)
Note:	*p<0.1; **p<0.05; ***p<0.01

Although one can't tell from the regression table that something is wrong (Table 17-1), the residual plot indicates the data do not form a shapeless cloud. Not only is there an obvious pattern to the residuals, but we are systematically overpredicting and underpredicting the observations as we move along the x axis. The residual plot indicates we've violated at least one of the assumptions associated with OLS, suggesting we might need to use a different estimator.

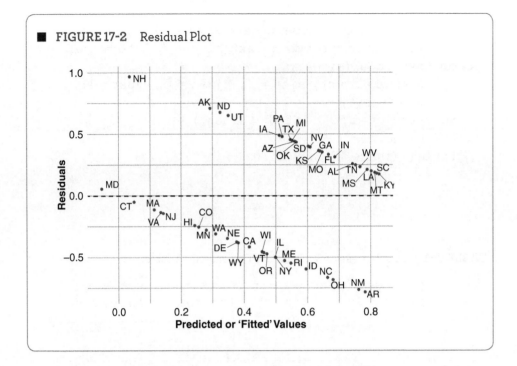

FIGURE 17-2 Residual Plot

> **KNOWLEDGE CHECK: Explain how logistic relationships violate the Gauss-Markov assumptions.**

3. Which of the following Gauss-Markov assumptions are violated in the context of fitting a line to a logistic relationship?

    a. The errors have constant variance.

    b. The errors are independent.

    c. The errors sum to zero.

    d. The residuals sum to zero.

4. When fitting a linear model to a logistic relationship, what will the residual plot reveal?

    a. A fan-shape relationship

    b. A formless cloud

    c. Underprediction and overprediction

    d. Significant outliers

Working With Logged Odds

Since an important assumption of the Gauss-Markov theorem is violated, we need to look for alternatives. Fortunately, if we conceptualize our problem in terms of the **logged odds** of voting, we can use the linear regression framework. To calculate the logged odds, we first need the odds ratio: the probability (Pr) of success over failure. For example, if there's an 80% chance that it will rain today, there is a 20% chance that it won't.

$$\text{Odds ratio} = \frac{\Pr(\text{rain})}{\Pr(1-\text{rain})}$$

The odds ratio would be the following:

$$\text{Odds ratio} = \frac{(.8)}{(1-.8)} = 4$$

The odds ratio would be 4. To calculate the logged odds, we simply take the log of 4 in this example: ln(4).

We use logged odds because if we transform the odds ratio by logging it, we now have a linear function to work with. The independent variables in a logistic model can be expressed as a linear combination of the logged odds. In the logistic regression, the coefficients of interest can be interpreted similarly to their OLS alternatives, only now we're expressing units in logged odds. For example, the constant tells us what the logged odds of voting are when income equals zero, and the coefficient on income indicates how much the logged odds for voting changes with a one-unit change in income.

The only problem, however, is that logged odds are not an intuitive unit of measurement. Converting logged odds to predicted probabilities provides clarity. Fortunately, R makes it easy to convert logged odds to predicted probabilities, generating graphics that show what the likelihood of an event is over the range of our independent variables.

Let's define a model (*model.glm*) with whether Trump won the state's electoral college vote (*trumpwin*) as the dependent variable and the median household income (*medinc*) as the independent variable (for presentation purposes, I divide *medinc* by $1,000 so that the units are thousands of dollars). The *glm* command in R will produce a regression table similar to what we've seen before (Table 17-2). However, the coefficients in the regression table refer to logged odds. For example, the regression table indicates that a unit change in x (an increase of $1,000 in median income) is associated with a .2 decrease in the logged odds of Trump winning the state.

Before we continue, there are a couple of things to note in Code Chunk 17-4. First, we use the command *glm()* instead of *lm()* since we're performing a nonlinear regression. The *glm* stands for "generalized linear model." Also note that in the regression command, I specify the option *family = binomial*. This specifies a logistic regression.

Code Chunk 17-4

```
states$inc1000 <- states$medinc/1000

model.glm <- glm(trumpwin ~ inc1000, data=states,
         family=binomial)

stargazer(model.glm, title = "Table 17-2: Logistic Regression of Trump Winning State", header = FALSE, type = "text")
```

TABLE 17-2 Logistic Regression of Trump Winning State

	Dependent variable:
	trumpwin
inc1000	−0.231***
	(0.066)
Constant	12.187***
	(3.414)
Observations	50
Log Likelihood	−23.298
Akaike Inf. Crit.	50.597
Note:	*p<0.1; **p<0.05; ***p<0.01

Unfortunately, even the most regular visitors to Las Vegas would have trouble explaining the substantive significance of that result. Consequently, we need to convert logged odds to predicted probabilities: the probability that Trump will win the state.

> **KNOWLEDGE CHECK: Calculate a logistic regression.**

5. Which of the following best describe logged odds?
    a. We can interpret the coefficients as a linear function of logged odds.
    b. They are easy to interpret.
    c. Their substantive importance is easy to grasp.
    d. Logged odds give the predicted probability of an event occurring.
6. Using the *states* data, regress the death penalty on the homicide rate (*murderrate*). Using logged odds, interpret the coefficient.

Working With Predicted Probabilities

To make the logistic enterprise more intuitive, convert the regression output from logged odds to **predicted probabilities**. Predicted probabilities are simply the probability of an event occurring. For example, we now calculate the probability of Trump winning at a specific level of income. If we go this route, however, the size of the change in the dependent variable depends on where we are in the distribution. In other words, when we ask what is the effect of income on Trump winning, our answer in the framework of predicted probabilities has to be "it depends" since we're converting a straight line (logged odds) to an S-shaped curve. Let's start with a simple bivariate regression, proceed to multiple regression, then move to an example.

Bivariate Logistic Regression

In our previous example, the relationship between Trump winning a state depended on the state's level of wealth. Our motivation for using logistic regression rested on the idea that an extra $1,000 at the low or high end of income matters less than in the middle—hence the S-shaped curve. Adding $1,000 to median household income in states where income is relatively high or low will be associated with relatively little change. The biggest gain in the probability of Trump winning happens somewhere toward the middle of the distribution.

Next we calculate predicted values for the dependent variable at different levels of the independent variable, convert those predicted values to predicted probabilities, then graph them to understand the relationship. Put simply, we take the predicted values in logged odds form (a line) and convert them to predicted probabilities (an S-shaped curve). All we're doing in this case is using the same model and estimation. However, once we've obtained the estimates, we're converting them from logged odds to predicted probabilities to help gain a better understanding of the results. To understand the substantive significance of a logistic regression, we follow these steps:

1. Obtain predicted values (in logged odds) by using the estimates from the GLM regression.

2. Convert the estimated logged odds of voting into the predicted probabilities of voting using the following:

$$\text{Probability of Trump Winning} = \frac{e^{\widehat{Y}_i}}{\left(1+e^{\widehat{Y}_i}\right)}$$

3. Graph the predicted probabilities

Observe the plot in Figure 17-3 for the bivariate logistic regression of Trump winning a state (*trumpwin*) on median household income in thousands (*inc1000*). Fortunately, the 'visreg' package makes these calculations automatically, providing a nice graph of the predicted probabilities as we range over any independent variable of interest. Code Chunk 17-5 is straightforward. First, I define an object that estimates (in logged odds) the relationship between the dichotomous variable *trumpwin* and *inc1000*, the income variable I created in Code Chunk 17-4. Note that the option *family = binomial* is used, which tells R to use logistic regression. I then plot the relationship using the *visreg()* command, which is an easy way to visualize any regression results. All that is required is to input the model name *trump.glm*, indicate which independent variable you want to graph (place it in quotes), and specify *scale = "response"*, which tells R to use predicted probabilities. I also graph the relationship without a confidence interval (*se = FALSE*) and place the figure in the ggplot framework (*gg = TRUE*).

Code Chunk 17-5

```
trump.glm <- glm(trumpwin ~ inc1000, data = states, family = binomial)
visreg(trump.glm, "inc1000", header = FALSE, scale = "response",
       band=FALSE, gg=TRUE) +
  theme_minimal() +
  theme(plot.title = element_text(size = 8, face = "bold"),
        axis.title = element_text(size = 8, face = "bold")) +
  annotate("segment", x = 50, xend = 50, y = 0, yend = .66,
           col = "grey", linetype = "dashed") +
  annotate("segment", x = 40, xend = 50, y = .66, yend = .66,
           col = "grey", linetype = "dashed") +
  annotate("segment", x = 60, xend = 60, y = 0, yend = .15,
           col = "grey", linetype = "dashed") +
  annotate("segment", x = 40, xend = 60, y = .15, yend = .15,
           col = "grey", linetype = "dashed") +
  annotate("text", x = 38, y = .66, parse = F,
           label = '.66') +
  annotate("text", x = 38, y = .15, parse = F,
           label = '.15') +
  ggtitle("Figure 17-3: Trump's Chances Decreased With Income")
```

Notice the curvature of the line; the greatest decline in the predicted probabilities of voting for Trump are around the middle of the distribution. Also note the ticks on the bottom and top of the graph (those are called the rug). They indicate the individual data points used in the regression. We can clearly see that Trump's chances of winning a state decline as the state's

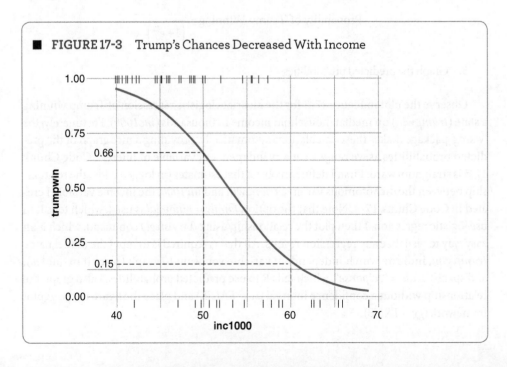

FIGURE 17-3 Trump's Chances Decreased With Income

median household income increases. Specifically, we can see that when income is $50,000, the probability of winning the state is at 66%. When income is at $60,000, the predicted probability is at 15%. Now that we've described this technique in the bivariate context, let's explore the multivariate case.

Multivariate Logistic Regression

Analysts display results from logistic regressions in a number of different ways. In the multivariate framework, choosing specific values of one independent variable and plotting the predicted probabilities over the range of another is helpful. To illustrate, let's add to the bivariate model we've already estimated (the logistic regression of Trump's electoral college win on median income). In this example, we might want to control for population density—Trump did particularly well with rural voters. Let's see what our predictions look like when we consider the relationship between Trump's probability of winning, median household income, and the population density of the state (Figure 17-4).

There are just a few adjustments from the previous code chunk. First, we add the variable *density* to the logistic regression (Code Chunk 17-6). Then, in the *visreg()* command, we indicate that we want to view the relationship between the probability of Trump winning and income *by* the density variable. The *visreg()* command makes this very easy by calculating the interquartile range of the density variable and providing a line for the lower quartile, the median, and the upper quartile.

Code Chunk 17-6

```
trump1.glm <- glm(trumpwin ~ inc1000 + density,
                  data = states, family = binomial)

visreg(trump1.glm, "inc1000", by="density", overlay = TRUE,
       band = FALSE, scale = "response", legend = TRUE, gg=TRUE) +
  theme_minimal() +
  theme(plot.title = element_text(size = 8, face = "bold"),
        axis.title = element_text(size = 8, face = "bold")) +
  annotate("segment", x = 38, xend = 50, y = .325, yend = .325, col =
"grey", linetype = "dashed") +
  annotate("segment", x = 38, xend = 50, y = .795, yend = .795, col =
"grey", linetype = "dashed") +
  annotate("text", x = 36, y = .795, parse = F,
           label = '.795') +
  annotate("text", x = 36, y = .325, parse = F,
           label = '.325') +
  annotate("segment", x = 50, xend = 50, y = 0, yend = .795,
           col = "grey", linetype = "dashed") +

  ggtitle("Figure 17-4: Trump's Chances Decreased as Income Rose") +
  xlab("Median Household Income in Thousands") +
  ylab("Predicted Probability of Trump Winning State")
```

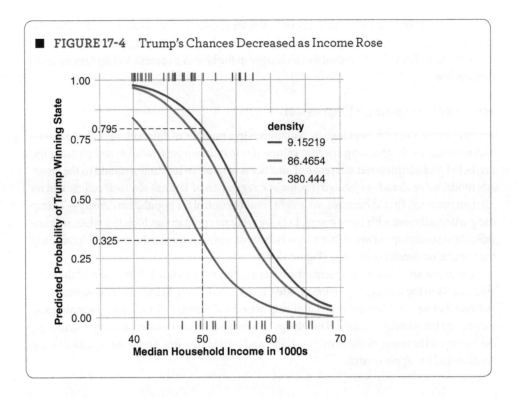

FIGURE 17-4 Trump's Chances Decreased as Income Rose

As can be observed in Figure 17-4, Trump's probability of winning a state varies significantly depending on income and population density. Consider Trump's probability of winning when median household incomes are at $50,000. Among the more densely populated states (the blue line), his probability of winning is 32.5%. In the most sparsely populated states (the red line), his probability of winning is 79.5%! Notice how that difference closes at both the low end (below $45,000) and the high end (above $60,000).[1]

Example: Obama's Electoral College Wins in 2012

Let's do the same exercise, only now we consider income, education, and population density's influence on Obama's chances in the 2012 election (Code Chunk 17-7). The data we'll be working with are the binary dependent variable *obamawin*, which indicates whether Obama

Code Chunk 17-7

```
Obama.glm <- glm(obamawin ~ inc1000 + hsdiploma + density,
                 data = states, family = binomial)

stargazer(Obama.glm, title =
        "Table 17-3: Probability of Obama Winning the
            State (2012)", header = FALSE, type = "text")
```

---

[1] The default for the *visreg* command creates three lines based on the 25th, 50th, and 75th percentile cases of the independent variable specified. In the example illustrated in Figure 17-4, the 25th, 50th and 75th percentile values for density are 9.2, 86.5, and 380.4, respectively.

TABLE 17-3 Probability of Obama Winning the State (2012)

```
===============================================================
                                     Dependent variable:
                                  ---------------------------
                                           obamawin
---------------------------------------------------------------
inc1000                                     0.205*
                                           (0.107)

hsdiploma                                   0.169
                                           (0.152)

density                                    0.019***
                                           (0.007)

Constant                                  -26.480**
                                           (11.539)
---------------------------------------------------------------
Observations                                  50
Log Likelihood                              -17.642
Akaike Inf. Crit.                           43.284
===============================================================
Note:                            *p<0.1; **p<0.05; ***p<0.01
```

won the state in the 2012 election. The independent variables we're interested in are income (*inc1000*), a measure of education (*hsdiploma*), and population density (*density*). The regression results are reported in Table 17-3.

Since we want to see how Obama's chances change as income changes while accounting for different values of education and population density, we need to know something about the *hsdiploma* variable: what values are most interesting? From a summary of *hsdiploma*, we find that the lowest quartile is equal to 82.53, the median is 86.1, and the upper quartile is 90.4. We use those three values to define the three lines we want to examine. As noted earlier, the *visreg()* command in R does this automatically (Code Chunk 17-8).

Code Chunk 17-8

```
visreg(Obama.glm, "inc1000", by = "hsdiploma", overlay = TRUE,
       scale = "response", band = FALSE, gg=TRUE) +
  theme_minimal() +
  theme(plot.title = element_text(size = 8, face = "bold"),
        axis.title = element_text(size = 8, face = "bold")) +
  annotate("segment", x = 38, xend = 50, y = .245, yend = .245, col =
              "grey", linetype = "dashed") +
```

```
annotate("segment", x = 38, xend = 50, y = .69, yend = .69, col =
    "grey", linetype = "dashed") +
annotate("text", x = 36, y = .69, parse = F, label = '.690') +
annotate("text", x = 36, y = .245, parse = F, label = '.245') +
annotate("segment", x = 50, xend = 50, y = 0, yend = .70, col =
    "grey", linetype = "dashed") +
ggtitle("Figure 17-5: Obama's Chances Increased as Income Rose") +
xlab("Median Household Income in Thousands") +
ylab("Predicted Probability of Obama Winning State")
```

We see that as income increases, Obama's chances of winning increase (Figure 17-5). However, once a certain level of income is reached, his chances of winning do not increase at the same rate. What is more, we see that the level of a state's education matters. An income of $50,000 is associated with a probability of Obama winning 24.5% when education levels are relatively low (the red line). For more highly educated states (the blue line), the probability of winning at $50,000 is 69%. Note that the relationship between income and the probability of winning changes with an increase in income *and* that it also depends on the level of education. In other words, in the multivariate logistic world, the relationship between any given variable and the probability of success depends not only on the value of the independent variable of interest but also on the level of the other independent variables.

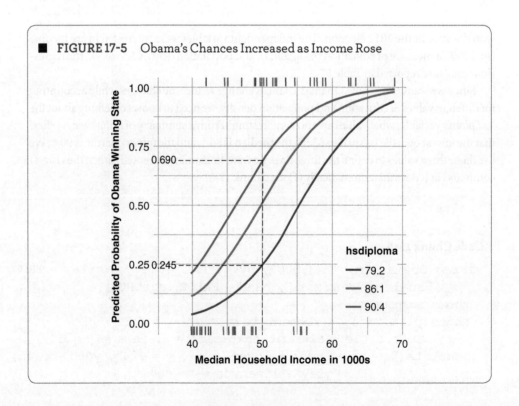

FIGURE 17-5 Obama's Chances Increased as Income Rose

> **KNOWLEDGE CHECK: Explain how to convert logged odds to predicted probabilities.**

7. What's the mathematical formula that converts the predicted value in logged odds into a predicted probability?
8. How do you interpret results that are given in predicted probabilities?
   a. A unit change in x is associated with a change in y's probability.
   b. A unit change in x is associated with a change in y's odds.
   c. A unit change in x is associated with probability y.
   d. At x, it is the probability of y.
9. Using the *states* data, what is the probability that a state will have the death penalty when the median household income is $50,000 at the lowest quartile of the variable democrat?

Model Fit With Logistic Regression

As with OLS, in the logistic regression framework we're interested in how well our model fits the data. Unfortunately, there is no direct comparison to the R^2 statistic that we get in the OLS framework. What we have instead are a number of pseudo R^2 statistics that help give a sense of how well our model fits the data. The command is *logitR2()*, which produces a number of different statistics (Pollock, 2014). The *logitR2()* command gives **McFadden's R^2**, the **Cox-Snell R^2**, and the **Nagelkerke R^2**. The three measures are similar to each other and give a sense of how well the model fits the data. The output from the *logitR2()* command also generates a chi-square test. The **chi-square test** indicates whether the model provides an advance over what is called the null-deviance or know-nothing model (model with no independent variables).

We can also use the chi-square test to investigate whether models with fewer independent variables outperform those with more independent variables. Using the *pchisqC()* command, specify the two models you wish to compare and give the difference in the number of independent variables each has. The command returns a p-value that indicates whether the full model offers an improvement over the reduced model. If the p-value is .05 or smaller, we can say that the full model is a significant improvement over the model with fewer independent variables. This allows an explicit test of whether the addition or subtraction of variables improves the model. To deepen our understanding, let's practice with an example.

Example: Obama, Income, and Education

To demonstrate both the measures of fit and chi-square test, let's estimate two different models based on voting, income, and education (Code Chunk 17-9). Model 1 simply regresses a binary dependent variable—whether Obama won the state or not—against median income. Model 2 regresses the same binary dependent variable against median income and education. In this exercise, we are less interested in the substantive significance of the model and more interested in evaluating the model's fit.

Code Chunk 17-9

```
model1.glm <- glm(obamawin ~ inc1000, data = states, family = binomial)
model2.glm <- glm(obamawin ~ inc1000 + hsdiploma, data = states,
                        family = binomial)

object <- logitR2(model1.glm)
object1 <- logitR2(model2.glm)

knitr::kable(object, format = "pandoc",
         caption = "Table 17-4: Measures of Fit for
             Logit Model 1", digits = 4)
knitr::kable(object1, format = "pandoc",
         caption = "Table 17-5: Measures of Fit
             for Logit Model 2", digits = 4)
```

It appears that adding education (*hsdiploma*) to the model has relatively little impact on the regression; the measures of fit stay roughly the same (Tables 17-4 and 17-5). Does the inclusion of *hsdiploma* help improve the fit of the model appreciably? To answer that, we use the *pchisqC* function from Pollock (2014) in Code Chunk 17-10 to perform the test.

Code Chunk 17-10

```
pchisqC(model1.glm$dev, model2.glm$dev, 1)
```

```
[1] "0.863444"
```

The number produced by the test is .863, which is well above the number .05 that marks the 95% level of significance. We therefore cannot reject the null hypothesis: adding education has no appreciable impact on our ability to explain why some states voted for Obama and others did not. In other words, the education variable does not seem to add much to our model.

TABLE 17-4 Measures of Fit for Logit Model 1

N	CHI2	DF	P	COX	NAG	RL2
50	18.6762	1	0	0.3117	0.4165	0.2707

TABLE 17-5 Measures of Fit for Logit Model 2

N	CHI2	DF	P	COX	NAG	RL2
50	18.7058	2	1e-04	0.3121	0.417	0.2711

Receiver Operating Characteristic Curves and the Area Under the Curve

While the various pseudo R^2 measures provide nice numerical summaries of fit, a useful way to visualize the fit of any logistic model involves **receiver operating characteristic (ROC) curves** and the associated **area under the curve (AUC)**. The rather technical-sounding names come from their origins in signal detection theory (radar). The underlying logic for their use in logistic regression is straightforward. The ROC curve and the AUC are based on the association of how many cases our model guessed right versus how many cases it guessed wrong. In the previous example, a positive case would be Obama winning a state. A good outcome for our model would be correctly predicting how many states Obama actually won. Similarly, success is defined by correctly predicting a loss. Comparing the outcomes of those two questions allows us to construct the ROC curve and the AUC.

In this example, we need to make some important distinctions within the states we predicted Obama would win and within the states we predicted he would lose. Let's start with our predictions on winning. For states we predicted he'd win and he actually won, we call those cases **true positives**. For states we predicted he'd win and he actually lost, we call those cases **false positives**. Now let's turn to our negative predictions. When we predicted correctly the cases he would lose, we call those **true negatives**. When we predicted incorrectly the cases he would lose (he actually won them), we call those **false negatives**. Table 17-6 helps delineate the four different situations within the context of Obama winning a state: true positive, false positive, true negative, and false negative.

Before we continue, it's important to point out that while ROC curves originated in signal detection theory, you might recognize these terms when evaluating medical tests. I mention medicine because there are two additional terms we need to understand that are best explained in the context of medical tests. The two additional terms are **sensitivity** and **specificity**. I think it's helpful to link the definitions of these terms from the frame of medical testing (e.g., patients and disease) to the analytics of our question regarding Obama's electoral victories. The sensitivity of a test is the probability of a positive test given that the patient has the disease. Sensitivity is also known as the *true positive rate* or the probability of detection. In our case, it is the probability that we would predict Obama won the state and he actually did. To be concrete, in our example, sensitivity would be calculated as follows:

$$\text{Sensitivity} = \frac{\text{Number of true positives}}{\text{Total number of states Obama won}}$$

Specificity is the probability of a negative test given that the patient is well. You might also see the term *false positive rate*, which is (1 − specificity). The false positive rate is also known

TABLE 17-6 Categories Used to Construct ROC Curves

	STATES ACTUALLY WON	STATES ACTUALLY LOST
PREDICTED WINS	True Positive	False Positive
PREDICTED LOSSES	False Negative	True Negative

as the probability of the false alarm. In our example, specificity is the probability of predicting Obama would lose the state and he actually lost it. Specificity would be calculated as follows:

$$\text{Specificity} = \frac{\text{Number of true negatives}}{\text{Total number of states Obama lost}}$$

Think of the logistic regression model as a medical test where a negative result indicates the patient is healthy and a positive result indicates the patient is sick. It's a little counterintuitive but when we say a patient tests positive and is actually sick, that is the same as our model predicting Obama will win the state and he actually wins (a true positive). When a patient tests negative and they are healthy, that is the same as our model predicting Obama will lose the state and he actually loses (a true negative).

The definition of the ROC curve can be somewhat confusing because it is defined in two different ways (but they mean the same thing). In some texts, you'll see the ROC curve defined by plotting the true positive rate against the false positive rate. In other texts, you'll see the ROC curve is defined as sensitivity plotted against 1 – specificity. Just remember that sensitivity is the same as the true positive rate and 1 – sensitivity is the same as the false positive rate. The R package used here ('pROC') plots sensitivity against specificity but reverses the x axis (the same as 1 – specificity).

With these definitions, we're ready to proceed. I construct a density plot that shows the cases Obama actually lost (red) and the states he actually won (blue). I plot those according to the predicted probability that Obama won the state.

To generate the graph, I first have to get the data in order. I collect the predicted probabilities from a simple bivariate model, a logistic regression of *states$obamawin* against *states$inc1000*. I then create two different data sets (*testwin* and *testlose*) that separate the cases for when Obama actually wins the cases and when he loses. Note that I include the name of the package 'dplyr' before the *filter()* and *select()* commands since they are masked. I then combine them into an object called *varlist*. Once in *varlist* I use the *melt()* command to stack the cases on top of each other so that I can plot the two density plots. I call that data set *newvar*. For plotting purposes, I also change the variable *L1* in *newvar* to a factor (Code Chunk 17-11).

Code Chunk 17-11

```
mtest.glm <- glm(obamawin ~ inc1000, data = states, family = binomial)

states$pred.glm <- predict(mtest.glm, type = "response")

testwin <- states %>% dplyr::filter(obamawin == 1) %>%
              dplyr::select(pred.glm)

testlost <- states %>% dplyr::filter(obamawin == 0) %>%
              dplyr::select(pred.glm)

varlist <- list(testwin, testlost)

newvar <- melt(varlist)

newvar$L1 <- as.factor(newvar$L1)
```

Once I've generated the data, I use the *ggplot()* command to plot the graph (Code Chunk 7-12). Remember, these densities represent the real outcomes of the 2012 election. The red represents the states Obama lost and the blue represent the states Obama won. I also established three thresholds for illustration (the dashed vertical lines), one at 25%, 50%, and 75%. Figure 17-6 indicates there are some cases that the model predicted Obama would win easily (high probability of winning) but he lost. Similarly, there are some cases our model was certain he'd lose (low probability of winning) but he won.

Code Chunk 17-12

```
ggplot(newvar,aes(x=value, fill=L1)) +
  geom_density(alpha = .7) +
  ggtitle("Figure 17-6: Predicted Probability That Obama
                   Won, Colored by Actual Wins/Losses") +
  theme_minimal() +
  theme(plot.title = element_text(size = 8, face = "bold"),
        axis.title = element_text(size = 8, face = "bold"),
        legend.position = "bottom") +
        scale_fill_manual(values = c("#0000bf", "#bf0000"), name = "",
                          labels = c("States Actually Won",
                                     "States Actually Lost")) +
  xlab("Predicted Probability of Obama Winning the State") +
  ylab("Density") +
  geom_vline(xintercept = .50, linetype = "dashed") +
  geom_vline(xintercept = .25, linetype = "dashed") +
  geom_vline(xintercept = .75, linetype = "dashed")
```

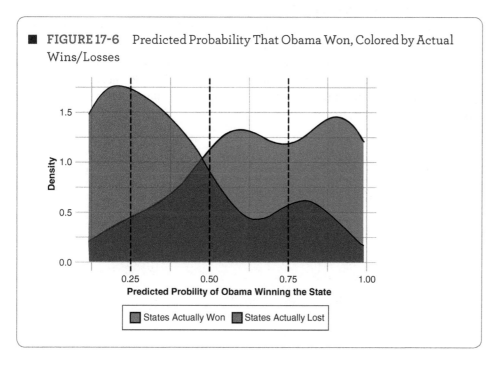

FIGURE 17-6 Predicted Probability That Obama Won, Colored by Actual Wins/Losses

The ROC curve is constructed by calculating the sensitivity and specificity of the test at different thresholds (we will use three thresholds to illustrate). At each threshold we take the percentage of positive cases that our model correctly predicted (*true positives*) and plot them against the percentage of states he lost that we correctly predicted (*true negatives*). Remember, the true positives represent sensitivity and the true negatives represent 1 − specificity. Again, to simplify, rather than use "1 − specificity" on the x axis, the graph reverses the axis (it decreases in value as we move from left to right).

$$\text{Coordinates of ROC Curve} = (\text{Specificity}, \text{Sensitivity})$$

At each threshold we take the number of states Obama won that we predicted correctly and divide by the number of states he won (sensitivity). We plot that against specificity: the number of states we predicted Obama would lose divided by the number of states he lost.

At the 25% threshold, we would have correctly predicted 96% of the states he won. In other words, 96% of the blue-shaded distribution lies to the right of the 25% threshold. We then calculate the percentage of the states we correctly predicted he would lose, which is 39%. In other words, we calculate the percentage of the red-shaded distribution (the states lost) that lies to the left of the 25% threshold. These are the cases we predicted Obama would lose and he did. We now have the two values we need to plot a point on the ROC curve (.39, .96).

To make the calculations, I'm simply taking each distribution and calculating the area to the left or right of the threshold. I provide the code and the numbers for those calculations in Code Chunk 17-13.

You'll see that I take each separate distribution (*testwin* and *testlost*) and find the area to the right of the threshold when calculating sensitivity and the left when I'm calculating specificity. I first define *d_fun* as the empirical cumulative distribution function *ecdf()*, which plots the coordinates of the density function so that we can calculate the percentage of each of the distributions. I then define *x0* as the threshold we want to use, then apply *d_fun* with that argument which gives the area to the left of the threshold. When calculating sensitivity, we want the area of the *testwin* distribution to the right of the threshold so we subtract *d_fun(x0)* from 1. When calculating specificity, we want the area to the left in the '*testlost*' distribution of the threshold so we skip the last step. As an exercise, use the code in Code Chunk 17-13 to calculate the sensitivity and specificity for the 50% threshold and see if it matches the value represented in the ROC curve in Figure 17-6.

Code Chunk 17-13

```
d_fun <- ecdf(testwin$pred.glm)
x0 <- .25
d_fun(x0)
1 - d_fun(x0)


d_fun <- ecdf(testlost$pred.glm)
x0 <- .25
d_fun(x0)
```

At the 75% threshold, the percentage of the blue-shaded distribution that lies to the right is 44%. At that threshold, we would correctly predict 44% of the states Obama won. To the left of the 75% threshold, we correctly predicted 83% of Obama's losses. Consequently, the x and y coordinates of the ROC curve for the 75% threshold are (.83, .44).

Now that we've made all of the calculations, it's important to stand back and understand the bigger picture. Think of our logistic regression model as a medical test in which we want to know if the patient has the disease. The stronger the medical test (the better our model), the more those density plots will be distinct from each other (the less overlap). While there is some overlap in this case, we'll see how that translates into what the ROC curve looks like in addition to its AUC.

Now that I have illustrated how to calculate a few points on the ROC curve, we'll let R do the rest of the work. There are several packages in R that will draw ROC curves. I chose to use 'pROC' because it can render figures in ggplot and it has an easy interface in drawing multiple ROC curves. For now, let's stick with our simple illustrative model of the bivariate logistic regression of *obamawin* on *inc1000*.

I first define the regression object (*roc.list*) and then I use the *ggroc* command to generate the plot. The rest of the code should look familiar, helping me make the usual refinements to the plot (Code Chunk 17-14).

Code Chunk 17-14

```r
roc.list <- roc(obamawin ~ inc1000, data = states)

g.list <- pROC::ggroc(roc.list)

g.list + theme_minimal() +
  theme(plot.title = element_text(size = 8, face = "bold"),
        axis.title = element_text(size = 8, face = "bold")) +
  geom_segment(aes(x = 1, xend = 0, y = 0, yend = 1),
                            color="grey", linetype="dashed") +
  geom_point(aes(.40, .961), col = "#bf0000") +
  geom_point(aes(.7826, .8148), col = "#bf0000") +
  geom_point(aes(.8260, .4444), col = "#bf0000") +
  annotate("text", x = .6, y = .6, label = "AUC = .839") +
  annotate("text", x = .38, y = .93, label = ".25", col = "#bf0000") +
  annotate("text", x = .75, y = .81, label = ".50", col = "#bf0000") +
  annotate("text", x = .79, y = .44, label = ".75", col = "#bf0000") +
  ggtitle("Figure 17-7: ROC Curve for Income")
```

The ROC curve for income's impact on Obama's electoral success indicates that when we've plotted the corresponding sensitivity and specificity values, roughly 84% of the area lies underneath the curve (Figure 17-7). The closer the curve is to the dotted line, the less successful our variable is at predicting wins and losses. In fact, the dashed line represents the point at which it's a 50/50 call, a coin flip. Curves that reach to the upper-left corner represent more successful models.

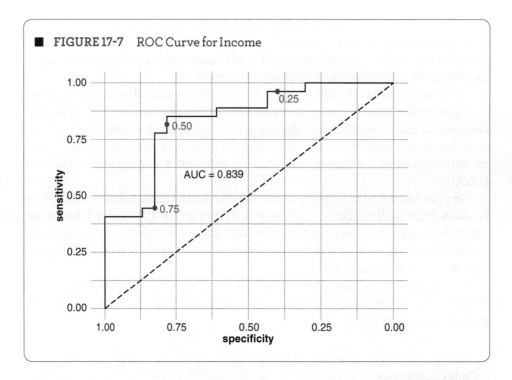

FIGURE 17-7 ROC Curve for Income

To review, the better our logistic regression model, the more separation there is between the density plots of our two predicted outcomes (whether we're examining electoral wins and losses or the occurrence of a disease). The more separation between those outcomes, the more area there will be underneath the curve. Large AUC values imply the ROC curve is getting closer and closer to the upper-left corner of the graph, providing a nice visual accounting of how well our model fits the data.

Now that we have a good understanding of the ROC curve and associated AUC, let's extend the example to the full model that uses three variables (income, education, and population density) to predict whether Obama win's the state's electoral votes. We established how well income helps predict Obama's wins in 2012. Of course, income can't be the only variable. Let's extend the model by adding two additional variables: *hsdiploma* and *density*.

Producing ROC curves for multiple variables in a regression is easy in R. In Code Chunk 17-15, I need only to specify the logistic regression and then plot.

Code Chunk 17-15
```
roc.list <- roc(obamawin ~ inc1000 + hsdiploma + density,
                data = states)

g.list <- pROC::ggroc(roc.list)

g.list +
  theme_minimal() +
  theme(plot.title = element_text(size = 8, face = "bold"),
        axis.title = element_text(size = 8, face = "bold")) +
```

```
geom_segment(aes(x = 1, xend = 0, y = 0, yend = 1),
                    color="grey", linetype="dashed") +
ggtitle("Figure 17-8: ROC Curves for 'inc100',
                    'hsdiploma', and 'density'")
```

As Figure 17-8 indicates, *inc1000* (represented by the red line) generally outperforms the ROC curves for *hsdiploma* and *density*. We know this because a larger part of the ROC curve for income approaches the upper-left corner of the graph. Although density does outperform income in one region (when specificity is greater than .80), the income curve is generally closer to the upper-left corner than are the other curves. In short, this provides a quick and easy way to evaluate the different variables in a logistic regression.

> **Art and Practice of Data Visualization**
> **CALCULATE THE AUC**
>
> While the *ggroc()* command in the 'pROC' package provides a quick and easy way to get a sense of how the different variables are actually performing, it offers little in terms of details. If you want to calculate the AUC and provide labels for different levels of thresholds, the *roc()* command is useful. However, it doesn't work with ggplot.

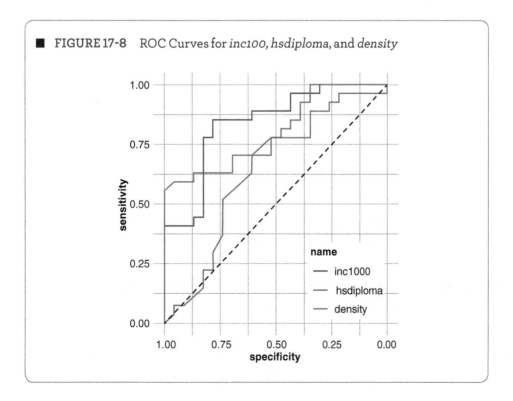

FIGURE 17-8   ROC Curves for *inc100, hsdiploma,* and *density*

> **KNOWLEDGE CHECK: Evaluate the ability of the logistic model to predict outcomes.**

10. Which are tools available for diagnostics in logistic regression?
    a. R^2 statistics
    b. Pseudo R^2 statistics
    c. Receiver operating characteristic curve
    d. Area under the curve

11. Which best describes the chi-squared test?
    a. It compares one model with another.
    b. It tests the different variables in a model.
    c. It records the number of successful predictions.
    d. It records the number of unsuccessful predictions.

12. Which of the following represents sensitivity?
    a. True positives
    b. The share of positive cases that were correctly predicted
    c. True negatives
    d. The share of negative cases that were correctly predicted

13. Which of the following represents specificity?
    a. True positives
    b. The share of positive cases that were correctly predicted
    c. True negatives
    d. The share of negative cases that were correctly predicted

14. Using logistic regression, regress whether a state has a "stand your ground" law on *hsdiploma* and *evangel*. Produce the ROC curves for those variables and answer the following:
    a. Which variable in the model is a more successful test?
    b. Does *hsdiploma* help us distinguish between the two kinds of states?
    c. Does *evangel* help us distinguish between the two kinds of states?

SUMMARY

We learned in this chapter that there are a number of interesting questions which involve a binary dependent variable. We also learned that the OLS framework does not do a good job of estimating the relationship between independent variables and binary dependent variables. Although we can easily generate coefficients in a logistic regression model, we need to do some extra work in order to understand the substantive relationship between our independent variables of interest and our dependent variable.

Specifically, in the logistic regression framework, the relationship between one independent variable and the dependent variable depends on the values of the other variables in the model (they are interactive). As a result, we have to spend a little more time to calculate the substantive impact. There are packages (we used 'visreg') that help produce these graphs. Performing diagnostics for logistic models also requires some additional work. The pseudo R^2 statistic helps demonstrate how well a model performs, but it does not provide the same information as does the R^2 statistic in OLS. ROC curves and AUC calculations provide nice summaries of how well our logistic models perform. To understand their underlying analytics, however, we first needed to understand sensitivity and specificity. Equipped with a complete picture of how to calculate, interpret, and diagnose regressions in the logistic framework, you now have a wider range of tools to conduct your own data analysis.

COMMON PROBLEMS

- *Sensitivity and specificity.* These terms get confused and have a relatively short shelf-life in the memory of people who don't use them often. Sensitivity is all about detecting a disease when it actually exists (a true positive). Specificity is the opposite, the ability to rule out a disease when it doesn't exist (a true negative). Medical tests, or logistic regressions, vary on how well they exceed at both.

- *Logged odds versus predicted probabilities.* Logged odds are the quickest way to interpret the coefficients from a logistic regression. A one-unit change in the independent variable is associated with an x change in the logged odds of a particular outcome. In other words, we interpret it very much like the slope of a line since the log of the odds ratio (the logged odds) puts a dichotomous dependent variable and a continuous independent variable in a linear framework. Predicted probabilities take the logged odds and transform them using the following expression:

$$\text{Predicted Probability} = \frac{e^{\hat{Y}_i}}{\left(1+e^{\hat{Y}_i}\right)}$$

where $\hat{Y}_i$ is the model's prediction for the ith case (expressed in logged odds). Since predicted probabilities transform a linear function into a non-linear one, we cannot interpret the coefficients as before. Instead, we choose a value of x and calculate the probability of the outcome happening at that value.

- *ROC and AUC.* While understanding how the ROC curve is derived requires some patience, the curve itself provides an intuitive way to visualize the fit of a logistic model. The closer the curve is to the diagonal line that intersects the graph, the worse the prediction. The diagonal line represents the accuracy of a coin flip. The closer the line reaches to the upper-left corner, the better the fit and the more area that will lie underneath the curve. While the AUC provides a nice simple numerical summary (varying between .50 and 1), the ROC provides somewhat more detail and can be helpful when looking at models with more than one variable.

REVIEW QUESTIONS

1. When do we use logistic regression?
2. What Gauss-Markov conditions are violated when applying OLS to a binary dependent variable?
3. Why are logged odds not so useful?
4. How are logged odds transformed to produce predicted probabilities?
5. What is the shape of the logistic curve? Why?
6. How is fit measured in the logistic framework?

7. How do we evaluate whether one model is better than another in the logistic framework?
8. How does one interpret coefficients from a logistic regression?
9. How should one present predicted probabilities?
10. What does a residual plot indicate when performing OLS on a binary dependent variable?

PRACTICE ON ANALYSIS AND VISUALIZATION

1. Under what circumstances do we use a logistic regression?
   a. When the dependent variable is logged
   b. When the dependent variable is categorical
   c. When the independent variables are dichotomous
   d. When the dependent variable is dichotomous

2. Which of the following questions imply using a logistic model?
   a. What explains why some teams win the Super Bowl?
   b. Will there be a football season this year?
   c. Why do people vote?
   d. Why do some people commit murder?

3. Why do models with dichotomous variables as dependent variables violate Gauss-Markov?
   a. The errors will have nonconstant variance.
   b. The errors will not sum to zero.
   c. The errors will not be independent.
   d. There will always be an omitted variable.

4. Using the *states* data, run a regression of death against *stuspend*, *murderrate*, *inc*, and *democrat*. Interpret the coefficients on *stuspend* and *murderrate* using logged odds.

5. Run the same regression and graph the relationship between the death penalty and *murderrate* at different levels of student spending using predicted probabilities.

6. Calculate the pseudo R^2 measures from the regression.

7. Using the chi-square test, does the *democrat* variable significantly improve the model?

8. Produce the ROC curve for the *stuspend* variable. Is it a useful predictor?

9. What is the AUC for the *stuspend* variable?

10. Plot the ROC curves for the full model. What variable best explains whether some states adopt the death penalty?

ANNOTATED R FUNCTIONS

The following functions appear in this chapter. They are listed in order of their first appearance (with the code chunk number in parentheses) and annotated here to give a very brief description of their use. Some are not stand-alone functions and only work in combination with other commands. As a reminder, the code in every chapter will work properly if executed in the order it appears. Proper execution also depends on typing the author-defined *libraries()* command, which loads the required R packages.

ggplot(): defines the basic structure of a plot (usually the x and y variables). (17-1)

aes(): the aes (called "aesthetics") function is used in ggplot to define the basic structure of the plot, which often includes the variables you want to use and any shapes or colors. (17-1)

geom_point(): draws points on a ggplot grid. (17-1)

theme_minimal(): specifies a minimalist style for ggplot. (17-1)

theme(): specifies font, size, and so forth in a ggplot. (17-1)

geom_smooth(): draws lines or curves to a scatter plot. (17-1)

geom_text_repel(): labels points in a scatter plot in the ggplot framework. (17-1)

ggtitle(): provides the title for a ggplot. (17-1)

ylab(): labels the y axis in ggplot. (17-1)

xlab(): labels the x axis in ggplot. (17-1)

lm(): linear model or regression command. (17-2)

stargazer(): generates a regression table with a specified model. (17-2)

resid(): when applied to an object described by a linear model, it will generate the residuals. (17-3)

predict(): when applied to an object described by a linear model, it will generate the predicted values. (17-3)

geom_hline(): produces a horizontal line in a figure. (17-3)

glm(): command for generalized linear models. For our purposes, it is the command to execute logistic regression. (17-4)

visreg: graphics package that plots predicted values from various models. Has a feature that creates ggplot-based graphics. (17-5)

annotate: allows the user to place words and lines within figures. (17-5)

logitR2(): author-defined function that produces pseudo R^2 statistics from a logistic regression. (17-9)

pchisq(): outputs the p-value from a chi-square test that compares two different models. Used for diagnostics in logistic regression. (17-10)

filter(): command from the 'dplyr' package that selects rows from a data set according to a criterion. (17-11)

select(): command from the 'dplyr' package that selects columns from a data set according to a criterion. (17-11)

melt(): command that takes two variables and stacks them on top of each other. It converts a data set in wide form and puts it in long form. (17-11)

geom_density(): a ggplot layer that draws a density plot. (17-12)

scale_fill_manual(): allows the user to fill shapes in ggplot with specified colors. (17-12)

geom_vline(): draws a vertical line in a ggplot. (17-12)

ecdf(): calculates the coordinates of a cumulative distribution function for a variable. (17-13)

roc(): command that takes a logistic model and calculates the coordinates to produce a ROC curve. (17-14)

ggroc(): converts a ROC curve drawn in base R to one rendered in ggplot. (17-4)

geom_segment(): draws a line in a ggplot graph. (17-14)

ANSWERS

KNOWLEDGE CHECK

1. a, d
2. b, d
3. b
4. c
5. a
6. Each additional murder per 100,000 population is associated with a .26 increase in the logged odds of a state instituting the death penalty.
7. $\dfrac{e^{\hat{Y}_i}}{\left(1+e^{\hat{Y}_i}\right)}$

8. d

9. between 70% and 80%

10. b, c, d

11. a

12. b

13. d

14. *evangel* (a), not much (b), yes (c)

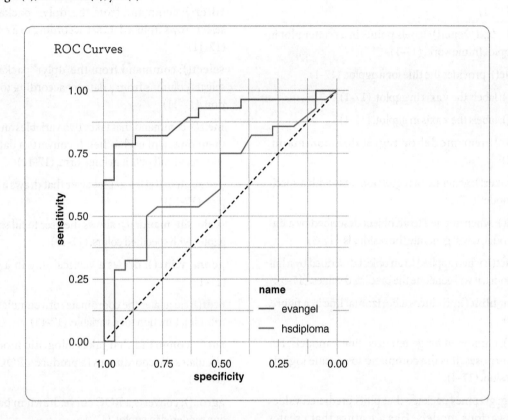

PRACTICE ON ANALYSIS AND VISUALIZATION

1. d

2. a, b, c, d

3. c

4. For each additional dollar spent on a student, the logged odds of having a death penalty decrease by .0005. For each homicide per 100,000 people, the logged odds of having a death penalty increase by .24.

5. predicted probabilities

```
death.glm <- glm(death ~ stuspend + murderrate + inc + democrat,
data=states, family=binomial)

visreg(death.glm, "murderrate", by="stuspend", overlay=TRUE, band=FALSE,
scale = "response", gg=TRUE) +
  theme_minimal()
```

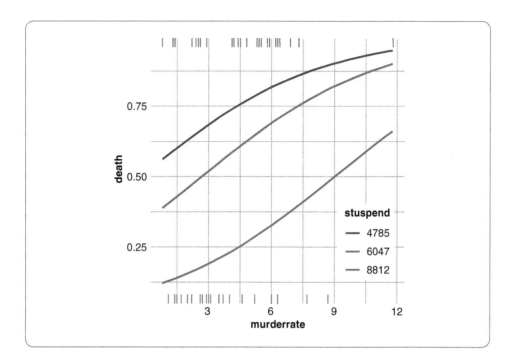

6. Pseudo R^2 values are Cox-Snell (.359), Nagelkerke (.481), and McFadden (.325).

    logitR2(death.glm)

```
   n   Chi2 Df           p       Cox       Nag       RL2
1 50 22.25937  4 0.0001779462 0.3592958 0.4813941 0.3245138
```

7. The p-value for the chi-square statistic is .086, indicating that adding the *democrat* variable does not significantly improve the model.

```
modelq7.glm <- glm(death ~ stuspend + murderrate + inc, data=states, fami-
ly=binomial)
modelq7a.glm <- glm(death ~ stuspend + murderrate + inc + democrat,
data=states, family=binomial)

object <- logitR2(modelq7.glm)
object1 <- logitR2(modelq7a.glm)

pchisqC(modelq7.glm$dev, modelq7a.glm$dev, 1)

        [1] "0.086357"
```

8. ROC for *stuspend*

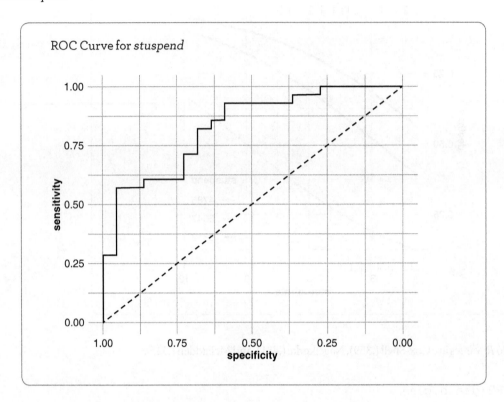

9. AUC for *stuspend* is .8279

```
auc(roc.list)

Area under the curve: 0.8279
```

10. ROC curves for model. *Stuspend* outperforms the others (it has the largest AUC).

```
aucinc <- roc(death ~ inc, data = states)

auc(aucinc)

Area under the curve: 0.7752
```

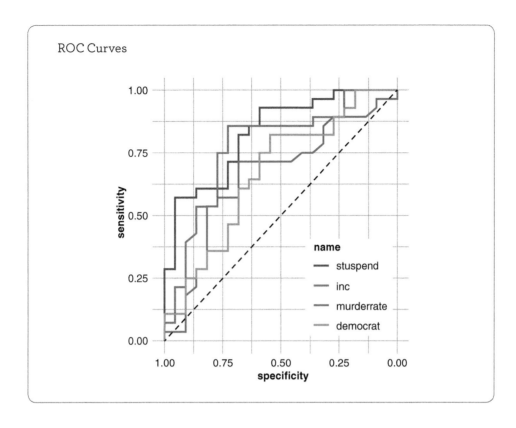

Access digital resources, including datasets, at
http://edge.sagepub.com/brownstats1e.

Appendix
Developing Empirical Implications

Overview

In exploratory data analysis (EDA), we can't rely on classical statistical methods to evaluate the accuracy of our results since we've tested different models, included and excluded variables in our model, and dropped or added certain cases. In this process, one should never hang their hat or bet the farm on a single regression estimate. While a comprehensive approach to confirmatory analysis can give us some confidence that we've actually found something in the data, there are a variety of means at our disposal to determine whether a discovery has truly been made. In this appendix, I list and briefly discuss some of those approaches. In a sense, this formalizes what we do as data analysts, making the process explicit for the novice.

The appendix begins by introducing a simple framework to generate empirical implications: the "if-then" statement. This simple exercise in logic can generate the most profound questions. The if-then formulation is then used to interrogate the left side of the equation (the dependent variable), the right side of the equation (the independent variables), specific observations, and the causal mechanisms that link the two sides of the equation under examination. This appendix concludes with the acknowledgment that the process never ends, requiring the analyst to determine when enough is enough!

Developing Empirical Implications

Key to the EDA enterprise is developing if-then statements after we've examined views of the data or generated estimates from a model. We're looking for additional circumstantial evidence to see whether what we observe in a plot or regression is something real and not the result of statistical chance, bad measurement, or sloppy theorizing. This appendix offers some suggestions on how to go about the enterprise. Just like rookie detectives can learn from a seasoned veteran, these pointers show how the data analyst can go about solving the crime.

Testing Additional Dependent Variables

Suppose we find there is a strong negative correlation between homicide rates and education. Specifically, we find that the homicide rate in a state is tightly correlated with the percentage of the state's population with a high school degree. If our theory, which now seems to be borne out by some evidence, is that education provides individuals with the means to avoid conflict, we might expect that other types of conflict can be avoided by high school graduates.

> In this case, we would make the following statement: if high school education reduces homicide rates, then other assault crimes should also vary with the population's level of education.

Provided we have the data or can get them, we can test this new hypothesis. If it pans out (i.e., the percentage of a state's population with a high school diploma is associated with other assault crimes), then we have more circumstantial evidence in our case. If there is no relationship between high school graduation and other assault crimes, then perhaps we've learned something interesting about different types of crime.

Testing Additional Independent Variables

Continuing with the homicide and high school education example, we can also test whether the same pattern exists between other independent variables and homicide.

> For example, if being more educated provides the means to avoid violent crimes, then we might expect that a larger percentage of the population having a college education might reduce homicide rates as well.

If we find there is a negative correlation between the percentage of a population with college degrees and the homicide rate, we have more evidence that education provides the means to avoid conflict. If there is no relationship, then perhaps we've discovered that the skills to avoid crime are learned in high school. Note that an important discovery is made no matter the result. Therefore, the discovery depends on a well-crafted question—a question generated by a clear analysis of the data.

Using Information on Cases

Making new discoveries does not always require collecting additional data or using different variables. Study your data more closely to gain insight. Here's where knowing more about your subject goes together with asking the right questions to make progress on a problem. Perhaps the negative relationship between education and homicide rates is purely a function of income, which happens to be strongly related to education. While we can always add income to a regression model to observe its impact on our results, we can also identify interesting cases in a scatter plot that could provide some clues.

Suppose that after having labeled all of the points with their state names, we find that Connecticut and Wyoming have very different homicide rates but very similar rates of high school graduation. We might be drawn to explore the role of income since the two states have very different median household incomes. Hypothetically, we might also note that Vermont has a much higher rate of homicide given its level of income and education. Given Vermont's permissive gun laws (hunting is popular in Vermont), we might be drawn to examine the role of gun laws in explaining homicide rates. All of these insights can be gained by a close and careful examination of a simple scatter plot.

Causal Mechanisms

As the final example, we can develop a number of hypotheses based on the empirical evidence we observe from our initial descriptions of the data. Again, consider that same negative correlation between the percentage of the population with high school degrees and homicide rates. From that initial observation, we can start to build a story (a theory) about why we think these two variables are related. Perhaps we think the more educated the population, the more income it will realize. Higher incomes characterize individuals whose leisure activity is more likely to involve golf, tennis, and swimming more than it will encompass hunting. With fewer individuals participating in gun-related activities, there may simply be fewer incidents of gun violence, reducing homicide rates.

While rather complex and drawn out, this line of reasoning allows us to test a number of different mechanisms. First, are higher rates of high school graduation associated with higher levels of income? Is there a relationship between types of leisure activity and gun violence? By systematically articulating the causal mechanisms we might think are at play, we can test the specific links that tie together the story we think lies behind the data.

If the mechanisms we test in this example don't pan out, we've either discovered that education doesn't matter that much or that it matters but not in the way we think. Perhaps the gun-owning strain of our theory is mistaken. Perhaps those with high school degrees can more readily identify relationships, places, and people to avoid. As policymakers, it would be important to know the specific causal links that connect homicide and education.

The Rabbit Hole

The process articulated above can be never-ending. There are always additional hypotheses that can be derived and tested from each additional test or view of the data.

Although what I've outlined here is not the most efficient process, it can be the most rewarding since it can lead to interesting discoveries: learning things about the world we didn't know, but thought we did. That said, it is important to know when to stop. Eventually, the paper needs to be written, the results need to be reported, and policy needs to be made. The more the analyst engages, however, in the back-and-forth between theory and evidence, the richer and more accurate the story.

Once practiced and developed, these skills translate to many different contexts. Making important discoveries aside, data analysis serves as an arena in which we can develop these critical skills of logic, creativity, and imagination.

Glossary

Chapter 1

R: a free (open-source) statistical programming language designed for statistical computing and graphics supported by the R Foundation for Statistical Computing.

S: a statistical program developed in the 1960s upon which R is based.

Package: additional code that can be installed and then loaded into R and RStudio that helps accomplish more specialized tasks.

R Markdown: a lightweight markup language that is designed to provide a record of the data analysis and a streamlined way to present the results in a professional manner. Once a markdown file is produced, it can generate a file in Microsoft Word, HTML, or LaTeX (PDF).

Function: in R, performs a series of predefined commands on an object.

Object: in R, can represent a number of different things. An object is the thing a function operates on and it can be data, figures, or a series of functions.

Code chunk: in R Markdown, a code chunk is the area where executable R code is placed (marked by a gray box). In this book code chunks are referred to as a series of R commands that are executed together.

Knit: knitting in R means rendering the R Markdown file into an HTML, Word, or PDF document.

YAML: code at the beginning of an R Markdown document that specifies document-wide features and characteristics of the document you want to generate.

R-script: a file you can create in RStudio that stores commands you've used in your analysis and want to store for future use or to help reproduce your analysis. R-scripts are less helpful for describing in English what analysis was conducted and they are less useful for presentation purposes (creating documents).

Variable: a list of numbers that record a unique characteristic for a collection of people, places, or things.

Data set: a collection of variables.

Chapter 2

Theory: a collection (or system) of hypotheses (ideas) used to explain something of interest. Theories have at least some empirical justification in that some evidence exists to back them up.

Hypothesis: a guess based on limited empirical data; the starting point for any investigation.

Model: an approximation of reality; a list of variables and how they relate to one another and to the phenomenon of interest.

Diagnostic: a view or summary of the data used to understand how model estimates were generated and whether model assumptions hold. Diagnostics are also used to identify outliers.

ggplot: a graphics package. The "gg" stands for the "grammar of graphics," a conceptual framework where different visual objects (lines, dots, colors, labels) are layered on top of each other to produce visualizations of data.

Scatter plot: a two-dimensional grid that shows how two variables relate by placing observations on that grid.

Smooth: a straight line or curve used to summarize the overall pattern in a scatter plot.

Linear: anything that can be described or summarized by a straight line.

Summarize: describing data, patterns, shapes, or the lack thereof can be illustrated with simple graphical representations (e.g., lines or curves) or with simple numerical values (e.g., the mean, median, or standard deviation) that describe the data. In statistical parlance, those shapes or numerical representations are said to "summarize" the data.

Positive association: the relationship between two or more variables that move in the same direction. As one increases, so do the others.

Bivariate regression model: a regression model with two variables, an independent variable (the cause) and the dependent variable (the effect).

Multiple regression model: a regression model with more than one independent variable.

Negative association: the relationship between two or more variables that move in the opposite direction. As one increases, the other decreases.

Stability: the amount of change registered in a model estimate based on changes to the variables or observations included in the estimation. It can also be the change registered in model estimates as a result of changes in the model's functional form.

Added variable plot: shows the relationship between an independent variable in the model and the dependent variable, holding all other variables in the model constant and having removed the effects of the other independent variables; otherwise known as a partial regression plot. The slope of the line in an added variable plot has the same slope as the corresponding coefficient in the regression.

Residual plot: the difference between the actual value and the predicted value from a regression model plotted against the model's dependent variable; a scatter plot with the residual on the y axis and the fitted (or predicted) value on the x axis. This is a helpful visualization to identify outliers and to evaluate whether OLS is the appropriate estimator.

Chapter 3

Data set: a collection (a series of columns and rows) of numbers or categories that specify amounts or characteristics for a group of persons, places, or things. Often, each column will represent a characteristic and each row will represent a specific person, place, or thing.

Continuous variable: a variable that records counts or amounts. Continuous variables can take on any value between the minimum and maximum.

Categorical variable: a variable indicating differences in kind. Categorical variables record distinct states of being or characteristics for people, places, or things.

Ordered categorical variable: a categorical variable whose categories can be ordered according to an underlying dimension (e.g., parties along a conservative-liberal spectrum; educational attainment from completing elementary school to receiving a PhD; different classes of crimes according to length of suggested prison sentence).

Shape: the distribution (location) of a variable's cases along the range of a variable.

Distribution: synonymous with shape in this instance.

Density plot: plot that provides a representation of a variable's distribution. It indicates where the majority of the cases lie between the maximum and minimum.

Skewed distribution: the shape of a variable that is not symmetrical. The shape of the variable's distribution above the mean is not the mirror image of that below the mean.

Histogram: a visualization that records the number of cases for each interval within a variable's range. Histograms also provide a useful view of a data's shape or distribution. Use histograms with continuous variables.

Range: the difference between the minimum and maximum values of a variable.

dplyr: a package in R designed to produce data tables according to a specific set of criteria. For example, a command can specify that only three variables of a data set be listed with the 10 highest values according to one of the variables. This package is very useful for data management and data description.

Tibble: a data construct in R that resembles a table of data. Its purpose is to allow easy manipulation of data sets in R and it allows 'dplyr' to manipulate data.

Validity: the degree to which a measure (a variable) represents the concept.

Reliability: the accuracy of a measure in terms of making comparisons both across individuals (e.g., Does the measure register important differences between individuals?) and across time (e.g., Will the measure register the same amount for an individual if repeated?).

Chapter 4

Mode: the category or value of a variable that contains the most observations, cases, or individuals.

Mean: the average value of a variable. The mean is calculated by summing all of the observations and then dividing by the number of observations.

Median: the middle case of a variable. The median is calculated by arranging all of the observations in a variable from lowest to highest values and picking out the middle case. In a variable with nine observations (an odd number), the fifth case represents the median. In a variable with 10 observations (an even number), simply take the average of the fifth and sixth cases.

Interquartile range: the distance between the 25th and 75th percentile cases of a variable; roughly thought of as the "middle half" of the variable.

Standard deviation: the average distance of a variable's cases from the mean. It is the square root of the mean of squared distances from the variable's mean. As such, it is expressed in the variable's original units.

Variance: the square of the standard deviation. It is the mean of the squared distances from the mean. As such, it is not expressed in original units and is less useful in providing the spread of a variable.

Chapter 5

Univariate description: a visualization of a single variable designed to reveal the central tendency and dispersion of a variable.

Frequency table: a table that records the number of observations contained in each category of a categorical variable. Use frequency tables with categorical variables.

Bar plot: a visualization of the frequency table. Each bar in the plot represents a category in a categorical variable. The bar indicates the number of observations in each category. Use bar plots with categorical variables.

Boxplot (box-and-whisker plot): a visualization of a variable's distribution. The "box" represents the interquartile range, while the "whisker" represent outlying values on either side of the interquartile range. Use boxplots with continuous variables.

Skewness: the degree to which a variable's distribution deviates from a normal distribution; the degree to which a variable's distribution is not symmetrical.

Stem-and-leaf plot: a visualization of the data that represents a horizontal orientation of a histogram. The stem represents the first numerical partition of a continuous variable, while the leaves represent the individual cases that exist in each partition. The partitions can represent different decimal places in a number. Use a stem-and-leaf plot with continuous variables when the number of observations is relatively small.

Bivariate description: a visualization of the data designed to reveal the relationship between two variables.

Negative association: a relationship between two variables in which an increase in one is associated with a decrease in the other.

Linear relationship: a relationship between two continuous variables in which an increase or decrease in one variable is associated with a constant increase or decrease in another.

Correlation: a single numerical summary of the relationship between two continuous variables. It is a number that varies between −1 (strong negative correlation) and 1 (high positive correlation). A strong correlation (close to 1 or −1) indicates that a movement in one variable is associated with a similar movement in another.

Mosaic plot: a visualization of two categorical variables that indicates the number of observations each category in one variable shares in common with the other. The thickness of the columns also indicates how many cases exist in the variable designated as X. Use mosaic plots when examining the relationship between two categorical variables.

Cross-tab: a numerical representation of the mosaic plot. Each cell reveals the number of observations each category of a variable shares in common with another. Cross-tabs also indicate the percentage of a variable's cases that exist in each cell. Use cross-tabs when examining the relationship between two categorical variables.

Bubble plot: a scatter plot where the observations are sized according to a third variable. While the third variable can be continuous or categorical, the variables on the x and y axes should be continuous.

Chapter 6

Transforming data: re-expressing data in forms that accord to our hypotheses or permit better visualizations of our data. Re-expressing the data can come in the form of using a mathematical formula (taking logs), changing a continuous variable into a categorical one, or changing the number of categories in a categorical variable.

Box-Cox ladder of transformations: a set of mathematical expressions we can use to transform continuous variables. The expressions form a progression of transformations that help facilitate our visual examination of the data.

Logarithmic transformations: the log transformation is perhaps the most common transformation of data. Taking logs re-expresses a number by calculating the exponent needed to produce the number according to a given base. The two most common bases are log to the base 10 or what is called the natural log (2.718).

Natural log: the natural log is 2.718. When re-expressing any number by taking the natural log, we are calculating the exponent needed to convert 2.718 to that number. For example, if we take the natural log of 10, we are calculating the exponent of 2.718 that produces $10 : 2.718^{(x)} = 10$. In this case, x would be 2.30. In other words, $2.718^{(2.30)} = 10$.

Log to the base 10: when taking logs to the base 10, we're simply re-expressing a number by calculating the exponent needed to convert 10 to that number. For example, if we take the log to the base 10 of 10, we are calculating the exponent of 10 that produces $10 : 10^{(x)} = 10$. In this case, x would be 1. In other words, $10^{(1)} = 10$.

Chapter 7

Chart clutter: not to be confused with too much information. Chart clutter comprises all of the extra features of a figure or table that are unnecessary because they convey no information. Chart clutter is usually associated with style, such as shaded axes, boxes around labels, strange fonts, and gray backgrounds.

Explanatory: visualizations that are explanatory should emphasize a single point; they are constructed to hit the reader right between the eyes with a key insight. In these figures, too much information should be avoided.

Exploratory: visualizations that produce information in a neutral way so that possible rival hypotheses are given equal weight.

Informational: visualizations that produce as much information per square inch as possible. One could think of these as reference points, plots that readers can go back to and study with care in order to more fully understand a relationship or the overall structure of the data (central tendency, spread, and where the cases lie).

Context: in storytelling, it is the beginning of the story. It introduces the characters and the situation they face. In presenting data, it introduces the question and the problem. It also builds the audience's intuition about the data and measures used in the exercise.

Causation: explains why two variables are correlated with each other. It tells the story as to why two things are related.

Reduced form: often causal explanations involve several mechanisms that lead from x to causing y and from y to causing z. The reduced form of that causal change would be the relationship between x and z.

Chapter 8

Population: a defined set of individuals or objects that represent the focus of interest. For example, if we want to know who might win a U.S. presidential election, we are interested in the population of registered voters.

Sample: a subset of the population.

Random sample: a sample taken from the population in which each individual or object has an equal chance of being chosen.

Sample bias: occurs whenever a sample taken from the population does not give each individual or object an equal chance of being chosen. For example, if the population of interest is a university's student population, taking a sample from a single class does not give every student at the university a chance to be drawn.

Sampling with replacement: after each individual or object is drawn from a population, they are placed back in the population so that they have an equal chance—along with all other individuals or objects—of being selected again. This ensures each individual or object has an equal chance of being selected and that each selection or draw is independent of each other.

Law of large numbers: if a certain chance experiment is repeated an unlimited number of times under exactly the same conditions and if the repetitions are independent of each other, then the fraction of times that a given event A occurs will converge with probability 1 to a number that is equal to the probability that A occurs in a single repetition of the experiment.

Sampling distribution: the distribution of a summary statistic (e.g., mean, median, standard deviation, etc.), obtained from a series of random samples.

Central limit theorem: establishes the following characteristics when taking random samples from a population: (1) the sampling distribution of means from a population will tend toward normality as the number of observations in each sample increases; (2) the sampling distribution of sums from a population will tend toward normality as the number of observations in each sample increases; and (3) when drawn from a normally distributed population, the sampling distribution will tend toward normality regardless of the number of observations.

Normal distribution: a symmetrical bell-shaped curve centered over the sample or population's mean.

Uniform distribution: a distribution in which the number of observations is constant over the range of the variable. A uniform distribution resembles the shape of a rectangle.

Exponential distribution: a distribution in which the number of observations is highest near a variable's minimum, gradually decreasing as the maximum is approached.

Bimodal distribution: the highest frequency of observations exists in two distinct regions over the range of the variable. It appears as if there are two "humps" in the distribution.

Standard normal distribution: a normal distribution that is perfectly symmetrical, centered over a mean of 0 with a standard deviation of 1. With 100% of the area lying underneath, 68% of the distribution lies between 1 standard deviation above and below the mean, 95% of the distribution lies between 2 standard deviations above and below the mean, and 99.75% of the area exists between 3 standard deviations above and below the mean. These properties provide the mechanics to calculate confidence intervals and perform hypothesis tests.

z-score: the difference between an observation and the sample mean, divided by the sample's standard deviation. The z-score standardizes whatever kind of thing we're measuring and puts it in units expressed as standard deviations. This allows us to determine how much of the distribution lies to the left or right of the observation.

Chapter 9

Confidence interval: sometimes referred to as the margin of error in electoral polls. The confidence interval represents the range of values around a sample statistic that includes the population parameter for a given level of confidence. The greater the level of confidence, the larger the interval.

Population proportion: the percentage of a population belonging to a specific category. In elections, the population proportion represents the fraction of the population that supports a particular candidate. The formula for the confidence interval is different for population proportions than population means.

Critical z-value: the distance in a standard normal distribution from the mean that represents a given area underneath the curve, measured in standard deviations. For example, if we choose the 90% level of confidence, we use 1.65 as the critical z-value since 90% of the area underneath the standard normal curve is ±1.65 standard deviations above and below the mean.

t-Distribution: analogous to the standard normal distribution except that its purpose is explicitly for small samples. Also, unlike the standard normal distribution, there are many different t-distributions, all based on how many observations are in the sample. Eventually, as the number of observations increases, the shape of the t-distribution approximates the standard normal distribution.

Degrees of freedom: the number of independent data points used to calculate a statistic. Consequently, when determining which t-distribution to use, we use the degrees of freedom (n – 1) rather than the number of observations (n). It is a concept that appears in many different contexts in statistics.

Bias: the difference between the mean of an estimate's sampling distribution and the population parameter. If the mean of an estimator's sampling distribution does not center on the population's mean, the estimator is biased.

Bessel's correction: since the sample standard deviation is biased when n is used in the denominator (on average it has a downward bias), subtracting 1 from the number of observations in the denominator accounts for the bias, producing an unbiased estimator.

Critical t-value: the distance in a t-distribution from the mean measured in standard deviations. For example, if we choose the 90% level of confidence, we use 1.697 as the critical t-value since 90% of the area underneath the standard normal curve is ±1.697 standard deviations above and below the mean of 0.

Sampling variance: when we draw a sample from a population and compute a statistic, it will rarely turn out to be an exact match with the population parameter. The difference between the sample statistic and the population parameter is called the sampling variance or the error variance.

Null hypothesis: when conducting hypothesis testing (comparing means from two samples), the null hypothesis is that there is no difference between the means. If the t-statistic indicates there is a significant difference between two means, we "reject" the null hypothesis.

Two-sample t-test: a statistical test that establishes a confidence interval around the difference between the means of two samples. It indicates the likelihood that the difference in means between two samples is not zero, the null hypothesis.

t-Ratio (for a two-sample t-test): the statistic calculated by taking the difference between the sample statistic and the null hypothesis (usually zero) and dividing by the square root of the pooled sample variance.

Chapter 10

Exploratory data analysis (EDA): a term coined by John Tukey to describe a process of analysis characterized by the back-and-forth between generating hypotheses and examining data. The classical approach forms hypotheses then tests them with data, mimicking what might go on in a laboratory. EDA recognizes that looking at the data first can help form better hypotheses. An important downside to consider with EDA, the back-and-forth approach can violate a central assumption in statistical theory (random selection of observations), rendering classical statistical tests less useful.

Feeling thermometer: a survey instrument used to capture a respondent's attitudes toward a person, place, or thing. Usually the respondent is asked to share their feelings—how much they like, or how warmly they feel—toward something or someone. The thermometer is designed to generate a continuous variable, often on a scale that ranges between 0 and 100.

Jitter plot: identifies the individual cases of a boxplot but separates (jitter) the points so that they can be identified. Although boxplots are an extremely valuable way to visualize data, the box can hide the details that the jitter plot helps reveal.

Chapter 11

Controlled comparison: a bivariate view of two variables designed to show how their relationship changes with respect to a third. They are called "controlled" comparisons since we say they show the relationship between two variables, controlling for a third. They are synonymous with accounting for a third variable in an argument about the causal relationship between two variables.

Dichotomous variable: a variable that takes on two possible values. If expressed numerically, the variable usually records a 0 or a 1 depending on whether a condition, character, or trait is present. Categorical variable that are treated as either/or are also dichotomous (e.g., college educated or not, male or female, win or loss). Dichotomous variables are also called "dummy" variables.

Chapter 12

Linear regression: in the context of two variables (the focus in this chapter), a linear regression fits a line to a scatter plot that summarizes the relationship between the two variables.

Substantive significance: refers to the magnitude of the coefficients in a regression model: the change in y associated with a unit change in x.

Statistical significance: refers to the level of certainty (or confidence) we have that the estimated coefficient is not equal to zero. An estimate is statistically significant if we can reject the possibility that it is the result of random chance.

Slope: the change in y associated with a unit change in x.

Intercept: the point where the regression line crosses the y intercept. In some cases it reveals important information, in others it is less helpful.

Coefficient: the estimated slope of the regression line.

Predicted value: the regression line itself. Our prediction of y given the value of x.

Residual: the difference between an observation's actual value and its predicted value ($y_i - \hat{y}_i$).

R^2 statistic: the explained variance of y [$\Sigma(\hat{y}_i - \bar{y})$] divided by its total variance y [$\Sigma(y - \bar{y})$]. It is a number that ranges from 0 to 1 indicating how much of the variation in y is explained by the variable(s) included in the model.

Total deviation (in the context of R^2): the difference between Y_i and $\bar{Y}$.

Explained deviation (in the context of R^2): the difference between the predicted value $\hat{Y}_i$ and the mean of $\bar{Y}$.

Unexplained deviation (in the context of R^2): the difference between the observed value Y_i and the predicted value $\hat{Y}_i$.

t-Ratio (for a regression coefficient): a measure of statistical significance. It is the regression coefficient divided by its standard error. It indicates how confident we can be that the true slope of the line describing the relationship between two variables is not zero. Absolute values greater than 2 indicate we can be confident that the true slope is not zero by conventional standards of statistical significance.

Chapter 13

Multiple regression analysis: fits a line to summarize the relationship between two variables (a dependent variable and independent variable), controlling for other independent variables.

Additive multiple regression model: a strictly linear model that assumes the independent variables in the model are independent of each other.

Adjusted R^2 statistic: a version of the R^2 statistic that accounts for the number of predictors in the model, increasing only if the added predictors improve the fit of the model. It is interpreted the same way as the R^2 statistic.

Empirical implication: a hypothesized relationship that follows logically from an existing empirical relationship.

Chapter 14

Dummy variable: a variable consisting of zeros and ones that indicates whether the case can be classified as belonging to one class or another. In R, dummy variables can simply contain two different strings indicating whether the case belongs to one category or another (e.g., democracy or dictatorship, male or female, Republican or Democrat).

Additive model: a multiple regression model that assumes there is no relationship between the independent variables. Mathematically, an additive model is an equation with multiple terms which are added to each other:
$y = a + b + c$.

Interactive model: a multiple regression model that does not assume independence between all of the independent variables. Mathematically, an interactive model is an equation with multiple terms, some of which are multiplied together:
$y = a + b + c + a \times c$.

Reference category: the category in a dummy variable that is assigned the value of zero.

Chapter 15

Efficiency: a criterion used to compare estimators or statistics. An estimator is deemed more efficient if its sampling distribution has less variance. If the sampling distribution from one statistic or estimator has less variance than another, it means on average it is a more accurate estimate of the population parameter.

Consistency: the accuracy of an estimator increases as the number of observations increases.

Gauss-Markov theorem: states that if three important conditions are met, the OLS estimator is the best linear unbiased estimator of a line. The three conditions are the following: (1) the errors from the model sum to zero, (2) the errors are independent of each other, and (3) the errors have constant variance.

Error: the difference between the actual value of an observation and the prediction generated by the true model. Since we really never know exactly what the true model is, the error is a theoretical construct.

Chapter 16

Outlier: a term many use generically to describe an observation that stands out from the rest of the data. In this chapter we gave it a more specific definition: an observation that stands out from the rest of the data in terms of y.

Leverage: an observation that stands out from the rest of the data in terms of the independent variables (the x axis).

Influence: the impact any given observation has on a number of different summaries or estimates from a regression model.

Cook's distance (Cook's D): an influence measure based on the combination of the residual and leverage. It measures the impact of an observation on the model estimates. It follows that Cook's D measures how much all of the predicted values change when the ith observation is removed from the regression.

dfbeta: an influence measure that registers the impact an observation has on each regression coefficient.

Chapter 17

Logistic regression: a regression model that fits a logistic curve to data in order to explain dichotomous (binary) outcomes.

Logged odds: the log of the odds ratio. The odds ratio is the probability of x divided by the probability of 1 − x.

Predicted probability: the probability of an outcome (usually denoted by a binary variable registering "1") predicted by a logistic regression model. In a logistic model, the predicted probabilities follow an S-shaped curve. Similar to the predicted value $\hat{y}_i$ in an OLS regression, a predicted probability exists for each value of an independent variable. For a specific x, we say that there is a probability that the outcome coded "1" will occur.

McFadden's R^2: a pseudo R^2 statistic that is based on showing the improvement of the model when independent variables are added to a simple model with only the constant. It is the most conservative of the three pseudo R^2 measures generated by the *logreg2()* command.

Cox-Snell R^2: a pseudo R^2 statistic that is based on showing the improvement of the model when independent variables are added to a simple model with only the constant. It is more conservative than the Nagelkerke R^2 but not as conservative as McFadden's R^2.

Nagelkerke R^2: a pseudo R^2 statistic that is based on showing the improvement of the model when independent variables are added to a simple model with only the constant. It provides the least conservative estimate of all the pseudo R^2 measures.

Chi-square test: a test statistic that takes the difference between the null deviance (from the model with fewer independent variables) and the residual deviance (from the full model). A p-value is then calculated for the chi-square number, which indicates whether the full model is an improvement over the model with fewer independent variables. P-values lower than .05 indicate the full model is an improvement at conventional levels of significance.

Receiver operating characteristic (ROC) curve: graphs the relationship between the sensitivity and specificity of a test or model over different thresholds. It provides an indication of how well the model (or test) separates the two outcomes of a logistic regression.

Area under the curve (AUC): provides a numerical summary of the area that lies underneath the ROC curve. The higher the number, the closer it is to 1 and the further away it is from .5, means the model does a good job of explaining why you observe one outcome instead of the other.

True positive: a positive result on a test when the outcome is positive. Analogous to a medical exam where a patient tests positive and is indeed sick.

False positive: a positive result on a test when the outcome is, in fact, negative. Analogous to a medical exam where patient tests positive but is not really sick.

True negative: a negative result on a test when the outcome is negative. Analogous to a medical exam where a patient tests negative and is, in fact, not sick.

False negative: a negative result on a test when the outcome is positive. Analogous to a medical exam where a patient tests negative but is actually ill.

Sensitivity: the probability of achieving a true positive. Analogous to the probability of detecting a disease in a patient when it is present.

Specificity: the probability of achieving a true negative. Analogous to the probability of achieving a negative result when the disease is not present.

References

Chapter 1

Gaubatz, K. T. (2015). *A survivor's guide to R*. SAGE.

Chapter 2

Alesina, A. (2003). The size of countries: Does it matter? *Journal of the European Economic Association, 1*(2–3), 301–316.

Alesina, A., Devleeschauwer, A., Easterly, W., Kurlat, S., & Wacziarg, R. (2003). Fractionalization. *Journal of Economic Growth, 8*(2), 155–194.

"China Invents the Digital Totalitarian State." (2016 December 17). *The Economist*. https://www.economist.com/briefing/2016/12/17/china-invents-the-digital-totalitarian-state

Cukier, K., & Mayer-Schönberger, V. (2013). *Big data: A revolution that will transform how we live, work, and think*. Eamon Dolan/Houghton Mifflin Harcourt.

Jones, C. (2015). *The facts of economic growth*. National Bureau of Economic Research.

Lazer, D., Kennedy, R., King, G., & Vespignani, A. (2014). The parable of Google flu: Traps in big data analysis. *Science, 343*(6176), 1203–1205.

Lloyd, P., & Lee, C. (2018). A review of the recent literature on the institutional economics analysis of the long-run performance of nations. *Journal of Economic Surveys, 32*(1), 1–22.

Paulos, J. A. (1995). *A mathematician reads the newspaper*. Basic Books.

Pollock, P. H. (2014). *An R companion to political analysis*. SAGE/CQ Press.

Przeworski, A., Alvarez, R. M., Alvarez, M. E., Cheibub, J. A., Limongi, F., & Neto, F. P. L. (2000). *Democracy and development*. Cambridge University Press.

"Violent Crime Is Down in Chicago." (2018 May 5). *The Economist*. https://www.economist.com/united-states/2018/05/05/violent-crime-is-down-in-chicago

Chapter 3

de Fezensac, M. (1852). *A journal of the Russian campaign of 1812*. Parker, Furnirall & Parker.

Wickham, H., & Grolemund, G. (2017). *R for data science*. O'Reilly.

Chapter 4

Cassidy, J. (2009). *How markets fail: The logic of economic calamities*. Farrar, Strauss, and Giroux.

Fisher, M., & Keller, J. (2017 November 7). What explains US mass shootings? International comparisons suggest an answer. *New York Times*.

Chapter 5

Lucas, R. E., Jr. (1988). On the mechanics of economic development. *Journal of Monetary Economics, 22*(1), 3–42.

Ruger, W. P., & Sorents, J. (2009 February). *Freedom in the 50 states: An index of personal and economic freedom*. George Mason University Mercatus Center.

Tufte, E. R. (1997). *Visual explanations*. Graphics Press LLC.

Chapter 6

Binswanger, M. (2006). Why does income growth fail to make us happier? Searching for the treadmills behind the paradox of happiness. *Journal of Socio-Economics, 35*(2), 366–381.

Goldstone, J. A., Bates, R. H., Epstein, D. L., Gurr, T. R., Lustik, M. B., Marshall, M. G., Ulfelder, J., & Woodward, M. (2010). A global model for forecasting political instability. *American Journal of Political Science, 54*(1), 190–208.

Przeworski, A., Alvarez, R. M., Alvarez, M. E., Cheibub, J. A., Limongi, F., & Neto, F. P. L. (2000). *Democracy and development*. Cambridge University Press.

Tukey, J. W. (1977). *Exploratory data analysis*. Addison-Wesley.

Chapter 7

Duarte, N. (2010). *Resonate: Present visual stories that transform audiences*. Wiley.

"Economists Are Rethinking the Numbers on Inequality." (2019 November 28). *The Economist*. https://www.economist.com/briefing/2019/11/28/economists-are-rethinking-the-numbers-on-inequality

Knaflic, C. N. (2015). *Storytelling with data: A data visualization guide for business professionals*. Wiley.

Piketty, T. (2014). *Capital in the twenty-first century*. Belknap Press.

Piketty, T., & Saez, E. (2003). Income inequality in the United States, 1913–1998. *Quarterly Journal of Economics*, *118*(1), 1–41.

Piketty, T., Saez, E., & Zucman, G. (2018). Distributional national accounts: methods and estimates for the United States. *Quarterly Journal of Economics*, *133*(2), 553–609.

Piketty, T., & Zucman, G. (2014). Capital is back: Wealth-income ratios in rich countries 1700–2010. *Quarterly Journal of Economics*, *129*(3), 1255–1310.

Putnam, R. (1993). *Making democracy work: Civic traditions in modern Italy*. Princeton University Press.

Strunk, W., & White, E. B. (1979). *The elements of style* (3rd ed.). Macmillan Publishing Company.

Tufte, E. R. (2006). *Beautiful evidence*. Graphics Press LLC.

Vogler, C. (2007). *The writer's journey: Mythic structure for writers* (3rd ed.). Michael Wiese Productions.

Chapter 8

Galton, F. (1889). *Natural inheritance*. Macmillan.

Stigler, S. M. (2016). *The seven pillars of statistical wisdom*. Harvard University Press.

Tijms, H. (2004). *Understanding probability: Chance rules in everyday life*. Cambridge University Press.

Chapter 9

Schumacker, R. E. (2015). *Learning statistics using R*. SAGE.

Chapter 10

Knaflic, C. N. (2015). *Storytelling with data: A data visualization guide for business professionals*. Wiley.

Tukey, J. W. (1977). *Exploratory data analysis*. Addison-Wesley.

Chapter 11

Karstedt, S. (2006). Democracy, values, and violence: Paradoxes, tensions, and comparative advantages of liberal inclusion. *Annals of the American Academy of Political and Social Science*, *605*(1), 50–81.

Karstedt, S. (2013). *Legitimacy in non-democratic regimes* (J. Tankebe & A. Liebling eds.). Oxford University Press.

LaFree, G., & Tseloni, A. (2006). Democracy and crime: A multilevel analysis of homicide trends in forty-four countries, 1950–2000. *Annals of the American Academy of Political and Social Science*, *605*(1), 25–49.

Chapter 12

Tufte, E. R. (1974). *Data analysis for politics and policy*. Prentice-Hall.

Chapter 13

Cukier, K., & Mayer-Schönberger, V. (2013). *Big data: A revolution that will transform how we live, work, and think*. Eamon Dolan/Houghton Mifflin Harcourt.

Greene, W. H. (2000). *Econometric analysis* (4th ed). Prentice-Hall.

Chapter 16

Fox, J., & Weisberg, S. (2011). *An R companion to applied regression* (2nd ed). SAGE.

Chapter 17

Pollock, P. H. (2014). *An R companion to political analysis*. SAGE/CQ Press.

Index

Abortion rates
   EDA, 322–328
   income and, 223–227
   multiple regression model, 432–434
   religion and, 405–407
Added variable plots, 56–57, 529–530
Additive model, 461
Additive multiple regression
    model, 430
Adjusted R^2 statistic, 437
aes(), 36
Allaire, J. J., 2
alpha option, 109, 503
Analysis of covariance (ANCOVA)
    model, 474
Area under the curve (AUC), 553–559, 561
Arguments, 33
   controlled comparisons, 370
   multiple regression, 430–431

Bar plot, 135–136
basicStats(), 88
Bernoulli, J., 241
Bessel's correction, 288
Best linear unbiased estimator (BLUE), 492
Bias, 286
Biased estimator, 487
Big data, 30–31
Bimodal distribution, 257
Bivariate data description, 131, 146–169
   boxplots, 154–159
   bubble plot, 163, 165–169
   cross-tab, 162–163
   dependent/independent variables
    and, 149
   mosaic plot, 159–161
   scatter plot, 148–154
Bivariate logistic regression, 545–547
Bivariate regression, 396–398,
    405–415, 450
   dummy variables, 461–463
   ELF and homicide rate, 409–411
   GDP per capita and voter turnout,
    412–415
   homicide rate and political instability,
    412, 413 (figure), 413 (table)
   income and stand your ground laws,
    539–541
   leverages from, 518–522
   multiple regression *vs.*, 416
   Muslim feeling thermometer, 462–463
   R^2 statistic, 399
   religion and abortion rates, 405–407
   religion and gun laws, 408–409
   residual plots, 498–500
Bivariate regression models, 54
   results stability from, 56
Box-and-whisker plot, 136–139
Box-Cox ladder of transformations,
    194–201
   GDP per capita, 195–196
   GDP per capita and human capital
    index, 198–201
   GDP per capita and women legislators,
    196–198
Boxplots
   bivariate, 154–159
   gay thermometer and party
    identification, 154–155
   girls-to-boys educational achievement,
    155–157
   homicide rate, 138–139
   Muslims and Obama feeling
    thermometer, 158
   Trump feeling thermometer by age,
    183–184
   univariate, 136–139
   voting turnout, 136–137
Bubble plot, 163, 165–169
   democracy and infant mortality,
    166–168, 169 (figure)
   women legislators, education and
    evangelism, 165, 166 (figure)

Campbell, J., 228
Categorical comparisons, 310–314, 348
   party identification and attitudes
    toward police, 310–313
   religion and politics, 313–314
Categorical variables, 72
   association between continuous
    variables and, 309
   comparison of, 310–314
   continuous variables to, 182–187
   continuous variable *vs.*, 94, 170, 330
   controlled comparisons, 349–362
   cross-tab, 162–163
   mosaic plot, 159–161
   relationship between, 309
   same variable name, 182
Cat plot, 221, 222 (figure)
Causal story, 231–233
Causation, 231–233, 300–301
Central limit theorem, 246–259, 271
   implications, 247–252
   observations from normal
    distribution, 251–252
   property of, 253
   sampling distribution of means as
    normal, 248–250
   sampling distribution of sums as
    normal, 250–251
   sampling from distributions, 253–259
   standard normal distribution, 259–264
Central tendency, 99, 124
   mean, 102–106
   measurement, 100–108
   median, 106–108
   mode, 100–102
   outliers and, 100
Chart clutter, 207–209, 234
Chi-square test, 551
Clutter, 207–209, 234
CO_2 emission and GDP per capita,
    179–181
Code chunk, 8, 13, 14 (figure), 15
   execution, 16 (figure)
   options, 15
Coefficient, 395
   dummy variables/interactions, 479
   multiple regression, 449
Comparisons, 307–330
   categorical variables, between,
    310–314, 348
   continuous and categorical variables,
    between, 315–317
   continuous variables, between,
    318–321, 348
   controlled, 347–371
   gender and education, 318–319
   gender and policymaking, 319–321
   Obama feeling thermometer, 315–317
   party identification and attitudes
    toward police, 310–313

purpose of, 308
religion and politics, 313–314
Conceptual simplicity, 116
Confidence intervals, 272
  age of population, 278–280
  confidence levels, 297–299
  federal government bias, 275–277
  hours studied per week, 277–278
  income and support for Trump, 296–297
  incomes, 293–295
  Karl Rove meltdown, 274–275
  large samples, with, 272–280
  means comparison, 292–297
  population mean, 277–280
  population proportion, 272–277
  race and attitudes toward police, 295–296
  sample standard deviation, 284–289
  scientists, attitudes toward, 278
  small samples, 289–291
  t-distribution, 280–289
  winning elections, 272–274
Consistency, 487
Console pane, RStudio, 18
Context, 231
Continuous variables, 71–72
  association between categorical variable and, 309
  categorical variables, to, 182–187
  categorical variable vs., 94, 170, 330
  comparison of, 318–321, 348
  controlled comparisons, 363–369
  relationship between, 309–310
Controlled comparisons, 347–371
  arguments and, 370
  categorical variables, 349–362
  comparison vs., 348
  continuous variables, 363–369
  definition, 348
  homicide rates, 365–369
  immigration, attitudes toward, 349–355
  infant mortality, 363–365
  police, attitudes toward, 349–355
Cook's distance (Cook's D), 522–524
Correlation, 152
Cox-Snell R^2, 551
Critical thinking, 24, 61
Critical t-value, 289, 302
Critical z-value, 273, 302
Cross-tab, 162–163
  gender and party ID, 161 (table)
  mosaic plot advantages over, 159
  terrorism fear and party ID, 163, 164 (table)
Curves, 152

Data
  displaying. See Data presentation
  exploration vs. presentation, 62
  range of, 82–83
  sets, 19, 69–70
  shape of, 78–81
  transform/transforming. See Transform/transforming data
Data analysis, 29–62
  big data, 30–31
  components of, 33–34
  data description and hypotheses formulation, 33, 35–53
  diagnostics, 34, 56–60
  marketable skill, as, 31
  model building and estimation, 33, 54–55
  motivations, 30–32
  public good, as, 31–32
  quantitative, 31
  question, generation, 61
  surveys, by, 31
Data description, 33, 48, 67–94
  additional clues, revealing, 88–90
  advantages, 67
  bivariate, 131, 146–169
  data sets, 69–70
  hypotheses formulation from, 33–53
  measurement, 91–93
  puzzles vs. questions, 84–85
  question formulation, 85–88
  range of data, 82–83
  shape of data, 78–81
  time/effort and, 77–83
  univariate, 131–145
  variables, 69–70
  See also Bivariate data description; Univariate data description
Data presentation, 205–233
  data exploration vs., 62
  graphs, 233
  story. See Story/storytelling
  style, elements of, 206–227
Data visualization
  audience, 220–222, 234
  causation, 231–233
  clutter, 207–209, 234
  documentation, 229–230
  explanatory, 222–225
  exploratory, 222–225
  focus, sharpening, 209–211
  graphs, 233
  informational, 226
  purpose of, 222–227, 233–234, 330
  style, elements of, 206–227
  words and pictures integration, 211–214
  worthy pictures, 215–220
db_diag(), 523

Degrees of freedom, 282–284, 302, 404
Democracy
  foreign aid, GDP per capita and, 474–478
  GDP per capita and, 184–186
  homicide rates and GDP per capita by, 367–369
  infant mortality and, 166–168, 169 (figure)
  region, by, 316–317
de Moivre, A., 247
Density plot, 78–79, 244
Dependent variable, 149
  independent variable vs., 171
descr, 134
Describing data. See Data description
Description, data. See Data description
Detection, probability of, 553
dfbetas, 522, 524–528
Diagnostics, 34, 56–60, 485–507, 513–531
  added variable plots, 529–530
  errors, 492
  Gauss-Markov assumptions, 492–497
  leverages, 517–522
  measures of influence, 522–528
  outliers, 514–516
  regression analysis, in, 486
  residual plots, 58–60, 62, 498–506
  stability, results, 56–58
  statistics/estimators, properties of, 487–491
Dichotomous variable, 367–368, 461
Dispersion, 99, 124
  interquartile range, 114–115
  measurement, 112–121
  range, 113–114
  standard deviation, 115–121
Distribution of data, 78
Documentation, story, 229–230
dplyr, 86
Duarte, N., 205–206, 222
Dummy variables, 459
  additive model, 461
  bivariate regression, 461–463
  coefficient on, 479
  definition, 460
  interactive model, 461
  multiple regression, 464–467

EDA. See Exploratory data analysis (EDA)
Education
  Bernie Sanders, income and, 470–474
  boxplots of educational achievement, 155–157
  expenditure, 3
  gender and, 318–319
  homicide rate and, 438–448

income, gender and, 468–470
infant mortality and, 150–152
Obama, income and, 551–552
party affiliation and, 91–92
party identification *vs.* attitudes toward immigration and, 356–358
voter turnout and, 394–398
women legislators, evangelism and, 165, 166 (figure)
Efficiency, 487
Efficient estimator, 487
ELF. *See* Ethnolinguistic fractionalization (ELF)
ELF and GDP, 36–42
clutter, 207–209
Code Chunk, 36, 38–39, 41
industrial countries and, 38, 39 (figure)
low GDP per capita countries, 39–41
negative relationship, 37, 41
nonlinear relationship, 38, 40 (figure), 153 (figure)
scatter plot, 37 (figure), 152–153, 217–220
words and pictures integration, 211–214
Empirical cumulative distribution function, 556
Empirical implication, 445, 569–571
Error(s), 492
messages, 71
residual *vs.*, 507
Estimators, properties of, 487–491, 507
Ethnolinguistic fractionalization (ELF)
homicide rate and, 409–411
infant mortality and, 363–365, 366 (figure)
Explained deviation, 399
Explanatory data visualization, 222–225
Exploration, data, 62, 307–308
Exploratory data analysis (EDA), 307–308, 428, 569
abortion rates, 322–328
Exploratory data visualization, 222–225
Exponential distribution, 79 (figure), 255
Extreme values, 94, 100, 106, 109, 122
range and, 113

False alarm, probability of, 553–554
False negatives, 553
False positive rate, 553–554
False positives, 553
Feeling thermometer, 315. *See also* Muslim feeling thermometer; Obama feeling thermometer; Trump feeling thermometer

File folder, 6
freq(), 134
Frequency table, 134–135
Functions, R, 4–5
format, 20

Galton, F., 247
gather(), 120
Gauss-Markov assumptions, 492–497
constant variance errors, 495–497
independently distributed errors, 494–495
logistic regression violation of, 539–542
residual plots and, 500–506
zero error, 493–494
Gauss-Markov theorem, 492
GDP per capita, 37
abortion rate and, 223–227
Bernie Sanders, education and, 470–474
education, gender and, 468–470
ELF and. *See* ELF and GDP
ELF *vs.* infant mortality and, 363–365, 366 (figure)
homicide rates and, 365–369
Obama winning state and, 548–550
stand your ground laws and, 539–541
Trump winning state and, 543–548
voter turnout and, 412–415
women's suffrage and. *See* Women's suffrage and GDP
Gender
education and, 318–319
income, education and, 468–470
infant mortality and, 318–319
party identification and, 159–161
party identification *vs.* attitudes toward police and, 349–355
policymaking and, 319–321
ggplot, 36
ggplot2, 36
ggroc(), 559
Goodness of fit. *See* R^2 statistic
Grand Army's March, 67–69

head(), 69
Histograms, 80, 139–143
Homicide rates
boxplot, 138–139
education and, 438–448
ELF and, 409–411
GDP per capita and, 365–369
mean, 104–106
median, 106–108
political instability and, 412, 413 (figure), 413 (table)
Human capital and GDP per capita, 48–50

Hypotheses, 33
theory *vs.*, 62
Hypotheses formulation, 33
data description, from, 33–53
ELF, 36–42
human capital, 48–50
political stability, 50–53
women's suffrage, 42–48

If-then statement, 34
Impatience, 206
Income. *See* GDP per capita
Independent variable, 149
dependent variable *vs.*, 171
Infant mortality, 80–81
democracy and, 166–168, 169 (figure)
education and, 150–152
ELF and, 363–365, 366 (figure)
gender and, 318–319
income and, 88–90
interquartile range, 114–115
range, 114
Influence, 522. *See also* Measures of influence
Informational data visualization, 226
installD(), 9, 24
Interactions, 468–478
Bernie Sanders, education and income, 470–474
coefficient on, 479
foreign aid, GDP per capita and democracy, 474–478
graphing, 479
income, education and gender, 468–470
Interactive model, 461
Intercept, 393–398
Interquartile range, 114–115
standard deviation *vs.*, 122

Jitter plots, 316

kable(), 70
Knaflic, C. N., 205–207
Knitting, 8
Know-nothing model, 551

Law of Frequency of Error, 247
Law of large numbers, 241–246, 253
definition, 242
visualization, 242–246
Leverages, 517–522
bivariate regression, from, 518–522
outliers, measures of influence *vs.*, 514, 532
residual plots and, 517–518
libraries(), 10–11, 133

library(), 10
Linear model function, 395
Linear regression, 391–416
  advantages, 392
  definition, 392
  intercept, 393–398
  R^2 statistic, 399–402
  slope, 393–398
  statistical significance, 303, 392, 403–404
  voter turnout and education, 394–398
Linear relationship, 41, 152
Lines, 152
lm(), 54
Logarithmic transformations, 194–195
  variables with zeros, 195
Logged odds, 542–544
  predicted probabilities *vs.*, 561
Logistic curve, 538
Logistic regression, 537–561
  AUC, 553–559
  bivariate, 545–547
  Gauss-Markov assumptions, violation of, 539–542
  logged odds, 542–544
  model fitting with, 551–559
  multivariate, 547–548
  Obama, income and education, 551–552
  predicted probabilities, 544–550
  questions and problems, 538
  R^2 statistic, 551
  ROC curves, 553–559
  stand your ground laws, 539–542
  substantive significance, 545
  Trump winning state, 543–544
logitR2(), 551
Log to the base 10, 194
Lower-left pane, RStudio, 18
Lower-right pane, RStudio, 18

Mathematician Reads the Newspaper, A (Paulos), 31
McFadden's R^2, 551
Mean, 102–106, 124
  amount of cash in hand, 102–104
  calculation of, 102–106
  definition, 102
  homicide rates, 104–106
  mathematical formula, 103
  median *vs.*, 109–112
  population, 277–280
Measures of influence, 522
  Cook's distance (Cook's D), 522–524
  dfbetas, 522, 524–528
  outliers, leverages *vs.*, 514, 532
Median, 106–108
  definition, 106
  homicide rates, 106–108

mean *vs.*, 109–112
median(), 107
Mode, 100–102
  definition, 100
  mode(), 107
  party identification, 102–103
Model building and estimation, 33, 54–55
Models, 33
Mosaic plot, 159–161
  advantage over cross-tabs, 159
  axis labels, 315
  party ID and terrorism fear, 162–163
  party identification and attitudes toward immigration, 356–362
  party identification and attitudes toward police, 311–312, 349–355
  party identification and gender, 159–161
  religion and politics, 313–314
Multiple regression, 427–449
  abortion rates, 432–434
  additive multiple regression model, 430
  bivariate regression *vs.*, 416
  coefficients, 449
  definition, 428
  dummy variable, 464–467
  estimates interpretation, 435–437
  homicide rate and education, 438–448
  interactions in, 468–478
  models and arguments, 430–431
  models, theory and evidence, 431–434
  overview, 428–429
  R^2 statistic, 437, 450
  statistical significance, 436–437
  substantive significance, 435–436
Multiple regression models, 54
  additive, 430
Multivariate logistic regression, 547–548
Muslim feeling thermometer, 462–463

Nagelkerke R^2, 551
Natural log, 194
Negative association, 55, 152
Normal curve, 247
Normal distribution, 79 (figure), 90, 137, 144 (figure), 248
  sampling distribution as, 248–250
  standard, 259–264
  t-distribution *vs.*, 282
  voter turnout, 80
Null-deviance, 551
Null hypothesis, 293

Obama feeling thermometer
  Muslims and, 158
  party identification and, 315–317

Obama winning state
  predicted probabilities, 548–550, 554–555
  ROC curves and AUC, 556–559
Object, R, 4–5
Observations *vs.* samples, 267
Odds ratio, 543
olsrr, 523, 529
Ordered categorical variables, 72–77
Ordinary least squares (OLS), 397, 485–507
Outliers, 514–516
  Idaho, 516
  leverages, measures of influence *vs.*, 514, 532
  residual plots and, 515

Packages, 132–133
  definition, 2
  dplyr, 86
  installing, 9–10, 11 (figure)
  loading, 10–11
  visreg, 57
Partial regression plots, 529–530
Party identification, 86–87
  abortion rate and, 321–328
  attitudes toward immigration and, 356–362
  attitudes toward police and, 311–312, 349–355
  bar plot, 215–216
  employment and, 135–136
  frequency table, 134–135
  gay feeling thermometer and, 154–155
  gender and, 159–161
  mode calculation, 102–103
  Obama feeling thermometer and, 315–317
  terrorism fear and, 162–163
  Trump feeling thermometer and, 315–316
Patience, 62–63
pchisqC(), 551
Pearson's R correlation, 152–154
Pipe operator, 86
Police, attitudes toward
  party identification and, 311–312, 349–355
  race and, 295–296
Political stability hypothesis, 50–53
  Code Chunk, 50–51
  GDP per capita and, 50, 51 (figure)
  nonlinear relationship, 52 (figure)
Pollock, P. H., 552
Population mean, 277–280
  population proportion *vs.*, 302
Population parameters, 239
  sample statistic *vs.*, 266

Population proportion, 272–277
  population mean *vs.*, 302
Populations, 238–239, 266
Positive association, 52
Predicted probabilities, 544–550
  bivariate logistic regression, 545–547
  logged odds *vs.*, 561
  multivariate logistic regression, 547–548
  Obama winning state, 548–550, 554–555
  Trump winning state, 543–548
Predicted value, 396
Presenting data. *See* Data presentation
Probability of detection, 553
Probability of false alarm, 553–554
Probability theory, 237–266
  central limit theorem, 246–259
  law of large numbers, 241–246
  populations, 238–239, 266
  random samples, 238, 240–241
  sample bias, 240–241
  samples, 238–239
Przeworski, A., 184
Pseudo R^2 statistic, 551, 553
Puzzles *vs.* questions, 84–85

Quantitative data analysis, 31
Questions
  formulation, 85–88
  puzzles *vs.*, 84–85
Quetlet, A., 247

R
  blind colors in, 216
  definition, 2
  dummy variables in, 465
  error messages, 71
  functions, 4–5
  interface, 2
  object, 4–5
  object-oriented language, as, 5, 21
R^2 statistic
  adjusted, 437
  linear regression, 399–402
  logistic regression, 551
  multiple regression, 437, 450
Random digit dialing, 241
Random samples, 238, 240–241, 267
Range, 113–114
  data, 82–83
  interquartile, 114–115
Receiver operating characteristic (ROC) curves, 553–559, 561
Reduced form of causal relationship, 232
Reference category, 462, 465

Regression analysis. *See* Bivariate regression; Linear regression; Logistic regression
Regression diagnostics, 486
Reliability, measurement, 91–93
Religion
  abortion rate and, 405–407
  gun laws and, 408–409
  party identification *vs.* attitudes toward immigration and, 360–362
  politics and, 313–314
Residual, 34, 396, 492
  errors *vs.*, 507
  income and stand your ground laws, 539–541, 542 (figure)
Residual plots, 58–60, 62, 498–506, 513
  bivariate regression and, 498–500
  Gauss-Markov assumptions and, 500–506
  leverages and, 517–518
  outliers and, 515
R Markdown, 2
  advantage, 2, 11
  Cheat Sheet, 13, 14 (figure)
  features of, 13, 15
  file, 8
  R-scripts *vs.*, 18
  RStudio navigation with, 12–18
ROC curve. *See* Receiver operating characteristic (ROC) curves
R-scripts, 18
RStudio, 2
  application, 7 (figure)
  code, 3
  data file, 8, 9 (figure)
  load command, 8, 9 (figure)
  lower-left pane (console), 18
  lower-right pane, 18
  navigation with R Markdown, 12–18
  packages, installing, 9–10, 11 (figure)
  panes, 18
  upper-left pane, 18
  upper-right pane (global environment), 18

S, 2
Sample(s), 238–239
  bias, 240–241
  observations *vs.*, 267
  random, 238, 240–241, 267
Sample statistic, 239
  population parameter *vs.*, 266
Sampling distribution, 242, 266–267
  bimodal distribution, from, 257–259
  exponential distribution, from, 255–257
  mean as normal distribution, 248–250
  sums as normal distribution, 250–251
  uniform distribution, from, 253–255

Sampling variance, 292
Sampling with replacement, 240
Scatter plot, 37, 148–154, 330
  democracy and infant mortality, 166, 167 (figure)
  education and infant mortality, 150–152
  ELF and GDP, 37 (figure), 152–153, 217–220
  ELF and infant mortality, 363–365, 366 (figure)
  GDP per capita and homicide rates, 365–369
  gender and infant mortality, 318–319
  income and stand your ground laws, 539, 540 (figure)
  labeling in, 171
  Pearson's R correlation, 152–154
  points label, 321
  political knowledge and voting turnout, 148, 149 (figure)
Schumacker, R. E., 297
select(), 69
Sensitivity, 553–554, 561
Shape of data, 78–81
Skewed distribution, 79, 94, 138, 144 (figure)
Skewness, 140–143
Slope, 393–398
Smooth, 38
Specificity, 553–554, 561
S-shaped curve, 538
Stability, 56–58
Standard deviation, 115–121
  amount of cash in students hand, 118–121
  calculation, 116–118
  conceptual simplicity, 116
  definition, 115
  interquartile range *vs.*, 122
  mathematical formula, 117
  sd(), 116
  unit of measurement, 116
Standard error, 273–274, 276, 404, 407
Standard normal distribution, 259–264
  critical z-scores and, 263–264
Stand your ground laws, 539–542
stargazer(), 54–55, 395
Statistical inference, 300–301
Statistical significance
  levels of, 437
  linear regression, 303, 392, 403–404
  multiple regression, 436–437
  substantive significance *vs.*, 416
Statistics
  golden rule, 250
  properties of, 487–491
Stem-and-leaf plot, 143–145

Story/storytelling
  causation, 231–232
  causation to action, from, 232–233
  context, setting, 231
  credible storyteller, 229–230
  documentation, 229–230
  elements of, 228–229
  hero's journey, 228
Strunk, W., 206–207
Student success, 82–83
Style, elements of, 206–227
  audience, 220–222, 234
  clutter, 207–209, 234
  focus, sharpening, 209–211
  purpose, 222–227, 233–234
  words and pictures integration, 211–214
  worthy pictures, 215–220
Substantive significance, 405
  linear regression, 392
  logistic regression, 545
  multiple regression, 435–436
  statistical significance *vs.*, 416
Summarization, 48
Summation sign, 103–104, 124
Sum of squared deviation, 123

t-distribution, 280–291, 302
  degrees of freedom, 282–284
Theories, 33, 54
  hypotheses *vs.*, 62
Tibble, 86
Total deviation, 399
Transform/transforming data, 175–202
  Box-Cox ladder of, 194–201
  categories, changing, 187–193
  CO_2 emission and GDP per capita, 179–181
  continuous to categorical variables, 182–187
  democracy and GDP per capita, 184–186
  practical reasons, 178–181
  theoretical reasons, 176–177
  Trump feeling thermometer and family income, 188–193
  Trump feeling thermometer by age, 183–184
t-ratio, 437
  regression coefficient, 403–404
  two-sample t-test, 293
True negatives, 553
True positive rate, 553–554
True positives, 553
Trump feeling thermometer
  age, by, 183–184
  family income and, 188–193
  party identification and, 315–316
Trump winning state
  logistic regression, 543–544
  predicted probabilities, 545–548
t-test, 293, 302
Tufte, E., 207, 215, 226, 229
  Challenger disaster, 133–134
  displaying data, 205–206
  Grand Army's March, 67–69
Tukey, J., 194, 307–308
Two-sample t-test, 293

Unbiased estimator, 487
Unexplained deviation, 399
Uniform distribution, 253
Unit of measurement, 116
Univariate data description, 131–132
  bar plot, 135–136
  boxplots, 136–139
  five views of, 133–145
  frequency table, 134–135
  histogram, 139–143
  stem-and-leaf plot, 143–145
Upper-left pane, RStudio, 18

Validity, measurement, 91–92
Variables, 19, 69–70
  categorical. *See* Categorical variables
  continuous. *See* Continuous variables
  dependent, 149
  dichotomous, 367–368, 461
  dummy. *See* Dummy variables
  independent, 149
  ordered categorical, 72–77
  range of, 113–114
  transforming, 194
  types of, 71–77
Variance, 123
visreg(), 57, 548
Vogler, C., 228
Voter/voting turnout
  boxplots, 136–137
  education and, 394–398
  GDP per capita and, 412–415
  normal distribution, 80
  political knowledge and, 148, 149 (figure)

White, E. B., 206
Women's suffrage and GDP, 42–48
  bivariate model, 55
  Code Chunk, 42–45, 47
  gap during World War II, 45–46
  GDP, large variation in, 46
  industrial countries and, 42–43, 44 (figure)
  negative relationship, 46, 48
  result stability, 57–58
  sub-Saharan Africa and, 42–43, 44 (figure), 47

YAML, 13, 14 (figure)

Z-scores, 262
  critical, 263–264